# Nature-Inspired Intelligent Techniques for Solving Biomedical Engineering Problems

Utku Kose
*Suleyman Demirel University, Turkey*

Gur Emre Guraksin
*Afyon Kocatepe University, Turkey*

Omer Deperlioglu
*Afyon Kocatepe University, Turkey*

A volume in the Advances in Bioinformatics and Biomedical Engineering (ABBE) Book Series

IGI Global
DISSEMINATOR OF KNOWLEDGE

Published in the United States of America by
IGI Global
Medical Information Science Reference (an imprint of IGI Global)
701 E. Chocolate Avenue
Hershey PA, USA 17033
Tel: 717-533-8845
Fax: 717-533-8661
E-mail: cust@igi-global.com
Web site: http://www.igi-global.com



Library of Congress Cataloging-in-Publication Data

Names: Kose, Utku, 1985- editor. | Guraksin, Gur Emre, 1980- editor. | Deperlioglu, Omer, 1967- editor.
Title: Nature-inspired intelligent techniques for solving biomedical engineering problems / Utku Kose, Gur Emre Guraksin, and Omer Deperlioglu, editors.
Description: Hershey PA : Medical Information Science Reference, [2018] | Includes bibliographical references and index.
Identifiers: LCCN 2017032791| ISBN 9781522547693 (h/c) | ISBN 9781522547709 (eISBN)
Subjects: | MESH: Biomedical Engineering--methods | Artificial Intelligence | Models, Biological | Bionics
Classification: LCC R857.M3 | NLM QT 36 | DDC 610.28--dc23 LC record available at https://lccn.loc.gov/2017032791

This book is published in the IGI Global book series Advances in Bioinformatics and Biomedical Engineering (ABBE) (ISSN: 2327-7033; eISSN: 2327-7041)

British Cataloguing in Publication Data
A Cataloguing in Publication record for this book is available from the British Library.

All work contributed to this book is new, previously-unpublished material. The views expressed in this book are those of the authors, but not necessarily of the publisher.

For electronic access to this publication, please contact: eresources@igi-global.com.

# Advances in Bioinformatics and Biomedical Engineering (ABBE) Book Series

Ahmad Taher Azar
Benha University, Egypt

ISSN:2327-7033
EISSN:2327-7041

## Mission

The fields of biology and medicine are constantly changing as research evolves and novel engineering applications and methods of data analysis are developed. Continued research in the areas of bioinformatics and biomedical engineering is essential to continuing to advance the available knowledge and tools available to medical and healthcare professionals.

The **Advances in Bioinformatics and Biomedical Engineering (ABBE) Book Series** publishes research on all areas of bioinformatics and bioengineering including the development and testing of new computational methods, the management and analysis of biological data, and the implementation of novel engineering applications in all areas of medicine and biology. Through showcasing the latest in bioinformatics and biomedical engineering research, ABBE aims to be an essential resource for healthcare and medical professionals.

## Coverage

- Chemical Structures
- Genomics
- Nucleic Acids
- Data Analysis
- Neural Engineering
- Finite Elements
- Health Monitoring Systems
- Therapeutic Engineering
- Protein Structure
- Tissue Engineering

IGI Global is currently accepting manuscripts for publication within this series. To submit a proposal for a volume in this series, please contact our Acquisition Editors at Acquisitions@igi-global.com or visit: http://www.igi-global.com/publish/.

The Advances in Bioinformatics and Biomedical Engineering (ABBE) Book Series (ISSN 2327-7033) is published by IGI Global, 701 E. Chocolate Avenue, Hershey, PA 17033-1240, USA, www.igi-global.com. This series is composed of titles available for purchase individually; each title is edited to be contextually exclusive from any other title within the series. For pricing and ordering information please visit http://www.igi-global.com/book-series/advances-bioinformatics-biomedical-engineering/73671. Postmaster: Send all address changes to above address.

## Titles in this Series

*For a list of additional titles in this series, please visit: www.igi-global.com/book-series*

*Applying Big Data Analytics in Bioinformatics and Medicine*
Miltiadis D. Lytras (Deree - The American College of Greece, Greece) and Paraskevi Papadopoulou (Deree - The American College of Greece, Greece)
Medical Information Science Reference • ©2018 • 465pp • H/C (ISBN: 9781522526070) • US $245.00 (our price)

*Comparative Approaches to Biotechnology Development and Use in Developed and Emerging Nations*
Tomas Gabriel Bas (University of Talca, Chile) and Jingyuan Zhao (University of Toronto, Canada)
Medical Information Science Reference • ©2017 • 592pp • H/C (ISBN: 9781522510406) • US $205.00 (our price)

*Computational Tools and Techniques for Biomedical Signal Processing*
Butta Singh (Guru Nanak Dev University, India)
Medical Information Science Reference • ©2017 • 415pp • H/C (ISBN: 9781522506607) • US $225.00 (our price)

*Handbook of Research on Computational Intelligence Applications in Bioinformatics*
Sujata Dash (North Orissa University, India) and Bidyadhar Subudhi (National Institute of Technology, India)
Medical Information Science Reference • ©2016 • 514pp • H/C (ISBN: 9781522504276) • US $230.00 (our price)

*Applying Business Intelligence to Clinical and Healthcare Organizations*
José Machado (University of Minho, Portugal) and António Abelha (University of Minho, Portugal)
Medical Information Science Reference • ©2016 • 347pp • H/C (ISBN: 9781466698826) • US $165.00 (our price)

*Biomedical Image Analysis and Mining Techniques for Improved Health Outcomes*
Wahiba Ben Abdessalem Karâa (Taif University, Saudi Arabia & RIADI-GDL Laboratory, ENSI, Tunisia) and Nilanjan Dey (Department of Information Technology, Techno India College of Technology, Kolkata, India)
Medical Information Science Reference • ©2016 • 414pp • H/C (ISBN: 9781466688117) • US $225.00 (our price)

*Big Data Analytics in Bioinformatics and Healthcare*
Baoying Wang (Waynesburg University, USA) Ruowang Li (Pennsylvania State University, USA) and William Perrizo (North Dakota State University, USA)
Medical Information Science Reference • ©2015 • 528pp • H/C (ISBN: 9781466666115) • US $255.00 (our price)

IGI Global
DISSEMINATOR OF KNOWLEDGE

701 East Chocolate Avenue, Hershey, PA 17033, USA
Tel: 717-533-8845 x100 • Fax: 717-533-8661
E-Mail: cust@igi-global.com • www.igi-global.com

# Editorial Advisory Board



# List of Reviewers

# Table of Contents

# Detailed Table of Contents



Artificial intelligence has a remarkable effect on many different fields with its flexible and comprehensive solution approaches to solve real-world problems. In this context, the field of biomedical engineering has also been affected by employment of different artificial intelligence-based techniques. This chapter aims to give a theoretical discussion on using nature-inspired artificial intelligent techniques for obtaining intelligent applications within biomedical engineering. As it is known, techniques within the field of artificial intelligence are inspired from nature. So, it is a good approach to focus on nature-inspired techniques for discussing intelligent biomedical engineering research works. Readers will have a chance to understand some ways of using artificial intelligence for achieving better results in biomedical engineering and the related developments associated with this field.



Swarm intelligence is a branch of computational intelligence where algorithms are developed based on the biological examples of swarming and flocking phenomena of social organisms such as a flock of birds. Such algorithms have been widely utilized for solving computationally complex problems in fields of biomedical engineering and sociology. In this chapter, two different swarm intelligence algorithms, namely the jumping frogs optimization (JFO) and bacterial foraging optimization (BFO), are explained in detail. Further, a synergetic algorithm, namely the coupled bacterial foraging/jumping frogs optimization algorithm (BFJFO), is described and utilized as a tool for control of the heroin epidemic problem.

Managing medical information and knowledge is becoming an increasing problem for healthcare professionals. Medical science that contains ever-increasing amounts of knowledge, such as the medical history of a patient, medical data about diseases, diagnosis and treatment methods, should be necessarily a science of information. The real problem faced by patients and healthcare providers is finding and using relevant knowledge at the right time. In this context, in the middle of 1950s, intelligent computer systems, called clinical decision support systems (CDSS), were introduced as a new concept. CDSS is defined as an active intelligent system that can help medical experts to make decisions by taking specific recommendations. Also, it provides decisions based on resolving patient-specific information and related medical truths. The objective of this chapter is to focus on these systems and explain relations with the field of artificial intelligence methods, approaches, or techniques in this manner.



Artificial immune systems inspired by the natural immune system are used in problems such as classification, optimization, anomaly detection, and error detection. In these problems, clonal selection algorithm, artificial immune network algorithm, and negative selection algorithm are generally used. This chapter aims to solve the problem of correct identification and classification of patients using negative selection (NS) and variable detector negative selection (V-DET NS) algorithms. The authors examine the performance of NSA and V-DET NSA algorithms using three sets of medical data sets from Parkinson, carotid artery doppler, and epilepsy patients. According to the obtained results, NSA achieved 92.45%, 91.46%, and 92.21% detection accuracy and 92.46%, 93.40%, and 90.57% classification accuracy. V-DET NSA achieved 94.34%, 94.52%, and 91.51% classification accuracy and 94.23%, 94.40%, and 89.29% detection accuracy. As can be seen from these values, V-Det NSA yielded a better result. Artificial immune system emerges as an effective and promising system in terms of problem-solving performance.



In this chapter, the differential diagnosis of erythematous diseases was determined using data mining and machine learning algorithms. In this chapter, data mining and its application to differential diagnosis of erythematous squamous diseases were discussed. A dermatology dataset from UCI Machine Learning Repository was used for the study. The dataset consists of 366 data items with 34 attributes. Initially, feature selection was made, and then classification was performed by using various algorithms. The number of attributes has been reduced from 34 to 19 as a result of the integration of the correlation-based filter methods and various heuristic search methods. The evaluation results show that Naive Bayes has 100% success rate in classification of psoriasis, seborrheic dermatitis, lichen planus, rose disease, chronic dermatitis, and pityriasis rubra pilaris diseases with 19 attributes selected with feature extraction algorithms.

One of the most popular applications of artificial intelligence within the medical field is developing medical diagnosis systems. Because artificial-intelligence-based techniques are able to use pre-data and instant data flow for making predictions, it is an easy task to design intelligent systems that can give advice to people or perform diagnosis-based decision making. So, it has been an important research interest to design and develop intelligent systems, which are able to make diagnoses for medical purposes. In this sense, the objective of this chapter is to introduce a general medical diagnosis system that can be used for detecting diseases. In detail, the system employs artificial neural networks and swarm-intelligence-based techniques to form a general framework of intelligent diagnosis. The chapter briefly focuses on the infrastructure of the system and discusses its diagnosis potential.



The aim of this chapter is to classify normal and extra systole heart sounds using artificial intelligence methods. Initially, both heart sounds have been passed from Butterworth, Chebyshev, Elliptic digital filter in specific frequency values to remove noise. Afterwards, features of heart sounds have been obtained for classification. For this process, wavelet transform and Mel-frequency cepstral coefficients (MFCC) methods have been applied. Training and test data have been created for classifier by taking means and standard deviation of gained feature. Support vector machine (SVM) and artificial neural network (ANN) methods have been used for classification of these heart sounds. Using wavelet and MFCC features, classification success of SVM has been obtained as 93.33% and 100%, respectively. Using wavelet and MFCC features, classification success of ANN has been obtained as 83.33% and 90%, respectively.



Bacteria counting process and bacterial counting by eye can be performed by manual counting in which a specialist counts the bacteria with naked eye or the automatic counting that utilizes computer-based techniques. The counting process becomes exhausting, long, and incorrect when the counting is performed by specialist. Therefore, there is a need for an efficient and more reliable method for the bacteria counting process performed automatically. In this chapter, a semi-automatic method has been designed for bacterial counting from the images of petri plates and photomicrograph. For this purpose, the image processing procedure has been applied to the images. In the image processing procedure, watershed transform was used to separate the adjacent bacteria and thus a more accurate counting was performed. The efficiency of the presented method has been shown for many different images, and the counting results of the proposed method have been compared with counting results of an expert. The method has been shown to be a successful and promising work for bacterial counting.

# Foreword

Today, Artificial Intelligence has an important role on improving people's life standards. In time, this scientific field has improved more and more and affected almost all fields including even art, and literature. So, we can see that a future standing over intelligent technologies is not something like a science-fiction. Here, one of the most important thing to consider is what such intelligent technologies has triggering factors on the background. One cannot deny that the 'nature' is one of them.

Nature has always directed the humankind to improve great effects of the sciences. We cannot think about anything having no connection with some factors – elements taking place in the nature. Planes have been invented by inspiring from birds. Many inventions that we use in daily life has inspirations from physical and / or chemical laws. So, thanks to such infinite examples, we can understand the role of nature on Artificial Intelligence. When we focus more on details, we can see that the nature-inspired intelligent techniques are important research sub-field of Artificial Intelligence.

It is whether a Machine Learning technique or not, a typical Artificial Intelligence technique has many inspirations from natural factors. For example, Artificial Neural Networks find its meaning in brain and neurons while optimization techniques are often associated with swarm behaviors and functions resulting to collaborative actions that are inspirations for collaborative problem solution ways. Some remarkable ones of them are:

- Connective mechanism among different natural actors.
- Interactions between the nature and living organisms.
- Problem solving actions taken by natural elements – actors.
- Features and mechanisms that appeared itself in time over the nature.
- Adaptive mechanisms shown by living organisms in the nature.
- Adaptive mechanisms appeared in an evolutionary process within the nature.
- Many other things that will surprise us in a near or far future.

At the final, it is possible to indicate that nature had had many contributions to the development of nature-inspired techniques and it also has many hidden things to appear in the future. Let's see what we will be see in the intersection of all technological developments – improvements occurring.

This book: 'Nature-Inspired Intelligent Techniques for Solving Biomedical Engineering Problems' is briefly one side of application ways related to intelligent techniques. As a vital factor in people's life, biomedical has a remarkable role on shaping the future. So, it had found its way in Artificial Intelligence as like many other fields having more active roles in our life. Changing from decision support mechanisms to predictive roles for better medical data, nature-inspired techniques have covered Biomedical

Engineering problems for a long time of period. Today, we can see applications done via from simpler to more advanced techniques for solving simple but time taking problems or too much complicated big problems including several sub-problem-steps to be overcome first. At this point, this valuable book study employing different chapters with many different problem types is a 'mind opening' material for our scientific community.

I would like to thank to the editors Dr. Kose, Dr. Guraksin, and Dr. Deperlioglu for their valuable contribution to the scientific literature. I hope everyone interested in this labyrinths of nature-inspired techniques will improve their awareness on the associated research topics and the book will be also a valuable references work for anyone, who is at just starting steps towards use of intelligent solutions on Biomedical Engineering oriented problems. Enjoy!

*Ali Hakan Isik*
*Mehmet Akif Ersoy University, Turkey*
*July 2017*

**Ali Hakan Isik** *was born in Burdur, Turkey, in 1980. He received B.S. and M.Sc. degree in electrical-electronics engineering from the Gazi University, Ankara, Turkey. He received Ph. D. degree in the field of Electronics and Computer Education department from Gazi University, in 2012. He worked for Turk Telekom Corporation which is incumbent telecommunication operator in Turkey and it has 17 million subscribers. He also worked as a system administrator in Gazi University Distance Education system at the Institute of Informatics. He is currently Assoc. Prof. Dr. in the department of Computer Engineering in Mehmet Akif Ersoy University as being the head of the department. His current research interests include expert systems, distance education, telemedicine, signal processing, computer education, and computer science.*

# Foreword

There are many words to say about effects of Artificial Intelligence in our life. Since its first appearance in the scientific literature, 'science of making intelligent systems' has taken many steps ahead and reached to an unstoppable level of innovations and developments. When we think about just our daily life, we can see that intelligent systems take more role in our works or practical actions and there are many differences among years after especially 2000, for the use of intelligent approaches, methods, and techniques. Currently, we cannot think about any field that is not associated with Artificial Intelligence. In this book, the scope was chosen as Biomedical Engineering.

Thanks to rapid technological developments, many different engineering fields have appeared in time. Biomedical Engineering is one of these fields and it has a remarkable role on improving medical oriented studies by combining different aspects from different disciplines. Because it is too important to employ effective and efficient solutions within especially medical oriented works, today's world is highly interested in achieving always the better for Biomedical Engineering. This book of 'Nature-Inspired Intelligent Techniques for Solving Biomedical Engineering Problems' in your hand employs many different, valuable chapters providing information about recent Artificial Intelligence based applications for Biomedical Engineering problems. As it is known, there is a great hunger for such published results for understanding the associated literatures better. When I heard first about nature-inspired techniques, I often think about optimization algorithms designed to provide intelligent mechanisms to solve optimization problems. But when you read this reference book, you will see that the inspiration from the nature is for more than just optimization and has an important role on improving the Biomedical Engineering field. I think that this book and the chapters included over here will be a good alternative for all interested readers and researchers, who are at the beginning of the objective research subject or highly enrolled in the related field with associated, similar research works discussed under the chapters.

Thank you very much to the editors and Drs. Kose, Guraksin, and Deperlioglu for providing such a unique work for the scientific community. Looking forward to seeing their new contributions.

*José Antonio Marmolejo*
*Panamerican University. Mexico*
*July 2017*

**José Antonio Marmolejo** *is a Professor at Panamerican University, Mexico. His research is on operations research, large-scale optimization techniques, computational techniques and analytical methods for planning, operations, and control of electric energy and logistic systems. He received his Doctorate in Operations Research (Hons) at National Autonomous University of Mexico. At present, He has the third highest country-wide distinction granted by the Mexican National System of Research Scientists for scientific merit (SNI Fellow, Level 1). He is a member of the Network for Decision Support and Intelligent Optimization of Complex and Large Scale Systems and Mexican Society for Operations Research. He has co-authored research articles in science citation index journals, conference proceedings, presentations, books and book chapters.*

# Preface

## INTRODUCTION

The humankind has a complex role on shaping the World with new discoveries, and inventions. Although there are many problems about not protecting the peace, and the natural balance of the World, humankind will always the main actor of determining the future as something similar to a utopia or dystopia. In this context, the life and its standards have been changed with new discoveries, and inventions leading to use of new technologies. Today, one cannot deny that there are lots of technological tools and approaches directing us to experience a different type of life according to people lived before in the past centuries. Majority of general ideas about that changed experience of the life is associated with an improvement within the modern life.

It can be clearly said that rapid developments and improvements within different technologies have remarkable roles on improving our modern life. In this way, it is always possible to obtain effective solutions for especially real-world based problems. If we can just have a second to observe the environment surrounding us, we can see there are many different examples of technology use towards solving real-world based problems. At this point, it is also important that mathematical dynamics of the related solution approaches are generally based on natural dynamics. We can say that nature has a remarkable role on thinking about solutions and designing mathematical structures in the way of developing effective scientific, computational approaches, methods, or techniques. As the humankind causes eliminating of some natural sources because of its unstoppable passions, the nature has always given many sources to us in order to understand it and solve the problems with inspirations or discoveries within its dynamics.

In time, this connection between nature and science has made it possible to introduce new research fields. Artificial Intelligence is also one of these fields and it employs many sub-research interest appeared in time thanks to great diversity of solution ways. In detail, nature-inspired techniques are one the remarkable solutions ways that can be examined in detail and they can be used in the context of effective and efficient intelligent solution approaches widely. Because of their great potential on forming effective algorithmic solution ways, nature-inspired techniques have always been popular among Artificial Intelligence based research works. It is also remarkable that these techniques can be used with other Artificial Intelligence techniques in order to form hybrid solution models. Eventually, it is all Artificial Intelligence having a great potential of changing the way of traditional solutions and proving to solve all the problems that have never solved before.

The related inspiration from nature can be both from human or living organism systems or iteratively occurred, intelligent-like dynamics in the nature. Generally, especially cooperative behaviors by swarms in the nature had enabled researcher to design and develop different types of algorithmic techniques inspired from swarms like bird, bee, fish swarms, so it has been possible to discuss about a new research interest called as Swarm Intelligence. On the other hand, it is also possible to classify such nature-inspired techniques by focusing on some mechanisms inspired from the theory of Evaluation. It is also remarkable that there has been always a focus on improving already designed nature-inspired techniques in order to fit them better to the problems encountered.

Real-world based problems have always been important elements to evaluate success of such techniques. Research works in the associated literature have shown that especially mathematical features of nature-inspired techniques can be adapted to all considered problems from different disciplines. Eventually, flexible and improvable mathematical features of nature-inspired techniques have made it easier to employ them in different disciplines in order to obtain effective solutions for the newly encountered problems. From this perspective, it can be thought that nature-inspired techniques are important actors on keeping great potential of the Artificial Intelligence field for the future.

*Nature-Inspired Intelligent Techniques for Solving Biomedical Engineering Problems* aims to provide a general view on applications of nature-inspired techniques (Artificial Intelligence) in Biomedical Engineering, discuss about potentials of the Artificial Intelligence field from an engineering perspective, and finally improve awareness on the power of Artificial Intelligence and nature-inspired techniques on ensuring effective and efficient solutions for encountered biomedical based problems. As it can be understood from the objectives, the mission of the book is to provide a valuable, quality reference work by focusing on an enough spectrum of the biomedical world meeting with the Artificial Intelligence. The book is a great opportunity of international audience from both Artificial Intelligence and Biomedical Engineering to provide their results on applications of nature-inspired techniques, enable international audience to improve their knowledge and awareness on the associated literatures, and keep a unique reference work.

The book is briefly a research document intersecting both Artificial Intelligence and Biomedical Engineering. In this sense, it can be expressed that it focuses on aiding interested people, academicians, researchers – scientists within the Artificial Intelligence (and of course also sub-areas like nature-inspired techniques, Machine Learning, Swarm intelligence...etc.) and the covered sub-fields of the Biomedical Engineering. Furthermore, the book can be used as also an educational material to examine latest improvements within the covered fields and understand the use of nature-inspired techniques (Artificial Intelligence) in this context.

The chapters in the book include mostly use of nature-inspired techniques and reporting obtained results on problems solutions. In this context, used techniques differ from Swarm Intelligence (i.e. Particle Swarm Optimization, Artificial Bee Colony, Ant Colony Optimization) techniques, Evolutionary (i.e. Genetic Algorithms) algorithms, general nature-inspired techniques (i.e. Artificial Neural Networks) and even newly introduced hybrid approaches formed by more than one techniques. In this context, it is also possible to said that this book is a source for general chapters providing views on the recent literature, future works, analysis of the subject area, recent specific research orientations...etc. Eventually, a reference work on multidisciplinary application potentials of nature-inspired techniques (Artificial Intelligence) is obtained, thanks to efforts from valuable scientific audience.

Considering the chapters here, there have been a good diversity of scope. At this point, we can express proudly that the desired coverage listed as follows can be examined among the pages of this book by the interested readers:

- General nature-inspired Artificial Intelligence techniques (Artificial Neural Networks, Genetic Algorithms, etc.) and their usage in problems of Biomedical Engineering.
- Nature-inspired, Artificial Intelligence optimization algorithms (Particle Swarm Optimization, Ant Colony Optimization, Artificial Bee Colony, etc.) and their usage within optimization problems of Biomedical Engineering.
- Swarm Intelligence based, nature-inspired system applications.
- Knowledge-based, nature-inspired system applications.
- Intelligent problem solutions associated with medical diagnosis approach supported by nature-inspired techniques.
- Applications of intelligent hybrid systems formed by nature-inspired techniques and other Artificial Intelligence techniques.
- Discussions on the most recent literature of nature-inspired techniques over Biomedical Engineering problems.
- Discussions on the future of nature-inspired techniques as associated with the problems of Biomedical Engineering.
- Discussions on the relations between Artificial Intelligence and other alternative, strong research fields.
- Applications of newly developed nature-inspired techniques within Biomedical Engineering oriented problems – studies.
- Alternative, software-based, intelligent solution systems for applying to specific problems related to Biomedical Engineering.
- Focus on ethical issues regarding to applying intelligent solutions within Biomedical Engineering problems.

## ORGANIZATION OF THE CHAPTER

In the content, there are 15 chapters about research works that have been performed based on both nature-inspired techniques and Biomedical Engineering oriented problems. The related chapters can be explained briefly as follows:

### Chapter 1

Artificial Intelligence has a remarkable effect on many different fields with its flexible and comprehensive solution approaches to solve especially real world problems. In this context, the field of Biomedical Engineering has also been affected by employment of different Artificial Intelligence based techniques. In this chapter, it is aimed to give a theoretical discussion on using Nature-Inspired Artificial Intelligent techniques for obtaining intelligent applications within Biomedical Engineering. As it is known, techniques appeared within the field of Artificial Intelligence are inspired from the nature. So, it is a good approach to focus on Nature-Inspired techniques for discussing about intelligent Biomedical Engineering research works. In detail of this chapter, the readers will have a chance to understand some ways of running Artificial Intelligence for achieving better results in Biomedical Engineering and the related developments associated with this field.

## Chapter 2

Swarm intelligence is a branch of computational intelligence where algorithms are developed based on the biological examples of swarming and flocking phenomena of social organisms such as a flock of birds. Such algorithms have been widely utilized for solving computationally complex problems in fields of biomedical engineering and sociology. In this chapter, two different swarm intelligence algorithms namely the Jumping Frogs Optimization (JFO) and Bacterial Foraging Optimization (BFO) are explained in detail. Further, a synergetic algorithm namely the Coupled Bacterial Foraging/Jumping Frogs Optimization Algorithm (BFJFO) is described and utilized as a tool for control of the heroin epidemic problem.

## Chapter 3

Managing medical information and knowledge is becoming an increasing problem for healthcare professionals. Medical science that contains ever-increasing amounts of knowledge, such as the medical history of a patient, medical data about diseases, diagnosis and treatment methods, should be necessarily a science of information. The real problem faced by patients and health care providers is finding and using relevant knowledge at the right time. In this context, in the middle of 1950s, intelligent computer systems, called Clinical Decision Support Systems (CDSS), were introduced as a new concept. CDSS is defined as an active intelligent system that can help medical experts to make decisions by taking specific recommendations. Also, it provides decisions based on resolving patient specific information and related medical truths. Objective of this chapter is to focus on these systems and explain relations with the field of Artificial Intelligence methods, approaches or techniques in this manner.

## Chapter 4

Artificial immune systems inspired by the natural immune system are used in problems such as classification, optimization, anomaly detection, error detection. In these problems, Clonal Selection Algorithm, Artificial Immune Network Algorithm and Negative Selection Algorithm are generally used. This study aims to solve the problem of correct identification and classification of patients using Negative Selection (NS) and Variable Detector Negative Selection (V-DET NS) algorithms. Here we examined the performance of NSA and V-DET NSA algorithms using three sets of medical data sets from Parkinson, Carotid Artery Doppler and Epilepsy patients. According to the obtained results, NSA achieved 92.45%, 91.46% and 92.21% detection accuracy and 92.46%, 93.40% and 90.57% classification accuracy. V-DET NSA achieved 94.34%, 94.52% and 91.51% classification accuracy and 94.23%, 94.40% and 89.29% detection accuracy. As can be seen from these values, V-Det NSA yielded a better result. Artificial immune system emerges as an effective and promising system in terms of problem solving performance.

## Chapter 5

In this study, the differential diagnosis of erythematous diseases was determined using data mining and machine learning algorithms. In the study, dermatology dataset which was shared in UCI Machine Learning Repository was used. The dataset consists of 366 data items with 34 attributes. Initially, feature selection was made and then, classification was performed by using various algorithms. The number of attributes has been reduced from 34 to 19 as a result of the integration of the correlation based filter methods and various heuristic search methods. The evaluation results show that Naive Bayes has 100% success rate in classification of psoriasis, seborrheic dermatitis, lichen planus, rose disease, chronic dermatitis and pityriasis rubra pilaris diseases with 19 attributes selected with feature extraction algorithms.

## Chapter 6

One of the most popular application ways of Artificial Intelligence within medical field is developing medical diagnosis systems. Because Artificial Intelligence based techniques are able to use pre-data and instant data flow for making predictions, it is an easy task to design intelligent systems that can give advices to people or perform diagnosis based decision makings. So, it has been an important research interest to design and develop intelligent systems, which are able to make diagnosis for medical purposes. In this sense, objective of this chapter is to introduce a general medical diagnosis system that can be used for detecting diseases. In detail, the system employs Artificial Neural Networks and Swarm Intelligence based techniques to form a general framework of intelligent diagnosis. The chapter briefly focuses on infrastructure of the system and discusses about its diagnosis potentials.

## Chapter 7

Aim of this chapter is to classify normal and extra systole heart sounds using by artificial intelligence methods. Initially, both of heart sounds have been passed from Butterworth, Chebyshev, Elliptic digital filter in specific frequency values to remove noise. Afterwards, features of heart sounds have been obtained for classification. For this process, Wavelet transform and Mel-frequency Cepstral Coefficients (MFCC) methods have been applied. Training and test data have been created for classifier by taking means and standard deviation of gained feature. Support Vector Machine (SVM) and Artificial Neural Network (ANN) methods have been used for classification of these heart sounds. Using Wavelet and MFCC features, classification success of SVM has been obtained as 93,33% and 100% respectively. Using Wavelet and MFCC features, classification success of ANN has been obtained as 83,33% and 90% respectively.

## Chapter 8

Bacteria counting process and bacterial counting by eye can be performed by manual counting in which a specialist counts the bacteria with naked eye or the automatic counting that utilizes the computer based techniques. The counting process becomes exhausting, long and incorrect when the counting performed by specialist. Therefore, there is a need for an efficient and more reliable methods for the bacteria counting process performed automatically. In this study, a semi-automatic method has been designed for bacterial counting from the images of petri plates and photomicrograph. For this purpose, the image processing procedure has been applied to the images. In the image processing procedure, watershed transform was used for to separate the adjacent bacteria and thus a more accurate counting was performed. The efficiency of the presented method has been shown for many different images, and the counting results of the proposed method has been compared with counting results of an expert. The method has been shown to be a successful and promising work for bacterial counting.

## Chapter 9

Artificial Intelligence is widely enrolled in different types of real world problems. In this context, developing diagnosis based systems is one of the most popular research interest recently. Considering medical service purposes, using such systems have enabled doctors and other individuals taking roles in medical services to take instant, efficient expert support from computers. One cannot deny that intelligent systems are able to make diagnosis over any type of disease. That just depends on decision making infrastructure of the formed intelligent diagnosis system. In the context, of the explanations, this chapter introduces a diagnosis system formed by Support Vector Machines (SVM) trained by Vortex Optimization Algorithm (VOA). As a continuation of previously done works, the research considered here aims to diagnose diabetes. The chapter briefly gives information about details of the system and findings reached to after using the developed system.

## Chapter 10

Recently the field of Artificial Intelligence encounters with some ethical questions as associated with especially future of humankind. Although it is a common question asked for years, the existence of humankind against badly configured intelligent systems, is more important nowadays. As a result of rapid developments in intelligent systems and their increasing role in our life, there is a remarkable anxiety about dangerous Artificial Intelligence. Because of that, some research interests gathered under some topics like Machine Ethics, Future of Artificial Intelligence, and even Existential Risks are taking researchers' interest too much. As associated with this state, objective of this chapter is to examine ethic factors in using intelligent systems for Biomedical Engineering oriented purposes. The chapter will firstly give essential information about the background and then consider possible scenarios that may require ethical adjustments during design and development of Artificial Intelligence oriented systems for Biomedical Engineering problems.

## Chapter 11

Time series analysis has a wide application interest in Artificial Intelligence oriented research studies. Because it is easy to run Machine Learning based solutions directly over time series, it has been a popular approach to use alternative types of intelligent systems to analyze time series. Regarding to such works, time series prediction is known as a remarkable topic as followed by researchers from even different fields. Objective of this chapter is to provide an alternative work by using Artificial Neuro-Fuzzy Inference System trained by the League Championship Algorithm, which is an optimization algorithm from the associated literature. As the application objective, Electroencephalogram (EEG) time series have been tried to be predicted by using the designed ANFIS-LCA approach. The chapter briefly introduces details about the approach and reports findings from the performed prediction operations.

## Chapter 12

Currently, medical diagnosis has a strong relation with the Artificial Intelligence oriented approaches. Because it is practical to employ intelligent mechanisms over some input data-expert knowledge and design effective solution ways, even Biomedical Engineering field is interested in taking support from Artificial Intelligence. If applications in this manner are taken into consideration, we can see that medical diagnosis have a big percentage. In the sense of the explanations, objective of this study is to use Genetic Algorithm (GA) for diagnosing headache diseases. As a popular and essential technique benefiting from evolutionary mechanisms, GA can deal with many different types of real world problems. So, it has been chosen as the solution way - algorithm over the headache disease detection problem, which shapes the research framework of the study. The paper content gives information about the performed diagnosis application and the results.

## Chapter 13

Along with the rise of Artificial Intelligence (AI), there are many different research fields taking much more place in our life. Because of growing amount of data and needs for immediate access to information for dealing with the problems, different types of research fields take place within the scientific community. Internet of Things (IoT) is one of them and it enables devices to communicate with each other machines in order to form a general network of physical, working devices. Objective of this chapter in this manner is to provide a general discussion over using nature-inspired techniques of AI to form the future of Biomedical Engineering build over IoT. Because it is often thought that the medical services of the future will be based on autonomous machines supported with AI and IoT, discussing about such topic by considering Biomedical Engineering applications will be good for the related literature.

## Chapter 14

Deep Learning (DL) is a new method of Machine Learning, as running over Artificial Neural Networks, which have structure above of the standards to deal with big amount of data. That is generally because of increasing amount of data, input data sizes, and of course more complexity of objective real-world problems. Performed research studies in the associated literature show that the DL currently has a good performance among considered problems and it seems as a strong solution way for more advanced problems of the future. In this context, this chapter aims to provide some essential information about DL and its applications within the field of Biomedical Engineering. The chapter was organized as a reference source for enabling readers to have idea about relation between DL and Biomedical Engineering.

## Chapter 15

Cloud Computing is an effective way of using hardware and software oriented resources at optimum levels. Thanks to this technology, it is also possible to share big amount of resources effectively and accurately among target users. Because it is a rapidly growing technology, one cannot deny that it has remarkable relations with alternative research fields having great potential and application scope. It is clear that Artificial Intelligence is one of these fields. As associated with both these research fields - technologies, purpose of this chapter is to examine Artificial Intelligence based Biomedical Engineering works supported - connected with Cloud Computing. Because it has a vital importance with applications regarding to the medical - health problems, Biomedical Engineering needs support from the most recent technologies and research fields in this manner. So, the chapter has provided a view over the intersection of these three research fields as trying to improve awareness among interested readers.

## CONCLUSION

Generally, this book comes with a great opportunity for the scientists and researchers in the field of Artificial Intelligence and Biomedical Engineering to disseminate their original, quality, novel and innovative research work findings and obtained results to be read by the global research scholars. Here, we strongly suggest to benefit from all roots of the book to have idea about the intersection of Artificial Intelligence and Biomedical Engineering. In addition to the chapters and suggestions by their valuable authors, we have also listed a wide list of additional references at the end of the book. It is another strong suggestion to improve the knowledge and have some ideas about which scientific works to be examined considering the whole scope of the book.

As the editors, we sincerely thank all the contributors for their valuable efforts and kind interests in enabling this book to be a global reference for the scientists, engineers, researchers, and academicians as well as anyone, who are interested in both Artificial Intelligence (more specifically nature-inspired techniques) and also Biomedical Engineering fields. We are looking forward to being serving for the scientific community with newer projects.

*Utku Kose*
*Suleyman Demirel University, Turkey*

*Gur Emre Guraksin*
*Afyon Kocatepe University, Turkey*

*Omer Deperlioglu*
*Afyon Kocatepe University, Turkey*

# Acknowledgment

As the editors of this valuable reference book, we would like to take this great opportunity to sincerely thank the following friends and colleagues for their kind help and valuable support on the book content: Ahmet Arslan, Aydin Cetin, Harun Uguz, Pandian Vasant, and Tuncay Yigit. Their fabulous ideas, suggestions, feedback, and also constructive comments for the improvement of the organization and quality of the book are gratefully acknowledged.

Furthermore, we sincerely thank the group at IGI Global in Hershey PA, USA, for their great help and excellent support on this project. In particular, special thanks to Ms. Jan Travers, Ms. Katie Shearer, and Ms. Marianne Caesar of IGI Global for their great and valuable cooperation.

*Utku Kose*
*Suleyman Demirel University, Turkey*

*Gur Emre Guraksin*
*Afyon Kocatepe University, Turkey*

*Omer Deperlioglu*
*Afyon Kocatepe University, Turkey*

# Introduction

Intelligence has been defined in many different ways by many scientists working on it. Most recently, many psychologists have developed very different theories about the concept of intelligence, as in the theories of multiple intelligences. Intelligence is information of the person' s learning, understanding and the ability to deal with the environment. These specific abilities can be listed as follows:

- The ability to changes in the current environment or to adapt to another environment.
- The capacity to obtain knowledge and information.
- Intelligence and abstract thinking capacity.
- Understanding associations.
- Assessment and judgment skills.
- Original and productive thinking ability.

It is possible to add many more abilities to the list above. But in general, it is to investigate the ability of the person the learning and practice and their relationship with the environment. In medical application, the concept of intelligent emerges in many different ways and many different applications. Continuous improvement in the world of information technology, the methods of diagnosis, consultations and treatments has changed naturally. Information technology has been playing a very important role in biomedical engineering and new technological solutions have emerged into existing problems of medical applications. Information technology has been playing a very important role in biomedical engineering. New technological solutions have especially emerged into existing problems of medical decisions. With the emergence of advanced computers medical reasoning and inference systems began to change. Firstly, the concept of medical record with the popularization of the Internet has been changed to electronic records. Then everything started to get clever name, the classic insights was completely changed such as smart equipment, smart devices, interactive software, intelligent systems, tele-medicine with mobile devices.

The concept of intelligent medical environment, health workers can easily reach, can be used safely, and are able to solve problems of medical consultation, diagnosis and treatment. It is quite different from traditional medical environments. It does not cover only clinical data, individual patient data, and inference events, but also adapts to the user requests.

As understood from the above description, intelligent medical environments are some kind of software solutions, which are designed to facilitate effective medical decision support through the exploration of complex problems. Generally, intelligent medical environments are mentioned together with the concepts of decision support systems and knowledge based systems. Intelligent medical systems in studies many different names and many different approaches have been given. Some of these are listed as follows: An intelligent information access system assisting, a clinical decision support system, a web-based clinical decision support system, an intelligent diagnosis system, an adaptive clinical decision support system, an adaptive and intelligent hospital systems etc.

Intelligent medical systems are an interactive computer software designed to assist healthcare workers in their decision-making tasks. Intelligent medical systems employ artificial intelligence and knowledge based techniques to help clinicians inference and decision making. In these systems, the ability of clinicians to decision making from clinical data can be selected as a suitable platform and different possibilities can be offered. So, these systems provide a more efficient medical environment to clinicians with different levels of expertise and inference abilities, by providing different environments. In Intelligent medical systems design, many methods or approaches such as Bayesian networks, artificial neural networks, genetic algorithms and rule-based approaches are used.

Artificial intelligence, which has become a subdivision of computer science, has found many fields of application such as education, medical, commercial, and industrial applications. In particular, artificial intelligence techniques, which are more effective in solving problems in nature, are inspired by nature. Researchers are trying to develop new computer techniques by simulating existing systems in nature such as ant colony optimization, particle swarm optimization, artificial neural networks, vortex optimization and artificial immune system to solve complex problems and develop the most appropriate solution. There is a need for multidisciplinary rigorous work in order to optimize the algorithms, techniques, or approaches inspired by nature to computer environments. Recently, many algorithms have been developed that are inspired by nature to solve numerical and combinational optimization problems. These algorithms mimic the way the real biological system works such as Particle Swarm Optimization, Ant colony Optimization, Bacterial Foraging Optimization, River Formation Dynamics, Stochastic Diffusion, artificial neural networks, vortex optimization and artificial immune system. At this point, we can also indicate that the famous meta-heuristics and nature-inspired methods and techniques, which have been just mentioned and also the other ones such as other Swarm Intelligence techniques, Artificial Neural Networks, Support Vector Machines, Artificial Neuro-Fuzzy, Genetic Algorithm, Wavelet, Machine Learning, and even Cloud Computing techniques as out of Artificial Intelligence have been extensively and rigorously applied in solving the real world problems in Biomedical Engineering.

It is an extremely need for forming a reference work, which focuses on Intelligent medical systems, containing application of nature-inspired Artificial Intelligence techniques. Generally, it is possible to mention that such reference works would be indicators of the current and cutting edge research development in the research field of Artificial Intelligence. In more detail, such works will eliminate the gap about decision support systems that form the basis of adaptive and interactive health systems. Because of their ability to remove distance and time constraints, web-based systems and related techniques are supported at a high rate by means of intelligence medical environment. So, these topics are remarkable ones to be examined by the researchers. In addition, the future of biomedical engineering will be determined with the employment of intelligent systems within all medical fields. So, alternative works in the literature give us a chance to see the current potential of nature-inspired Artificial Intelligence techniques in medical application activities and enables us to think about the future better. Biomedical Engineering problems are currently very famous across the planet and researchers from all over the world very keen to solve these problems by using nature-inspired intelligent techniques. Forming reference works in this manner would be a wonderful contribution to the body of the knowledge in the research interests of Biomedical Engineering as being useful and fruitful research reference for the academicians, medical operators, decision makers, post-graduate students, industrialists, managers and executives across the globe.

By taking the related reasons indicated briefly under the previous paragraphs, this book in which you will just lose yourself within the next pages – chapters, aims to have a role on contributing to the related literature as being a specific focus to the nature-inspired techniques. Because of that, its title includes the words of 'nature-inspired'. We suggest readers the following points while reading the content:

- Follow the provided references and the additional readings,
- Take notes of important points,
- Consider what has been done in each work and what type of open problems may be remaining,
- What is waiting for us regarding to the future of intelligent Biomedical Engineering.

*Omer Deperlioglu*
*Afyon Kocatepe University, Turkey*

*Pandian Vasant*
*Universiti Teknologi Petronas, Malaysia*

# Chapter 1
# Towards an Intelligent Biomedical Engineering With Nature-Inspired Artificial Intelligence Techniques

**Utku Kose**
*Suleyman Demirel University, Turkey*

## ABSTRACT

*Artificial intelligence has a remarkable effect on many different fields with its flexible and comprehensive solution approaches to solve real-world problems. In this context, the field of biomedical engineering has also been affected by employment of different artificial intelligence-based techniques. This chapter aims to give a theoretical discussion on using nature-inspired artificial intelligent techniques for obtaining intelligent applications within biomedical engineering. As it is known, techniques within the field of artificial intelligence are inspired from nature. So, it is a good approach to focus on nature-inspired techniques for discussing intelligent biomedical engineering research works. Readers will have a chance to understand some ways of using artificial intelligence for achieving better results in biomedical engineering and the related developments associated with this field.*

## INTRODUCTION

Many different fields have effective roles on shaping the humankind's life. Even since ancient eras, humans have had the chance of using knowledge and abilities to change the world and their life, thanks to some discoveries and inventions. In time, specializations have appeared as a result of increasing needs for newer discoveries and inventions. As a result, developments and improvements determining current and future state of the humankind have been performed and classified under different fields. Today, although some main scientific fields like Physics, Chemistry, Mathematics has still great importance in scientific developments, some new specializations formed thanks to combinations of developments in such fields have started to take our interests. It is clear that especially rise of computer and electronic technologies

DOI: 10.4018/978-1-5225-4769-3.ch001

have taken the triggering roles in rapid appearance of newer fields. Today, Artificial Intelligence is one of these fields and it generally has a great importance in developing intelligent systems, which are able to simulate intelligent behaviors shown by humans or other living organisms. Also, considering the current state of Artificial Intelligence, we can say that it is associated with all things in the nature if they have something intelligently used even they belong to lifeless objects. As the future has many alternative scientific fields like Big Data (Mayer-Schonberger & Cukier, 2013; Zikopoulos & Eaton, 2011), Internet of Things (Weber & Weber, 2010; Xia et al., 2012), it is certain that the future of humankind will be based on collaboration of these and many alternative fields. Here, Artificial Intelligence seems to be having a remarkable place with its flexible and effective solution approaches, methods, and techniques for every kind of real-world based problems.

When we focus on connections of the Artificial Intelligence with other fields of the modern life, it is possible to see many different research work orientations changing according to structure of different fields. Especially applied fields have great potential to employ intelligent solutions designed over techniques of Artificial Intelligence. In detail, engineering oriented fields are important to be considered in this manner. Currently, the field of Biomedical Engineering is one of the trendiest fields in which intelligent solution approaches are widely used. Dealing with the applications of engineering oriented studies within medical and biology to solve healthcare problems (Bronzino, 1999; Bronzino & Peterson, 2014; Enderle & Bronzino, 2012), Biomedical Engineering is an important field needing support from innovative solution ways and the field of Artificial Intelligence has done that easily according to the latest developments done so far. This situation can be also a result of instant needs for effective solutions for healthcare services and solutions done on the background. Because studies associated with Artificial Intelligence has proved the role of intelligent solution mechanisms for solving problems of even the most complicated fields have always resulted in successful and effective outputs at the end, popularity of Artificial Intelligence based applications in Biomedical Engineering has increased day by day. At this point, it would be a good way to improve awareness among anyone, who are interested in employment of intelligent solutions for Biomedical Engineering problems.

In this chapter, it is aimed to give a theoretical discussion on using nature-inspired Artificial Intelligent techniques for obtaining intelligent applications within Biomedical Engineering. As it is known, techniques appeared within the field of Artificial Intelligence are inspired from the nature. So, it is a good approach to focus on nature-inspired techniques for discussing about intelligent Biomedical Engineering research works. In detail of this chapter, the readers will have a chance to understand some ways of running Artificial Intelligence for achieving better results in Biomedical Engineering and the related developments associated with this field. From a general perspective, the chapter has written as a starting reference for the works focused on intelligent solutions for Biomedical Engineering problems.

Considering the subject of the chapter, remaining sections are organized as follows: The next section is devoted to some essential information on how Artificial Intelligence can be employed within different types of applied studies – problems. After that, the third section explains some nature-inspired techniques that can be run among the related problems. Following the third section, the fourth section provides information about essential ways of running nature-inspired techniques for solving Biomedical Engineering oriented studies. Next, the fifth section focuses some ideas regarding to the future and the chapter is ended with final discussions under the last – sixth section.

## APPLICATION POTENTIALS OF ARTIFICIAL INTELLIGENCE

Before focusing on the concept of intelligent Biomedical Engineering and its causes, it is better to explain why that field of Artificial Intelligence can be easily employed within different fields. In detail, these explanations will give us some ideas about flexible sides of the field of Artificial Intelligence.

### Elements of Artificial Intelligence

Artificial Intelligence is generally build over some essential elements, which are associated with different sides of intelligent problem solution ways. These elements are some type of building blocks that enable us to understand better about details of Artificial Intelligence. The related elements can be explained briefly as follows:

- **Problems:** Artificial Intelligence is actually a scientific approach that was appeared as a result of real-world based problems. Because there have been some unsolved problems needing innovative solutions or already problems needing more effective solutions, intelligent mechanisms simulating something from the nature has found its way to be employed through problems.
- **Math and Logic:** Typical Artificial Intelligence approaches, methods, and techniques employ math and logic based, algorithmic solution steps to achieve effective, efficient, and flexible solutions for encountered problems. Because all solution mechanisms of intelligent systems employ strong logical structures and Math on the background, it has been always possible to use Artificial Intelligence in different problems, adapt it for newly encountered problems if it is necessary, and also design and develop newer intelligent methods or techniques that can be specifically run for different problems.
- **Simulation:** One of the most important element of Artificial Intelligence is its simulation capabilities. In this context, simulation means not directly simulating something but inspiring from some aspects of nature to simulate them under algorithmic structures supported with Math and logic. Nature is associated with many other actors like humans or natural dynamics.
- **Human – Nature:** Starting points of the field of Artificial Intelligence are human and the nature and it seems that the future of this field will be always associated with them. At this point, it possible to express that the human has great importance on shaping first essentials of Artificial Intelligence by enabling researchers to think about possibilities on simulating human behaviors and thinking styles for getting effective intelligent computational algorithms. In detail, even human brain and many other organism systems have enabled Artificial Intelligence to derive new techniques. Later, other living organisms and even lifeless objects in the nature have been inspiration sources for developing new intelligent techniques. At the final, we can accept the human and the associated objects of the nature as under another element completing the general structure of the Artificial Intelligence.

The mentioned elements can be examined in more detail, but when we talk about just them uniquely, it is possible to understand what the concept of Artificial Intelligence is like to be. According to that, it is possible to give an equation of Artificial Intelligence formed by human_nature as included by the function of simulation and the function of math_logic to be applied for a problem (Equation 1).

*Figure 1. Elements of Artificial Intelligence*

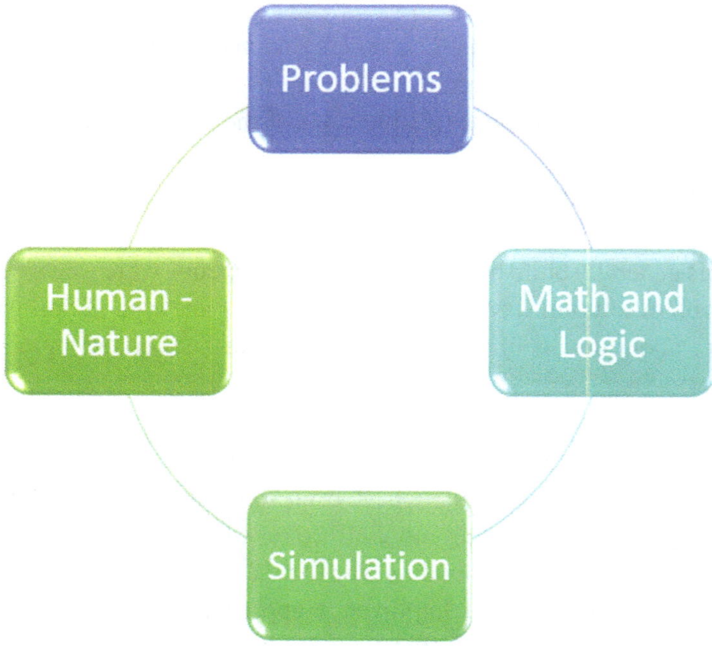


$$AI = math\_logic(simulation(human\_nature)) \quad (1)$$

## Artificial Intelligence Approaches – Methods – Techniques

Because the field of Artificial Intelligence has been an active field within time, there have been many different variations under it. In time, some more alternative research ways have appeared under same problem applications. So, it has been necessary to use a classification scheme under Artificial Intelligence. At this point, we can firstly examine Artificial Intelligence based solutions under some approaches, which are actually associated with problem types in which intelligent solutions are widely used. Under the approaches, we can see some specific methods of Artificial Intelligence and finally, methods include specific techniques, which everyone generally know more about them / their names (Figure 2). As an example, adaptive control or prediction can be accepted as an approach whereas Swarm Intelligence (Kennedy, 2006), Machine Learning (Alpaydin, 2014), or Cybernetics (Wiener, 1961) are some methods under them. Some well-known intelligent solution mechanisms called as Fuzzy Logic (Klir & Yuan, 1995; Zadeh, 1996), Artificial Neural Networks (Principe & Fancourt, 2013; Yegnanarayana, 2009), or Genetic Algorithms (Holland, 1995; Holland, 2012) are techniques of the field of Artificial Intelligence.

## Applying Intelligent Solutions in Problems

Artificial Intelligence based techniques can be applied in different real-world based problems by running different approaches. At this point, it is important to determine how to employ an intelligent solution within a specific problem. In detail, it is possible to use an intelligent solution mechanism by using it i.e.

*Figure 2. Approach – method – technique organization in Artificial Intelligence*

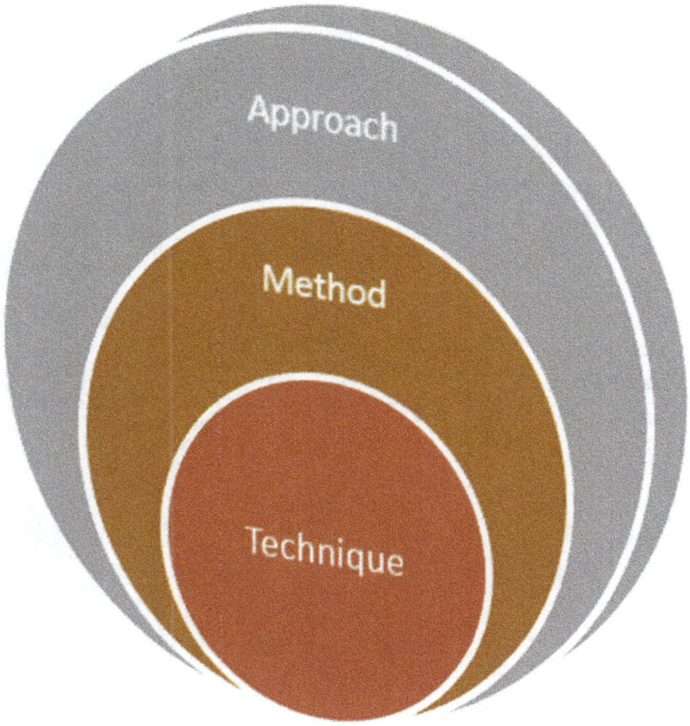


directly over the problem or achieving something hybrid to get a desired solution flow. Different types of solution application approaches followed with the Artificial Intelligence can be explained briefly as follows:

- **Direct Application:** For solving unsolved problems or providing alternative and / or better solutions for problem, a common way is to use an Artificial Intelligence technique directly over the problem.
- **Hybrid Application:** Sometimes, it can be not too possible to solve a complex problem by using just one Artificial Intelligence technique. In such times, researchers generally form a hybrid system formed by two or more techniques to achieve the desired solution flow. In the literature, there are many different types of hybrid Artificial Intelligence based systems and this approach is very trendy nowadays (Abraham & Pedrycz, 2008; Cheng & Lien, 2012; de Carvalho et al., 2014; Corchado et al., 2012; Hedayat et al., 2009; Hudson & Cohen, 2000; Nikolopoulos, 1997; Spelt et al., 1991; Toro et al., 2013; Wermter, 2000).
- **Using as a Step:** For improving quality and effectiveness of solutions in different fields, it is also possible to use Artificial Intelligence techniques as a step among the solution flow. This is achieved by employing single technique or a hybrid structure within an already used solution flow designed before.
- **Using as Pre or Post Support:** For some problems, it is possible to use intelligent techniques to prepare inputs of the problems for the solution process or analyze – evaluate outputs at the final. This approach can be accepted as also the type of 'using as a step' but it has different meanings in detail as its constant place as 'before' or 'after' support.

*Figure 3. Application approaches of intelligent solutions within problems*

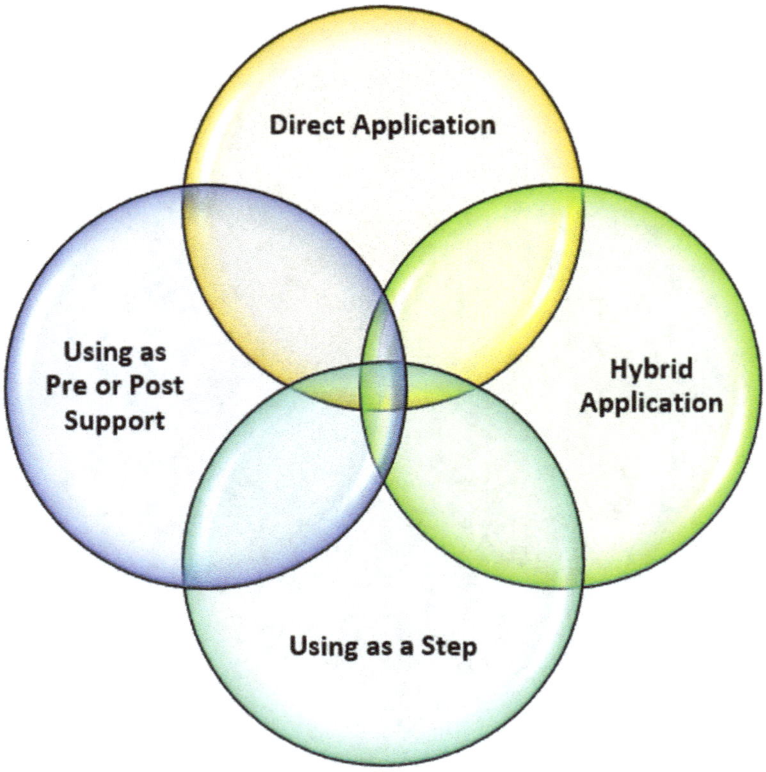


## Flexibility of Artificial Intelligence

One important question widely asked about Artificial Intelligence is that 'how it can be such flexible to be applied in different fields even for complex problems?'. This question can be answered from different perspectives with many different 'making-sense' reasons. Here, some ideas by the author can be expressed as follows:

- **Learning Abilities:** Under Machine Learning, Artificial Intelligence has many different techniques that can run 'training' process to learn something from the data – environment and act according to the training processes – experiences for next problems. In detail, learning types of Artificial Intelligence techniques are classified under three types called as 'supervised learning', 'unsupervised learning', and 'reinforcement learning' respectively (Caruana & Niculescu-Mizil, 2006; Coates et al., 2011; Kaelbling et al., 1996; Sutton & Barto, 1998). But the main mechanisms of such learning processes as changing the actions of techniques make it easier to adapt techniques to different types of problems included under different fields.
- **Effective Use of Math and Logic:** Artificial Intelligence techniques are based on unique algorithmic structures supported with Math and logic. Because use of mathematical equations with well-designed logical flows make it successful to achieve something flexible that can direct itself to different scenarios according to the objective problem and making the general structure flexible enough.

- **Similarity to the Common Reasoning Followed by the Human and Nature:** Artificial Intelligence techniques employ simple logical reasoning flows, which are meaningful for us and the natural dynamics occurred in the nature. Because this reasoning can be modified easily, it is possible to be flexible for any type of Artificial Intelligence technique.
- **Common Sides of Different Problems:** Out of intelligent solutions, it is also possible to indicate that many different real-world based problems actually have some common sides because some scientific fields like Math and logic has always very similar approaches in even different problems. Because of such common sides, it is easier to use similar solution ways for problems from different fields. This can be another reason explaining the flexibility of Artificial Intelligence but of course, Artificial Intelligence based techniques often provide better solutions according to traditional solution ways, which is a brief explanation for multidisciplinary potential of the Artificial Intelligence.

Application potentials have made Artificial Intelligence an important multidisciplinary problem solution approach for different fields. On the other hand, some techniques have attracted researchers to focus more on nature-inspired techniques that can be examined in detail under the field of Artificial Intelligence. Although all intelligent techniques can actually be associated with aspects of the nature, some of them are more understandable if we consider not only human but also other objects taking part in the nature.

## NATURE-INSPRED ARTIFICIAL INTELLIGENCE TECHNIQUES

If we consider both humans and the remaining nature in a common framework, it can be seen that the whole Artificial Intelligence techniques are somehow nature oriented solution ways. But in this chapter, it is more focused on the techniques coming from the natural factors out of humans. Actually, if we think about some techniques like Artificial Neural Networks, it is possible to see that it is related to not only human brain or neurons but also to brains and neurons of other living organisms. So, because such living organisms are part of the nature, we can include such techniques in the same scope (Also, it is a better classification to include them under bio-inspired techniques in the context of nature-inspired techniques). But for example Fuzzy Logic (Klir & Yuan, 1995; Zadeh, 1996) is generally accepted as inspired from human thinking, so it is out of scope of the nature-inspired techniques here.

### Artificial Neural Networks

Artificial Neural Networks (ANN) is generally known as a technique inspired from human brain. But here, it is considered within the scope of the nature, because it simulates the interaction mechanism of biological neurons, which are part of not only humans but also living organisms having brains and neural systems as being part of also nature. ANN are briefly parallel and distributed data processing systems, which were developed to be used for simulating neurons and in this way, they are able to learn from the environment in order to improve the process performance through learning (Deperlioglu & Kose, 2011). ANN are maybe the most popular Artificial Intelligence technique as often used by researchers to solve problems in a wide scope of i.e. classification, prediction, control, pattern recognition, and associative memory (Carpenter, 1989; Cochocki & Unbehauen, 1993; Elmas, 2003; Miller et al., 1995; Ripley, 1994). In a structural manner, ANN include some type of processing objects connected to each other over connections having weight values. Called as artificial neurons, each of these objects comes with input(s), weight value(s), summation function, transfer function, and finally an output. As default, all inputs received by an artificial neuron are summed and went through the transfer function. If the output is input for another artificial neuron, then this new value is evaluated by the next artificial neuron (Anderson & McNail, 1992; Deperlioglu & Kose, 2011; Ugur & Kinaci, 2010; Yegnanarayana, 2009). From a general perspective, the artificial neurons are connected with some other ones to build input, hidden, and output layers according to the objective problem structure and principles for some known ANN model structures (i.e. Multi-Layer Perceptron, Hopfield, Elman) - (Deperlioglu & Kose, 2011; Elmas, 2003; Yegnanarayana, 2009). A typical Multi-Layer Artificial Neural Networks model structure is shown in Figure 4 (Ahmad & Saleem, 2014).

## Genetic Algorithms

Genetic Algorithms (GA) is another popular technique used often by researchers in even different disciplines. Generally, GA is a typical algorithmic solution introduced by Holland (1995) and based on the formulation by Fisher (1958) for expressing a mathematical structure for gene spreading among generations. As default, GA employs some particles coded specially to represent some genes in the context of chromosomes and these particles go through some logical mechanisms inspired from crossover, mutation, and natural selection, which are some key aspects of the Theory of Evaluation. It is though in the GA algorithmic flow that successful particles (individuals) should receive crossover operation to spread their genes in newly created generations (particles – individuals). Furthermore, mutation, which allows changing specific genes in the chromosome codes of chosen particles is used another mechanism

*Figure 4. A typical multi-layer Artificial Neural Network structure*
*Ahmad & Saleem, 2014.*

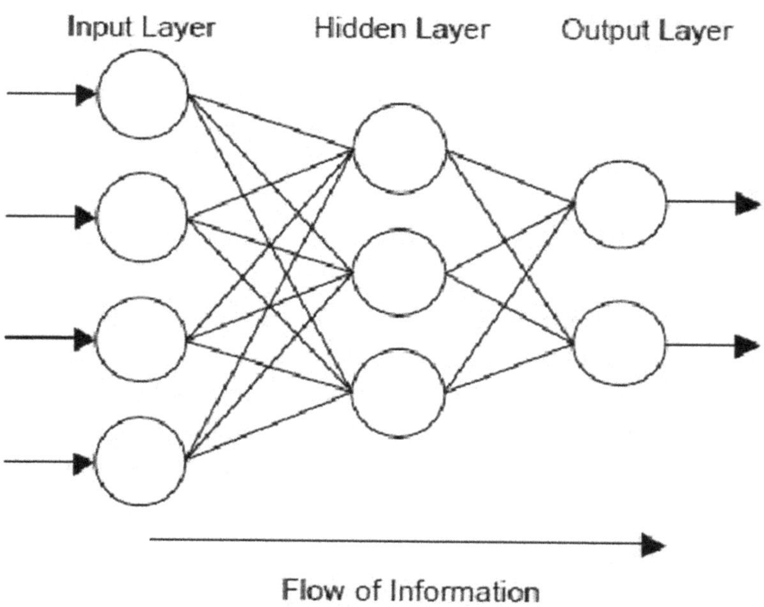

within the algorithmic problem solution process (Holland, 2012). If it is examined from a general view, an algorithmic simulation of natural selection is seen through a typical solution flow of the GA. Because of its wide use, GA is briefly explained under a separate title but it is also a Swarm Intelligence oriented technique. Some essential information regarding to Swarm Intelligence and details for a few techniques – algorithms are provided under the next sub-title. Figure 5 shows the algorithmic flow in a flow chart for the default GA (Shyr, 2010).

## Swarm Intelligence Oriented Techniques

In the literature, we can see that there are many different Swarm Intelligence oriented techniques (Brownlee, 2011; Yang, 2010b). So, it is not possible to cover all of them here. In order to understand what is Swarm Intelligence is for and which typical features and functions do they employ, a brief introduction information is given as follows:

*Figure 5. A flow chart for the Genetic Algorithms*
*Shyr, 2010.*

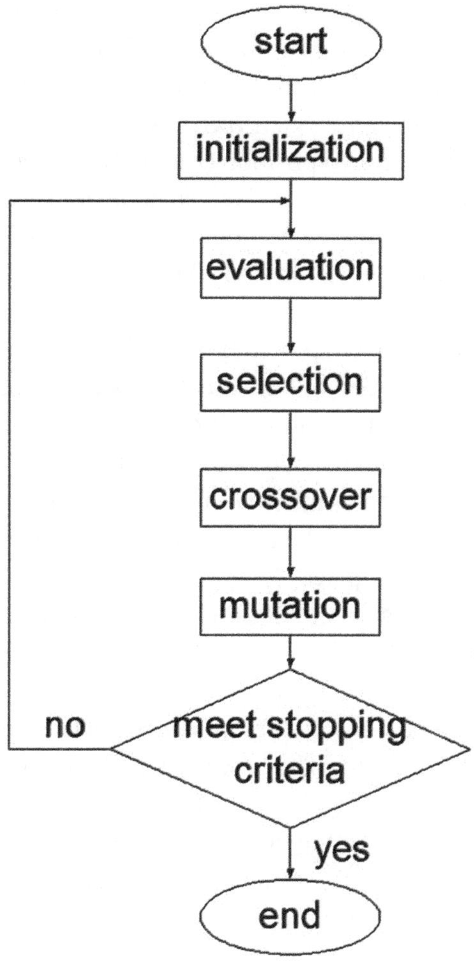

As firstly introduced in the context of cellular robotic systems by Beni and Wang (1993), Swarm Intelligence is related to natural and artificial systems having members, who can show intelligent, collaborative behaviors without being controlled by a central authority (Blum & Li, 2008; Engelbrecht, 2006; Garnier et al., 2007; Kose, 2015). As inspired from the nature, systems within Swarm Intelligence employ members – particles that have specific features and functions to use collaborative advantages to overcome an encountered problem. Swarms showing intelligent behaviors – actions as taking part in the nature are groups of animals or other living organisms like birds, fishes, ants, bees, bacteria and even lifeless objects like rivers, trees…etc. (Blum & Li, 2008; Garnier et al., 2007; Parpinelli & Lopes, 2011).

By inspiring from swarms in the nature, the Swarm Intelligence approach in the context of Artificial Intelligence focuses on techniques – algorithms, which are able to solve especially optimization problems by benefiting from particles with some parameters (features) and actions (methods) to interact with the solution space of the problem and other particles through some algorithmic steps.

In order to give essential information, a few of the remarkable Swarm Intelligence algorithms – techniques can be explained briefly as follows (Kose, 2015):

- **Particle Swarm Optimization (PSO) Algorithm:** Particle Swarm Optimization (PSO) algorithm is a simple technique as first introduced by Kennedy, Eberhart and Shi (Eberhart & Kennedy, 1995; Shi, & Eberhart, 1998). It briefly focuses on social behaviors, actions, and movements of organisms in a bird flock or fish school (Kennedy, 1997). PSO algorithm employs a swarm of candidate solutions (particles) that is spread to the solution space of the objective problem. In detail, these particles are started to move within the solution space according to some mathematical functions and the final it is aimed to find some optimums on requirements of the problem. For each particle, the movements are affected by their own best-known position within the search space as well as the entire swarm's best known position (Kennedy & Eberhart, 1995; Kennedy, 1997; Shi & Eberhart, 1998). From a general perspective, calculated positions direct the whole population for next movements to find appropriate solution(s) for the considered problem. These solution steps are repeated within a loop. General algorithm of the default PSO is presented in Figure 6 (Tarique & Gabbar, 2013).
- **Cuckoo Search (CS) Algorithm:** Cuckoo Search (CS) algorithm is one of the most successful nature-inspired technique – algorithm as based on the Swarm Intelligence. As introduced by Yang and Deb (2009), it inspires from the obligate brood parasitism of some cuckoo species. Briefly the principle is associated with that the cuckoos lay their eggs in the nests of other host birds and some host birds may engage direct conflict with the intruding cuckoos. At this point, if the host bird discovers some eggs not its own, then it either throws these eggs away or abandons the nest to build a newer nest on a different place (Yang, & Deb, 2009). In detail, the following conditions are accepted within CS (Yang, & Deb, 2009): (1) Each egg in a nest is a solution and a cuckoo egg means a new solution. (2) Objective is to use new and seems better solutions to replace worse solutions placed in the nests. (3) Through the solution process, each nest includes one egg or multiple eggs according to the problem structure. (4) Each cuckoo lays one egg at a time and dumps its egg in a randomly chosen nest. (5) The best nests including high quality of eggs are saved for the next generation. (6) Number of available hosts nests is a constant value. (7) The egg laid by a Cuckoo is seen by the host bird according to a probability value. In the context of the explanations on the problem solution flow, Figure 7 shows a pseudo-code for the CS (Yang, & Deb, 2009):

*Figure 6. General algorithm of the Particle Swarm Optimization*
*Tarique & Gabbar, 2013.*

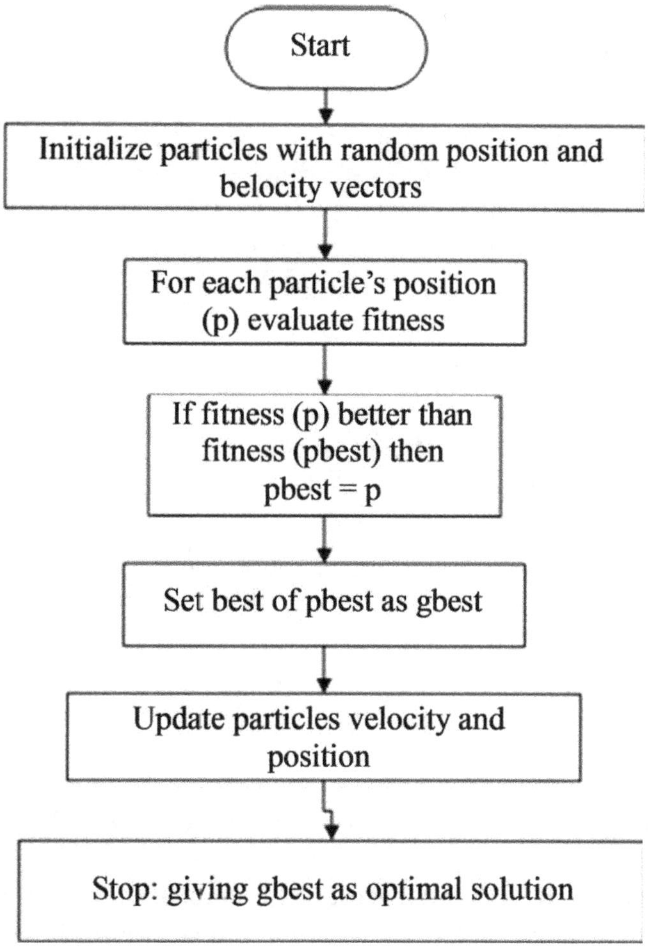


- **Artificial Bee Colony (ABC) Algorithm:** Artificial Bee Colony (ABC) algorithm is introduced by Karaboga (2005) as simulating the behaviors of bees for optimization purposes. As general, the algorithm details are inspired from the roles of honey bees in a swarm. At this point, there are three types of bees – particles in the algorithm. They are called as onlooker bees, scouts, and worker bees respectively. In default ABC algorithm, a colony of onlooker bees looks for rich food sources as optimum values for the considered problem. Here, the problem is represented as a problem of finding the best parameter vector minimizing / maximizing the objective function(s). The bees – particles randomly discover a population of initial solution vectors and try to iteratively improve the related vectors thanks to the neighbor search mechanism (Karaboga, 2010). The mechanism expressed here is about that each worker bee goes to its food source and then comes back to the swarm for performing a dance. A worker bee, whose food source was abandoned, is transformed into a scout and following that, it starts looking for finding a new food source. Meanwhile, onlooker bees see the performed dances by the worker bees and choose food sources

*Figure 7. Pseudo-code for the Cuckoo Search*
*Yang & Deb, 2009.*

**begin**
   *Objective function* $f(\mathbf{x})$, $\mathbf{x} = (x_1, ..., x_d)^T$
   *Generate initial population of*
      $n$ *host nests* $\mathbf{x}_i$ $(i = 1, 2, ..., n)$
   **while** *(t <MaxGeneration) or (stop criterion)*
      *Get a cuckoo randomly by Lévy flights*
         *evaluate its quality/fitness* $F_i$
      *Choose a nest among n (say, j) randomly*
      **if** $(F_i > F_j)$,
            *replace j by the new solution;*
      **end**
      *A fraction* $(p_a)$ *of worse nests*
            *are abandoned and new ones are built;*
      *Keep the best solutions*
            *(or nests with quality solutions);*
      *Rank the solutions and find the current best*
   **end while**
   *Postprocess results and visualization*
**end**

according to influence of the performed dances (Karaboga, 2005, 2010). Figure 8 presents all of these algorithmic flows briefly (Karaboga, & Akay, 2009).

- **Ant Colony Optimization (ACO) Algorithm:** As introduced by Dorigo (1992; Colorni et al., 1991), Ant Colony Optimization (ACO) algorithm is a combinatorial optimization technique as focused on a search for the most optimal path in some kind of solution space. In detail, a graph based data model is used for determining solution spaces for the problems that can be solved by ACO. ACO is simply accepted as a simulation approach designed for the behavior of ants seeking a path between the colony and a food source found. So, in order to understand more about the algorithm structure of the ACO, it is better to examine natural aspects of ant behaviors. Typical natural behaviors of an ant are as follows: (1) initially, an ant wanders randomly and when it detects a food source, it returns back to the colony as laying down the 'pheromone'. Other ants are able to find paths according to the pheromone, so they can use the path to reach to the food. The more food is reached, the more pheromone is laid down to improve effect of the path towards food (Bachir et al., 2012). A pseudo-code for the default ACO is presented in Figure 9 (Dorigo & Blum, 2005).
- **Artificial Immune Systems (AIS) and the Clonal Selection Algorithm (CLONALG):** Almost all complex living organisms including humans and animals have biological immune systems. As being part of the biology and also acceptable to be included within the nature, techniques inspired from immune systems can be examined here within nature-inspired techniques. At this point, Artificial Immune System (AIS) is an approach based on simulating the biological im-

*Figure 8. Algorithmic flow of the Artificial Bee Colony*
*Karaboga, & Akay, 2009.*

```
Initialize Population
repeat
   Place the employed bees on their food sources
   Place the onlooker bees on the food sources depending on their nectar amounts
   Send the scouts to the search area for discovering new food sources
   Memorize the best food source found so far
until requirements are met
```

*Figure 9. Pseudo-code for the Ant Colony Optimization*
*Dorigo & Blum, 2005.*

```
input: An instance P of a CO problem model 𝒫 = (S, f, Ω).
InitializePheromoneValues(𝒯)
s_bs ← NULL
while termination conditions not met do
   𝔖_iter ← ∅
   for j = 1, ..., n_a do
      s ← ConstructSolution(𝒯)
      if s is a valid solution then
         s ← LocalSearch(s)          {optional}
         if (f(s) < f(s_bs)) or (s_bs = NULL) then s_bs ← s
         𝔖_iter ← 𝔖_iter ∪ {s}
      end if
   end for
   ApplyPheromoneUpdate(𝒯, 𝔖_iter, s_bs)
end while
output: The best-so-far solution s_bs
```

mune system, which protects the living organism against any antigenic factor. It is possible to indicate that a typical AIS operates over some predefined algorithms designed by considering functions or features of the biological immune system (Tezel, & Kose, 2011). For example, Clonal Selection algorithm (CLONALG) is one of these algorithms as introduced to the literature by de Castro and Von Zuben (2002). CLONALG briefly inspires from the clonal selection mechanism, which is about mechanisms of an adaptive immune response given to an antigenic stimulus. The mechanism consists of the following phases: (1) The related cells recognizing existent antigens are selected to proliferate against potential future antigen presence. (2) The selected cells are also taken into an affinity maturation phase and their affinities to the antigens are improved in this way for getting better protection (Tezel, & Kose, 2011). Figure 10 shows the algorithmic flow of the CLONALG (Ulutas, & Konak, 2011).

*Figure 10. Algorithmic flow of the Clonal Selection Algorithm*
*Ulutas & Konak, 2011.*

Step 1: *Initialization*: Randomly initialize a population of individuals ($N$).
Step 2: *Evaluation*: Given a set of patterns to be recognized ($P$), for each pattern, determine its affinity (match) with each element of the population.
Step 3: *Selection and cloning*: Select a number ($n$) of the best highest affinity elements of $N$ and generate copies of these individuals proportionally to their affinity with the antigen.
Step 4: *Hypermutation*: Mutate all the copies with a rate proportional to their affinity with the input pattern.
Step 5: *Receptor editing*: Add the mutated individuals to the population and reselect a number ($d$) of the maturated (optimized) individuals as memory.
Step 6: Repeat steps 2–5 until a termination criterion is met.

Readers wanting to know more about alternative algorithms are referred to the algorithms – techniques listed as follows:

- Flower Pollination Algorithm (Yang, 2012),
- Bat Algorithm (Yang & Hossein Gandomi, 2012),
- Firefly Algorithm (Yang, 2010a),
- Differential Evaluation Algorithm (Storn & Price, 1997),
- Vortex Optimization Algorithm (Kose & Arslan, 2015),
- Krill Herd Algorithm (Gandomi & Alavi, 2012),
- Imperialist Competitive Optimization Algorithm (Atashpaz-Gargari & Lucas, 2007),
- Intelligent Water Drops Algorithm (Shah-Hosseini, 2009),

## ACHIEVING INTELLIGENT BIOMEDICAL ENGINEERING WITH NATURE-INSPIRED TECHNIQUES

By taking the explained nature-inspired techniques into consideration or examining their details, it is possible to predict how intelligent Biomedical Engineering can be achieved. This is also associated with having some idea about what type of problems that Biomedical Engineering is interested in. It can be clearly expressed that there are billions of different research works currently taking place in the literature as providing solutions for different Biomedical Engineering problems, with the support by Artificial Intelligence and nature-inspired techniques.

Considering the associated literature some popular research types on achieving intelligent Biomedical Engineering can be expressed brief as follows:

- **Intelligent Diagnosis:** Because it is an important approach to diagnose diseases effectively, there is a remarkable interest in designing and developing diagnosis systems based on nature-inspired techniques. Such systems can be also in the form of Expert Systems (Giarratano & Riley, 1998) or nature-inspired systems can support infrastructure of Expert Systems. Some examples of applications are: Bose et al., 2005; de Carvalho Gomes & Maia, 2016; Makas & Yumusak, 2013; Marinakis et al., 2009; Neelima & Babu, 2016; Thangavel & Karnan, 2005.

- **Optimization Problems:** Sometimes problems within Biomedical Engineering may require optimization based mathematical models to solve the problem. This can be for classifying biomedical data for further investigations or calculating some variables important for the objective problem that can be considered as a Biomedical Engineering problem. Some examples of such works are: Aler et al., 2012; Balestra et al., 2010; Bursa & Lhotska, 2007; Fidanova & Roeva, 2013; Wang et al., 2014.
- **Classification:** Some Biomedical Engineering problems may require classifying data considered in the objective problem. In that case, nature-inspired techniques that are capable of classifying can be used accordingly. Readers are referred to Al-Muhaideb & El Bachir Menai, 2013; Bertolazzi et al., 2008; Mousavirad & Ebrahimpour-Komleh, 2014; Suganya & Sumathi, 2015 as some example of such works.
- **Signal Processing:** In Biomedical Engineering, processing data is always an important research way considered by the researchers. In this context, processing biomedical signals is also an important aspect within Biomedical Engineering problems. At this point, Artificial Intelligence and nature-inspired techniques are often used for signal processing purposes. Some examples in this manner are: Cagnoni, 2014; Escalona-Vargas et al., 2014; Gacek & Pedrycz, 2011; Nait-Ali, 2009; Nakib et al., 2009.
- **Image Processing:** As another data processing approach, also image processing has a remarkable place among research works in Biomedical Engineering. Especially processing over medical image data requires effective solution ways to achieve desired objectives easily. In this sense, Artificial Intelligence and nature-inspired techniques are often preferred because of their rapid and effective solutions for image processing problems. Some research works from the literature as regarding to this research approach are: Kockanat & Karaboga, 2017; Kaur & Mann, 2016; Mesejo et al., 2015; Rogai et al., 2016; Valsecchi et al., 2017.
- **Hardware/Robotic Systems:** In addition to soft computing based solutions, Artificial Intelligence and nature-inspired techniques can find their way in hardware systems or robotic solutions. At this point, Artificial Intelligence is employed within the software side of the system as controlling the whole machine or supporting some problem steps while the machine is working on the objective problem. Some examples are: Broekens et al., 2009; Camarillo et al., 2004; De Momi & Ferrigno, 2010; Taylor et al., 2008; Wada et al., 2008.
- **Other Uses:** Out of the mentioned intelligent Biomedical Engineering approaches, there are many different types of uses of Artificial Intelligence in Biomedical Engineering problems. These can be associated with control systems or biomechanics (Chawla & Duhan, 2014) as the field of Biomedical Engineering has many sub-fields under it. Furthermore, there is also a possibility of new uses of intelligent systems that may appear in the near future as alternative solutions for Biomedical Engineering problems and the humankind still should discover – find them.

## FUTURE RESEARCH DIRECTIONS

In addition to the expressed application ways of nature-inspired techniques to achieve intelligent Biomedical Engineering solutions, it is possible to also have some predictions regarding to future research works, which may be done in a near future or at least some decades later. Considering the experiences and the current state of the literature, some ideas can be expressed briefly as follows:

- It will be possible to develop fully automated surgery robots and there will no need for humans to take part in surgeries. This will also reduce time of surgeries by improving also success rates and effectiveness.
- Thanks to intelligent software systems that will take part in even mobile devices, it will be possible to get immediate medical expert support by requiring no need to go to a i.e. hospital.
- There will be innovative changes in the fields associated with medical – health services thanks to Artificial Intelligence based research works and in this way, educational processes will also receive revolutionary changes at the final.
- Considering the education side more, it can be expressed that the future of education will be open with autonomous systems as including also educational processes for the field of medical – health. At this point, both theoretical and applied side of the education for medical – health can be achieved by using intelligent systems.
- There may be a possibility of eliminating all diseases if the Artificial Intelligence based approaches, methods, and techniques can be used always in the advantages of the humankind.
- Nature will be always an inspiration factor for designing and developing new Artificial Intelligence techniques to solve Biomedical Engineering problems.
- There are many different problem types under the Biomedical Engineering and it can be indicated that there are some gaps in using Artificial Intelligence effectively in all of different problems types. In the future, it can be said that there will be no gap thanks to increasing use of intelligent solution ways within Biomedical Engineering problems.

## CONCLUSION

The field of Biomedical Engineering is currently in a great improvement loop of innovations thanks to research efforts supported by Artificial Intelligence. In detail, nature-inspired techniques have always had a great potential to derive unique problem solution mechanisms from the nature. In this sense, this chapter has provided some essential information on role of nature-inspired Artificial Intelligence techniques to achieve intelligent solutions for Biomedical Engineering studies. It is clear that some of the mentioned solutions expressed under this chapter can be associated with also some other fields. But this situation is an important sign of how Artificial Intelligence can be flexible enough to cover many different fields and shape the future of humankind. It is also important to have real-time, effective and efficient solutions for such an important field associated with healthcare services and dealing with lives of living organisms. It seems that the Artificial Intelligence is strong enough to improve effectiveness and success of such healthcare services by providing innovative solutions thanks to techniques derived from natural dynamics – mechanisms – features. But it can also be seen that the future has many different innovations, improvements, and new developments, which are still background of a mist. So, research efforts towards understanding and improving the literature of intelligent Biomedical Engineering are key points to be keeping up-to-date for the future.

## REFERENCES

Abraham, A., & Pedrycz, W. (2008). *Hybrid Artificial Intelligence Systems. Lecture Notes in Artificial Intelligence, 5271. Retrieved from* http://crina.softcomputing.net/hais2008-preface.pdf

Ahmad, Z., & Saleem, A. (2014). Impact of Governance on Human Development. *Pakistan Journal of Commerce & Social Sciences*, *8*(3).

Al-Muhaideb, S., & El Bachir Menai, M. (2013). Hybrid metaheuristics for medical data classification. *Hybrid Metaheuristics*, 187-217.

Aler, R., Vega, A., Galván, I. M., & Nebro, A. J. (2012). Multi-objective metaheuristics for preprocessing EEG data in brain–computer interfaces. *Engineering Optimization*, *44*(3), 373–390. doi:10.1080/0305215X.2011.641542

Alpaydin, E. (2014). *Introduction to machine learning*. MIT Press.

Anderson, D., & McNeill, G. (1992). Artificial neural networks technology. *Kaman Sciences Corporation*, *258*(6), 1–83.

Atashpaz-Gargari, E., & Lucas, C. (2007). Imperialist competitive algorithm: an algorithm for optimization inspired by imperialistic competition. In *Evolutionary computation, 2007. CEC 2007. IEEE Congress on* (pp. 4661-4667). IEEE. doi:10.1109/CEC.2007.4425083

Bachir, B., Ali, A., & Abdellah, M. (2012). Multiobjective optimization of an operational amplifier by the ant colony optimisation algorithm. *Electrical and Electronic Engineering*, *2*(4), 230–235. doi:10.5923/j.eee.20120204.09

Balestra, G., Foglietta, C., & Gaetano, L. (2010). A Decision Support System based on Meta-heuristics and MCDA for Healthcare Service and Technology Optimal Location. *Studies in Health Technology and Informatics*, *160*, 1470–1470.

Beni, G., & Wang, J. (1993). Swarm intelligence in cellular robotic systems. In *Robots and Biological Systems: Towards a New Bionics?* (pp. 703–712). Berlin: Springer. doi:10.1007/978-3-642-58069-7_38

Bertolazzi, P., Felici, G., Festa, P., & Lancia, G. (2008). Logic classification and feature selection for biomedical data. *Computers & Mathematics with Applications (Oxford, England)*, *55*(5), 889–899. doi:10.1016/j.camwa.2006.12.093

Blum, C., & Li, X. (2008). Swarm intelligence in optimization. In *Swarm Intelligence* (pp. 43–85). Springer Berlin Heidelberg. doi:10.1007/978-3-540-74089-6_2

Bose, J., Jaya, J., Karnan, M., & Thanushkodi, K. (2005). Metaheuristic algorithms for detection of ovarian cancer. *Proceeding of Image Processing and Pattern Recognition*, 35.

Broekens, J., Heerink, M., & Rosendal, H. (2009). Assistive social robots in elderly care: A review. *Gerontechnology (Valkenswaard)*, *8*(2), 94–103. doi:10.4017/gt.2009.08.02.002.00

Bronzino, J. D. (1999). *Biomedical engineering handbook* (Vol. 2). CRC Press.

Bronzino, J. D., & Peterson, D. R. (2014). *Biomedical engineering fundamentals*. CRC Press.

Brownlee, J. (2011). *Clever algorithms: nature-inspired programming recipes*. Jason Brownlee.

Bursa, M., & Lhotska, L. (2007). Ant Colony Inspired Clustering in Biomedical Data Processing. In *3rd European Symposium on Nature-inspired Smart Information Systems*. Aachen: NiSIS.

Cagnoni, S. (2014). Evolutionary image analysis and signal processing. In *Proceedings of the Companion Publication of the 2014 Annual Conference on Genetic and Evolutionary Computation* (pp. 795-818). ACM.

Camarillo, D. B., Krummel, T. M., & Salisbury, J. K. Jr. (2004). Robotic technology in surgery: Past, present, and future. *American Journal of Surgery, 188*(4), 2–15. doi:10.1016/j.amjsurg.2004.08.025 PMID:15476646

Carpenter, G. A. (1989). Neural network models for pattern recognition and associative memory. *Neural Networks, 2*(4), 243–257. doi:10.1016/0893-6080(89)90035-X

Caruana, R., & Niculescu-Mizil, A. (2006). An empirical comparison of supervised learning algorithms. In *Proceedings of the 23rd international conference on Machine learning* (pp. 161-168). ACM. doi:10.1145/1143844.1143865

Chawla, M., & Duhan, M. (2014). Applications of recent metaheuristics optimisation algorithms in biomedical engineering: A review. *International Journal of Biomedical Engineering and Technology, 16*(3), 268–278. doi:10.1504/IJBET.2014.065807

Coates, A., Ng, A., & Lee, H. (2011, June). An analysis of single-layer networks in unsupervised feature learning. In *Proceedings of the fourteenth international conference on artificial intelligence and statistics* (pp. 215-223). Academic Press.

Cochocki, A., & Unbehauen, R. (1993). *Neural networks for optimization and signal processing*. John Wiley & Sons, Inc.

Colorni, A., Dorigo, M., & Maniezzo, V. (1991). *Distributed optimization by ant colonies. In Actes de la première conférence européenne sur la vie artificielle* (pp. 134–142). Paris, France: Elsevier.

Corchado, E., Graña, M., & Woniak, M. (2012). Editorial: New trends and applications on hybrid artificial intelligence systems. *Neurocomputing, 75*(1), 61–63. doi:10.1016/j.neucom.2011.06.021

de Carvalho Gomes, F., & Maia, J. G. R. (2016). Learning Optimal Decision Lists as a Metaheuristic Search for Diagnosis of Parkinson's Disease. In *International Workshop on Machine Learning, Optimization and Big Data* (pp. 391-401). Springer. doi:10.1007/978-3-319-51469-7_32

de Castro, L. N., & Von Zuben, F. J. (2002). Learning and optimization using the clonal selection principle. *IEEE Transactions on Evolutionary Computation, 6*(3), 239–251. doi:10.1109/TEVC.2002.1011539

De Momi, E., & Ferrigno, G. (2010). Robotic and artificial intelligence for keyhole neurosurgery: The ROBOCAST project, a multi-modal autonomous path planner. *Proceedings of the Institution of Mechanical Engineers. Part H, Journal of Engineering in Medicine*, *224*(5), 715–727. doi:10.1243/09544119JEIM585 PMID:20718272

Deperlioglu, O., & Kose, U. (2011). An educational tool for artificial neural networks. *Computers & Electrical Engineering*, *37*(3), 392–402. doi:10.1016/j.compeleceng.2011.03.010

Dorigo, M. (1992). *Optimization, learning and natural algorithms* (PhD Thesis). Politecnico di Milano, Italy.

Eberhart, R., & Kennedy, J. (1995). A new optimizer using particle swarm theory. In *Micro Machine and Human Science, 1995. MHS'95., Proceedings of the Sixth International Symposium on* (pp. 39-43). IEEE. doi:10.1109/MHS.1995.494215

Elmas, C. (2003). *Artificial neural network – Theory, architecture, training, implementation*. Ankara, Turkey: Seçkin Press. (In Turkish)

Enderle, J. D., & Bronzino, J. D. (2012). *Introduction to biomedical engineering*. Academic press.

Engelbrecht, A. P. (2006). *Fundamentals of computational swarm intelligence*. John Wiley & Sons.

Escalona-Vargas, D. I., Lopez-Arevalo, I., & Gutiérrez, D. (2014). Multicompare tests of the performance of different metaheuristics in EEG dipole source localization. *The Scientific World Journal*. PMID:24757424

Fidanova, S. S., & Roeva, O. N. (2013). Metaheuristic techniques for optimization of an E. coli cultivation model. *Biotechnology, Biotechnological Equipment*, *27*(3), 3870–3876. doi:10.5504/BBEQ.2012.0136

Fisher, R. A. (1958). *The genetic theory of natural selection*. Oxford University Press.

Gacek, A., & Pedrycz, W. (Eds.). (2011). *ECG signal processing, classification and interpretation: a comprehensive framework of computational intelligence*. Springer Science & Business Media.

Gandomi, A. H., & Alavi, A. H. (2012). Krill herd: A new bio-inspired optimization algorithm. *Communications in Nonlinear Science and Numerical Simulation*, *17*(12), 4831–4845. doi:10.1016/j.cnsns.2012.05.010

Garnier, S., Gautrais, J., & Theraulaz, G. (2007). The biological principles of swarm intelligence. *Swarm Intelligence*, *1*(1), 3–31. doi:10.1007/s11721-007-0004-y

Giarratano, J. C., & Riley, G. (1998). *Expert systems*. PWS Publishing Co.

Hedayat, A., Davilu, H., Barfrosh, A. A., & Sepanloo, K. (2009). Optimization of the core configuration design using a hybrid artificial intelligence algorithm for research reactors. *Nuclear Engineering and Design*, *239*(12), 2786–2799. doi:10.1016/j.nucengdes.2009.08.027

Holland, J. (1995). *Hidden order: How adaptation builds complexity*. Basic Books.

Holland, J. H. (2012). Genetic algorithms. *Scholarpedia, 7*(12), 1482. doi:10.4249/scholarpedia.1482

Hudson, D. L., & Cohen, M. E. (2000). *Neural networks and artificial intelligence for biomedical engineering*. New York: IEEE.

Kaelbling, L. P., Littman, M. L., & Moore, A. W. (1996). Reinforcement learning: A survey. *Journal of Artificial Intelligence Research, 4*, 237–285.

Karaboga, D. (2005). *An idea based on honey bee swarm for numerical optimization* (Vol. 200). Technical report-tr06, Erciyes University, Engineering Faculty, Computer Engineering Department.

Karaboga, D. (2010). Artificial bee colony algorithm. *Scholarpedia, 5*(3), 6915. doi:10.4249/scholarpedia.6915

Karaboga, D., & Akay, B. (2009). A comparative study of artificial bee colony algorithm. *Applied Mathematics and Computation, 214*(1), 108–132. doi:10.1016/j.amc.2009.03.090

Kaur, B., & Mann, P. S. (2016). Ant Colony based Metaheuristic Technique for Color Image Segmentation. *International Journal of Computers and Applications, 153*(1).

Kennedy, J. (1997). The particle swarm: social adaptation of knowledge. In *Evolutionary Computation, 1997., IEEE International Conference on* (pp. 303-308). IEEE. doi:10.1109/ICEC.1997.592326

Kennedy, J. (2006). Swarm intelligence. In Handbook of nature-inspired and innovative computing (pp. 187-219). Springer US.

Klir, G., & Yuan, B. (1995). *Fuzzy sets and fuzzy logic* (Vol. 4). Prentice hall.

Kockanat, S., & Karaboga, N. (2017). Medical Image Denoising Using Metaheuristics. In *Metaheuristics for Medicine and Biology* (pp. 155–169). Springer Berlin Heidelberg. doi:10.1007/978-3-662-54428-0_9

Kose, U. (2015). Present state of swarm intelligence and future directions. In *Encyclopedia of Information Science and Technology* (3rd ed.; pp. 239–252). IGI Global. doi:10.4018/978-1-4666-5888-2.ch023

Kose, U., & Arslan, A. (2015). On the Idea of a New Artificial Intelligence Based Optimization Algorithm Inspired From the Nature of Vortex. *BRAIN. Broad Research in Artificial Intelligence and Neuroscience, 5*(1-4), 60–66.

Makas, H., & Yumusak, N. (2013). A comprehensive study on thyroid diagnosis by neural networks and swarm intelligence. In *Electronics, Computer and Computation (ICECCO), 2013 International Conference on* (pp. 180-183). IEEE. doi:10.1109/ICECCO.2013.6718258

Marinakis, Y., Dounias, G., & Jantzen, J. (2009). Pap smear diagnosis using a hybrid intelligent scheme focusing on genetic algorithm based feature selection and nearest neighbor classification. *Computers in Biology and Medicine, 39*(1), 69–78. doi:10.1016/j.compbiomed.2008.11.006 PMID:19147127

Mayer-Schonberger, V., & Cukier, K. (2013). *Big Data: A Revolution That Will Transform How We Live*. Hachette.

Mesejo, P., Valsecchi, A., Marrakchi-Kacem, L., Cagnoni, S., & Damas, S. (2015). Biomedical image segmentation using geometric deformable models and metaheuristics. *Computerized Medical Imaging and Graphics*, *43*, 167–178. doi:10.1016/j.compmedimag.2013.12.005 PMID:24480648

Miller, W. T., Werbos, P. J., & Sutton, R. S. (Eds.). (1995). *Neural networks for control.* MIT press. doi:10.1007/978-3-642-57760-4

Mousavirad, S. J., & Ebrahimpour-Komleh, H. (2014). Population-Based Feature Selection for Biomedical Data Classification. In *Data Mining and Analysis in the Engineering Field* (pp. 296–326). IGI Global. doi:10.4018/978-1-4666-6086-1.ch016

Naït-Ali, A. (Ed.). (2009). *Advanced biosignal processing*. Springer Science & Business Media. doi:10.1007/978-3-540-89506-0

Nakib, A., Naït-Ali, A., Van Wassenhove, V., & Siarry, P. (2009). Parametric Modeling of Some Biosignals Using Optimization Metaheuristics. In *Advanced Biosignal Processing* (pp. 291–305). Springer Berlin Heidelberg. doi:10.1007/978-3-540-89506-0_14

Neelima, E., & Babu, M. P. (2016). MAGED: Metaheuristic Approach on Gene Expression Data: Predicting the Coronary Artery Disease and the Scope of Unstable Angina and Myocardial Infarction. *Global Journal of Computer Science and Technology, 16*(4).

Nikolopoulos, C. (1997). *Expert systems: introduction to first and second generation and hybrid knowledge based systems*. Marcel Dekker, Inc.

Parpinelli, R. S., & Lopes, H. S. (2011). New inspirations in swarm intelligence: A survey. *International Journal of Bio-inspired Computation*, *3*(1), 1–16. doi:10.1504/IJBIC.2011.038700

Principe, J. C., & Fancourt, C. L. (2013). Artificial neural networks. Handbook of Global Optimization, 2, 363-386.

Ripley, B. D. (1994). Neural networks and related methods for classification. *Journal of the Royal Statistical Society. Series B. Methodological*, 409–456.

Rogai, F., Manfredi, C., & Bocchi, L. (2016). Metaheuristics for specialization of a segmentation algorithm for ultrasound images. *IEEE Transactions on Evolutionary Computation*, *20*(5), 730–741. doi:10.1109/TEVC.2016.2515660

Shah-Hosseini, H. (2009). The intelligent water drops algorithm: A nature-inspired swarm-based optimization algorithm. *International Journal of Bio-inspired Computation*, *1*(1-2), 71–79. doi:10.1504/IJBIC.2009.022775

Shi, Y., & Eberhart, R. (1998). A modified particle swarm optimizer. In *Evolutionary Computation Proceedings, 1998. IEEE World Congress on Computational Intelligence., The 1998 IEEE International Conference on* (pp. 69-73). IEEE. doi:10.1109/ICEC.1998.699146

Shyr, W. J. (2010). Parameters Determination for Optimum Design by Evolutionary Algorithm. In *Convergence and Hybrid Information Technologies*. InTech. doi:10.5772/9638

Spelt, P. F., Knee, H. E., & Glover, C. W. (1991). Hybrid artificial intelligence architecture for diagnosis and decision-making in manufacturing. *Journal of Intelligent Manufacturing*, *2*(5), 261–268. doi:10.1007/BF01471174

Storn, R., & Price, K. (1997). Differential evolution–a simple and efficient heuristic for global optimization over continuous spaces. *Journal of Global Optimization*, *11*(4), 341–359. doi:10.1023/A:1008202821328

Suganya, P., & Sumathi, C. P. (2015). A novel metaheuristic data mining algorithm for the detection and classification of parkinson disease. *Indian Journal of Science and Technology*, *8*(14). doi:10.17485/ijst/2015/v8i14/72685

Sutton, R. S., & Barto, A. G. (1998). Reinforcement learning: An introduction: Vol. 1. *No. 1*. Cambridge, MA: MIT Press.

Tarique, A., & Gabbar, H. A. (2013). Particle swarm optimization (PSO) based turbine control. *Intelligent Control and Automation*, *4*(02), 126–137. doi:10.4236/ica.2013.42018

Taylor, R. H., Menciassi, A., Fichtinger, G., & Dario, P. (2008). Medical robotics and computer-integrated surgery. In *Springer handbook of robotics* (pp. 1199–1222). Springer Berlin Heidelberg. doi:10.1007/978-3-540-30301-5_53

Tezel, G., & Kose, U. (2011). Headache diagnosis by using the clonal selection algorithm. In *6th International Advanced Technologies Symposium 2011* (pp. 144-148). Elazig, Turkey: Firat University.

Thangavel, K., & Karnan, M. (2005). Computer aided diagnosis in digital mammograms: Detection of microcalcifications by meta heuristic algorithms. *GVIP Journal*, *5*(7), 41–55.

Toro, C. H. F., Meire, S. G., Gálvez, J. F., & Fdez-Riverola, F. (2013). A hybrid artificial intelligence model for river flow forecasting. *Applied Soft Computing*, *13*(8), 3449–3458. doi:10.1016/j.asoc.2013.04.014

Ugur, A., & Kinaci, A. C. (2010). A web-based tool for teaching neural network concepts. *Computer Applications in Engineering Education*, *18*(3), 449–457. doi:10.1002/cae.20184

Ulutas, B. H., & Konak, S. (2011). A review of clonal selection algorithm and its applications. *Artificial Intelligence Review*, *36*(2), 117–138. doi:10.1007/s10462-011-9206-1

Valsecchi, A., Bermejo, E., Damas, S., & Cordón, O. (2017). Metaheuristics for Medical Image Registration. Handbook of Heuristics, 1-22.

Wada, K., Shibata, T., Musha, T., & Kimura, S. (2008). Robot therapy for elders affected by dementia. *IEEE Engineering in Medicine and Biology Magazine*, *27*(4), 53–60. doi:10.1109/MEMB.2008.919496

Wang, L., Ni, H., Yang, R., Pappu, V., Fenn, M. B., & Pardalos, P. M. (2014). Feature selection based on meta-heuristics for biomedicine. *Optimization Methods & Software*, *29*(4), 703–719. doi:10.1080/10556788.2013.834900

Weber, R. H., & Weber, R. (2010). *Internet of things* (Vol. 12). New York: Springer. doi:10.1007/978-3-642-11710-7

Wermter, S. (2000). *Hybrid neural systems (No. 1778)*. Springer Science & Business Media. doi:10.1007/10719871

Wiener, N. (1961). *Cybernetics or Control and Communication in the Animal and the Machine* (Vol. 25). MIT Press. doi:10.1037/13140-000

Xia, F., Yang, L. T., Wang, L., & Vinel, A. (2012). Internet of things. *International Journal of Communication Systems, 25*(9), 1101–1102. doi:10.1002/dac.2417

Yang, X. S. (2010a). Firefly algorithm, Levy flights and global optimization. *Research and Development in Intelligent Systems, 26*, 209-218.

Yang, X. S. (2010b). *Nature-inspired metaheuristic algorithms*. Luniver Press.

Yang, X. S. (2012). *Flower pollination algorithm for global optimization*. UCNC.

Yang, X. S., & Deb, S. (2009). Cuckoo search via Lévy flights. In *Nature & Biologically Inspired Computing, 2009. NaBIC 2009. World Congress on* (pp. 210-214). IEEE.

Yang, X. S., & Hossein Gandomi, A. (2012). Bat algorithm: A novel approach for global engineering optimization. *Engineering Computations, 29*(5), 464–483. doi:10.1108/02644401211235834

Yegnanarayana, B. (2009). *Artificial neural networks*. PHI Learning Pvt. Ltd.

Zadeh, L. A. (1996). Fuzzy logic= computing with words. *IEEE Transactions on Fuzzy Systems, 4*(2), 103–111. doi:10.1109/91.493904

Zikopoulos, P., & Eaton, C. (2011). *Understanding big data: Analytics for enterprise class hadoop and streaming data*. McGraw-Hill Osborne Media.

## ADDITIONAL READING

Bar-Cohen, Y., & Breazeal, C. L. (2003). *Biologically inspired intelligent robots* (Vol. 122). Spie Press. doi:10.1117/3.2068093

Beekman, M., Sword, G. A., & Simpson, S. J. (2008). Biological foundations of swarm intelligence. *Swarm Intelligence*, 3–41.

Bell, N., & Oommen, B. J. (2017). A novel abstraction for swarm intelligence: Particle field optimization. *Autonomous Agents and Multi-Agent Systems, 31*(2), 362–385. doi:10.1007/s10458-016-9350-8

Beni, G. (2004). From swarm intelligence to swarm robotics. In *International Workshop on Swarm Robotics* (pp. 1-9). Springer, Berlin, Heidelberg.

Black, J. B., Perez-Pinera, P., & Gersbach, C. A. (2017). Mammalian Synthetic Biology: Engineering Biological Systems. *Annual Review of Biomedical Engineering, 19*(1), 249–277. doi:10.1146/annurev-bioeng-071516-044649 PMID:28633563

Brambilla, M., Ferrante, E., Birattari, M., & Dorigo, M. (2013). Swarm robotics: A review from the swarm engineering perspective. *Swarm Intelligence, 7*(1), 1–41. doi:10.1007/s11721-012-0075-2

Chakraborty, A., & Kar, A. K. (2017). Swarm intelligence: A review of algorithms. In *Nature-Inspired Computing and Optimization* (pp. 475–494). Springer International Publishing. doi:10.1007/978-3-319-50920-4_19

Cheng, M. Y., & Lien, L. C. (2012). Hybrid Artificial Intelligence–Based PBA for Benchmark Functions and Facility Layout Design Optimization. *Journal of Computing in Civil Engineering, 26*(5), 612–624. doi:10.1061/(ASCE)CP.1943-5487.0000163

Chi, M. T. (2014). *The nature of expertise*. Psychology Press.

Chiong, R. (Ed.). (2009). *Nature-inspired algorithms for optimisation* (Vol. 193). Springer. doi:10.1007/978-3-642-00267-0

Coello, C. C., Dhaenens, C., & Jourdan, L. (Eds.). (2009). *Advances in multi-objective nature inspired computing* (Vol. 272). Springer.

Cruz, C., González, J. R., Pelta, D. A., Krasnogor, N., & Terrazas, G. (Eds.). (2010). *Nature Inspired Cooperative Strategies for Optimization (NICSO 2010)* (Vol. 284). Springer.

de Carvalho, M. P. A. C., Wozniak, J. S. P. M., & Corchado, H. Q. E. (2014). Hybrid Artificial Intelligence Systems. *9th International Conference, HAIS 2014*, Salamanca, Spain. Springer.

Dorf, R. C. (2016). *Sensors, Nanoscience, Biomedical Engineering, and Instruments: Sensors Nanoscience Biomedical Engineering*. CRC press.

Dorigo, M. (2016). Ten years of swarm intelligence. *Swarm Intelligence, 10*(4), 245–246. doi:10.1007/s11721-016-0130-5

Eberhart, R. C. (1998). Overview of computational intelligence [and biomedical engineering applications]. In *Engineering in Medicine and Biology Society, 1998. Proceedings of the 20th Annual International Conference of the IEEE*(Vol. 3, pp. 1125-1129). IEEE.

Enderle, J. D., & Bronzino, J. D. (2012). *Introduction to biomedical engineering*. Academic press.

Fister, I., Jr., Yang, X. S., Fister, I., Brest, J., & Fister, D. (2013). A brief review of nature-inspired algorithms for optimization. *arXiv preprint arXiv:1307.4186.*

Foster, K. R., Koprowski, R., & Skufca, J. D. (2014). Machine learning, medical diagnosis, and biomedical engineering research-commentary. *Biomedical Engineering Online, 13*(1), 94. doi:10.1186/1475-925X-13-94 PMID:24998888

Gang, L., Wenyu, Y., Ling, L., Qilian, Y., & Xuemin, Y. (2000). An artificial-intelligence approach to ECG analysis. *IEEE Engineering in Medicine and Biology Magazine, 19*(2), 95–100. doi:10.1109/51.827412 PMID:10738667

Gao, S., & Yang, J. Y. (2006). *Swarm intelligence algorithms and applications* (pp. 112–117). Beijing: China Water Power Press.

Haugeland, J. (1989). *Artificial intelligence: The very idea*. MIT press.

Helms, M., Vattam, S. S., & Goel, A. K. (2009). Biologically inspired design: Process and products. *Design Studies*, *30*(5), 606–622. doi:10.1016/j.destud.2009.04.003

Hinchey, M. G., Sterritt, R., & Rouff, C. (2007). Swarms and swarm intelligence. *Computer*, *40*(4), 111–113. doi:10.1109/MC.2007.144

Hudson, D. L., & Cohen, M. E. (2000). *Neural networks and artificial intelligence for biomedical engineering* (pp. I–XXIII). New York: IEEE.

Karaboga, D., & Akay, B. (2009). A survey: Algorithms simulating bee swarm intelligence. *Artificial Intelligence Review*, *31*(1-4), 61–85. doi:10.1007/s10462-009-9127-4

Lei, W., & Qi, K. (2007). Nature-inspired Computation—Effective Realization of Artificial Intelligence. *Systems Engineering-Theory & Practice*, *27*(5), 126–134. doi:10.1016/S1874-8651(08)60034-4

Liu, J., & Tsui, K. C. (2006). Toward nature-inspired computing. *Communications of the ACM*, *49*(10), 59–64. doi:10.1145/1164394.1164395

Liu, Y., & Passino, K. M. (2000). Swarm intelligence: Literature overview. Department of electrical engineering, the Ohio State University.

Magin, R. L. (2017). Bioengineering and Cybernetics: A Modern Caduceus. *IEEE Pulse*, *8*(1), 44–47. doi:10.1109/MPUL.2016.2627461 PMID:28129142

Marrow, P. (2000). Nature-inspired computing technology and applications. *BT Technology Journal*, *18*(4), 13–23. doi:10.1023/A:1026746406754

Matarić, M. J. (1998). Behavior-based robotics as a tool for synthesis of artificial behavior and analysis of natural behavior. *Trends in Cognitive Sciences*, *2*(3), 82–86. doi:10.1016/S1364-6613(98)01141-3 PMID:21227083

Musib, M., Wang, F., Tarselli, M. A., Yoho, R., Yu, K. H., Andrés, R. M., & Dutton-Regester, K. et al. (2017). Artificial intelligence in research. *Science*, *357*(6346), 28–30. doi:10.1126/science.357.6346.28 PMID:28684488

Nayak, R., Jain, L., & Ting, B. (2001). Artificial neural networks in biomedical engineering: a review.

Onet, E. V., & Vladu, E. (2008). Nature inspired algorithms and Artificial Intelligence. *Journal of Computer Science and Control Systems*, (1), 66.

Ortuño, F. M., Valenzuela, O., Glösekötter, P., & Rojas, I. (2016). Main findings and advances in biomedical engineering and bioinformatics from IWBBIO 2015. *Biomedical Engineering Online*, *15*(1), 79. doi:10.1186/s12938-016-0187-9 PMID:27454414

Peng, Y., Zhang, Y., & Wang, L. (2010). Artificial intelligence in biomedical engineering and informatics: An introduction and review. *Artificial Intelligence in Medicine*, *48*(2), 71–73. doi:10.1016/j.artmed.2009.07.007 PMID:20045299

Pham, D. T., & Pham, P. T. N. (1999). Artificial intelligence in engineering. *International Journal of Machine Tools & Manufacture*, *39*(6), 937–949. doi:10.1016/S0890-6955(98)00076-5

Rennard, J. P. (Ed.). (2006). *Handbook of research on nature-inspired computing for economics and management*. IGI Global.

Ryan, M. L. (1991). *Possible worlds, artificial intelligence, and narrative theory*. Indiana University Press.

Sandell, H. S., & Bourne, J. R. (1985). Expert systems in medicine: A biomedical engineering perspective. *Critical Reviews in Biomedical Engineering*, *12*(2), 95–129. PMID:3884272

Shadbolt, N. (2004). Nature-inspired computing. *IEEE Intelligent Systems*, *19*(1), 2–3. doi:10.1109/MIS.2004.1265875

Vassiliadis, V., & Dounias, G. (2009). Nature–Inspired Intelligence: A review of selected methods and applications. *International Journal of Artificial Intelligence Tools*, *18*(04), 487–516. doi:10.1142/S021821300900024X

Waldner, J. B. (2013). *Nanocomputers and swarm intelligence*. John Wiley & Sons.

Weiss, G. (Ed.). (1999). *Multiagent systems: a modern approach to distributed artificial intelligence*. MIT press.

Yang, X. S., Cui, Z., Xiao, R., Gandomi, A. H., & Karamanoglu, M. (Eds.). (2013). *Swarm intelligence and bio-inspired computation: theory and applications*. Newnes. doi:10.1016/B978-0-12-405163-8.00001-6

Ye, Z. (2005). Artificial-intelligence approach for biomedical sample characterization using Raman spectroscopy. *IEEE Transactions on Automation Science and Engineering*, *2*(1), 67–73. doi:10.1109/TASE.2004.840071

## KEY TERMS AND DEFINITIONS

**Artificial Intelligence:** (1) Artificial intelligence is a software-based system that can perform functions such as perception, learning, reasoning, thinking, problem solving, inference, and decision making, as similar to human. (2) Artificial intelligence is a scientific area in which it is tried to simulate human intelligence by employing a computer or an autonomous system.

**Biomedical Engineering:** Biomedical engineering is a type of engineering field that combines principles, research works, of both medicine, biology, and some other engineering fields like computer engineering, electronic engineering, and machine engineering to design and develop healthcare-oriented systems.

**Expert Systems:** Expert systems are a type of artificial intelligence technique that employs some expert-based rules on the background to solve real-world-based problems, which need some expert support to have idea about solution ways and/or running them effectively.

**Nature-Inspired Techniques:** Nature-inspired techniques are a type of intelligent problem-solving technique employed within the field of artificial intelligence and which are inspired by the dynamics and behaviors in nature, including humans, other living organisms, and even other natural objects.

**Swarm Intelligence:** Swarm intelligence is a research approach under artificial intelligence focused on techniques, algorithms inspired from swarm-based living or lifeless objects of nature, to employ some algorithmic solution steps for optimization.

# Chapter 2
# Coupling of Optimization Algorithms Based on Swarm Intelligence:
## An Application for Control of Heroin Addiction Epidemic

**Kamalanand Krishnamurthy**
*Anna University, India*

**Mannar Jawahar Ponnuswamy**
*Anna University, India*

## ABSTRACT

*Swarm intelligence is a branch of computational intelligence where algorithms are developed based on the biological examples of swarming and flocking phenomena of social organisms such as a flock of birds. Such algorithms have been widely utilized for solving computationally complex problems in fields of biomedical engineering and sociology. In this chapter, two different swarm intelligence algorithms, namely the jumping frogs optimization (JFO) and bacterial foraging optimization (BFO), are explained in detail. Further, a synergetic algorithm, namely the coupled bacterial foraging/jumping frogs optimization algorithm (BFJFO), is described and utilized as a tool for control of the heroin epidemic problem.*

## INTRODUCTION

Computational artificial intelligence can be described as the computational techniques built using the foundations of the intelligent behavior exhibited in nature. The central aim of computational intelligence is to design, build, and experiment with systems that perform tasks commonly viewed as requiring intelligence (Poole & Mackworth, 2010). Such computational techniques based on the intelligent behavior exhibited in nature are highly useful for solving certain computationally demanding problems and is highly useful for decision making. The field of artificial intelligence has developed rapidly as the power of computers has increased (Rossini, 2000).

DOI: 10.4018/978-1-5225-4769-3.ch002

The task of finding the global optimum in a multi parameter and multiple state problems is extremely difficult since the search space increases. In such cases, it is necessary to adopt highly efficient algorithms for scanning the entire search space for the elite solution. Further, there is a need for avoiding the premature convergence to local minima. Such problems can be overcome using novel optimization strategies known as swarm intelligence algorithms. Swarm intelligence algorithms are versatile population-based optimization techniques based on random search principles, similar to evolutionary algorithms (Kamberaj, 2014). Heuristic and random search algorithms are useful in obtaining approximations to the optimal solutions in a bounded time (Rebollo-Ruiz & Graña, 2014).

Swarm Intelligence is an innovative distributed intelligent paradigm for solving optimization problems that originally took its inspiration from the biological examples of swarming and flocking phenomena. This discipline is inspired by the behavior of social creatures and organisms such as frogs, a flock of birds, swarm of bacteria etc. In general, this is done by mimicking the behavior of the biological creatures within their swarms and colonies (Nedjah & Mourelle, 2006).

Swarm intelligence is a novel and innovative branch of meta-heuristics developed by mimicking or imitating the behavioral pattern of natural insects or organisms (Zhang, Lee, Chan, Choy & Wu, 2015). In other words, swarm Intelligence is the property of a system in which the collective behavior of simple quasi-independent agents which locally interact with their environment will lead to global intelligent behavior. Swarm intelligence techniques have been widely used for solving hard estimation problems since they are robust, flexible, fault tolerant, scalable, and highly parallelizable (Da San Martino et al. 2006; Kennedy & Eberhart, 2001; Consoli et al. 2008). Swarm Intelligence techniques provide powerful tools to help doctors to analyze biomedical datasets and medical images (Min, Liu, He, Gong, Fong, Xu, & Wong, 2016; Podgorelec & Kokol, 2001; Garlapati, Vundavilli & Banerjee, 2010; Jalali-Heravi & Ebrahimi-Najafabadi, 2011; Bursa, Lhotska, Chudacek, Spilka, Janku, & Hruban, 2015).

Some of the common algorithms based on swarm intelligence are:

1. **Particle Swarm Optimization:** A stochastic optimization technique that draws inspiration from the behavior of a flock of birds or the collective intelligence of a group of social insects for finding the optimal solution (Biswas, Dasgupta, Das & Abraham, 2007; Kamalanand & Mannar Jawahar, 2014; Kamalanand & Mannar Jawahar, 2015; Kamalanand & Mannar Jawahar, 2016)
2. **Bacterial Foraging Optimization:** The algorithm mimics the foraging strategies of the E. Coli bacteria for finding the optimal solution (Biswas, Dasgupta, Das & Abraham, 2007; Rajinikanth & Latha, 2011; Rajinikanth & Latha, 2012; Kamalanand & Mannar Jawahar, 2014; Kamalanand & Mannar Jawahar, 2015; Kamalanand & Mannar Jawahar, 2016)
3. **Bee Colony Optimization:** This algorithm utilizes the individual cognitive abilities and self-organization of a colony of honey bees for problem solving (Ari, Yenke, Labraoui, Damakoa, & Gueroui 2016)
4. **Ant Colony Optimization:** The behavior of real ants which find the shortest path to arrive at the food sources is mimicked. This process is performed through a chemical known as pheromone which is secreted by the ants on to the ground which increases the probability that other ants will follow the same path (Zheng, Yin, Fu, Fu, & Liu 2014).
5. **Firefly Algorithm:** This algorithm mimics the behavior of the Fire flys for searching the optima in a search space (Min, Liu, He, Gong, Fong, Xu, & Wong 2016; Yang 2010; Raja, Manic & Rajinikanth, 2013; Rajinikanth & Couceiro, 2015; Zahedi, Akbari, Shokouhifar, Safaei & Jalali, 2016; Yang, 2009).

6. **Cuckoo Algorithm:** The Cuckoo algorithm searches the optima by simulating the behavior of a Cuckoo laying eggs (Min, Liu, He, Gong, Fong, Xu, & Wong 2016; Yang & Deb 2009).
7. **Bat Algorithm:** It searches the optima in search space by simulating bats that sense distance via echolocation (Min, Liu, He, Gong, Fong, Xu, & Wong 2016; Yang 2010).
8. **Cat Swarm Optimization:** An algorithm which mimics the behavior of cats such as their natural instinct, strong curiosity, alertness and superior hunting skills (Chu, Tsai & Pan, 2006; Panda, Pradhan & Majhi, 2011; Pradhan & Panda 2012; Tsai, Pan, Chen, Liao & Hao, 2008).
9. **Jumping Frogs Optimization:** An algorithm that mimics the ability of frogs to jump from one rock to another, in a pond (Consoli et al. 2008; Garcia & Perez 2008; Kamalanand & Mannar Jawahar, 2015).
10. **Swallow Swarm Optimization:** The swallow swarm optimization algorithm is inspired by the collective behavior and movement of swallows along with the interaction between flock members (Neshat & Sepidname, 2015; Kaveh, Bakhshpoori & Afshari, 2014; Neshat, Sepidnam, & Sargolzaei, 2013).
11. **Termite Colony Optimization:** This algorithm is a population based optimization technique which is derived from the intelligent behaviors of termites and its hill building capability. The algorithm uses a decision making model which mimics the ability of termites to adjust their movement trajectories (Hedayatzadeh, Salmassi, Keshtgari, Akbari & Ziarati, 2010; Hoolimath, Kiran & Reddy, 2012).
12. **Symbiotic Organism Search (SOS):** This algorithm utilizes the effect of symbiotic interaction of organisms utilized for their survival and propagation in their ecosystem (Guha, Roy, & Banerjee, 2017; Duman, 2016).
13. **Bird Mating Optimizer (BMO):** An algorithm inspired by mating strategies of bird species during mating season, for solving continuous optimization problems (Askarzadeh, 2014; Zhu, Huang & Lu, 2017).

By studying the behavioral and self-organizing pattern exhibited by a group of organisms, it is possible to develop new and novel swarm intelligence techniques (Bursa, Lhotska, Chudacek, Spilka, Janku, & Hruban, 2015). Even though swarm intelligence algorithms have proved their efficiency in finding the optimal solution of hard optimization problems, in some cases the algorithms may fail because of the inability to search the entire search space. This problem occurs in parameter estimation problems where there are a large number of parameters to be estimated from few observations. In such cases, it is necessary to use hybrid or coupled swarm algorithms. The coupled or hybrid algorithms use two or more algorithms in conjunction with each other. The coupled algorithm utilizes the features of all the individual algorithms coupled together. Hence, such coupling results in a synergetic effect and it is possible to search almost the entire search space using these algorithms (Kamalanand & Jawahar, 2012, 2013, 2015).

The United Nations Office on Drugs and Crime (UNODC) estimated that in 2011 a total of 14.0 million people injected drugs worldwide. Further, opioid use has gone up in parts of Asia (East and South-East Asia, as well as Central and West Asia) and Africa since 2009 (UNODC 2013). Heroin users are at high risk for addiction. It is estimated that about 23 percent of individuals who use heroin become de-pendent on it. The spread of heroin habituation and addiction appears like the phenomena of epidemics, including rapid diffusion and clear geographic boundaries (Wang et al. 2011). Illicit drug use has extracted significant costs on societies and governments. The problem has reached epidemic proportions in most countries despite the widespread counter measures and campaigns organized at national and international levels (Agrawal et al. 2004).

Mathematical modelling is very useful to predict how classes of drug takers behave, and provide a suggestion for the treatment strategies (Wang et al. 2011). The mathematical models used to represent various dynamical aspects of sociological and medical processes can explain apparently complex problems. It is in the context of modeling, that engineers can apply their knowledge and use established techniques to help scientific discovery and formulate guidelines for clinical tests, treatment planning and practical measurements (Xia 2003; Filter 2005). The objective of this study is to devise a mathematical methodology for control of the heroin epidemic problem using mathematical model based approach along with tools based on swarm intelligence.

In this chapter, two different swarm intelligence based optimization algorithms namely, the Jumping Frogs Optimization (JFO) and Bacterial Foraging Optimization (BFO) is described in detail. Further, a coupled algorithm namely the coupled Bacterial Foraging/Jumping Frogs Optimization (BFJFO) is presented. Finally, the BFJFO algorithm along with the White and Comiskey mathematical model of heroin epidemic is utilized as a tool for control of the heroin epidemic problem.

## BACKGROUND

In recent years, several researchers have utilized the efficiency of swarm intelligence algorithms in a variety of fields such as mathematical modeling, sociology, economics, biology, medicine, engineering etc. and have reported improved performance and high problem-solving ability of such algorithms. Kang et al. (2016) have utilized swarm intelligence algorithms such as the particle swarm optimization and artificial bee colony algorithm for selection of the hyper-parameters of $\nu$-Support Vector Machine ($\nu$-SVM) model. The authors have demonstrated that these algorithms are preferable ways for determining the SVM hyper-parameters. Martinez-Torres and Olmedilla (2016) have applied Particle Swarm Optimization (PSO) algorithm in the field of social network analysis. The authors have proposed the use of swarm intelligence for optimizing the problem of obtaining the optimum discriminant functions for distinguishing between innovation and non-innovation solvers. Also, the authors have concluded that, the PSO algorithm provides the optimum coefficient values for the discriminant functions. Sabir and Ali (2016) have utilized swarm intelligence algorithms in the fields of alternative fuels and solar power, by designing an optimal PID controller through swarm intelligence algorithms for sun tracking system. The authors have demonstrated a faster convergence rate, small variance and standard deviation of design parameters obtained for solar tracking with respect to azimuth and tilt angle.

Giagkos and Wilson (2014) have examined swarm intelligence based routing protocols and have proposed a new bee-inspired routing protocol for providing multi-path routing in wireless ad hoc networks of mobile nodes. The authors have demonstrated that the application of swarm intelligence offers a high level of adaptability and efficiency under various network conditions which allows the protocol to outperform conventional approaches. Chaurasia and Singh (2015) have applied a grouping-based artificial bee colony algorithm coupled with a local search for grouping the wireless network cells into contiguous areas called registration areas for minimizing the cost of managing the location of mobile devices. The authors have compared the proposed approach based on swarm intelligence with the available approaches and have demonstrated the superiority of the swarm approach in terms of both solution quality and running time. Li et al. (2015) have proposed a novel swarm intelligence optimization algorithm, known as the modified shuffled frog leaping algorithm to compensate the wavefront aberration in free space optical communication systems.

Niu et al. (2016) have applied swarm intelligence algorithms in two scheduling problems in container terminal namely the Yard Truck Scheduling Problem (YTSP) and the integrated Yard Truck Scheduling Problem and Storage Allocation Problem (YTS–SAP). The authors have adopted two algorithms based on swarm intelligence technique namely the particle swarm Optimization (PSO) and the bacterial colony optimization (BCO), for obtaining the problem solution. Gandomi et al. (2015) have applied recent swarm intelligence techniques namely the accelerated particle swarm optimization (APSO), firefly algorithm (FA), and cuckoo search (CS) in civil engineering applications. The authors have utilized the swarm algorithms for cantilever retaining wall design which is an important optimization task in structural and civil engineering. Further, the authors have concluded that these algorithms are highly useful and best suited for solving optimization problems in civil engineering. Bhattacharyya and Raj (2016) have applied swarm intelligence algorithms for the effective co-ordination of Flexible Ac Transmission System (FATS) devices with other existing Var sources. The authors have compared the use of Simple Particle Swarm Optimization (SPSO), Adaptive Particle Swarm Optimization (APSO) and Evolutionary Particle Swarm Optimization (EPSO) for the optimal setting of the Var sources and FACTS devices. Krause et al. (2015) have created a self-organizing sensor network with the capability of overcoming sensor failures for online monitoring of bioprocesses through multivariate predictive sensors. The authors have utilized a swarm intelligence algorithm for decision making purpose in identifying control trajectories for biotechnological processes.

Lazzús et al. (2016) have presented a hybrid swarm intelligence algorithm for parameter estimation of a chaotic system. The authors have implemented a hybrid approach based on particle swarm optimization with ant colony optimization (PSO–ACO) for estimating the parameters of a three-dimensional Lorentz chaotic system. The authors have further compared the estimation results obtained from other algorithms with the proposed hybrid algorithm and have shown that the proposed hybrid PSO–ACO is a highly efficient tool for parameter estimation of nonlinear and chaotic systems.

Hetmaniok et al. (2015) have compared two swarm intelligence algorithms namely the artificial bee colony and ant colony algorithms for solving the two-dimensional inverse Stefan problem. Inverse problem is a problem which involves reconstruction of some parameters of a mathematical model using measured values. The authors have demonstrated that the adopted algorithms are highly efficient for inverse problems. Rebollo-Ruiz and Graña (2014) have proposed a new swarm intelligence algorithm namely the gravitational swarm intelligence for graph coloring algorithm. Lalwani et al. (2016) have introduced a set-based two-level particle swarm optimization algorithm having multiple swarms for finding secondary structure of Ribonucleic acid with high prediction accuracy. The authors have developed the algorithm in such a way that the first level of the algorithm aims on traversing the entire search space for the best solution of each swarm and the second level of the algorithm works at the global best solution of each swarm.

In recent years, several researchers have utilized coupled and hybrid swarm intelligence algorithms for solving problems pertaining to engineering and sciences. Chaves-González et al (2015) have proposed a multiobjective swarm intelligence approach for solving optimization problems. The authors have proposed a constrained multiobjective approach for solving software requirement problems. Kang et al (2016) have proposed a new clustering algorithm by combining a new swarm intelligence algorithm called mussels wandering optimization with the conventional K-means clustering method. The authors have demonstrated that the proposed algorithm makes efficient use of the global optimization capability of mussels wandering optimization and local search ability of K-means algorithm. The authors have further compared the efficiency of the proposed hybrid algorithm with the K-means algorithm and have concluded that the hybrid algorithm outperforms the K-means technique. Zahedi et al (2016) have proposed a swarm intelligence based fuzzy routing protocol which uses fuzzy logic to overcome the uncertainties occurring in the wireless sensor environment. Further, the authors have optimized the fuzzy rule base using a hybrid swarm intelligence algorithm namely the hybrid Firefly Algorithm and Simulated Annealing (FASA) since the firefly algorithm has good global exploration ability among the search space and the simulated annealing has a very good local exploration capability due to its local-based search mechanism and single solution evolution. Neshat and Sepidname (2015) have presented a new hybrid algorithm namely the Fuzzy Adaptive Swallow Swarm Optimization (FASSO) for the simultaneous computation of multimodal functions. The authors have demonstrated that this hybrid method can overcome the weak points of the swallow swarm optimization algorithm.

Meshram et al (2016) have developed a content based image retrieval system using a hybrid Ant colony and Particle swarm Optimization (ACPSO) algorithm. The developed technique involves the hybridization of the ant colony optimization algorithm with the particle swarm optimization algorithm for use in image clustering. Further, the authors have compared the ACPSO algorithm with other algorithms and have demonstrated the improved efficiency of ACPSO algorithm in comparison with other algorithms. Kamalanand and Jawahar (2012) have utilized a coupled algorithm known as coupled jumping frogs/particle swarm optimization (JFPSO) for estimating the parameters of three-dimensional HIV model. Also, Kamalanand and Jawahar (2013) have used the coupled bacterial foraging/jumping frogs algorithm (BFJFO) for resource limited estimation of HIV-1 viral load from CD4 cell count measurements. Further, the authors have compared the efficiency of individual algorithms with coupled and hybrid algorithms and have demonstrated that the coupled algorithms are more efficient in finding the optimal solution when compared to individual swarm intelligence algorithms (Kamalanand & Mannar Jawahar, 2015).

## SWARM INTELLIGENCE BASED CONTROL OF HEROIN ADDICTION EPIDEMIC

### The Jumping Frogs Optimization

The jumping frogs optimization algorithm is a highly efficient stochastic optimization technique which is based on the particle swarm optimization algorithm (Garcia & Perez 2008). The JFO mimics the movement of a group of frogs jumping from rock to rock at a given time interval. The JFO functions without some of the significant characteristics of the PSO such as the movement, velocity and inertia.

In this technique, a swarm $S$ containing $n$ particles ($S$ = 1; 2;.....; n) whose positions $x_i$ evolves in the solution space, jumping from one solution to another. In each iteration, each particle has a random behavior and jumps to another solution in a fashion guided by the effect of some attractors.

JFO considers three attractors for the movement of each particle i: its own best position to date ($b_i$), the best position of its social neighborhood ($g_i$), interpreted as the best position obtained within the swarm in the current iteration, and the best position to date obtained by all the particles, which is called the global best position ($g*$). A jump approaching an attractor consists of changing a feature of the current solution by a feature of the attractor. Each particle is further allowed to have a random behavior by performing random jumps. A random jump consists of selecting at random a feature of the solution and changing its value (Consoli et al. 2008; Garcia & Perez, 2008; Kamalanand & Jawahar, 2015).

The initial positions of the swarm *S* are updated at each iteration. Considering the i-th particle of the swarm ($i \in$ S) and a generic iteration k, the update procedure to obtain the new position $x_{ik}$ from its previous position $x_{ik-1}$ is as follows. Position $x_{ik}$ is obtained by performing random jumps with respect to its current position $x_{ik-1}$ with probability $c_1$, improving jumps approaching $b_i$ with probability $c_2$, improving jumps approaching $g_i$ with probability $c_3$, and improving jumps approaching g* with probability $c_4 = (1-c_1-c_2-c_3)$. The number of jumps to be performed at each iteration is selected at random in the following way. Given the probability $c_i$, we iteratively choose a random number $\zeta$ with uniform distribution in [0,1]. If $\zeta < c_i$, we perform a new jump; otherwise the jumps stop. The update equation (Garcia & Perez 2008) can be expressed as:

$$x_{i+1} = c_1 x_i \oplus c_2 b_i \oplus c_3 g_i \oplus c_4 g^* \qquad (1)$$

At the end of this stage, a local search is applied to the resulting particle. Then all the attractors ($b_i$, $g_i$, $g^*$) are updated, and the same procedure is repeated for all the particles in the swarm. The entire algorithm continues until the user termination conditions are satisfied.

## JFO Algorithm

```
Inputs: Set of available measurements
Initialization:
-          Set initial value for each of the parameters to be estimated
-          Set the size n_s of the swarm S
begin
-          Generate the initial swarm S with positions at random:
X = [x_0, x_1, . . ., x_ns ] ← Generate-Swarm-At-Random;
-          Update the vector of the best positions B = [b_0, b_1, . . ., b_ns ] ← X;
-          Extract the best position among all the particles: g* ← Extract-The-
Best(S, X);
repeat
for i = 1 to n_s do
          if i = 1 then
-          Initialize the best position of the social neighborhood: g_i;
else
-          Update the best position of the social neighborhood i: g_i ← g_i-1;
end
-          Select at random a number between 0 and 1: ζ =Random(0, 1);
if ζ ∈ [0, 0.25] then
-          selected ← x_i;
else if ζ ∈ [0.25, 0.5] then
-          selected ← b_i;
else if ζ ∈ [0.5, 0.75] then
-          selected ← g_i;
```

```
else if ζ ∈ [0.75, 1] then
-          selected ← g*;
- Combine particle i and the selected particle: xi ← Combine(xi, selected);
- Local-Search(i, xi);
if |xi| < |bi| then
-          Update the best position of the given particle i: bi ← xi;
end
if |xi| < |gi| then
-          Update the best position of the social neighborhood of i: gi ← xi;
end
if |xi| < |g*| then
-          Update the global best position to date: g* ← xi;
end
end
until termination conditions;
- Update θ=θg*;
end
```

## Bacterial Foraging Optimization

The bacterial foraging optimization is based upon search and optimal foraging decision making capabilities of the *E.Coli* bacteria. The *E.coli* bacterium has a control system that enables it to search for food and try to avoid noxious or poisonous substances. The selection behavior of bacteria tends to eliminate poor foraging strategies and improve successful foraging strategies. This ability of the bacterium is mathematically utilized for obtaining the solution of an optimization problem.

The foraging dynamics of the bacteria can be described using four stages namely, swarming and tumbling via flagella, chemotaxis, reproduction and finally, elimination and dispersal. In the first stage, the bacteria move as a swarm using movements known as swimming and tumbling. Tumbling mode indicates a change in the future swim direction. A chemotaxis step is a set of consequence swim steps followed by tumble. The actual number of swim steps is determined by the environment. If the environment shows good nutrient concentration in the direction of the swim, the bacteria swim more steps. The end of the chemotactic step is determined by either reaching the maximum number of steps or by reaching a poor environment. When the swim steps are stopped, a tumble action takes place (Rajinikanth & Latha, 2012).

After chemotaxis, the population of bacteria which have had sufficient nutrients will reproduce or split n two. Next, an elimination event may occur when a local event such as an increase in temperature occurs, killing a population of bacteria that are currently in a region with a high concentration of nutrients. Finally, dispersal of bacteria may occur due to an event such as a sudden flow of water which disperses bacteria from one place to another. The event of dispersal may assist chemotaxis since dispersal may place bacteria near good food sources (Korani 2008; Kamalanand & Jawahar, 2015).

In mathematical terms, the coordinates of a bacterium represent an individual solution of the optimization problem. Such a set of trial solutions converges towards the optimal solution following the foraging group dynamics of the bacteria population (Chen et al. 2007; Korani 2008; Kamalanand & Jawahar, 2015).

In order to find the minimum of a fitness function, assume that θ is a coordinate in the solution space and *J(θ)* in an attractant-repellant profile, which represents the location of nutrients and noxious substances (Chen et al. 2007).

$$J(\theta), \theta \in \mathbb{R}^P \quad (2)$$

$$P(j,k,l) = \left\{ \theta^i(j,k,l) \mid i = 1,2,....,S \right\} \quad (3)$$

Equation (3) denotes the positions of each bacterium in the population. For the $i^{th}$ bacteria, it represents the bacteria at the $j^{th}$ chemotactic step, $k^{th}$ reproduction step and the $l^{th}$ elimination-dispersal event.

The evaluation of the fitness of each bacteria, requires the calculation of attractant and repellant coefficients using equation (4), and then combined with the fitness value in equation (5), where $J_{cc}$ represents the combined cell to cell attraction and repelling effects, and $\theta=[\theta_1, \theta_2,... \theta_P]^T$. The values *d, h* and *w* are the attractant and repellant coefficients.

$$\begin{aligned} J_{cc}\left(\theta, P(j,k,l)\right) &= \sum_{m=1}^{S} J_{cc}^{i}\left(\theta, \theta^i(j,k,l)\right) \\ &= \sum_{m=1}^{s}\left[-d_{attractant} \exp\left(-w_{attractant} \sum_{m=1}^{p} \left(\theta_m - \theta_m^{\ i}\right)^2\right)\right] \\ &+ \sum_{m=1}^{s}\left[-h_{repellant} \exp\left(-w_{repellant} \sum_{m=1}^{p} \left(\theta_m - \theta_m^{\ i}\right)^2\right)\right] \end{aligned} \quad (4)$$

$$J(\theta) = J(i,j,k,l) + J_{cc}(\theta, P) \quad (5)$$

Each bacterium in the population will tumble once, and the bacteria which have better fitness values after the tumble, will go on the swim step under the swim length limit $N_s$. The movement can be described by equation (4), and C(i) denotes the step size, and $\phi(j)$ is a random variable in [0, 1]. $C(i)\phi(j)$ is the moving length that the $i^{th}$ bacterium takes. The evaluation and movement process take turns to execute until the lifetime of the bacteria reach the limit, and then the reproduction process takes place. The least healthy bacteria eventually die while each of the healthier bacteria (those yielding lower value of the objective function) asexually split into two bacteria, which are then placed in the same location.

$$\theta^i(j+1,k,l) = \theta^i(j,k,l) + C(i)\phi(i) \quad (6)$$

Gradual or sudden changes in the local environment may kill a group of bacteria that are currently in a region with a high concentration of nutrient gradients. Events can take place in such a fashion that all the bacteria in a region are killed or a group is dispersed into a new location. To simulate this phenomenon in the algorithm, some bacteria are liquidated at random with a very small probability while the new replacements are randomly initialized over the search space (Chen et al. 2007; Korani 2008; Kamalanand & Mannar Jawahar, 2015).

### BFO Algorithm

**Step 1:** Initialization
**Step 2:** Evaluation
**Step 3:** Moving: Tumble / Swim
**Step 4:** Check: End of life?
  **Yes:** Go to **Step 5**
  **No:** Go to **Step 2**
**Step 5:** Reproduction
**Step 6:** Check: End of reproduction?
  **Yes:** Go to **Step 7**
  **No:** Go to **Step 2**
**Step 7:** Elimination
**Step 8:** Check: End of elimination?
  **Yes:** Go to **Step 9**
  **No:** Go to **Step 2**
**Step 9:** Report solution/Stop

## Coupled Bacterial Foraging/Jumping Frogs Optimization

The conventional PSO, JFO and BFO algorithms, when used separately, may not explore some parts of the search space containing high-quality solutions. However, when two such algorithms are used in conjunction with each other, provides more chances to grasp at least one elite solution at the very beginning of the search providing that an initial exploration is not repeated too many times (Panikhom, Sarasiri & Sujitjorn, 2010; Kamalanand & Jawahar, 2012; Kamalanand & Jawahar, 2013).

The BFJFO algorithm utilizes the capability of both the BFO and the JFO algorithm for finding the solution of the optimization problem. When both the algorithms are used in conjunction with each other, the chances of obtaining an elite solution is increased (Kamalanand & Mannar Jawahar, 2015).

### BFJFO Algorithm

**Step 1:** Set initial values
**Step 2:** Call BFO algorithm
**Step 3:** Transport the solution obtained from BFO to JFO as an initial solution
**Step 4:** Call JFO
**Step 5:** Check if termination condition is reached
  **Yes:** Report solution
  **No:** Transport the solution obtained from JFO to BFO as an initial solution:Go to **Step 2**
**Step 6:** Stop

## The White and Comiskey Model

The White and Comiskey heroin epidemic model assumes three classes of people namely, susceptibles (S), drug users (U), and drug users undergoing treatment (T). The susceptible class is composed of people who have never used the drug before. The drug users are the people who begin using the drug and the drug users undergoing treatment are those people who go into medical treatment to battle their addiction. The dynamics of the model can be described as follows:

- The susceptible class can either" smarten-up" (or die) and leave, or they can become drug users.
- The drug users can either continue using the drug," smarten-up" or die, or enter medical treatment.
- The drug users undergoing treatment can then either relapse back into the drug or" smarten-up" (or die).

The White and Comiskey Model can be described as a dynamical system consisting of three simultaneous non-linear ordinary differential equations parameterized using eight parameters, as described in Table 1 (White & Comiskey 2007).

$$\left.\begin{aligned} \frac{dS}{dt} &= \Lambda - \frac{\beta_1 US}{N} - \mu S \\ \frac{dU}{dt} &= \frac{\beta_1 US}{N} - \rho U + \frac{\beta_3 UT}{N} - (\mu + \delta_1)U \\ \frac{dT}{dt} &= \rho U - \frac{\beta_3 UT}{N} - (\mu + \delta_2)T \end{aligned}\right\} \quad (7)$$

*Table 1. The parameters of the White and Comiskey model and their description*

| Model Parameter | Description |
|---|---|
| $\Lambda$ | Number of individuals entering susceptible population |
| $\beta_1$ | Probability of becoming a drug user per unit time |
| $\beta_3$ | Probability that a drug user in treatment relapses back into untreated drug Abuse |
| $\rho$ | Proportion of drug users who enter treatment per unit time |
| $\mu$ | Natural death rate of general population |
| $\delta_1$ | Enhanced removal rate/death rate for drug users |
| $\delta_2$ | Enhanced removal rate/death rate for drug users seeking treatment |
| N | Total population |

## Proposed Control Strategies

### First Strategy

Let us consider that the control of heroin epidemic is performed using awareness programmes or campaigns as a control force. In the White and Comiskey model, the effect of this control force can be incorporated using a suitable forcing function ($k_1$) which has a negative effect on the number of drug users. Hence, the model along with this control force can be described as follows.

$$\left.\begin{aligned} \frac{dS}{dt} &= \Lambda - \frac{\beta_1 US}{N} - \mu S \\ \frac{dU}{dt} &= \frac{\beta_1 US}{N} - \rho U + \frac{\beta_3 UT}{N} - (\mu + \delta_1)U - k_1 \\ \frac{dT}{dt} &= \rho U - \frac{\beta_3 UT}{N} - (\mu + \delta_2)T \end{aligned}\right\} \tag{8}$$

### Second Strategy

Now, let us consider another control strategy where, rehabilitation is the control force. The effect of this control force can be incorporated using a suitable forcing function ($k_2$) which has a positive effect and increases the number of drug users undergoing treatment. The model along with this control force can be described as follows.

$$\left.\begin{aligned} \frac{dS}{dt} &= \Lambda - \frac{\beta_1 US}{N} - \mu S \\ \frac{dU}{dt} &= \frac{\beta_1 US}{N} - \rho U + \frac{\beta_3 UT}{N} - (\mu + \delta_1)U \\ \frac{dT}{dt} &= \rho U - \frac{\beta_3 UT}{N} - (\mu + \delta_2)T + k_2 \end{aligned}\right\} \tag{9}$$

### Estimation of Control Forces

The control forces $k_1$ and $k_2$ can be described using a proportional-Integral type of equation, as follows.

$$k_1(t) = C_{p1}\left(r_1(t) - U(t)\right) + C_{I1}\int\left(r_1(t) - U(t)\right)dt \tag{10}$$

$$k_2(t) = C_{p2}\left(r_2(t) - T(t)\right) + C_{I2}\int\left(r_2(t) - T(t)\right)dt \tag{11}$$

where, $C_{p1}$, $C_{I1}$, $C_{p2}$ and $C_{I2}$ are the parameters of the control forces. In the proposed strategies, in order to estimate the control forces (awareness programmes and rehabilitation programmes in terms of financial expenditures), the parameters of equations (10) and (11) need to be estimated.

In this work, the estimation of the parameters was performed using the Coupled Bacterial Foraging/ Jumping Frogs Optimization. The computer simulations of the proposed methodology were performed using Matlab 7 and the results are presented. Figure 1(a) shows the variation in the susceptible population as a function of time, when awareness programmes are used as a control force. It is seen that the susceptible population increases linearly with increase in time, since some of the drug users may enter the susceptible population due to awareness programmes. Figure 1(b) shows the variation in the population of drug users as a function of time, when awareness programmes are used as a control force. It is seen that the drug user population decreases exponentially with respect to time, since the some of the drug users enter the susceptible population or may enter rehabilitation due to the effect of the awareness programmes. The variation of the population of drug users undergoing treatment is shown as a function of time in Figure 1 (c), when awareness programmes are used as a control force. It is observed that the number of drug users under rehabilitation increases nonlinearly with respect to time. This increase is due to the fact that the drug users enter rehabilitation because of the awareness programmes.

Similarly, the variation of the susceptible population, the population of drug users and the population of drug users undergoing treatment is shown as a function of time in Figures 2(a), 2(b) and 2(c) respectively, when rehabilitation is used as a control force. It is seen that that the susceptible population increases linearly with respect to time, since the drug users undergoing treatment can recover and enter the susceptible population. Further, a linear decrease in the population of drug users is observed since the some of the drug users may enter rehabilitation or may die. Finally, it appears that the population of drug users undergoing treatment increases almost linearly due to the effect of the control force.

## Advantages and Limitations of Swarm Intelligence Based Optimization Algorithms

The main advantages of swarm intelligence techniques are robustness, flexibility, fault tolerance, scalability and high parallelizability. Due to these features, such algorithms have been utilized for solving hard optimization problems. Also, these techniques can be efficiently used for estimating several parameters of nonlinear mathematical models consisting of several state variables. Coupling and hybridization of such algorithms results in improved performance. Due to the interaction of two different algorithms, it is easier to find an elite solution and the coupled algorithm is able to efficiently search the entire search space without neglecting any region of the search space. Further, coupling of two different swarm intelligence algorithms unites the intelligent features of both the algorithms resulting in a more efficient technique for solving problems requiring intelligence.

The major limitation of swarm intelligence techniques is the time-consuming nature of such algorithms. Since these techniques are based on random search, several iterations along with a large population size is required for finding the optimal solution in the search space.

*Figure 1. The variation of susceptible population (a), the population of drug users (b) and the population of drug users undergoing treatment (c) (with awareness programmes as the control force), shown as a function of time*

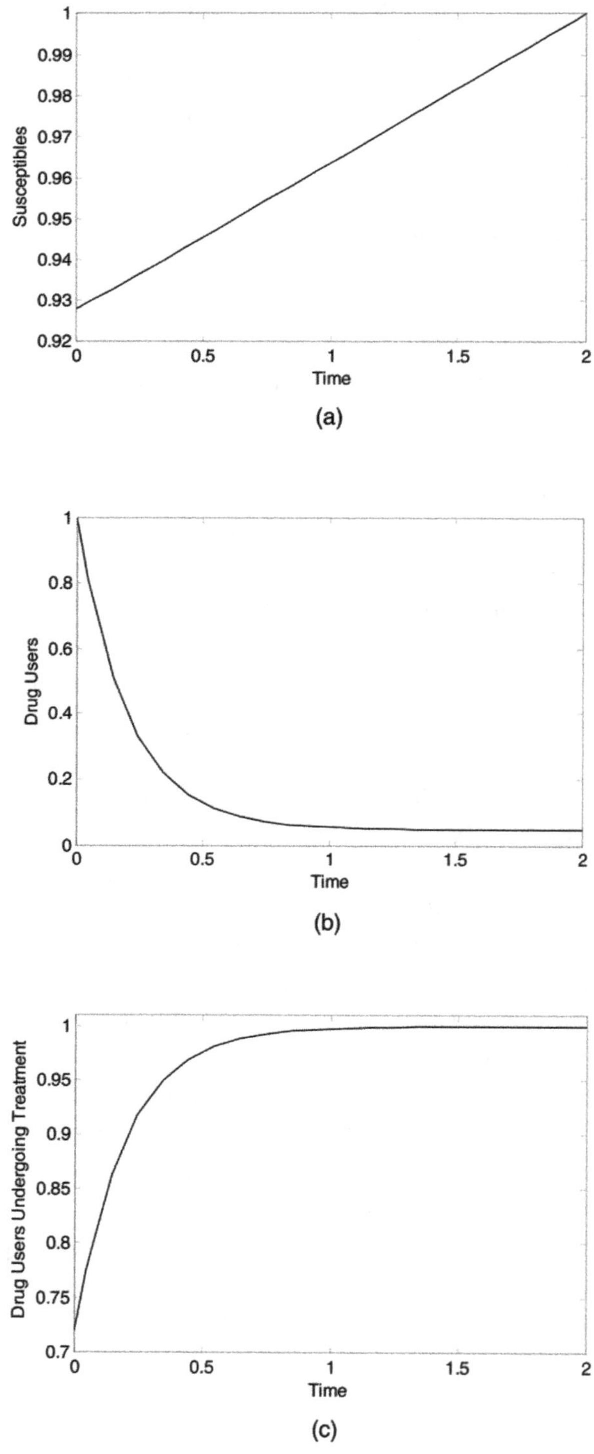

*Figure 2. The variation of susceptible population (a), the population of drug users (b) and the population of drug users undergoing treatment (c) (with rehabilitation as the control force), shown as a function of time*

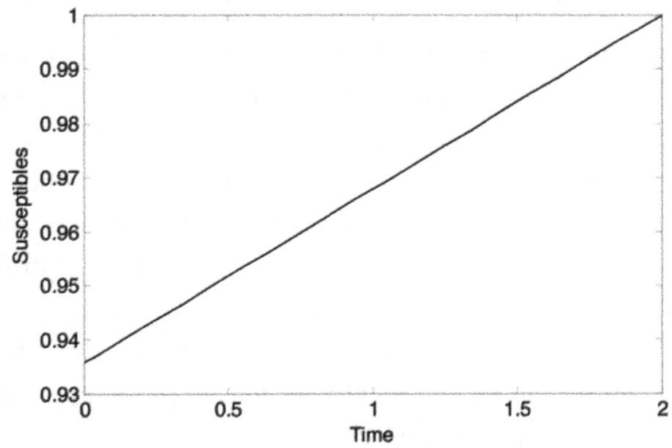


(a)

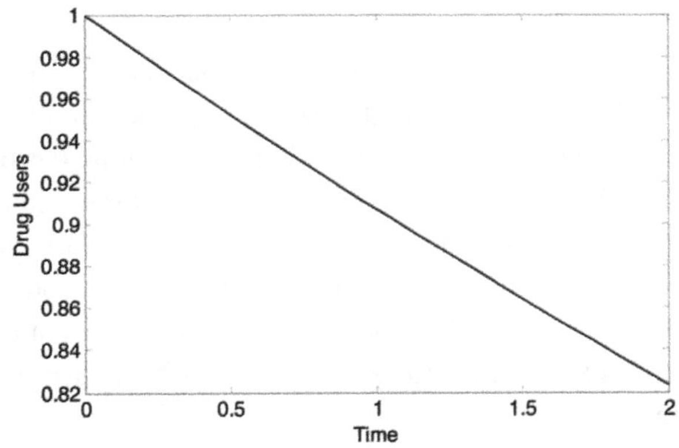


(b)

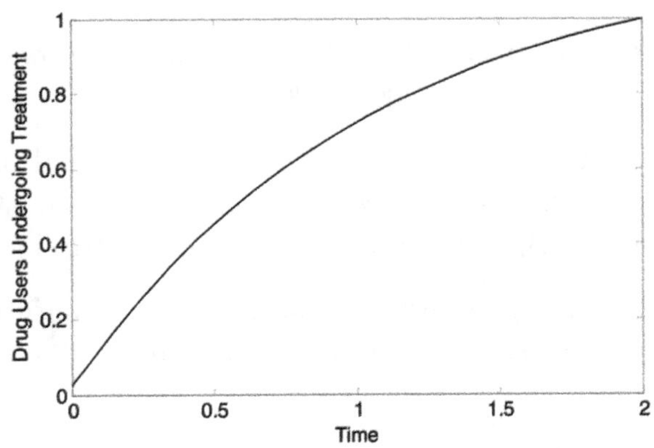


(c)

## FUTURE RESEARCH DIRECTIONS

The heroin epidemic problem is a global epidemic and is difficult to control. The proposed methodology may be utilized for analyzing and understanding the dynamics of the epidemic as well as the forces necessary to control the epidemic. In order to use the proposed methodology, data collection based on surveys for identification of the number of susceptible, drug users and the population of drug users undergoing rehabilitation is necessary for a particular geographical location. Further, for that closed area, the parameters of the White and Comiskey model of heroin epidemic need to be estimated using the collected data. For the parameter estimation problem, swarm intelligence techniques can effectively be used since these algorithms have shown their efficiency in several inverse problems. The model tailored for the particular geographical location can be utilized for estimating the control forces (awareness and rehabilitation programmes) required for controlling the epidemic. Such techniques can be highly helpful for governments to make assessments on the budget and expenditure involved towards conducting awareness programmes and establishing rehabilitation centers for reducing the drug users.

## CONCLUSION

Heroin dependency has become a global problem in recent years and control of the heroin epidemic is a major task for governments. In this work, a mathematical approach to the control of heroin epidemic has been proposed. The White and Comiskey model of heroin epidemic has been utilized and the effect of two different control strategies with awareness programmes and rehabilitation as control forces has been analyzed using computer simulations. Based on the computer simulations performed, it appears that the proposed strategy is efficient as a tool for computing the forces for control of the heroin epidemic. Results demonstrate that the mathematical model based approach in conjunction with the Coupled Bacterial Foraging/Jumping Frogs Optimization Algorithm may be an useful methodology for efficient control of the heroin epidemic.

## REFERENCES

Agrawal, M. R., Kaya, C. Y., & Filar, J. A. (2004). *A Class of Models and Budget Strategies for the Control of Heroin Epidemic in Australia*. Academic Press.

Ari, A. A. A., Yenke, B. O., Labraoui, N., Damakoa, I., & Gueroui, A. (2016). A power efficient cluster-based routing algorithm for wireless sensor networks: Honeybees swarm intelligence based approach. *Journal of Network and Computer Applications*, *69*, 77–97. doi:10.1016/j.jnca.2016.04.020

Askarzadeh, A. (2014). Bird mating optimizer: An optimization algorithm inspired by bird mating strategies. *Communications in Nonlinear Science and Numerical Simulation*, *19*(4), 1213–1228. doi:10.1016/j.cnsns.2013.08.027

Bhattacharyya, B., & Raj, S. (2016). Swarm intelligence based algorithms for reactive power planning with Flexible AC transmission system devices. *International Journal of Electrical Power & Energy Systems*, *78*, 158–164. doi:10.1016/j.ijepes.2015.11.086

Biswas, A., Dasgupta, S., Das, S., & Abraham, A. (2007). Synergy of PSO and bacterial foraging optimization—a comparative study on numerical benchmarks. In *Innovations in Hybrid Intelligent Systems* (pp. 255–263). Springer Berlin Heidelberg. doi:10.1007/978-3-540-74972-1_34

Bursa, M., Lhotska, L., Chudacek, V., Spilka, J., Janku, P., & Hruban, L. (2015). Information retrieval from hospital information system: Increasing effectivity using swarm intelligence. *Journal of Applied Logic*, *13*(2), 126–137. doi:10.1016/j.jal.2014.11.006

Chaurasia, S. N., & Singh, A. (2015). A hybrid swarm intelligence approach to the registration area planning problem. *Information Sciences*, *302*, 50–69. doi:10.1016/j.ins.2015.01.012

Chaves-González, J. M., Pérez-Toledano, M. A., & Navasa, A. (2015). Software requirement optimization using a multiobjective swarm intelligence evolutionary algorithm. *Knowledge-Based Systems*, *83*, 105–115. doi:10.1016/j.knosys.2015.03.012

Chen, T. C., Tsai, P. W., Chu, S. C., & Pan, J. S. (2007). A novel optimization approach: bacterial-GA foraging. In *Innovative Computing, Information and Control, 2007. ICICIC'07. Second International Conference on* (pp. 391-391). IEEE. doi:10.1109/ICICIC.2007.67

Chu, S. C., Tsai, P. W., & Pan, J. S. (2006). Cat swarm optimization. In *Pacific Rim International Conference on Artificial Intelligence* (pp. 854-858). Springer Berlin Heidelberg.

Consoli, S., Pérez, J. M., Darby-Dowman, K., & Mladenović, N. (2008). Discrete particle swarm optimization for the minimum labelling Steiner tree problem. In Nature inspired cooperative strategies for optimization (NICSO 2007) (pp. 313-322). Springer Berlin Heidelberg. doi:10.1007/978-3-540-78987-1_28

Da San Martino, G., Cardillo, F. A., & Starita, A. (2006). A new swarm intelligence algorithm for image alignment. In *Advances in Medical, Signal and Information Processing, 2006. MEDSIP 2006. IET 3rd International Conference On* (pp. 1-4). IET. doi:10.1049/cp:20060387

Duman, S. (2016). Symbiotic organisms search algorithm for optimal power flow problem based on valve-point effect and prohibited zones. *Neural Computing & Applications*, 1–15.

Filter, R. A. (2005). *Dynamic HIV/AIDS parameter estimation with applications* (Doctoral dissertation). University of Pretoria.

Gandomi, A. H., Kashani, A. R., Roke, D. A., & Mousavi, M. (2015). Optimization of retaining wall design using recent swarm intelligence techniques. *Engineering Structures*, *103*, 72–84. doi:10.1016/j.engstruct.2015.08.034

Garcia, F. J. M., & Perez, J. A. M. (2008). Jumping frogs optimization: a new swarm method for discrete optimization. *Documentos de Trabajo del DEIOC*, (3).

Garlapati, V. K., Vundavilli, P. R., & Banerjee, R. (2010). Evaluation of lipase production by genetic algorithm and particle swarm optimization and their comparative study. *Applied Biochemistry and Biotechnology*, *162*(5), 1350–1361. doi:10.1007/s12010-009-8895-2 PMID:20099046

Giagkos, A., & Wilson, M. S. (2014). BeeIP–A Swarm Intelligence based routing for wireless ad hoc networks. *Information Sciences*, *265*, 23–35. doi:10.1016/j.ins.2013.12.038

Guha, D., Roy, P., & Banerjee, S. (2017). Quasi-oppositional symbiotic organism search algorithm applied to load frequency control. *Swarm and Evolutionary Computation*, *33*, 46–67. doi:10.1016/j.swevo.2016.10.001

Hedayatzadeh, R., Salmassi, F. A., Keshtgari, M., Akbari, R., & Ziarati, K. (2010). Termite colony optimization: A novel approach for optimizing continuous problems. In *2010 18th Iranian Conference on Electrical Engineering* (pp. 553-558). IEEE. doi:10.1109/IRANIANCEE.2010.5507009

Hetmaniok, E., Słota, D., & Zielonka, A. (2015). Using the swarm intelligence algorithms in solution of the two-dimensional inverse Stefan problem. *Computers & Mathematics with Applications (Oxford, England)*, *69*(4), 347–361. doi:10.1016/j.camwa.2014.12.013

Hoolimath, P. K. G., Kiran, M., & Reddy, G. R. M. (2012). Optimized Termite: A bio-inspired routing algorithm for MANET's. In *2012 International Conference on Signal Processing and Communications (SPCOM)* (pp. 1-5). IEEE. doi:10.1109/SPCOM.2012.6290230

Jalali-Heravi, M., & Ebrahimi-Najafabadi, H. (2011). The use of ladder particle swarm optimisation for quantitative structure–activity relationship analysis of human immunodeficiency virus-1 integrase inhibitors. *Molecular Simulation*, *37*(15), 1221–1233. doi:10.1080/08927022.2011.586347

Kamalanand, K., & Jawahar, P. M. (2012). Coupled jumping frogs/particle swarm optimization for estimating the parameters of three dimensional HIV model. *BMC Infectious Diseases*, *12*(1), 1. PMID:22214291

Kamalanand, K., & Jawahar, P. M. (2013). Resource limited estimation of HIV-1 viral load from CD4 cell count using coupled bacterial foraging/jumping frogs algorithm. *The World Allergy Organization Journal*, *6*(S1), 189. doi:10.1186/1939-4551-6-S1-P189

Kamalanand, K., & Jawahar, P. M. (2014). A graphical user interface for resource limited estimation of HIV-1 viral load using swarm intelligence techniques. *Journal of Bioinformatics and Intelligent Control*, *3*(2), 110–116. doi:10.1166/jbic.2014.1073

Kamalanand, K., & Jawahar, P. M. (2015). Comparison of Swarm Intelligence Techniques for Estimation of HIV-1 Viral Load. *IETE Technical Review*, *32*(3), 188–195. doi:10.1080/02564602.2014.1000981

Kamalanand, K., & Jawahar, P. M. (2016). Comparison of particle swarm and bacterial foraging optimization algorithms for therapy planning in HIV/AIDS patients. *International Journal of Biomathematics*, *9*(02), 1650024. doi:10.1142/S1793524516500248

Kamberaj, H. (2014). Q-Gaussian swarm quantum particle intelligence on predicting global minimum of potential energy function. *Applied Mathematics and Computation*, *229*, 94–106. doi:10.1016/j.amc.2013.12.036

Kang, F., Xu, Q., & Li, J. (2016). Slope reliability analysis using surrogate models via new support vector machines with swarm intelligence. *Applied Mathematical Modelling*, *40*(11), 6105–6120. doi:10.1016/j.apm.2016.01.050

Kang, Q., Liu, S., Zhou, M., & Li, S. (2016). A weight-incorporated similarity-based clustering ensemble method based on swarm intelligence. *Knowledge-Based Systems*, *104*, 156–164. doi:10.1016/j.knosys.2016.04.021

Kaveh, A., Bakhshpoori, T., & Afshari, E. (2014). An efficient hybrid particle swarm and swallow swarm optimization algorithm. *Computers & Structures*, *143*, 40–59. doi:10.1016/j.compstruc.2014.07.012

Kennedy, J., Kennedy, J. F., Eberhart, R. C., & Shi, Y. (2001). *Swarm intelligence*. Morgan Kaufmann.

Korani, W. M., Dorrah, H. T., & Emara, H. M. (2009). Bacterial foraging oriented by particle swarm optimization strategy for PID tuning. In *Computational Intelligence in Robotics and Automation (CIRA), 2009 IEEE International Symposium on* (pp. 445-450). IEEE. doi:10.1109/CIRA.2009.5423165

Krause, D., Hussein, M. A., & Becker, T. (2015). Online monitoring of bioprocesses via multivariate sensor prediction within swarm intelligence decision making. *Chemometrics and Intelligent Laboratory Systems*, *145*, 48–59. doi:10.1016/j.chemolab.2015.04.012

Lalwani, S., Kumar, R., & Gupta, N. (2016). An efficient two-level swarm intelligence approach for RNA secondary structure prediction with bi-objective minimum free energy scores. *Swarm and Evolutionary Computation*, *27*, 68–79. doi:10.1016/j.swevo.2015.09.008

Lazzús, J. A., Rivera, M., & López-Caraballo, C. H. (2016). Parameter estimation of Lorenz chaotic system using a hybrid swarm intelligence algorithm. *Physics Letters. [Part A]*, *380*(11), 1164–1171. doi:10.1016/j.physleta.2016.01.040

Li, Z., Cao, J., Zhao, X., & Liu, W. (2015). Swarm intelligence for atmospheric compensation in free space optical communication—Modified shuffled frog leaping algorithm. *Optics & Laser Technology*, *66*, 89–97. doi:10.1016/j.optlastec.2014.08.012

Martinez-Torres, R., & Olmedilla, M. (2016). Identification of innovation solvers in open innovation communities using swarm intelligence. *Technological Forecasting and Social Change*, *109*, 15–24. doi:10.1016/j.techfore.2016.05.007

Meshram, S. P., Thakare, A. D., & Gudadhe, S. (2016). Hybrid Swarm Intelligence Method for Post Clustering Content Based Image Retrieval. *Procedia Computer Science*, *79*, 509–515. doi:10.1016/j.procs.2016.03.065

Min, X., Liu, L., He, Y., Gong, X., Fong, S., Xu, Q., & Wong, K. K. (2016). Benchmarking Swarm Intelligence Clustering Algorithms with Case study of Medical Data. *Computerized Medical Imaging and Graphics*. doi:10.1016/j.compmedimag.2016.06.005 PMID:27666794

Nedjah, N., & de Macedo Mourelle, L. (2006). *Swarm Intelligent Systems* (Vol. 26). Springer. doi:10.1007/978-3-540-33869-7_6

Neshat, M., Sepidnam, G., & Sargolzaei, M. (2013). Swallow swarm optimization algorithm: A new method to optimization. *Neural Computing & Applications*, *23*(2), 429–454. doi:10.1007/s00521-012-0939-9

Neshat, M., & Sepidname, G. (2015). A new hybrid optimization method inspired from swarm intelligence: Fuzzy adaptive swallow swarm optimization algorithm (FASSO). *Egyptian Informatics Journal*, *16*(3), 339–350. doi:10.1016/j.eij.2015.07.003

Niu, B., Xie, T., Tan, L., Bi, Y., & Wang, Z. (2016). Swarm intelligence algorithms for Yard Truck Scheduling and Storage Allocation Problems. *Neurocomputing*, *188*, 284–293. doi:10.1016/j.neucom.2014.12.125

Panda, G., Pradhan, P. M., & Majhi, B. (2011). IIR system identification using cat swarm optimization. *Expert Systems with Applications*, *38*(10), 12671–12683. doi:10.1016/j.eswa.2011.04.054

Panikhom, S., Sarasiri, N., & Sujitjorn, S. (2010). Hybrid bacterial foraging and tabu search optimization (btso) algorithms for lyapunov's stability analysis of nonlinear systems. *International Journal of Mathematics and Computers in Simulation*, *3*(4).

Podgorelec, V., & Kokol, P. (2001). Towards more optimal medical diagnosing with evolutionary algorithms. *Journal of Medical Systems*, *25*(3), 195–219. doi:10.1023/A:1010733016906 PMID:11433548

Poole, D. L., & Mackworth, A. K. (2010). *Artificial Intelligence: foundations of computational agents*. Cambridge University Press. doi:10.1017/CBO9780511794797

Pradhan, P. M., & Panda, G. (2012). Solving multiobjective problems using cat swarm optimization. *Expert Systems with Applications*, *39*(3), 2956–2964. doi:10.1016/j.eswa.2011.08.157

Raja, N. S. M., Manic, K. S., & Rajinikanth, V. (2013). Firefly algorithm with various randomization parameters: an analysis. In *International Conference on Swarm, Evolutionary, and Memetic Computing* (pp. 110-121). Springer International Publishing. doi:10.1007/978-3-319-03753-0_11

Rajinikanth, V., & Couceiro, M. S. (2015). RGB histogram based color image segmentation using firefly algorithm. *Procedia Computer Science*, *46*, 1449–1457. doi:10.1016/j.procs.2015.02.064

Rajinikanth, V., & Latha, K. (2011). Bacterial foraging optimization algorithm based PID controller tuning for time delayed unstable system. *The Mediterranean Journal of Measurement and Control*, *7*(1), 197–203.

Rajinikanth, V., & Latha, K. (2012). Controller parameter optimization for nonlinear systems using enhanced bacteria foraging algorithm. *Applied Computational Intelligence and Soft Computing*, *2012*, 22. doi:10.1155/2012/214264

Rajinikanth, V., & Latha, K. (2012). I-PD controller tuning for unstable system using bacterial foraging algorithm: A study based on various error criterion. *Applied Computational Intelligence and Soft Computing*, *2012*, 2. doi:10.1155/2012/329389

Rebollo-Ruiz, I., & Graña, M. (2014). An empirical evaluation of Gravitational Swarm Intelligence for graph coloring algorithm. *Neurocomputing*, *132*, 79–84. doi:10.1016/j.neucom.2013.03.064

Rossini, P. (2000). Using expert systems and artificial intelligence for real estate forecasting. In *Sixth Annual Pacific-Rim Real Estate Society Conference* (pp. 24-27). Academic Press.

Sabir, M. M., & Ali, T. (2016). Optimal PID controller design through swarm intelligence algorithms for sun tracking system. *Applied Mathematics and Computation*, *274*, 690–699. doi:10.1016/j.amc.2015.11.036

Tsai, P. W., Pan, J. S., Chen, S. M., Liao, B. Y., & Hao, S. P. (2008). Parallel cat swarm optimization. In *2008 International Conference on Machine Learning and Cybernetics* (Vol. 6, pp. 3328-3333). IEEE. doi:10.1109/ICMLC.2008.4620980

UNODC. (2013). *United nations office on drugs and crime - world drug report 2013*. United Nations Publication.

Wang, X., Yang, J., & Li, X. (2011). Dynamics of a heroin epidemic model with very population. *Applied Mathematics*, *2*(06), 732–738. doi:10.4236/am.2011.26097

White, E., & Comiskey, C. (2007). Heroin epidemics, treatment and ODE modelling. *Mathematical Biosciences*, *208*(1), 312–324. doi:10.1016/j.mbs.2006.10.008 PMID:17174346

Xia, X. (2003). Estimation of HIV/AIDS parameters. *Automatica*, *39*(11), 1983–1988. doi:10.1016/S0005-1098(03)00220-6

Yang, X. S. (2009, October). Firefly algorithms for multimodal optimization. In *International Symposium on Stochastic Algorithms* (pp. 169-178). Springer Berlin Heidelberg.

Yang, X. S. (2010). A new metaheuristic bat-inspired algorithm. In Nature inspired cooperative strategies for optimization (NICSO 2010) (pp. 65-74). Springer Berlin Heidelberg. doi:10.1007/978-3-642-12538-6_6

Yang, X. S. (2010). *Nature-inspired metaheuristic algorithms*. Luniver Press.

Yang, X. S., & Deb, S. (2009, December). Cuckoo search via Lévy flights. In *Nature & Biologically Inspired Computing, 2009. NaBIC 2009. World Congress on* (pp. 210-214). IEEE.

Zahedi, Z. M., Akbari, R., Shokouhifar, M., Safaei, F., & Jalali, A. (2016). Swarm intelligence based fuzzy routing protocol for clustered wireless sensor networks. *Expert Systems with Applications*, *55*, 313–328. doi:10.1016/j.eswa.2016.02.016

Zhang, S., Lee, C. K., Chan, H. K., Choy, K. L., & Wu, Z. (2015). Swarm intelligence applied in green logistics: A literature review. *Engineering Applications of Artificial Intelligence*, *37*, 154–169. doi:10.1016/j.engappai.2014.09.007

Zheng, W. B., Yin, H. T., Fu, J. F., Fu, P., & Liu, B. (2014). Range expansion of cooperative mobile wireless network based on swarm intelligence optimization. *Measurement*, *54*, 83–91. doi:10.1016/j.measurement.2014.04.003

Zhu, J. J., Huang, M., & Lu, Z. R. (2017). Bird mating optimizer for structural damage detection using a hybrid objective function. *Swarm and Evolutionary Computation*, *35*, 41–52. doi:10.1016/j.swevo.2017.02.006

## ADDITIONAL READING

Agar, M. H., & Wilson, D. (2002). Drugmart: Heroin epidemics as complex adaptive systems. *Complexity*, *7*(5), 44–52. doi:10.1002/cplx.10040

Alexander, B. K. (1987). The disease and adaptive models of addiction: A framework evaluation. *Journal of Drug Issues*, *17*(1), 47–66. doi:10.1177/002204268701700104

Aris, R. (2012). *Mathematical modelling techniques*. Courier Corporation.

Bailey, N. T. (1975). *The mathematical theory of infectious diseases and its applications*. Charles Griffin & Company Ltd.

Blum, W., Galbraith, P. L., Henn, H. W., & Niss, M. (2007). *Modelling and applications in mathematics education*. New York: Springer. doi:10.1007/978-0-387-29822-1

Brauer, F., Castillo-Chavez, C., & Castillo-Chavez, C. (2001). Mathematical models in population biology and epidemiology (Vol. 40, pp. xxiv+-416). New York: Springer.

Chao, P. A. N., Weiyu, J. I. A. O., & Tao, M. E. N. G. (2016). Optimal allocation of generalized power sources in active distribution network based on hybrid intelligent particle swarm optimization algorithm. *Power System Protection and Control, 44*(7), 69–71.

Dayan, P. (2009). Dopamine, reinforcement learning, and addiction. *Pharmacopsychiatry, 42*(S 01), S56-S65.

Diekmann, O., & Heesterbeek, J. A. P. (2000). *Mathematical epidemiology of infectious diseases: model building, analysis and interpretation* (Vol. 5). John Wiley & Sons.

Drucker, E., & Sidel, V. W. (1974). The communicable disease model of heroin addiction: A critique. *The American Journal of Drug and Alcohol Abuse, 1*(3), 301–311. doi:10.3109/00952997409011026 PMID:4467729

Du, K. L., & Swamy, M. N. S. (2016). Particle swarm optimization. In *Search and Optimization by Metaheuristics* (pp. 153–173). Springer International Publishing. doi:10.1007/978-3-319-41192-7_9

DuPont, R. L., & Greene, M. H. (1973). The dynamics of a heroin addiction epidemic. *Science, 181*(101), 716–722. doi:10.1126/science.181.4101.716 PMID:4724929

Frauenthal, J. C. (2012). *Mathematical modeling in epidemiology*. Springer Science & Business Media.

Greene, M. H. (1974). An epidemiologic assessment of heroin use. *American Journal of Public Health, 64*(12_Suppl), 1-10.

Greene, M. H., & Dupont, R. L. (1974). Heroin addiction trends. *The American Journal of Psychiatry, 131*(5), 545–550. PMID:4819047

Hritonenko, Y. (2003). *Applied Mathematical Modelling ofEngineering Problems* (Vol. 10). doi:10.1007/978-1-4419-9160-7

Hughes, P. H., & Crawford, G. A. (1972). A contagious disease model for researching and intervening in heroin epidemics. *Archives of General Psychiatry, 27*(2), 149–155. doi:10.1001/archpsyc.1972.01750260005001 PMID:5042822

Kaplan, C. D., Korf, D., & Sterk, C. (1987). Temporal and Social Contexts of Heroin-Using Populations An Illustration of the Snowball Sampling Technique. *The Journal of Nervous and Mental Disease, 175*(9), 566–574. doi:10.1097/00005053-198709000-00009 PMID:3655783

Kapur, J. N. (1988). *Mathematical Modelling*. New Age International.

Kolodny, A., Courtwright, D. T., Hwang, C. S., Kreiner, P., Eadie, J. L., Clark, T. W., & Alexander, G. C. (2015). The prescription opioid and heroin crisis: A public health approach to an epidemic of addiction. *Annual Review of Public Health, 36*(1), 559–574. doi:10.1146/annurev-publhealth-031914-122957 PMID:25581144

Kolodny, A., Courtwright, D. T., Hwang, C. S., Kreiner, P., Eadie, J. L., Clark, T. W., & Alexander, G. C. (2015). The prescription opioid and heroin crisis: A public health approach to an epidemic of addiction. *Annual Review of Public Health*, *36*(1), 559–574. doi:10.1146/annurev-publhealth-031914-122957 PMID:25581144

Liu, J., & Zhang, T. (2011). Global behaviour of a heroin epidemic model with distributed delays. *Applied Mathematics Letters*, *24*(10), 1685–1692. doi:10.1016/j.aml.2011.04.019

MacKintosh, D. R., & Stewart, G. T. (1979). A mathematical model of a heroin epidemic: Implications for control policies. *Journal of Epidemiology and Community Health*, *33*(4), 299–304. doi:10.1136/jech.33.4.299 PMID:536680

Massin, S. (2012). Is harm reduction profitable? An analytical framework for corporate social responsibility based on an epidemic model of addictive consumption. *Social Science & Medicine*, *74*(12), 1856–1863. doi:10.1016/j.socscimed.2012.02.006 PMID:22475402

Mavrovouniotis, M., Li, C., & Yang, S. (2017). A survey of swarm intelligence for dynamic optimization: Algorithms and applications. *Swarm and Evolutionary Computation*, *33*, 1–17. doi:10.1016/j.swevo.2016.12.005

Moharam, A., El-Hosseini, M. A., & Ali, H. A. (2016). Design of optimal PID controller using hybrid differential evolution and particle swarm optimization with an aging leader and challengers. *Applied Soft Computing*, *38*, 727–737. doi:10.1016/j.asoc.2015.10.041

Mousavi, M., Yap, H. J., Musa, S. N., Tahriri, F., & Dawal, S. Z. M. (2017). Multi-objective AGV scheduling in an FMS using a hybrid of genetic algorithm and particle swarm optimization. *PLoS One*, *12*(3), e0169817. doi:10.1371/journal.pone.0169817 PMID:28263994

Mulone, G., & Straughan, B. (2009). A note on heroin epidemics. *Mathematical Biosciences*, *218*(2), 138–141. doi:10.1016/j.mbs.2009.01.006 PMID:19563739

Narayanan, S. J., Thakur, P., & Bhatt, R. B. (2017). User Localization in an Indoor Environment Using Fuzzy Hybrid of Particle Swarm Optimization & Gravitational Search Algorithm with Neural Networks. In *Proceedings of Sixth International Conference on Soft Computing for Problem Solving: SocProS 2016* (Vol. 1, p. 286). Springer.

Pal, P. S., Choudhury, S., Ghosh, A., Kumar, S., Kar, R., Mandal, D., & Ghoshal, S. P. (2016). Social emotional optimization algorithm based identification of nonlinear hammerstein model. In *Communication and Signal Processing (ICCSP), 2016 International Conference on* (pp. 1633-1637). IEEE. doi:10.1109/ICCSP.2016.7754438

Prasad, D. R., Naganjaneyulu, P. V., & Prasad, K. S. (2016). A hybrid swarm optimization for energy efficient clustering in multi-hop wireless sensor network. *Wireless Personal Communications*, 1–13.

Redish, A. D. (2004). Addiction as a computational process gone awry. *Science*, *306*(5703), 1944–1947. doi:10.1126/science.1102384 PMID:15591205

Rossi, C. (2003). The role of dynamic modelling in drug abuse epidemiology. *Bulletin on Narcotics*, 33–44.

Samanta, G. P. (2011). Dynamic behaviour for a nonautonomous heroin epidemic model with time delay. *Journal of Applied Mathematics and Computing*, *35*(1-2), 161–178. doi:10.1007/s12190-009-0349-z

Skoullis, V. I., Tassopoulos, I. X., & Beligiannis, G. N. (2017). Solving the high school timetabling problem using a hybrid cat swarm optimization based algorithm. *Applied Soft Computing*, *52*, 277–289. doi:10.1016/j.asoc.2016.10.038

Strang, J., & Gossup, M. (Eds.). (1994). *Heroin addiction and drug policy: the British system*. Oxford: Oxford University Press.

Thangamani, C., & Chidambaram, M. (2017). A Novel Hybrid Genetic Algorithm with Weighted Crossover and Modified Particle Swarm Optimization. *Artificial Intelligent Systems and Machine Learning*, *9*(2), 25–30.

Unick, G. J., Rosenblum, D., Mars, S., & Ciccarone, D. (2013). Intertwined epidemics: National demographic trends in hospitalizations for heroin-and opioid-related overdoses, 1993–2009. *PLoS One*, *8*(2), e54496. doi:10.1371/journal.pone.0054496 PMID:23405084

Wang, G. G., Gandomi, A. H., Alavi, A. H., & Deb, S. (2016). A hybrid method based on krill herd and quantum-behaved particle swarm optimization. *Neural Computing & Applications*, *27*(4), 989–1006. doi:10.1007/s00521-015-1914-z

Yadav, A., Deep, K., Kim, J. H., & Nagar, A. K. (2016). Gravitational swarm optimizer for global optimization. *Swarm and Evolutionary Computation*, *31*, 64–89. doi:10.1016/j.swevo.2016.07.003

Yadav, B. Ch. S., Mathur, S., & Adamowski, J. (2016). Estimation of in-situ bioremediation system cost using a hybrid Extreme Learning Machine (ELM)-particle swarm optimization approach. *Journal of Hydrology (Amsterdam)*, *543*, 373–385. doi:10.1016/j.jhydrol.2016.10.013

## KEY TERMS AND DEFINITIONS

**Bacterial Foraging Optimization:** An algorithm that mimics the foraging capability of E. Coli bacteria for finding the optimal solution in an optimization problem.

**Computational Intelligence:** The learning ability of a computer through experimental observations.

**Coupled Algorithms:** Coupled algorithms are synergetic algorithms that utilize the power of two or more algorithms for finding the optimal solution.

**Dynamic Model:** A dynamic mathematical model represents the evolution of the state variables as a function of time.

**Heroin:** A highly addictive and illegal drug.

**Heroin Epidemic:** A widespread occurrence of heroin addiction in a community.

**Jumping Frogs Optimization:** A swarm intelligence algorithm that mimics the movement of a group of frogs jumping from rock to rock.

**Swarm Intelligence:** Algorithms based on swarming and flocking behavior exhibited by organisms and natural systems, useful for solving hard optimization problems.

**White and Comiskey Model:** A dynamic model of heroin epidemic that considers the susceptible population, population of drug users, and the population of drug users undergoing rehabilitation.

# Chapter 3
# Intelligent Techniques Inspired by Nature and Used in Biomedical Engineering

**Omer Deperlioglu**
*Afyon Kocatepe University, Turkey*

## ABSTRACT

*Managing medical information and knowledge is becoming an increasing problem for healthcare professionals. Medical science that contains ever-increasing amounts of knowledge, such as the medical history of a patient, medical data about diseases, diagnosis and treatment methods, should be necessarily a science of information. The real problem faced by patients and healthcare providers is finding and using relevant knowledge at the right time. In this context, in the middle of 1950s, intelligent computer systems, called clinical decision support systems (CDSS), were introduced as a new concept. CDSS is defined as an active intelligent system that can help medical experts to make decisions by taking specific recommendations. Also, it provides decisions based on resolving patient-specific information and related medical truths. The objective of this chapter is to focus on these systems and explain relations with the field of artificial intelligence methods, approaches, or techniques in this manner.*

## INTRODUCTION

Biomedical Engineering which is an important engineering area is interdisciplinary. Biomedical engineering is the application of the principles and problem-solving techniques of engineering to the fields of biology and health care. Biomedical engineers work with physician, clinicians, therapists, and other health care workers to develop systems, equipment, devices, and software in order to solve health care problems. In other words, the challenges created by the diversity and complexity of living systems require creative, knowledgeable, and imaginative people for working in same teams such as physicians, scientists, engineers, and even business professionals. Thus, they can produce an instrument to monitor, restore, and enhance normal body function. The biomedical engineer is ideally trained to work at the intersection of science, medicine, and mathematics to solve biological and medical problems. (IEEE EMB, 2015).

DOI: 10.4018/978-1-5225-4769-3.ch003

Biomedical engineers try to develop and evaluate systems and products such as artificial organs, prostheses (artificial devices that replace missing body parts), instrumentation, medical information systems, health management. Also, they interest to create new equipment or environments for such purposes as maximizing human performance, or providing non-invasive diagnostic tools using their knowledge of engineering (Biomedical Engineering, 2016). For this reason, biomedical engineers often use artificial intelligence to solve health care problems.

Generally, biomedical engineering is one direction of the growing field of medical or health care sciences, which develops the application of engineering, computer, and information sciences for the problems of health and life sciences. At each stage of medical practice, decision quality can have a significant impact. Human decision-making performance may be at the lowest level, and the problem may deteriorate as complexity increases. For these reasons, the importance of improving the medical decision support system is increasing, and the mostly use of these intelligent systems in all areas of medicine is becoming increasingly widespread. Decision-making and the use of artificial intelligence (AI) for medical decision-making is based on knowledge-intensive expert consultation systems which was introduced at the beginning of the 1970s. Meanwhile, sophisticated software environments are increasingly combining AI ideas and methods. Because at the same time they try to facilitate the task of building, validating and testing medical knowledge bases (Deperlioglu et al., 2015).

Jack Copeland describes artificial intelligence as "Artificial Intelligence (AI) is often defined as a science that does things that require intelligence when done by computers. AI has achieved some success in limited or simplified areas. However, AI last five years since its establishment, has been very slow progress and provide early optimism about reaching the level of human intelligence. It has led to the evaluation of deep difficulty of the problem." (Copeland, 2000). On the other hand, Clancey and Shortliffe provided this definition for medical AI in 1984:

*Medical artificial intelligence is primarily concerned with the creation of AI programs that make diagnosis and treatment recommendations. Unlike medical applications based on other programming methods, such as statistical and probabilistic methods, medical AI programs are based on the symbolic patterns of disease entities and their relationship to patient factors and clinical findings (Clancey & Shortliffe, 1984).*

In the medical field, there is too much available data, so it is difficult for physicians to work on very different situations for each encountered case. As general, there is often need for expert support during evaluation of large mass of information in order to perform the related operations. Therefore, physicians work in quite difficult conditions and they need help in their daily work. In order to facilitate physicians' work, examine large data sets, and perform any other tasks like finding similarities and differences among situations, software systems are also developed for decision support. After Lipkin and Hardy introduced the decision support system concept in 1958, Ledley and Lusted developed an approach to make inference automatically (Ledley & Lusted, 1959; Lipkin & Hardy, 1958). Such systems often use artificial intelligence components to do their tasks on evaluating big databases which is prepared by the experts. So, they are called as expert systems or knowledge-based systems. From 1950 till today, many expert systems or knowledge-based systems have been developed in different medical fields. As seen in Figure 1, there are many different types of medical tasks, which can be applied by expert systems. It is possible to list some of them as follows (Coiera, 2003).

- **Generating Alerts and Reminders:** An expert system incorporating a monitor or the similar device can stimulate changes in a patient's condition. In less acute cases, the laboratory can scan test results or drug orders, and then send reminders or warnings via an email system.
- **Diagnostic Assistance:** When inexperienced doctors should diagnose in a short-complicated time period, and the patient's condition is very serious, expert systems can support them to make diagnosis.
- **Therapy Critiquing and Planning:** Software systems can indicate deficiencies in current therapy program or prepare specific treatment programs for patients with exceptions. For example, a patient being treated for heart disease but also having diabetes can receive an appropriate treatment program, which was rigorously prepared by the expert system.
- **Agents for Information Retrieval:** It is possible for software systems to collect data for the relevant situation and transmit the appropriate functions that should be run. At this point, agents are some kind of elements, which are connected to the hospital database or the Internet and know what they are looking for in order to overcome a situation.
- **Image Recognition and Interpretation:** Many medical images can be automatically interpreted by software systems from X-ray plane up to more complex images such as angiograms, computed tomography CT and magnetic resonance imaging MRI scans. This is especially important for mass demonstrations. Because it is possible to encounter abnormal images that are difficult to analyze only through human capabilities.

Successful handling of a large number of complex problems in the real world has increased the popularity of intelligent algorithms and techniques that are inspired by nature. These are called Nature Inspired Intelligent (NII) techniques. Their successful applications of complex and difficult optimization problems emphasize the importance of artificial intelligence in the wider field. NII techniques benefit

*Figure 1. Some medical tasks, which can be applied by expert systems*

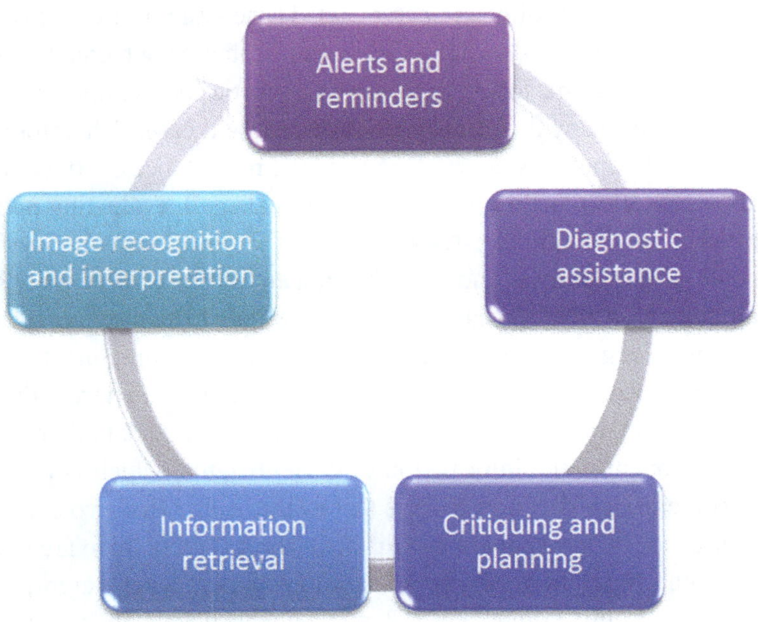

from the way biological systems deal with real-world situations. Especially when solving complex real-world situations, they mimic the way real biological systems work, such as the human brain, ant colony and the human immune system (Vassiliadis & Dounias, 2009). There are several methods, approaches or techniques that are inspired by nature, used to solve biomedical engineering problems such as Particle Survey Optimization (PSO), Ant Colony Optimization (ACO), DNA computation, artificial immune systems, and dice computation.

In this section, we briefly present a set of selected application which used in biomedical engineering fields by using NII approaches and explain the application steps of each process. Remaining content of the paper is organized as follows: The next chapter mentions the use of artificial intelligence in medical reasoning processes from the past to the present. The applications in medical reasoning and medical decision support systems, with artificial intelligence are briefly expressed. In the next section, the characteristics and the structures of expert systems and knowledge-based systems are explained. After this section, information was provided on the structure and characteristics of decision support systems used in medical consultation, diagnosis and treatment with artificial intelligence support. Later, opinions on decision-making applications and future expectations in the field of biomedical engineering were presented. Finally, it was finished with the comments in the last chapter.

## BACKGROUND

The developments in recent years will lead us to understand the future development of decision support technologies. The changes rapidly emerge in technology and in application environments. So, users become more sophisticated and more demanding. Organizations are becoming more complex but more agile and flexible. In addition, global factors and competitive factors are rapidly changing and affecting the design and use of these tools. Future surprises are anticipated but certain trends can be observed (Ramesh, et al, 2004).

Numerous intelligent systems have been developed to improve health care and provide better health facilities, reduce costs, and help to health care personnel. These systems are software applications that support medical decisions to facilitate decision-making in health care personnel with different models. Especially doctors haven't a lot of time to investigate the diagnostic and treatment methods applied in the past the medical profession for many different types of disease and cases. Therefore, they are generally called Biomedical Decision Support systems (BDSS) and the importance of BDDS is increasing day by day. This software approach offers suggestions to decision makers in evaluating patients with previous clinical data in addition to individual data. As general, it is a typical software system using artificial intelligence components such as learning and classification algorithms. This type of software systems is struggling to make inferences on what is the most appropriate class to be the closest in the case where the data from existing data or belong. Accordingly, BDDS has three main components. The first component is previously generated knowledge base of the experience that was obtained from the clinical studies for similar situations result. In fact, this component is considered as a large data base created by the expert medics. The second component is the information of current status, which is taken from the patient. The third component is the part of decision-makers that evaluates the first two components together by offering artificial intelligence inference using the different tools. Today there are used several methods such as fuzzy logic inference operations, artificial neural networks, genetic algorithms, support vector machines, and bee colonies in this component (Clancey & Shortliffe, 1984; Copeland, 2000; Miller,

1994; Stivaros et al., 2010). In literature, sometimes these BDSS are called adaptive or intelligent decision support systems (Chuang, & Yada, 1998; Papamichail, & French, 2005).

The main objective is to use artificial intelligence to prepare software would be smart assistants for physicians. Thus, the concept of intelligent assistants arises for BDSS to provide support at every stage that starting with the patient at hospital entrance and finishing with the patient treatment. In the next period, the effect of artificial intelligence will be very great on BDSSs. As Ramesh and his colleagues have said in their study: "There is strong evidence that the medical artificial intelligence can play a vital role in helping the clinician become effective in delivering health care in the 21st century. There is no doubt that these techniques will be used to reinforce and complement the future clinician's medical intelligence." (Nahar et al, 2012).

In the early 1970s, the first biomedical applications of artificial intelligence MYCIN had made and it has been very successful. In the 1980 s artificial intelligence applications has increased rapidly in the biomedical field and expert systems were performed such as PUFF, CADUCEUS and INTERNIST Sondak & Sondak, 1990). Today, many BDSSs have been developed to improve health care and provide better health facilities, reduce costs, and help to health care personnel. Intended use of the nature inspired intelligent techniques in BDSS can be listed as follows (Hoong, 1988; Ishak & Siraj, 2002):

- Provides an environment for the examination, organization, display and cataloging of medical information.
- Produces new tools for supporting activities of medical decision-making, training and research.
- Integrates activities in the fields of medical, computer, cognitive and other sciences.
- Offers a rich content discipline for future scientific medicine expertise.

Advancement in computer technologies set up a platform for development a better design of telemedicine applications. Especially mobile devices with the Internet have become the most important tools in medical applications (Ishak & Siraj, 2002). Bhargava and his colleagues said that the use of modern DSS should not have limits and be Web-based. They focus on the current use of Web-based DSS, on the trends in the installation of such systems and on the latest technological developments in this area (Bhargava et al., 2007).

Today, expert systems and knowledge-based systems are almost used in all medical units. These systems are developing day by day and prepared in user or task-oriented. For example, Song and colleagues have used support vector machines (SVM) in the decision-making process in their work. It was called embedded decision-making system and it has used SVM for classification in decision making system (Song et al., 2012). In another application using SVM, Ubeyli has reported that she combines a holistic view of automated diagnostic systems with spectral analysis techniques in the classification of electroencephalography (EEG) signals. In this paper, there are also descriptive and detailed information about the implementation of automated diagnostic systems and the extraction / selection of features from EEG signals (Ubeyli, 2009).

Decision support systems making use of artificial intelligence firstly was considered by only physician in supporting diagnostic process. Today, decision support systems include clinical studies, laboratory studies, consultation, diagnostic, and treatment phases and it supports all stages. Success strategies on clinical decision support systems have been explained by Mohamed Khalifa. In this study, the investigated proposals were collected under ten main topics to be addressed. These titles consist of the correct content of the BDSS, providing effective and reliable information, providing simple messages, giving users references, time savings, integration with clinical workflow, increasing system response and speed, adopting active and passive warning mechanisms, integration with other hospitals Information systems and the correct management of information of BDSS (Khalifa, 2014).

Based on the results of the clinical trials, the treatment methods carried out the work outlined by Valkenhoef and his colleagues. They explain the decision processes they have proposed and show how these are implemented in open source ADDIS software. ADDIS provides semi-automated meta-analysis, network based meta-analysis and utility-risk decision models, and visualization of all results. Figure 2 shows a screenshot from the ADDIS interface (Valkenhoefet al., 2013).

Also, Sikchi and his colleagues have showed the contribution of artificial intelligence in medical diagnosis by using fuzzy expert system (Sikchi et al., 2012).

In their study, Bahaa et al. have explained their artificial intelligence oriented solutions on how to find treatment needs and methods in orthodontic field. They said that:

*A valid problem assessment allows health care providers to determine treatment needs and priorities, such as improving health care towards greater financial accountability. Computer applications in different medical fields have received great interest and this interest has been further enhanced with artificial intelligence support (Bahaa et al., 2011).*

*Figure 2. A screenshot from the ADDIS interface*
*Valkenhoefet al., 2013.*

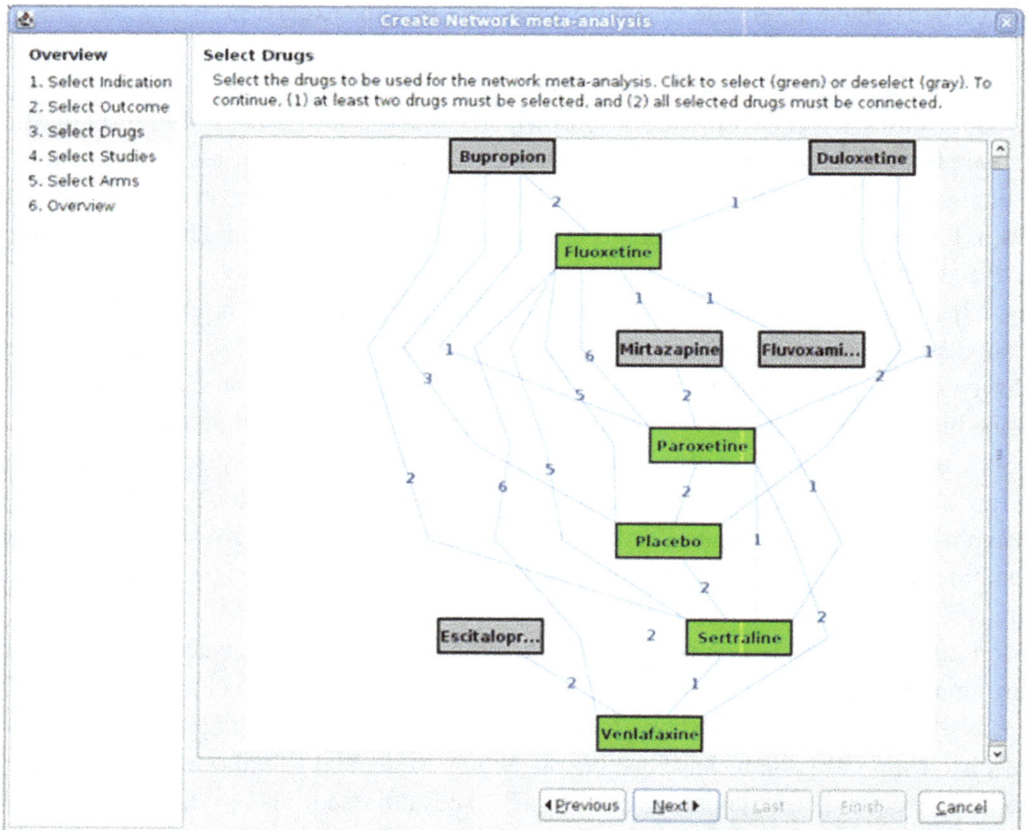

Yao and Kumar have made a clinical decision support system, which is called as CONFlexFlow. This system uses context and rules and tries to meet the goal of the research work, which introduces new approaches to designing clinical decision support systems. Undoubtedly, they demonstrated how flexible and adaptable clinical pathways can be designed taking into account medical information in the rules and have detailed contextual knowledge to achieve a high-quality result (Yao & Kumar, 2013).

Zhou and his colleagues have developed diagnosis systems for gastric cancer. In their paper, they have proposed a clinical decision support system based on bi-level belief rule based for diagnosing in gastric cancer. The proposed prototype CDSS consists of two layers, both modeled by the BRB. In addition, current methods for optimizing the BRB are sensitive to initial values and a new CSA strategy has been developed since local search capabilities are weak. It was also thought as a general method for optimizing BRB. The validity of the proposed CDSS prototype was tested using a range of real patient data and a comparative study of existing CDSSs. The results have shown that BBRB can produce better diagnostic performance (Zhou, 2013). As another application for the diagnosis of cancer, they have aimed to perform a comparative analysis of the computational intelligence algorithms for identifying breast cancer in its early stages. In this study, two types of data display are considered: a microarray-based and medical imaging based data display. In this study, contrary to previous researches, the imbalance of these data was taken into consideration (Shim et al., 2002).

## NATURE INSPIRED INTELLIGENT TECHNIQUES

Together with the technological developments in the present day, researchers in the field of biomedical have directed to develop computer software, which will provide expert support and help in making decisions for doctors. Naturally, these software which have reasoning, decision making, learning skills, are expected to benefit from expert opinions, and thus help the doctors inferring. When it comes to reasoning, decision making, learning skills, the concept of artificial intelligence comes to mind. Artificial intelligence, which has become a subdivision of computer science, has found many fields of application such as education, medical, commercial, and industrial applications. In particular, artificial intelligence techniques, which are more effective in solving problems in nature, are inspired by nature. Researchers are trying to develop new computer techniques by simulating existing systems in nature such as ant colony optimization, particle swarm optimization, artificial neural networks, vortex optimization and artificial immune system to solve complex problems and develop the most appropriate solution. As seen Figure 3, there is a need for multidisciplinary rigorous work in order to optimize the algorithms, techniques, or approaches inspired by nature to computer environments. However, it is possible to produce highly sophisticated intelligent computer software by working together with computer science, related scientific fields and engineering sciences. These natural inspiration techniques are used in many application areas to solve classification and optimization problems.

Recently, many algorithms have been developed that are inspired by nature to solve numerical and combinational optimization problems. These algorithms mimic the way the real biological system works. Accordingly, the algorithms in the six main categories that are inspired by nature are briefly described below (Gupta et al., 2011):

*Figure 3. Components of problem solving with the nature inspired intelligent techniques*

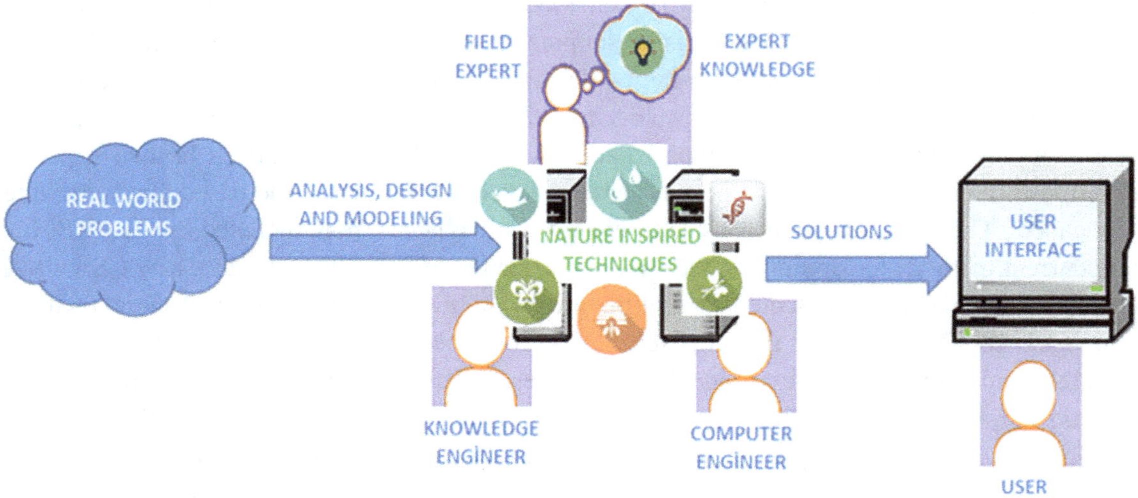


- **Swarm Intelligence:** These are generally algorithms that simulate the ability of living things such as birds, ants and bees to act together and solve problems. Researchers have been inspired by some features of these creatures, such as their ability to communicate with each other, organizational structures, and common reactions to certain events. Techniques most commonly encountered in artificial intelligence are usually in this class. Some examples of these algorithms are as Particle Swarm Optimization, Ant colony Optimization, Bacterial Foraging Optimization, River Formation Dynamics, Stochastic Diffusion Search etc.
- **Natural Evolution:** The most obvious example that can give to this group is Genetic Algorithms. Genetic Algorithms are a natural selection based adaptive search algorithms. They are based on evolutionary principles through natural selection; They use a specific group of individuals that produce selectors, mutations, and transitions in the presence of variation.
- **Biological Neural Network:** Natural intelligence, which exists in nature and most affected the scientists, is the human mind. The study of human mind and the ability of decision-making are attracting scientists. For this reason, natural intelligence, which is most wanted to be imitated or inspired, is human intelligence. At the very core of the idea of making intelligent machines lies human intelligence. The neurons in the human brain, the nerves surrounding the body, and the electrical signals that provide communication are the basis of human thinking, decision-making, and ability of movement. Artificial neural networks, which are simulations of this structure, are one of the most widely used methods today.

Artificial Neural Networks are well suited for prediction, analysis, forecasting and pattern identification and recognition of text, audio, video, speech and facial expressions (Gupta et al., 2011).

- **Molecular Biology:** The DNA structure where the information stored on it, discovered by Watson and Crick, and is the basic work field of biology. The DNA structure consists of two wrapped DNA strands. Each strand in the chain occurs from two components that make up the information as the sugar molecules, and the phosphate groups. DNA calculation is the best example of molecular biology methods. In general, it tries to solve real-world problems using biological molecules such as DNA and RNA.

- **Immune System:** The human immune system, with it's a lot of features such as durability, robustness and fault tolerance, is the most beautiful example for solving various real-world problems. Artificial Immune Systems are algorithms and systems that use the human immune system as a source of inspiration. The human immune system is a robust, decentralized, fault tolerant and adaptable system. Such features are highly desirable for the development of new computer systems (Gupta et al., 2011; Shanthi, 2009).
- **Biological Cell:** Membrane computation is the abstract computation approaches and models which is inspired by the structure and functioning of living cells. However, the cells are alone, but they are in an organization with the structure in which, it is a good example of abstraction. The models considered, called membrane systems (P systems), are parallel, and distributed computing models, processing multi sets of symbols in cell-like compartmental architectures (Gupta et al., 2011; Kennedy, 2001).

## DECISION SUPPORT SYSTEMS

Decision support systems are often referred to as "expert systems", because they have decision-making bases built by experts. Expert systems have three main components. The first component is the experiential knowledge base obtained as a result of clinical trials for similar situations. This component could also be thought as a large database, developed by clinical specialists. The second component is information, which is taken from the patient and the current case. The third component is the part of the decision makers that evaluates the first two components together and uses different artificial intelligence tools to make inferences. Today, many methods such as fuzzy logic, artificial neural networks, genetic algorithms, particle swarm optimization, ant colony optimization, DNA computation and bee colony are used for inference operations. Expert systems are used to support human decision making. Expert systems can be integrated into software and hardware products, thus providing an automated decision support system. The experts prefer specific information obtained from education, training, reading and experiences. An expert system tries to simulate human expertise by applying special field expertise (Deperlioglu & Polat, 2014). A block diagram that is showing the operation of an expert system is given in Figure 4. The components of an expert system can be listed as follows:

- **Knowledge Base:** Provides the necessary information to solve and understand problems.
- **User Interface:** Allows users to communicate with the system via the computer.
- **Inference Engine:** It uses the forward or backward chain. Develops methods for extracting and formulating results. It consists of the rules "IF ... THEN" to make an inference.
- **Workspace:** Keeps events, facts, rules and results in memory.
- **Description Facility:** Uses the facts and rules used to reach a conclusion.
- **Information Retrieval Facility:** Used to hold and load customized information.

Decision support systems use the current clinical information, together with patient information, to arrive at the conclusion by basically evaluating them as "if-then-the output is like this". The users put the result into practice if it is satisfactory. There are also decision support systems which interact with

*Figure 4. A block diagram that is showing the operation of an expert system*

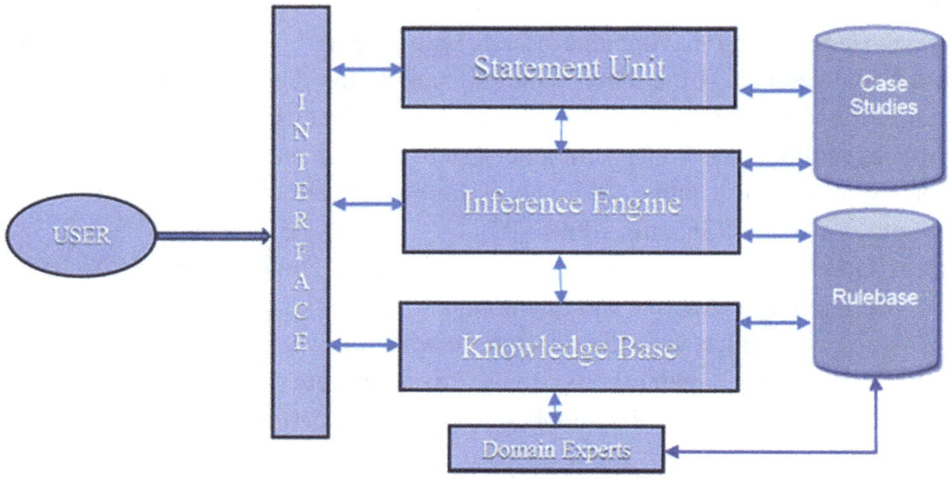


*Table 1. Classification of decision support systems*

| Type | Purpose |
| --- | --- |
| Passive systems | This decision support systems are the most common form. In this mode, the user must explicitly search the system for support. These are Knowledge management tools designed to provide environments for storing and retrieving medical information. |
| Semi-active systems | This system automatically triggers and transmits generally accepted information and procedural rules. It usually plays the role of ombudsman, stimulating a clinician in a particular clinical situation. These systems follow simple logic to achieve their intended behavior. |
| Active systems | This is triggered automatically and it can make decisions without clinician intervention. For example, requests for additional examinations based on health protocols, therapeutic examinations or clinical systems. |

the users by entering certain parameters. In this case, decision support systems can be classified into three groups as shown in Table 1 (Fraccaroa et al, 2015; Algarin, 2011; Deperlioglu et al, 2016).

- **Passive Systems:** Only provide medical data and information needed by physicians and health personnel, and they do not support the receipt of a specific case decision. They give all the information about the problem in the patient, but the evaluation and deduction are left to the decision makers completely. It also enhances the quality by gathering data and notes that may be needed for future decisions. They can also choose the intuitive ways of passing this knowledge. In such systems, the decision has been left entirely to the clinician.
- **Semi Passive Systems:** Are software designed to remind users of problems and diagnostic methods. For example, such a decision support system could alert the physician to the drug entry for an allergic patient or report abnormal conditions and adverse events that may occur in laboratory tests and medical imaging procedures. Thus, the diagnosis and treatment process for the negativities that may occur during the warning and reminder of the health personnel is fulfilled.

- **Active Systems:** This software offer recommendations to decision makers by evaluating previous clinical data as well as individual patient data. In general, using artificial intelligence components such as learning and classification algorithms, software is trying to deduce whether the incoming data is closest to the current data or which is the most appropriate class to belong to. While these systems also follow simple logic, they can use decision theory, cost-benefit analysis, and approximate set theory.

In expert system design, there are no conditions that require user information to be entered all the time. Sensors or receivers receive and display information from the devices, and then send the values to the expert system. Amendments or unusual values are treated as unusual indications and trigger a diagnostic system. In this case, other values are accessed and the account is entered. A warning system can generate a message as an automatic response and send it to an authority (Deperlioglu, 2016).

## NATURE INSPIRED INTELLIGENT TECHNIQUES FOR MEDICAL CONSULTATION, DIAGNOSIS, AND TREATMENT

Medical consultation is a process made by those who also have experience in general. This process requires the inference in very large data sets. Whether medical consultations are made by physicians or computer aided decision support system, it is a process that requires expert knowledge. The physicians don't have enough time to consult the experts for each situation. Likewise, there is no possibility to conduct large data sets research at every time. At this point, nature inspired intelligent techniques is employed in order to solve the related problem. Especially in the last 20 years, lots of research studies have been performed on this issue. MYCIN, INTERNIST, CASNET, PIP, and EXPERT can be given as examples for effectively usage of nature inspired intelligent techniques in decision support systems as expert consultation. As seen Figure 5, there is a change in medical consultation, diagnosis, and treatment and it makes the job of clinicians very easy. So, expert medical consultation is an expensive and a limited tool, but it is critical component in every health care system.

A Clinical Decision Support System (CDSS) is an interactive computer software designed to assist healthcare workers in their decision-making tasks. In CDSS design, many methods or approaches such as Bayesian networks, artificial neural networks, genetic algorithms and rule-based approaches are used. Many research efforts have been made to establish international health standards and to create a representation for shared rules for clinical practice. However, formal models of executable and flexible clinical workflows are few. In this respect, they are completely unique. The implementation of a clinical workflow depends largely on the structure of available medical knowledge, on available resources and on specific case reports. The Business Process Management Community (BPMN) has been working hard to design customized and flexible clinical processes using formal workflow languages that can be run by existing workflow engines. Figure 6 shows a simplified BPMN clinical workflow model (Yao & Kumar, 2013).

As seen in Figure 6, clinical workflows are highly dynamic, content sensitive, event-driven and knowledge-intensive (Yao & Kumar, 2013). For this reason, many authors have defined the properties of CDSS. Peek has listed 10 common properties in the scientific literature that are most often mentioned and generally valid. These properties are given in Figure 7.

*Figure 5. There is a change in medical consultation, diagnosis, and treatment*

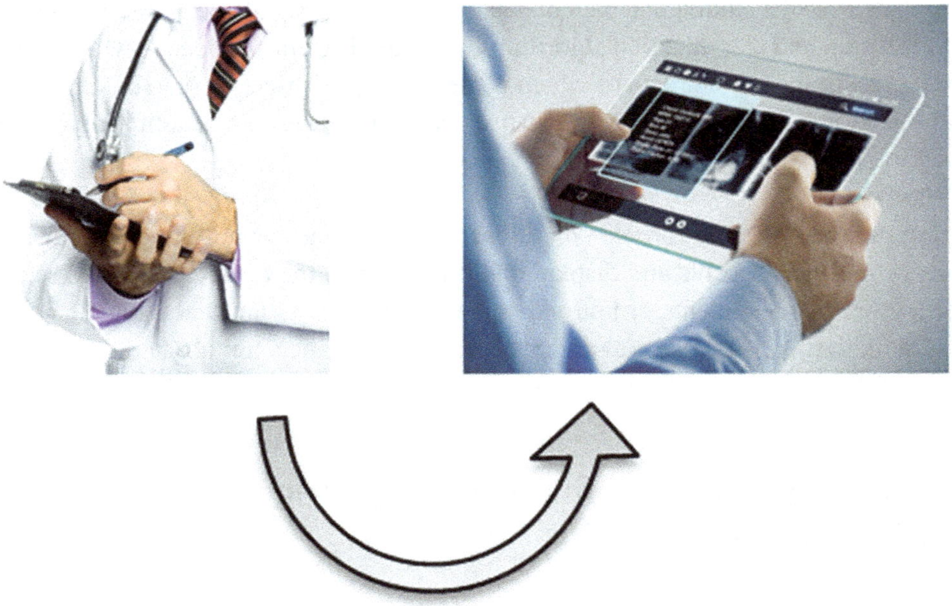

*Figure 6. A simplified BPMN clinical workflow model*
*Yao & Kumar, 2013.*

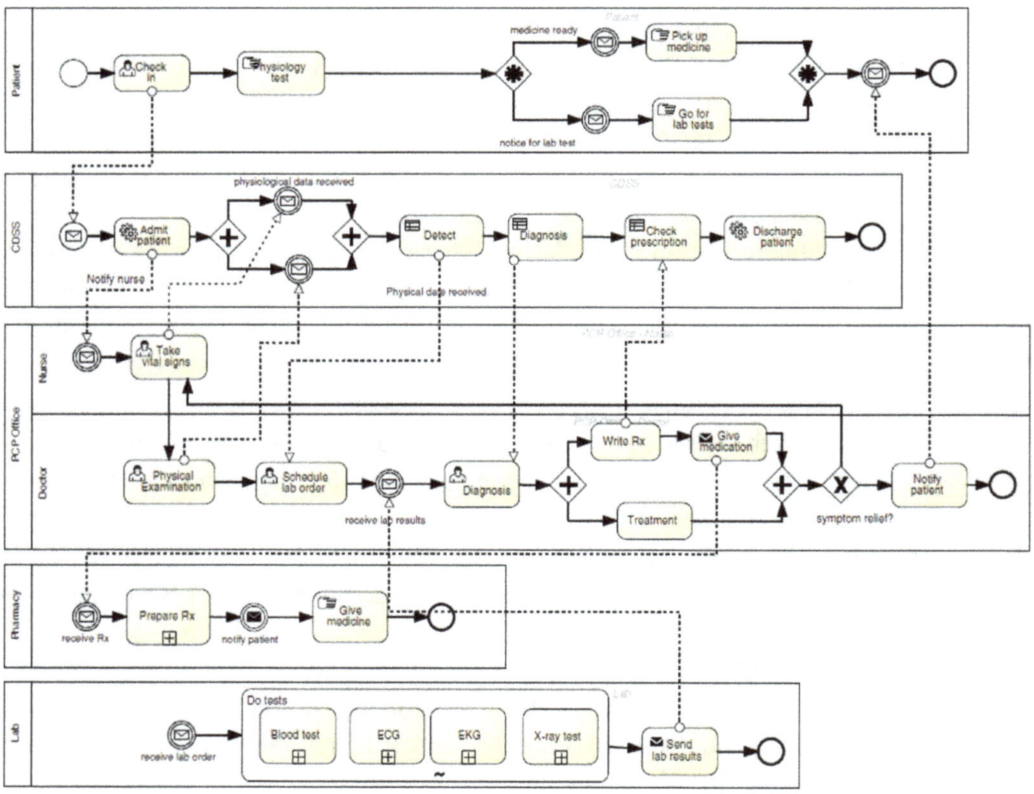

*Figure 7. Characterized CDSS properties*

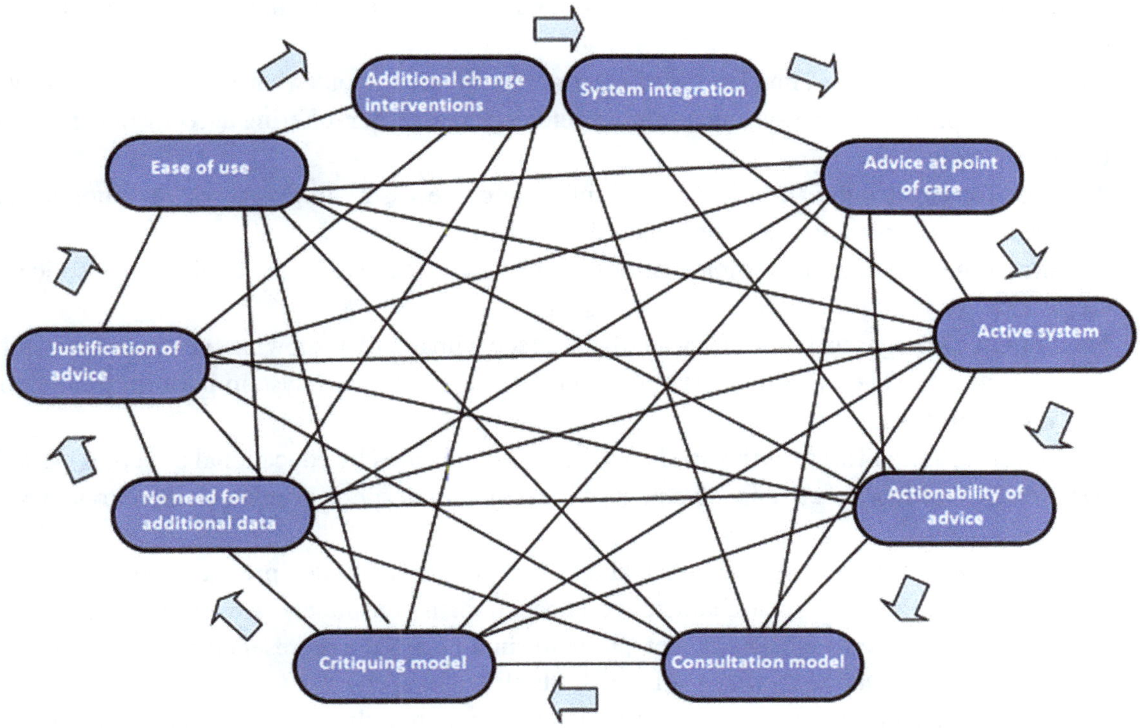


Here, it appears that the properties of CDSS are not independent. For example, active systems have more influence in decision-making behavior than passive systems, which require users to identify situations in which the recommendation will be useful and then make an explicit effort to reach the CDSS. However, when integrated with an electronic clinical information infrastructure, it is possible to create an active CDSS. The same is true so that no additional data entry is required from the user. Consultation and criticism are usually emerging as complementary models. Characterized CDSS properties by Peek are given below (Peek, 2011):

- **System Integration:** CDSS should work integrated with the clinical information infrastructure. For example, the CDSS should work in conjunction with electronic patient record systems, laboratory systems, medical imaging centers, and pharmacy departments.
- **Advice at Point of Care:** Decision support should be given when and where decisions are really taken. For example, the illness should be during a clinical examination or a multidisciplinary physician consultation.
- **Active System:** CDSS should automatically provide recommendations as part of the clinician's normal workflow, without waiting for the user to ask for help.
- **Actionability of Advice:** The CDSS provides advice on what to do for a patient rather than what is right for a patient. For example, CDSS tells physicians which tests need to be performed or which treatment should be initiated. However, CDSS does not comment on whether this treatment will cause some disturbances in the future.

- **Consultation Model:** CDSS serves as a consultant who produces user recommendations. It first accepts patient-specific data and evaluates the data. Then it offers recommendations on diagnosis or process management.
- **Critiquing Model:** The clinician has a preconceived notion about what is in a patient or what therapy is appropriate. The system acts as an inference researcher, offering agreements or alternative suggestions.
- **No Need for Additional Data Entry:** All information necessary to give advice is automatically removed from the patient's electronic records.
- **Justification of Advice:** Recommendations are proven to the user by explaining the justification steps or by providing evidence of basic research.
- **Ease of Use:** The system has a user-friendly interface with a screen showing the importance of the recommendation. The system is fast, so clinicians can save time or system perform transactions in a very short time.
- **Additional Change Interventions:** The CDSS is accompanied by educational direction, periodic performance feedback, active involvement of local opinion leaders, or other change strategies.

Combining many of the features described above, computer consultation programs began to exhibit depth and reasoning flexibility that characterize expert human consultants. However, the process of construction of these systems introduces new problems in terms of the representation, application and validity of medical information (Kulikowski, 1984; Shotliffe, 1984).

Clinical decision support systems are the first intelligent system that comes to mind when medical consultancy, diagnosis and treatment is mentioned. As previously defined, Clinical Decision Support Systems (CDSS) are computer-based health applications that are used to both support patient decision making and integrate clinical and patient information to create unique recommendations. CDSS not only increases the efficiency of the physician by increasing the physician's ability to process data and information, but also by providing computer advice to other health stakeholders in helping with clinical decision-making, drug and drug doses, laboratory results and diagnosis (Kawamoto et al., 2005). Thus, CDSS is computer software designed to meet the need for an adequate, exact and accurate data repository using a knowledge base built by a specialist clinician that performs many functions, from patient registration to clinician workflow components with full-computer. This integrated software covers patient care as well as a range of knowledge management tools (Kostic & Vasilyevic, 2011). Figure 8 shows an example of a medical decision-making system.

In-place processing of data at every workplace by medical doctors requires very large computer resources for calculations. On the other hand, the data can also be used for public health, prevention and disease prevention research. It can also be a valuable resource in the educational process. As shown in Figure 9, when health information systems are referred to, there is a system bias that covers all institutions providing health care services from the smallest clinics, family physician homes to large hospitals with the most complex systems. While health information systems offer these services, they only use computer networks (intranets) and general purpose computer networks (internet) that serve between the relevant institutions. As a result, it can be said that health information systems are a large computer network. The database on the large server systems and the workstations connected to them will record all the process from admission to treatment or until discharge from the hospital. In this network, there are many management tools such as the findings and reports of the physicians, the investigations made, the medications used, the treatment costs, personnel organization. Today, in many health systems, the

*Figure 8. General structure of medical decision making system*

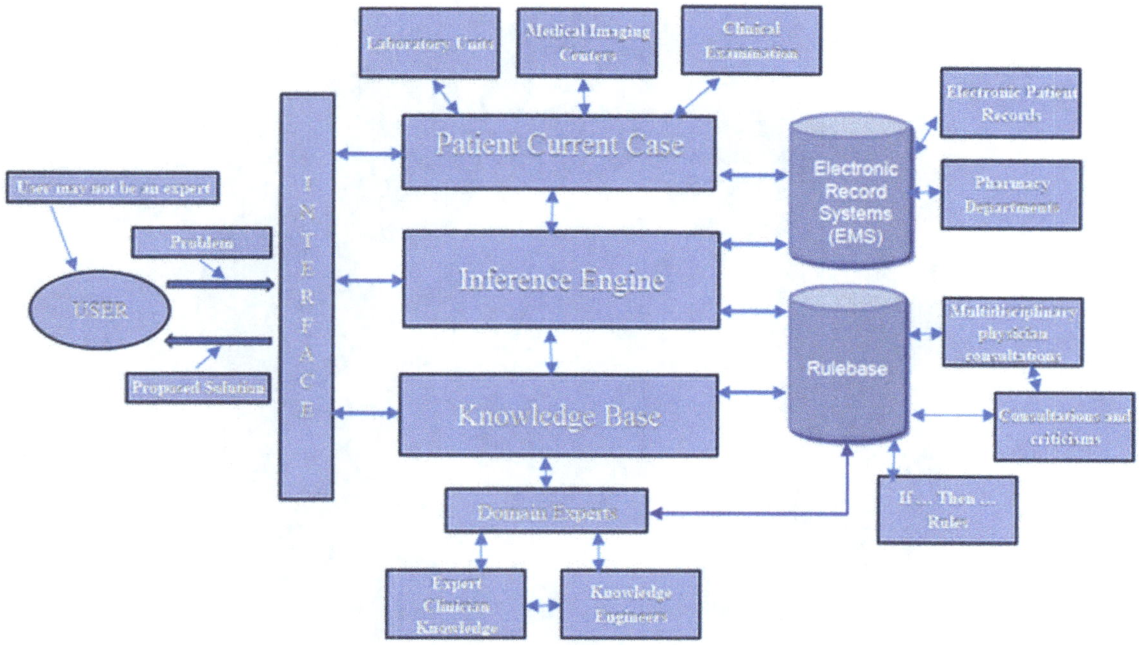


amount of data produced daily by Electronic Record Systems (ERS) can be estimated as hundreds of gigabytes. Unfortunately, the difference between the volume of generated information and the probability of effective use for practical diagnostic purposes are very large. (Deperlioglu & Polat, 2014; Jegelevicius et al., 2009).

Establishing systems to enable information sharing and decision support among clinicians across organizational and geographical boundaries is a complex but it is important task underlying the telehealth idea. In this context, the first place that comes to mind is decision support systems that can be accessed from anywhere. World Wide Web technologies have rapidly changed all the processes of Decision Support Systems such as design, development and implementation. In particular, Web technologies provide a new environment for sharing information about a new tool that provides decision support and decision support capabilities. The biggest breakthrough for CDSS developers is to use the Web as a "computer". Web-based Clinical Decision Support Systems, have qualified field knowledge on disease detection, treatment options and which drug should be used. The programs are based on the intellectual questionnaire created by the specialized experts and try to guess the best possibility accordingly. If the values found match the definitions, then the definitive result is obtained (Turban, 2011).

Figure 10 shows the partial architecture of clinical decision support systems. These systems help physicians to evaluate the patient in the best way, using up-to-date information and taking into account patient-specific information. Some of these programs are increasing diagnostic efforts by taking into account the basic clinical knowledge entered by clinicians. In addition, depending on the specific variables of the illness, special medication may be available and patient management and consultation functions are performed by comparing the patient's special information with the expert knowledge base. Some decision support systems can combine clinical and financial information repositories to assess service usage, assess cost components and assess clinical performance. As shown in Figure 11, Web-based

*Figure 9. Some health information systems*

*Figure 10. Partial architectural structure of web based clinical decision support systems*

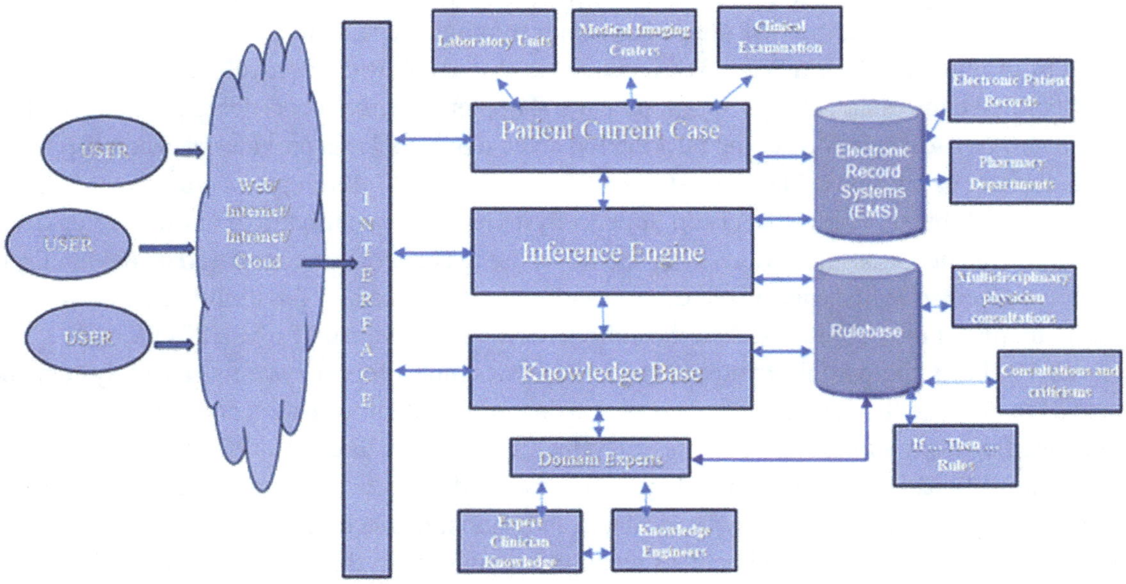

*Figure 11. Components of web based clinical decision support systems*

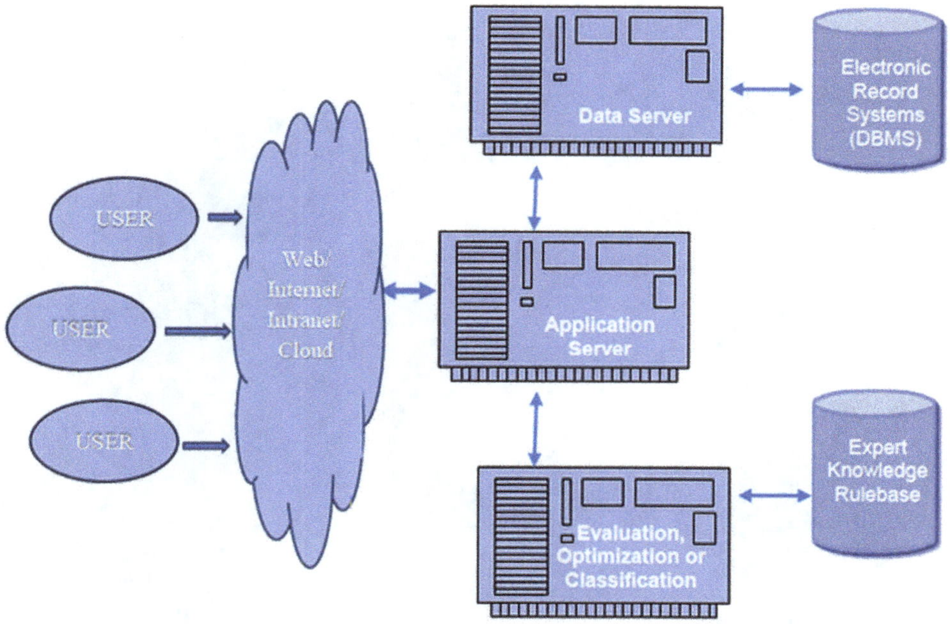


Clinical Decision Support Systems are interactive information systems that provide decision makers with modalities, information and data management tools in situations where semi-structured and unstructured situations or where a decision cannot be fully predicted. These systems aim to improve efficiency rather than improve decision making competence. The purpose of these systems is not to establish administrative provisions but to support them. Web-based Clinical Decision support systems offer physicians the ability to test alternate solutions and re-examine data during problem solving. Physicians using these systems formulate solution options for each problem and send them to the system. The system evaluates this recommendation by comparing it to physicians. The physician also chooses the best endorsement option among the assessed propositions, or prepares new options according to new information and presents them to the system for re-evaluation.

A lot of web-based CDSS has been developed until today. Some of these are summarized below. DXplain is a decision support system designed to produce a diagnostic list that outlines the clinical indication according to a number of clinical findings (signs, symptoms, laboratory data, examinations, etc.). DXplain, owned by Massachusetts General Hospital, can provide a list of more than 5,000 different clinical manifestations and any of the possible cases of the disease. The Early Referral Application (ERA), developed as an early referral practice for cancer patients, is an interactive decision support tool for supporting general practitioners in identifying cancer suspect patients. Isabel, one of the most well-known CDSSs, is a web-based diagnostic decision support system created by physicians in 2001 to provide diagnostic decision support. Isabel, who has been extensively studied, is also developing the cognitive skills of the clinician. It has also been observed that Isabel increases patient safety and quality of patient care (The Open Clinical Web site, 2016).

There are also studies in which new technologies are applied. For example, Dixon and his colleagues has developed a CDSS system based on cloud technology (Dixon, 2013). CDSSs have also been designed with the aim of strengthening decision-making with information research methods such as data min-

*Figure 12. Applications fields of clinical decision support systems*

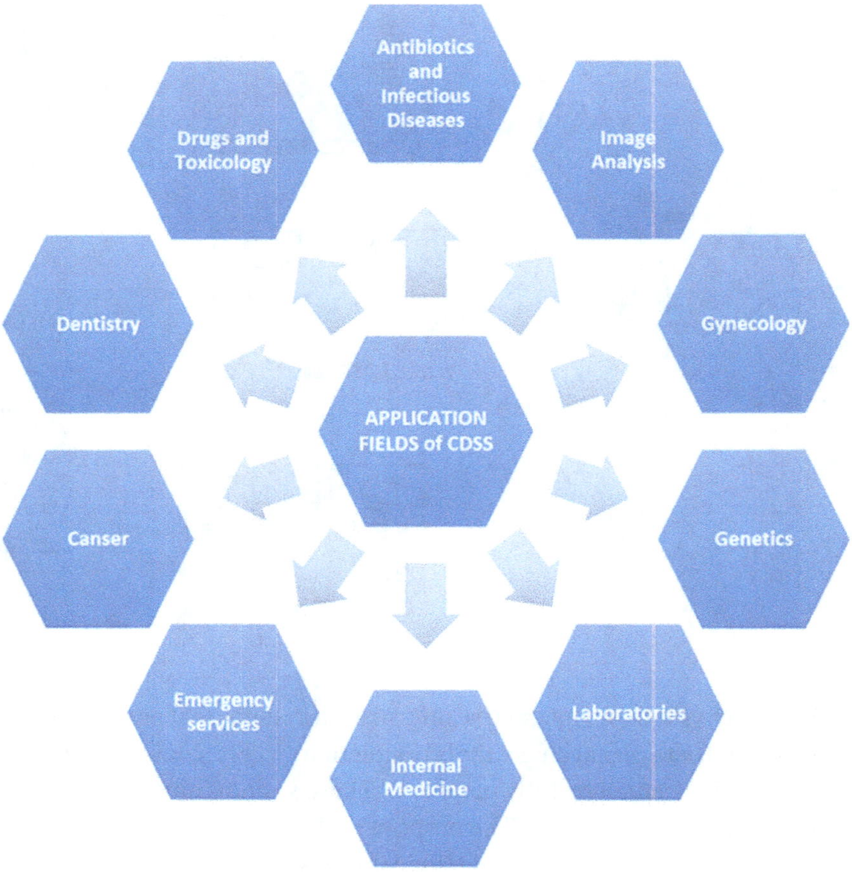


ing. Along with this, CDSS which can also work with mobile devices (such as tablets, smart phones) has been developed in parallel with the developing technology. For this reason, the direction of future development of CDSSs is also clear.

Until today, decision support systems have started to be used in almost all fields of medicine. Even, for many medical units, CDSSs have become indispensable. Figure 12 shows some medical units where CDSSs are commonly used.

Artificial intelligence techniques, which are integral with CDSS, are natural components of CDSS in all areas of medicine. Because medical inference based on artificial intelligence techniques. Among these techniques, the most preferred and most commonly used techniques are nature-inspired approaches.

## FUTURE RESEARCH DIRECTIONS

By taking the performed research works into account, it is also possible to think about some future possibilities on biomedical decision making supported by artificial intelligence. The foremost ideas by the authors can be indicated as follows:

- It is clear that the positive impact of artificial intelligence in biomedical decision making will continue in the future. But usage of artificial intelligence will be a common approach as a result of more employment of such decision-making technologies.
- It can be expressed that most of important tasks done by physicians will be taken over by software systems and physicians will be more focused on using such systems.
- Physicians in the future will be trained to design and develop expert systems or knowledge-based systems in order to enable computers to support them in decision making, performing surgeries, supporting sensitive tasks…etc.
- In the future, there will be new jobs or specialties that are related to employment of artificial intelligence in medical consultation, diagnosis and treatment.
- Decision support in medical field and it is application ways will be more and more important research interest of the future and because of this, new sub-fields that are focused on specific theoretical and applied sides of this research interest will be appeared.
- In the artificial intelligence field, new approaches, methods or techniques focused on only biomedical decision making applications will be introduced in the future.
- As a result of improved usage of artificial intelligence in medical consultation, diagnosis and treatment, new medical machines will be constructed and most of todays widely used medical systems/machines will be replaced by these new intelligent machines.
- Thanks to usage of more advanced artificial intelligence supported systems, serious diseases will be cured with early diagnosis and even such diseases will be no longer classified as serious, fatal ones.

The expressed ideas can be expended by thinking more about the current state. But the future will always have surprises.

## CONCLUSION

As mentioned above, in parallel to the developing computer and network technologies, the health sector benefits from a variety of fields such as information technology, management services, guidance, signal interpretation, laboratory services, record keeping systems. Clinical decision support systems are the leading systems used for this purpose.

When clinical decision support systems are referred to, naturally artificial intelligence methods come to mind. A Clinical Decision Support System (CDSS) is an interactive computer software designed to assist healthcare workers in their decision-making tasks. In CDSS design, many methods or approaches such as Bayesian networks, artificial neural networks, genetic algorithms and rule-based approaches are used. Successful handling of a large number of complex problems in the real world has increased the popularity of intelligent algorithms and techniques that are inspired by nature. These are called nature inspired intelligent techniques. Their successful applications of complex and difficult optimization problems emphasize the importance of artificial intelligence in the wider field.

Thanks to Web-based clinical decision support systems and electronic record systems, the accurate data in the electronic environment can be quickly shared according to the usage rights among all units of the health system. The sharing of information between departments such as polyclinics, laboratories and medical imaging departments and jointly informing each other through these sections, will ensure that transactions and decisions taken can be carried out more effectively and efficiently. Web-based clinical decision support systems will become more visual, more effective and efficient in the near future with the development of Web 3.0 techniques, cloud technologies, mobile devices and mixed nature inspired intelligent applications.

## REFERENCES

Algarin, A. D. R. (2011). *Clinical Decision Support Systems in Biomedical Informatics and Their Limitations.* Retrieved November 25, 2016, from http://www.engr.uconn.edu/~steve/Cse5810/delarosa.pdf

Bahaa, K., Noor, G., & Yousif, Y. (2011). *The Artificial Intelligence Approach for Diagnosis, Treatment and Modelling in Orthodontic.* INTECH Open Access Publisher. doi:10.5772/19597

Bhargava, H. K., Power, D. J., & Sun, D. (2007). Progress in Web-based decision support technologies. *Decision Support Systems, 43*(4), 1083–1095. doi:10.1016/j.dss.2005.07.002

Biomedical Engineering. (2016). *What is Biomedical Engineering?* Florida Institute of Technology. Retrieved November 20, 2016, from http://coe.fit.edu/biomedical-engineering/

Chuang, T. T., & Yada, S. B. (1998). The development of an adaptive decision support system. *Decision Support Systems, 24*(2), 73–87. doi:10.1016/S0167-9236(98)00065-7

Clancey, W. J., & Shortliffe, E. H. (1984). *Readings in Medical Artificial Intelligence: The First Decade.* Reading, MA: Addison Wesley.

Coiera, E. (2003). Guide to Medical Informatics, the Internet and Telemedicine (2nd ed.). London: Arnold.

Copeland, J. (2000). *What is Artificial Intelligence?* Retrieved November 20, 2016, from, http://www.alanturing.net/turing_archive/pages/reference articles/what is ai.html

Deperlioglu, O., Guraksin, G. E., & Kose, U. (2016). Web Based Clinical Decision Support Systems: Structures and Features. *Academic Informatics Conferences, AB 2016.*

Deperlioglu, O., Kose, U., & Guraksin, G. E. (2015). *Biomedical Decision Making: Reasoning and Decision Support with Artificial Intelligence.* 3rd International Congress on Natural and Engineering Sciences, ICNES 2015, Sarajevo, Bosnia Herzegovina.

Deperlioglu, O., & Polat, K. (2014). Biomedical Informatics Systems and Biostatistics. In M. H. Asyali, S. Kara, & B. Yilmaz (Eds.), *Fundamentals of Biomedical Engineering* (pp. 653–698). Ankara, Turkey: Nobel Academic Publishing.

Dixon, B. E., Dixon, B. E., Simonaitis, L., Goldberg, H. S., Paterno, M. D., Schaeffer, M., & Middleton, B. et al. (2013). A Pilot Study of Distributed Knowledge Management and Clinical Decision Support in the Cloud. *Artificial Intelligence in Medicine*, *59*(1), 45–53. doi:10.1016/j.artmed.2013.03.004 PMID:23545327

Fraccaroa, P., O'Sullivana, D., Plastirasa, P., & O'Sullivan, H., Dentonec, Biagioe, A., & Wellera, P. (2015). Behind the screens:Clinical decision support methodologies – A review. *Health Policy and Technology*, *4*, 29–38. doi:10.1016/j.hlpt.2014.10.001

Gupta, S., Bhardwaj, S., & Bhatia, P. K. (2011). A Reminiscent Study of Nature Inspired Computation. *International Journal of Advances in Engineering and Technology*, *1*(2), 117–125.

Hoong, N. K. (1988). Medical Information Science - Framework and Potential. In *International Seminar and Exhibition Computerization for Development-the Research Challenge* (pp. 191 – 198). Universiti Pertanian Malaysia.

IEEE EMB (Engineering in Medicine and Biology). (2015). *Designing a Career in Biomedical Engineering*. IEEE.

Ishak, W. H., & Siraj, F. (2002). Artificial intelligence in medical application: An exploration. *Health Informatics Europe Journal, 16.* Retrieved November 21, 2016, from, https://www.researchgate.net/publication/240943548

Jegelevicius, D., & Krisciukaitis, A. (2009). Network Based Clinical Decision Support System. *Proceedings of the 9th International Conference on Information Technology and Applications in Biomedicine.*

Kawamoto, K., Houlihan, C. A., Balas, E. A., & Lobach, D. F. (2005). Improving clinical practice using clinical decision support systems: A systematic review of trials to identify features critical to success. *British Medical Journal*, *330*(7494), 765–772. doi:10.1136/bmj.38398.500764.8F PMID:15767266

Kennedy, J., Eberhart, R. C. & Shi, Y. (2001). *Swarm Intelligence*. Morgan Kaufmann.

Khalifa, M. (2014). Clinical Decision Support: Strategies for Success. *Procedia Computer Science*, *37*, 422–427. doi:10.1016/j.procs.2014.08.063

Kostic, P., & Vasilyevic, Z. (2011). Knowledge Management System for Clinical Decision Support - Application in Cardiology. *19th Telecommunications forum TELFOR*, 1261-1264.

Kulikowski, C. A. (1984). Artificial Intelligence Methods and Systems for Medical Consultation. In W. J. Clancey & E. H. Shortliffe (Eds.), *Readings in Medical Artificial Intelligence: The First Decade.* Reading, MA: Addison Wesley.

Ledley, R. S., & Lusted, L. B. (1959). Reasoning foundations of medical diagnosis. *Science*, *130*(3366), 9–21. doi:10.1126/science.130.3366.9 PMID:13668531

Lipkin, M., & Hardy, J. D. (1958). Mechanical correlation of data in differential diagnosis of hematologic diseases. *Journal of the American Medical Association*, *166*(2), 113–125. doi:10.1001/jama.1958.02990020001001 PMID:13491316

Miller, R. A. (1994). Medical Diagnostic Decision Support Systems -Past, Present, and Future: A Threaded Bibliography and Brief Commentary. *Journal of the American Medical Informatics Association*, *1*(1), 8–27. doi:10.1136/jamia.1994.95236141 PMID:7719792

Nahar, J., Imam, T., Tickle, K. S., Ali, A. S., & Chen, Y. P. P. (2012). Computational intelligence for microarray data and biomedical image analysis for the early diagnosis of breast cancer. *Expert Systems with Applications*, *39*(16), 12371–12377. doi:10.1016/j.eswa.2012.04.045

Papamichail, K. N., & French, S. (2005). Design and evaluation of an intelligent decision support system for nuclear emergencies. *Decision Support Systems*, *41*(1), 84–111. doi:10.1016/j.dss.2004.04.014

Peek, N. (2011). Guideline-Based Decision Support Systems for Prevention and Management of Chronic Diseases. In C. S. Jao (Ed.), *Efficient Decision Support Systems – Practice and Challenges in Biomedical Related Domain* (pp. 67–86). Rijeka, Croatia: InTech. doi:10.5772/16882

Ramesh, A. N., Kambhampati, C., Monson, J. R. T., & Drew, P. J. (2004). Artificial intelligence in medicine. *Annals of the Royal College of Surgeons of England*, *86*(5), 334–338. doi:10.1308/147870804290 PMID:15333167

Shanthi, D., Sahoo, G., & Saravanan, N. (2009). Designing an Artificial Neural Network Model for the Prediction of Thrombo-embolic Stroke. *International Journals of Biometric and Bioinformatics*, *3*(1), 10–18.

Shim, J. P., Warkentin, M., Courtney, J. F., Power, D. J., Sharda, R., & Carlsson, C. (2002). Past, present, and future of decision support technology. *Decision Support Systems*, *33*(2), 111–126. doi:10.1016/S0167-9236(01)00139-7

Shortliffe, E. H., Buchanan, B. G., & Feigenbaum, E. A. (1984). Knowledge Engineering for Medical Decision Making: A Review of Computer-Based Decision Aids. In W. J. Clancey & E. H. Shortliffe (Eds.), *Readings in Medical Artificial Intelligence: The First Decade*. Reading, MA: Addison Wesley.

Sikchi, S. S., Sikchi, S., & Ali, M. S. (2012). Artificial intelligence in medical diagnosis. *International Journal of Applied Engineering Research*, *7*(11).

Sondak, V. K., & Sondak, N. E. (1990). New directions for medical artificial intelligent. *Computers & Mathematics with Applications (Oxford, England)*, *20*(4-6), 313–319. doi:10.1016/0898-1221(90)90336-I

Song, Z., Ji, Z., Ma, J. G., Sputh, B., Acharya, U. R., & Faust, O. (2012). A systematic approach to embedded biomedical decision making. *Computer Methods and Programs in Biomedicine*, *108*(2), 656–664. doi:10.1016/j.cmpb.2011.11.002 PMID:22136936

Stivaros, S. M., Gledson, A., Nenadic, G., Zeng, X.-J., Keane, J., & Jackson, A. (2010). Decision support systems for clinical radiological practice - towards the next generation. *The British Journal of Radiology*, *83*(995), 904–914. doi:10.1259/bjr/33620087 PMID:20965900

The Open Clinical Web Site. (n.d.). Retrieved November 28, 2016, from, http://www.openclinical.org/

Turban, E., Sharda, R. E., & Delen, D. (2011). *Decision Support and Business Intelligence Systems* (9th ed.). Prentice Hall.

Ubeyli, E. D. (2009). Decision support systems for time-varying biomedical signals: EEG signals classification. *Expert Systems with Applications, 36*(2), 2275–2284. doi:10.1016/j.eswa.2007.12.025

Valkenhoef, G. V., Tervonen, T., Zwinkels, T., Brock, B. D., & Hillege, H. (2013). ADDIS: A decision support system for evidence-based medicine. *Decision Support Systems, 55*(2), 459–475. doi:10.1016/j.dss.2012.10.005

Vassiliadis, V., & Dounias, G. (2009). Nature–Inspired Intelligence: A Review of Selected Methods and Applications. *International Journal of Artificial Intelligence Tools, 18*(04), 487–516. doi:10.1142/S021821300900024X

Yao, W., & Kumar, A. (2013). CONFlexFlow: Integrating flexible clinical pathways into clinical decision support systems using context and rules. *Decision Support Systems, 55*(2), 499–515. doi:10.1016/j.dss.2012.10.008

Zhou, Z.-G., Liu, F., Jiao, L.-C., Zhou, Z.-J., Yang, J.-B., Gong, M.-G., & Zhang, X.-P. (2013). A bi-level belief rule based decision support system for diagnosis of lymph node metastasis in gastric cancer. *Knowledge-Based Systems, 54*, 128–136. doi:10.1016/j.knosys.2013.09.001

## ADDITIONAL READING

Aleksovska-Stojkovska, L., & Loskovska, S. (2010). Clinical decision support systems: Medical knowledge acquisition and representation methods, *IEEE International Conference on Electro/Information Technology (EIT)*, pp. 1–6. doi:10.1109/EIT.2010.5612183

Anselma, L., Bottrighi, A., Molino, G., Montani, S., Terenziani, P., & Torchio, M. (2011). Supporting knowledge-based decision making in the medical context: The GLARE approach. *International Journal of Knowledge-Based Organizations, 1*(1), 42–60. doi:10.4018/ijkbo.2011010103

Bassa, A., Miguel, V., Cobos, A., Torremadé, E., Bergoñón, S., Crespo, C., & Espinosa, C. (2005). Impact of a clinical decision support system on the management of patients with hypercholesterolemia in the primary healthcare setting. *Disease Management & Health Outcomes, 13*(1), 65–72. doi:10.2165/00115677-200513010-00007

Bates, D., Kuperman, G., Wang, S., Gandhi, T., Kittler, A., Volk, L., & Middleton, B. et al. (2003). Ten Commandments for effective clinical decision support: Making the practice of evidence-based medicine a reality. *Journal of the American Medical Informatics Association, 10*(6), 523–530. doi:10.1197/jamia.M1370 PMID:12925543

Berner, E. S., & La Lande, T. J. (2007). Overview of clinical decision support systems. In Clinical decision support systems – theory and practice, (2nd ed.)., Berner E.S (ed.), Springer: New York. doi:10.1007/978-0-387-38319-4_1

Bishop, C. M. (2007). *Pattern Recognition and Machine Learning*. New York, USA: Springer.

Bonney, W. (2009). Is it appropriate, or ethical, to use health data collected for the purpose of direct patient care to develop computerized predictive decision support tools? *Studies in Health Technology and Informatics*, *143*, 115–121. PMID:19380924

Coiera, E., Westbrook, J. I., & Wyatt, J. C. (2006). The safety and quality of decision support systems. *Methods of Information in Medicine*, *45*(Suppl. 1), 20–25. PMID:17051290

Coppini, G., Diciotti, S., Falchini, M., Villari, N., & Valli, G. (2003). Neural networks for computer-aided diagnosis: Detection of lung nodules in chest radiograms. *IEEE Transactions on Information Technology in Biomedicine*, *7*(4), 344–357. doi:10.1109/TITB.2003.821313 PMID:15000360

Dayhoff, J. E., & DeLeo, J. M. (2001). Artificial neural networks: Opening the black box. *Cancer*, *91*(8Suppl.), 1615–1635. doi:10.1002/1097-0142(20010415)91:8+<1615::AID-CNCR1175>3.0.CO;2-L PMID:11309760

De Castro, L. N. (2007). Fundamentals of natural computing: An overview. *Physics of Life Reviews*, *4*(1), 1–36. doi:10.1016/j.plrev.2006.10.002

De Clercq, P. A., Blom, J. A., Korsten, H. H., & Hasman, A. (2004). Approaches for creating computer-interpretable guidelines that facilitate decision support. *Artificial Intelligence in Medicine*, *31*(1), 1–27. doi:10.1016/j.artmed.2004.02.003 PMID:15182844

Fuchsberger, C., & Miksch, S. (2002). *Asbru's execution engine: Utilizing guidelines for artificial ventilation of newborn infants. Technical report.* Vienna University of Technology, Institute of Software Technology and Interactive Systems.

Haykin, S. (2008). *Neural Networks and Learning Machines*. New Jersey, USA: Prentice Hall.

Holzinger, A. (2014). Medical Informatics [444.152, 2VO WS], Retrieved November 29, 2016, from http://genome.tugraz.at/medical_informatics.shtml

Holzinger, A. (2014). *Biomedical Informatics*. Switzerland: Springer International Publishing. doi:10.1007/978-3-319-04528-3

Hudson, D. L., & Cohen, M. E. (2000). *Neural Networks and Artificial Intelligence for Biomedical Engineering. The Institute of Electrical and Electronics Engineers, Inc.* New York: IEEE Press.

Jao, C. S. (Ed.). (2011). *Efficient Decision Support Systems – Practice and Challenges in Biomedical Related Domain*. Rijeka, Croatia: Intechopen. doi:10.5772/937

Kazemi, A., Fors, U. G., Tofighi, S., Tessma, M., & Ellenius, J. (2010). Physician order entry or nurse order entry? Comparison of two implementation strategies for a computerized order entry system aimed at reducing dosing medication errors. *Journal of Medical Internet Research*, *12*(1), e5. doi:10.2196/jmir.1284 PMID:20185400

Kim, G., Kim, Y., Lim, H., & Kim, H. (2010). An MLP-based feature subset selection for HIV-1 protease cleavage site analysis. *Artificial Intelligence in Medicine*, *48*(2-3), 83–89. doi:10.1016/j.artmed.2009.07.010 PMID:19945261

Kong, G., Xu, D.-L., & Yang, J.-B. (2008). Clinical decision support systems: A review on knowledge representation and inference under uncertainties. *International Journal of Computational Intelligence Systems*, *1*, 159–167.

Kwok, R., Dinh, M., Dinh, D., & Chu, M. (2009). Improving adherence to asthma clinical guidelines and discharge documentation from emergency departments: Implementation of a dynamic and integrated electronic decision support system. *Emergency Medicine Australasia*, *21*(1), 31–37. doi:10.1111/j.1742-6723.2008.01149.x PMID:19254310

Lin, S., Xiao, K., Huang, Y., Chiu, C., & Soo, V. (2010). Analysis of adverse drug reactions using drug and drug target interactions and graph-based methods. *Artificial Intelligence in Medicine*, *48*(2-3), 161–166. doi:10.1016/j.artmed.2009.11.002 PMID:19962282

Lisboa, P. J. (2002). A review of evidence of health benefit from artificial neural networks in medical intervention. *Neural Networks*, *15*(1), 11–39. doi:10.1016/S0893-6080(01)00111-3 PMID:11958484

Mahoney, C. D., Berard-Collins, C., Coleman, R., Amaral, J. F., & Cotter, C. M. (2007). Effects of an integrated clinical information system on medication safety in a multihospital setting. *American Journal of Health-System Pharmacy*, *64*(18), 1969–1977. doi:10.2146/ajhp060617 PMID:17823111

Negnevitsky, M. (2005). *Artificial intelligence: a guide to intelligent systems* (2nd ed.). Ma: Assison-Wesley.

Remzi, M., Anagnostou, T., Ravery, V., Zlotta, A., Stephan, C., Marberger, M., & Djavan, B. (2003). An artificial neural network to predict the outcome of repeat prostate biopsies. *Urology*, *62*(3), 456–460. doi:10.1016/S0090-4295(03)00409-6 PMID:12946746

Rodvold, D. M., McLeod, D. G., Brandt, J. M., Snow, P. B., & Murphy, G. P. (2001). Introduction to artificial neural networks for physicians: Taking the lid off the black box. *The Prostate*, *46*(1), 39–44. doi:10.1002/1097-0045(200101)46:1<39::AID-PROS1006>3.0.CO;2-M PMID:11170130

Santibanez, P., Chow, V., French, J., Puterman, M., & Tyldesley, S. (2009). Reducing patient wait times and improving resource utilization at British Columbia cancer agency's ambulatory care unit through simulation. *Journal of Health Care Management Science*, *12*(4), 392–407. doi:10.1007/s10729-009-9103-1 PMID:20058528

Sarkar IN, Butte AJ, Lussier YA, Tarczy-Hornoch P. & Ohno-Machado L. (2011).Translational bioinformatics: linking knowledge across biological and clinical realms. *Journal of the American Medical Informatics Association. Jul-Aug; 18(4),* 354-357.

Shortliffe, E. H. (2006). *Biomedical Informatics: Computer Applications in Health Care and Biomedicine* (3rd ed.). Springer. doi:10.1007/0-387-36278-9

Sistrom, C. L., Dang, P. A., Weilburg, J. B., Dreyer, K. J., Rosenthal, D. I., & Thrall, J. H. (2009). Effect of computerized order entry with integrated decision support on the growth of outpatient procedure volumes: Seven-year time series analysis. *Radiology*, *251*(1), 147–155. doi:10.1148/radiol.2511081174 PMID:19221058

Stefaniak, B., Cholewinski, W., & Tarkowska, A. (2003). Application of artificial neural network algorithm to detection of parathyroid adenoma. *Nuclear Medicine Review. Central & Eastern Europe*, *6*(2), 111–117. PMID:14737724

Sun, J., Han, L., & Zhao, Z. (2010). Gene- and evidence-based candidate gene selection for schizophrenia and gene feature analysis. *Artificial Intelligence in Medicine*, *48*(2-3), 99–106. doi:10.1016/j.artmed.2009.07.009 PMID:19944577

Taylor, A., Manatunga, A., & Garcia, E. V. (2008). Decision support systems in diuresis renography. *Seminars in Nuclear Medicine*, *38*(1), 67–81. doi:10.1053/j.semnuclmed.2007.09.006 PMID:18096465

Tehrani, F. T., & Roum, J. H. (2008). Intelligent decision support systems for mechanical ventilation. *Artificial Intelligence in Medicine*, *44*(3), 171–182. doi:10.1016/j.artmed.2008.07.006 PMID:18768304

Tewari, A., Issa, M., El Galley, R., Stricker, H., Peabody, J., Pow-Sang, J., & Menon, M. et al. (2001). Genetic adaptive neural network to predict biochemical failure after radical prostatectomy: A multi-institutional study. *Molecular Urology*, *5*(4), 163–169. doi:10.1089/10915360152745849 PMID:11790278

Tilbury, J. B., Van Eetvelt, P. W., Garibaldi, J. M., Curnow, J. S., & Ifeachor, E. C. (2000). Receiver operating characteristic analysis for intelligent medical systems— A new approach for finding confidence intervals. *IEEE Transactions on Biomedical Engineering*, *47*(7), 952–963. doi:10.1109/10.846690 PMID:10916267

Tleyjeh, M., Nada, H., & Baddour, L. M. (2006). Visualdx: Decision-support software for the diagnosis and management of dermatologic disorders. *Clinical Infectious Diseases*, 43. PMID:17029139

Tomatis, S., Bono, A., Bartoli, C., Carrara, M., Lualdi, M., Tragni, G., & Marchesini, R. (2003). Automated melanoma detection: Multispectral imaging and neural network approach for classification. *Medical Physics*, *30*(2), 212–221. doi:10.1118/1.1538230 PMID:12607839

Trafton, J., Martins, S., Michel, M., Lewis, E., Wang, D., Combs, A., & Goldstein, M. K. et al. (2010). Evaluation of the acceptability and usability of a decision support system to encourage safe and effective use of opioid therapy for chronic, noncancer pain by primary care providers. *Pain Medicine*, *11*(4), 575–585. doi:10.1111/j.1526-4637.2010.00818.x PMID:20202142

Tsay, J., Wu, B., & Jeng, Y. (2010). Hierarchically organized layout for visualization of biochemical pathways. *Artificial Intelligence in Medicine*, *48*(2-3), 107–117. doi:10.1016/j.artmed.2009.06.002 PMID:20036522

Vomweg, T. W., Buscema, M., Kauczor, H. U., Teifke, A., Intraligi, M., Terzi, S., & Thelen, M. et al. (2003). Improved artificial neural networks in prediction of malignancy of lesions in contrast-enhanced MR-mammography. *Medical Physics*, *30*(9), 2350–2359. doi:10.1118/1.1600871 PMID:14528957

Wang, H. Q., Zhu, H. L., Cho, W. C. S., Yip, T. T. C., Ngan, R. K. C., & Law, S. C. K. (2010). Low SCK. Method of regulatory network that can explore protein regulations for disease classification. *Artificial Intelligence in Medicine*, *48*(2-3), 119–127. doi:10.1016/j.artmed.2009.07.011 PMID:19962281

Wang, X., & Liotta, L. (2011). Clinical bioinformatics: A new emerging science. *Journal of Clinical Bioinformatics*, *1*(1), 1. doi:10.1186/2043-9113-1-1 PMID:21884620

Wetter, T. (2002). Lessons learnt from bringing knowledge-based decision support into routine use. *Artificial Intelligence in Medicine*, *24*(3), 195–203. doi:10.1016/S0933-3657(01)00103-8 PMID:11879989

Xu, R., Damelin, S., Nadler, B., & Wunsch, D. C. II. (2010). Clustering of high-dimensional gene expression data with feature filtering methods and diffusion maps. *Artificial Intelligence in Medicine*, *48*(2-3), 91–98. doi:10.1016/j.artmed.2009.06.001 PMID:19962867

Zeng, T., & Liu, J. (2010). Mixture classification model based on clinical markers for breast cancer prognosis. *Artificial Intelligence in Medicine*, *48*(2-3), 129–137. doi:10.1016/j.artmed.2009.07.008 PMID:20005686

Zhou, X., Chen, S., Liu, B., Zhang, R., Wang, Y., Li, P., & Yan, X. et al. (2010). Development of traditional Chinese medicine clinical data warehouse for medical knowledge discovery and decision support. *Artificial Intelligence in Medicine*, *48*(2-3), 139–152. doi:10.1016/j.artmed.2009.07.012 PMID:20122820

## KEY TERMS AND DEFINITIONS

**Artificial Intelligence:** (1) Artificial intelligence is a software that is specific to human intelligence and can perform functions such as perception, learning, reasoning, thinking, problem solving, inference, and decision making. (2) Artificial intelligence is an area of research that tries to simulate human intelligence using a computer or an autonomous system generated by human.

**Clinical Decision Support System:** Clinical decision support system is a medical artificial intelligence program that make inferences or suggestions by associating general symptoms of illnesses with the current state of the patient and clinical findings.

**Decision Support System:** Decision support systems are knowledge-based information systems that help people to make decisions and can make inferences, and work in interaction.

**Expert Systems:** An expert system tries to simulate human expertise by applying special field expertise. Expert systems can be integrated into software and hardware products, thus providing an automated decision support system.

**Medical Artificial Intelligence:** Medical artificial intelligence is primarily concerned with the creation of AI programs that make diagnoses and treatment recommendations.

**Nature-Inspired Techniques:** Nature-inspired techniques are naturally inspired artificial intelligence techniques such as genetic algorithm, artificial neural network, DNA computation, ant colonies. They are very effective in solving complex problems and achieving optimal solutions.

**Web-Based Clinical Decision Support System:** Web-based clinical decision support systems are clinical decision support systems that use web techniques for data exchange and have expert knowledge of the fields about disease detection, treatment options, which drugs should be used, etc.

# Chapter 4
# Performance of Negative Selection Algorithms in Patient Detection and Classification

**Orhan Bölükbaş**
*Aksaray University, Turkey*

**Harun Uğuz**
*Selcuk University, Turkey*

## ABSTRACT

*Artificial immune systems inspired by the natural immune system are used in problems such as classification, optimization, anomaly detection, and error detection. In these problems, clonal selection algorithm, artificial immune network algorithm, and negative selection algorithm are generally used. This chapter aims to solve the problem of correct identification and classification of patients using negative selection (NS) and variable detector negative selection (V-DET NS) algorithms. The authors examine the performance of NSA and V-DET NSA algorithms using three sets of medical data sets from Parkinson, carotid artery doppler, and epilepsy patients. According to the obtained results, NSA achieved 92.45%, 91.46%, and 92.21% detection accuracy and 92.46%, 93.40%, and 90.57% classification accuracy. V-DET NSA achieved 94.34%, 94.52%, and 91.51% classification accuracy and 94.23%, 94.40%, and 89.29% detection accuracy. As can be seen from these values, V-Det NSA yielded a better result. Artificial immune system emerges as an effective and promising system in terms of problem-solving performance.*

## INTRODUCTION

The artificial immune system is modeled considering the human immune system. Tissues and organs related to immune system in organism are thymus gland, bone marrow, lymph nodes, spleen and tonsils. There is no central organ in the immune system to coordinate the work of related tissues, organs, molecules and cells. The immune system recognizes foreign cells entering their body with their special cells and neutralizes them. The lymphocytes that are the basic immune cells are divided into two classes, the "T" and the "B" cells. While the "B" cells can recognize the antigens free in solution, the "T" cells

DOI: 10.4018/978-1-5225-4769-3.ch004

need the antigens to be made known by other helper cells. The natural immune system simply works like this: When a foreign molecule enters the body, special molecules (macrophages) wrap around this foreign molecule, allowing the other immune cells to see it. Helper "T" cells stimulate white blood cells when they see the specific molecules, and these cells multiply. The "B" cells begin to produce chemical substances called antibodies, and the antibodies combine with the antigen making it easier to destroy the antigen. The destructive "T" cells stimulated by the helper "T" cells enable the antigen to be destroyed. Natural immune system cells must be able to distinguish between their own cells and foreign cells and molecules (antigens) in order to perform the activities described above. The body's own cells are called "self" and the foreign cells are called "nonself". When "nonself" cells are destroyed, "self" cells are not damaged (de Castro & Timmis, 2002).

Inspired by the natural immune system, the generated artificial immune systems are used to achieve resolution in different types of problems. These are classification, pattern recognition, optimization, anomaly detection, computer and network security, error detection, time series analysis. Clonal Selection Algorithm (CSA), Artificial Immune Network Algorithm (AINET) and Negative Selection Algorithm (NSA) are generally used in these problems. The clonal selection algorithm is particularly used in multipurpose optimization and classification problems and the negative selection algorithm is used in anomaly detection, error checking, computer and network security problems.

The studies carried out in these areas are given below:

Forrest et al. (1994) investigated a computer security system based on the separation of "self" and "nonself" using an r-consecutive bit rule and a negative selection algorithm. The system operates according to the logic of the negative selection algorithm. A set of sensors is first created, and then the protected data is monitored by comparing with the generated sensors. If the number of common consecutive bits in two arrays is greater than or equal to a number r, then a match can be mentioned. The study also provided formulas to estimate the likelihood of a match between two random sequences and the likelihood of change-detection for different configurations of the system. According to the results obtained, as the number of arrays to be protected increases, the size of the sensors used does not need to increase. The probability of detection increases exponentially with the number of independent detection algorithms. The cost of creating a sensor increases exponentially with the size of the "self" set. In artificial immune systems, a basic type of sensor that includes all the useful features of the immune system has been described. Detectors are represented by bit-sequences in a binary Hamming shape-space. The detection event is accomplished by the matching process of the r-consecutive bits between the two sequences. It is the responsibility of the learning part of the system to mature lean sensors into memory sensors with negative selection. The established artificial immune system overlaps most important features of a classifier system. The recommended YBS detected all of the eight abnormal occurrences in a network system made up of 50 computers each of which has 100 sensors. While some systems in this area are giving millions of false alarms in a month, the proposed system gave two false alarms on average in a day.

Mori et al. (1997) have described a general autonomous distributed system for controlling a semiconductor production line. In their work, control of the production line is done with a set of agents (detector, mediator, inhibitor and restoration agents). Each agent is in contact with the production line and other agents. This relationship is based on the link in the vertebrate immune system. For example, detector agents correspond to B cells in the immune system and are used to detect certain defects in the system. Although the system has not been tried in practice, it has been claimed that the system will be successful in making real-time decisions and adapting to the changing environment.

Dasgupta and Forrest (1996) have developed an artificial immunity algorithm for tool error detection. This study was inspired by the negative-selection mechanism that allows the self (body elements) and nonself (foreign elements) cells of the immune system to separate each other. In this application, 'self' indicates the values of normal cutting operation and 'nonself' indicates any deviation beyond the allowable cutting force difference. The proposed algorithm was demonstrated in a simulation study for turning operations and in case of breakage of the tool tip, the algorithm's performance to detect this was determined. The algorithm detected breakdowns in the pen tip for all test cases.

Dasgupta and Forrest (1999) proposed a negative selection algorithm to detect variances in time series data. Cutting force values are changing with time. Values beyond the permissible deviation indicate 'non-selves' in the system. They applied a binary Hamming shape space to describe the elements of the system and applied an r consecutive bit rule to determine the degree of perception between the sensors and the decoded data. They gave the results for two sets of data in the study: the cutting dynamics of a turning operation and a synthetic signal. They demonstrated that the number of r-consecutive bits selected with the matching function affected the reliability of risk in the detection of errors

Taylor and Corne (2003) increased both the profit and performance of the company by early detecting the errors in cooling systems of supermarkets with NSA. This study was carried out by taking into account the time-dependent temperature values of the cooling cabins. In addition, in their study, by using Euclidean distance and r-consecutive bit calculations, their performance differences were also examined.

Boskovic and Mahra (2000) used the C-17 simulation aircraft in NASA to detect aircraft errors. This study was conducted by evaluating the data signals sent by the sensors on the body axis of the aircraft in flight. These signals were compared with the normal signals that should be taken and thus errors were detected.

Aydin, Karaköse and Akın (2009) conducted a fault diagnosis work using fuzzy logic and negative selection algorithm. In this study, the characteristics related to the broken rotor in the asynchronous motors were removed and fault diagnosis was performed by giving the direction of the spectrum change obtained by using the motor current signature analysis into the fuzzy logic system. In this study, stator failure, shaft bearing friction failure, and broken rotor failures occurring in asynchronous motors were successfully detected.

Akbal and Ergen (2007) performed attack detection and performance analysis on wireless networks. The proliferation of wireless networking systems and the development of mobile applications have increased the factors positing threat to network security. Firewalls, virus software, and protection programs used to protect the network are not enough to protect the system for long periods of time. The reason for this is the continuous increase of unwanted situations and the delay in taking precautions against them. The purpose of this study was to detect unwanted situations and to protect the network without any intervention and leading to any performance loss with YBS having the capacity to overcome this deficiency. In this study, attack detection in wireless networks was approached from a different perception and instead of conducting detection operation on all the users, the control operation was performed over the access point. When compared to classical methods, better and faster results were obtained.

Bendiab and Kholladi (2010) used a consultant NSA to perform skin detection from color and tissue characteristics. Computer images were used to detect skin color and tissue. These were images of someone's hand, face, etc. Classification with the characteristics obtained from these images was performed in NSA.

Aydın, Karaköse and Akın (2009) used a negative selection algorithm to detect broken rotor bar failures using artificial immune system-based fault detection to detect asynchronous motor failures. Fault related properties were obtained using the first difference filtering of a phase of the motor current and the Hilbert transformation. The negative selection input data were generated by obtaining the phase space of the signals of these properties through the nonlinear time series analysis method. The genetic algorithm was used to optimize and generate the detectors of the negative selection. With the maximum number of detectors, it was ensured to cover the maximum nonself space. The accuracy of the proposed method was verified using the simulation data obtained with the time-stepped combined end-effector state space.

In his doctoral studies, Zhou Ji (2006) used the data gathered from the blood tests of a group of 209 patients. Each patient had four different blood tests. These blood tests were used to screen for a rare genetic disorder. Of the participating patients, 75 were found to be the carriers of the disease and 134 were found to be normal. The carriers of the disease were considered to be "abnormal". As a result of this study, an average detection rate of 86.96% was achieved with NSA.

Ba-Karait et al. (2009) obtained a classification accuracy of 99.28% using a different method of NSA (SNS) from EEG data of epileptic patients.

Magna et al. (2016) achieved a maximum of 90.0% accuracy with the Adaptive Artificial Immune Networks (A2 INET) method.

Dong Li et al. (2016) used the data obtained from the blood tests of a group of 209 patients. These blood tests were used to screen for a rare genetic disorder. Of the participating patients, 75 were found to be the carriers of the disease and 134 were found to be normal. The carriers of the disease were considered to be "abnormal". As a result of this study, an average detection rate of 86.96% was achieved with V-det NSA.

Igawa and Ohashi (2009) achieved 75.6% classification accuracy using the Pima Indian Diabetes data set and the Artificial Negative Selection Classifier (ANSC) method.

For medical data, thyroid disease data set was applied in the performance analysis of AIRS system. This system reached 95.90% classification accuracy with 10-fold CV method (Kodaz et al., 2009).

The fuzzy appearance of the carotid artery Doppler signals makes physicians suspicious about the existence of diseases and sometimes causes false diagnosis. They technique gets around this problem using AIRS to decide and assist the physician to make the final judgment in confidence. AIRS has reached 99.29% classification accuracy using 10-fold cross validation (Latifoğlu et al., 2007).

Although several applications of AIS have been researched in recent years, very few of them have contributed to the research in the field of medicine. Therefore, this study not only contributes to the AIS field, but also contributes to the field of medicine.

This study addresses two AIS algorithms for the classification of medical data, Negative Selection and Variable Detector Negative Selection Algorithms. Algorithms were used for both detection and classification with testing limited to a data set for each problem. Binary classification is in the form of patient and normal.

The purpose of this study is to determine the success of the correct detection and classification of patients with the Negative Selection Algorithm; one of the Artificial Immune System (AIS) algorithms, using information from medical data. While the majority of research in this area has focused on statistical approaches, new studies have investigated artificial intelligence methods, especially neural networks and case-based logic. This study aims to deal with these problems using the relatively new AIS domain.

## ARTIFICIAL IMMUNE SYSTEM (AIS)

Artificial Immune Systems owe their development to models developed for understanding the innate immune system. The first studies in the field of AIS were conducted in 1980s. In the 1990s, the number of studies carried out increased rapidly and the AIS algorithms gradually begun to be applied to practical problems. New systems developed since 2000 and still being developed have proved to be superior over some other methods in some problems in terms of performance and thus made great contributions to the development of the field.

Application areas of Artificial Immune Systems can be summarized as follows: robotic, control, optimization, classification, clustering, artificial neural network approaches, anomaly detection, agent-based approaches, learning, inductive problem solving, pattern recognition, computer models, other approaches (Open Web server coordination, listing, data mining, classification systems, sensor-based diagnosis, development of gene libraries, self-identification etc.). In short, it is possible to use them in most of the areas where other artificial intelligence methods are used. However, their most popular application areas are classification and clustering algorithms, inconsistency detection, computer security and optimization applications. Although considerable success has been achieved in these areas of application, it is not yet possible to speak of AIS as a superior problem-solving tool. In order for AIS to have superiority over other methods, it must have features that are not available in other methods and it should show itself as an alternative way of solution in applications where these other methods are inadequate.

In order to design an artificial immune system, first it should be decided which elements of the immune system will be modeled to create it. Then the interactions of these elements with other elements and the environment should be estimated. Finally, it should be decided how the elements in the system should react to certain functions and situations.

As shown in Figure 1, the basis of every system to be constructed in the field of Artificial Immune Systems is built on the application area. An appropriate representation should be chosen according to the characteristics of the application area, and sensitivity criteria should be based on this representation. Immune algorithms emerging after the determination of the representation and sensitivity criteria determine the functions of the system elements. For example, a designer would like to create an artificial immune system for the optimization problem. For this purpose, he/she can choose to express *Antibodies* in the immune system as system units, and can express each of these *Antibodies* with the help of binary sequences. In this case, the type of representation selected by the designer is binary sequences and the sensitivity criteria should be determined considering this representation. As binary sequences should be taken into consideration for the sensitivity criteria, "Hamming" can be selected as measurement of distance. In the final and most important stage of constructing immune algorithm, the designer can model some mechanisms in the immune system considering the type of representation and sensitivity criterion. These models can use some mechanisms identical to the ones in immunity, as well as models inspired by these mechanisms (de Castro & Timmis,2002).

In the field of Artificial Immunity Systems, various methods have been developed for each of the above-mentioned operations. The shape-space approach that Perelson and Oster introduced in 1979 for the representation of system building units is the most widely accepted and used approach (Perelson & Oster, 1979). The interaction between Antibody and Antigen is one of the most preferred modellings in the process of using the sensitivity criterion.

*Figure 1. Layered structure of artificial immune system*

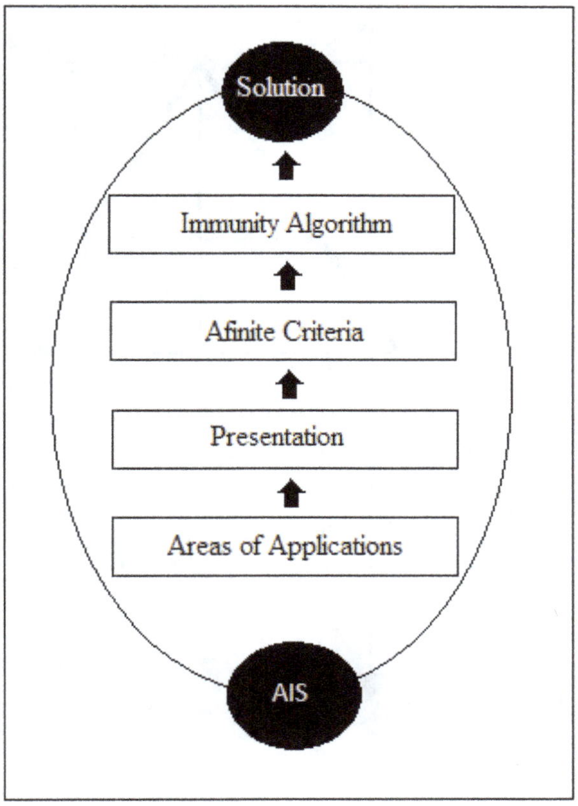


## Shape Space Representation

In order to recognize antigens entering the system, the population of all the cells is called the immune source or immune memory. If there is no protection against the antigens that enter our bodies, the memory is not complete. The immune memory must be complete to recognize all the antigens.

Complementarity is essential in the realization of an interaction in the immune system. In other words, if an antigen and an antibody complement each other, we can say that the interaction is excessive. The concept of shape space has been tried to be modeled with complementarity feature. Antigens cannot be taken directly in shape space representation. Instead, reverse is taken. Therefore, the more the distance between sensors is, the more the complementarity is.

According to Figure 2, the more the distance between antigen and antibody molecules is, the more the interaction is. By taking the inverse of the point where the antigens are present, complementary representation occurs on the shape space.

In Figure 3, the distance between the A antigen and C antibody is complementarily displayed. According to this figure, the interaction between A and C is more intense than the interaction between B and A. In the immune system, for lymphocytes to recognize antigens, the stimuli coming from these structures should be above a certain threshold level. This feature is expressed with a method called recognition circle in the shape space. Each self cell shown in this method has a threshold level and recognition circle. Self-cells can recognize the antigens present in this circle.

*Figure 2. Shape space representation of antigens and antibodies*

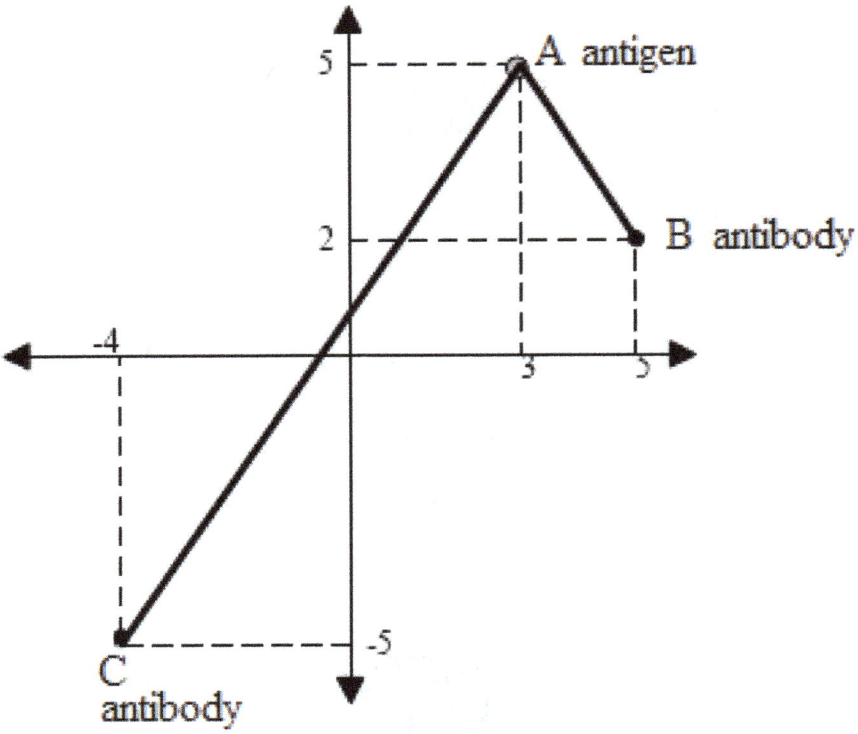


*Figure 3. Complementary representation of antigens and antibodies on the shape space*

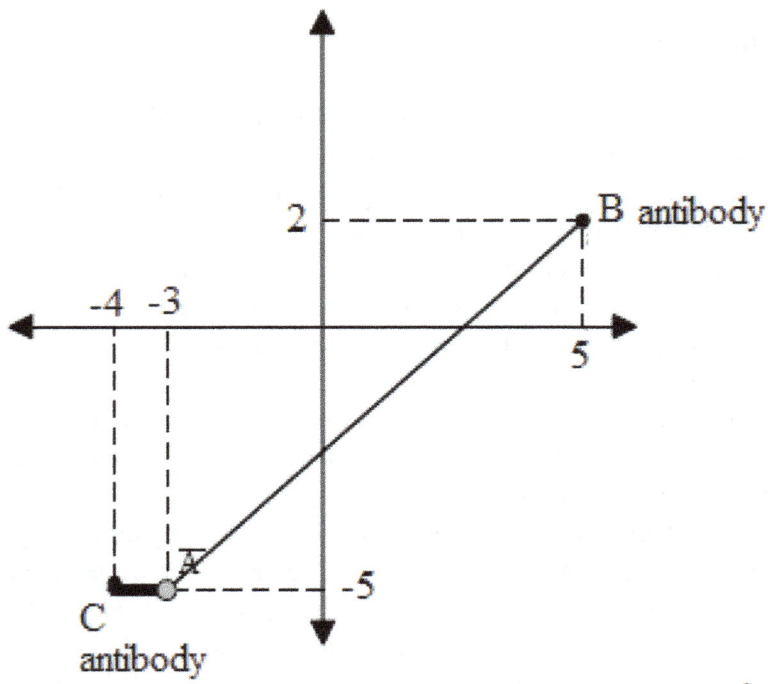

In Figure 4, three self-cells and antigens are shown on the recognition circle. For each cell, the threshold level of the recognition circle is shown with *e*. *Ve* is the determined circle. When we look at Figure 4, we can see that there are 2 antigens in *H1* recognition circle, there is no antigen in *H2* recognition circle and there is 1 antigen in *H3* recognition circle. In a shape space, antigen types are represented with *V* volume. When an antibody population is formed in the *V* volume, the resulting antibody population may recognize all antigens.

Mathematically, distances between an antigen and an antibody molecule can be calculated by different methods. If the vectors representing the antigen and the antibody are with real values, the calculation can be made using the Euclidean or Manhattan distance criteria (Kodaz,2007).

Antijen: $Ag = \{Ag_1,Ag_2,Ag_3,,,,,,,,,Ag_l\}$

Antikor: $Ab = \{Ab_1,Ab_2,Ab_3,,,,,,,,,Ab_l\}$

Euclidean distance between Ab and Ag:

$$D = \sqrt{\sum_{i=1}^{L}\left(Ab_i - Ag_i\right)^2}$$

Manhattan distance between Ab and Ag:

*Figure 4. Representation of antigens with recognition circle*

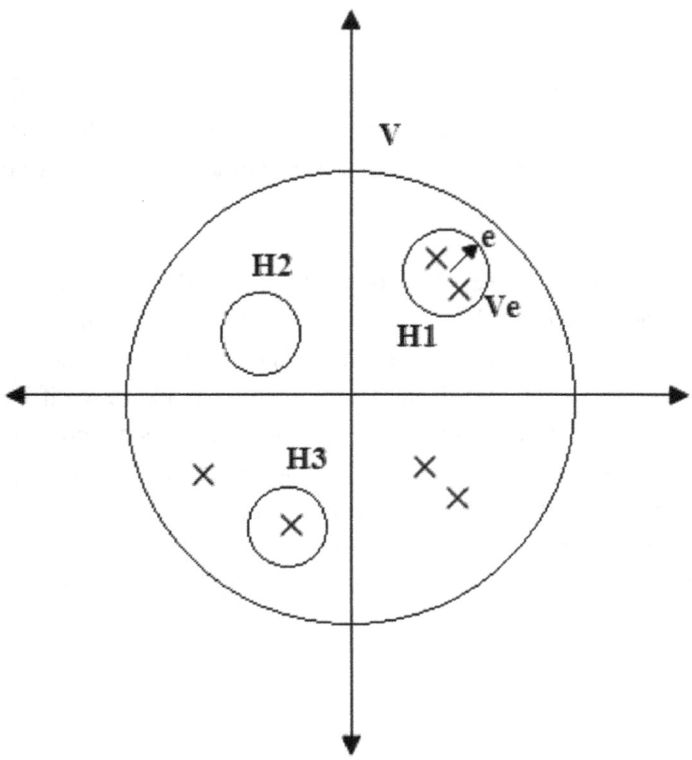

$$D = \sqrt{\sum_{i=1}^{L} \left| Ab_i - Ag_i \right|}$$

## ARTIFICIAL IMMUNE ALGORITHMS

In an artificial immune system design, artificial immune algorithms, which model the immune mechanisms that will form the framework of the system, should be established after the determination of the type of representation selected and sensitivity criteria in accordance with the structure of the field of application. During the stage of the formation of immune algorithms, clonal selection algorithms, negative selection algorithms and immune networks are created for different fields of application and are widely used methods.

### Negative Selection Algorithm

If we look at it from an information processing perspective, the negative selection presents an alternative paradigm to perform pattern recognition. This paradigm works with the sense of storing information about the complementary sets (nonself) of the patterns to be recognized (self). In the literature, negative selection algorithm has been proposed to be used with applications related to computer and network attacks, time series prediction, shape control and cutting and abnormality detection problems such as tolerance to hardware fault (Forrest et al., 1994).

First, the set of models to be protected is determined and is called 'self-set' (P). Based on the negative selection algorithm, a set of sensors (detectors) (M) responsible for identifying elements that do not belong to the 'self-set' are generated.

1. Generates random candidate elements *(C)* using the same representation acceptance.
2. Compares the elements in *C* with the elements in *P*. If there is a match; for example, one element of *P* is recognized by one element of *C*, dismiss this element of *C*. Vice a verse, keep this element of *C* in the detector set.

After the formation of the detector set (M), the next stage of the algorithm is the monitoring of the system for the existence of nonself models. In this case, a P* set was constructed for those to be protected. This set is made up of P set and other new models or a completely new set. For all elements of the detector set corresponding to the nonself models, a check is made to see if it recognizes an element of the P * set. If there is a match, it means that a nonself model has been recognized and reaction should start. The result of nonself detection varies depending on the event evaluation problem and is outside the scope of model recognition of the negative selection algorithm.

Negative Selection Algorithm is shown in Figure 6.

The codes of this algorithm were written in the Matlab program according to the flow chart in Figure 5 and the algorithm in Figure 6.

*Figure 5. Negative Selection Algorithm*
*Ji, 2006.*

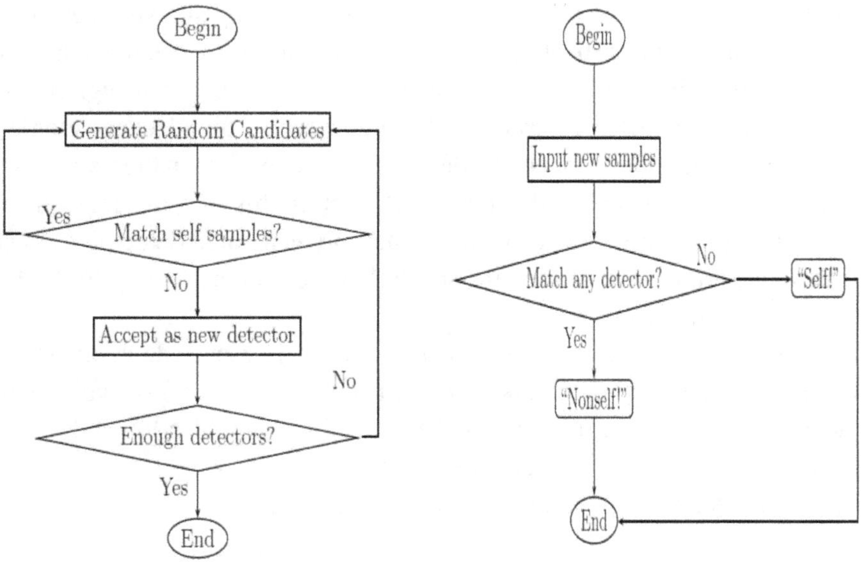


*Figure 6. Detector generation algorithm in negative detection using constant-sized detectors*

```
Require: {Generate pre-defined number of constant-sized detectors}

S     : set of self samples
m     : number of detectors
rs    :self Radius

1: D ← Ø
2: repeat                    {repeat until re-defined number is reached}
3:     x ← random sample from [0, 1]n
4:     for all every si in S = {si, i = 1, 2, · · · } do
5:           d ← Euclidean distance between si and x
6:           if d ≤ rs then
7:|                go to 2          {When recognized as self, try a different point.}
8:           end if
9:     end for
10: D ← D U {x}               {If not matched by any self samples, add to the detector set.}
11: until |D| = m
```

## V-Detector Negative Selection Algorithm

V-detector is an anomaly detection algorithm and application. That is, it is a general-purpose classification algorithm. It is an algorithm inspired by negative selection algorithm. Negative selection algorithms are mainly based on the negative selection process in the thymus, in which immature T cells that recognize self cells are eliminated before the rest can mature and are deployed into the immune system. The negative selection algorithm generates a set of detectors by removing any detector candidate that matches samples on its own as described. Matching rule is generally based on the Euclidean distance between the data to be tested and the detectors. Matching is generally defined as a distance within a certain threshold. In fact, detectors are in the form of hyper-sphere and the threshold is actually the radius of the detector. Ji and Dasgupta (2006) strengthened the negative selection algorithm by allowing detectors to have some variable properties. In real valued application which uses the Euclidean distance matching rule varying the radius of the detectors is an obvious choice considering that the non-self regions to be covered by detectors are very likely to be in different scales. In this algorithm, a candidate detector is generated, and its correspondence with the other detectors in the detector set and the data in the training set is calculated. Depending on its compliance, it is included in the detector set. In this way, the detector set changes until the condition is satisfied. In Figure 7, the change of randomly generated detector set is shown.

The codes of this algorithm were written in the Matlab program according to the flow diagram in Figure 8 and the algorithm in Figure 9.

Due to the structure of the algorithm several major loops were required, an outer *while loop* which exits when the maximum number of detectors have been generated. A *for loop* which loops through the current set of detectors of each randomly generated x and a *for loop* which loops through each member of the self sample for each randomly generated detector x.

V-detector Negative Selection algorithm is shown in Figure 9.

## PATIENTS DETECTION AND CLASSIFICATION PROBLEM

The anomaly-based disease detection problem can be seen as a learning task that tries to induce a general function that can distinguish normal and abnormal samples from an educational set. However, in a many anomalies detection problem, only normal samples are used for training. This shows that it is different from a classical classification algorithm.

*Figure 7. Variable detectors gives a graphic representation*

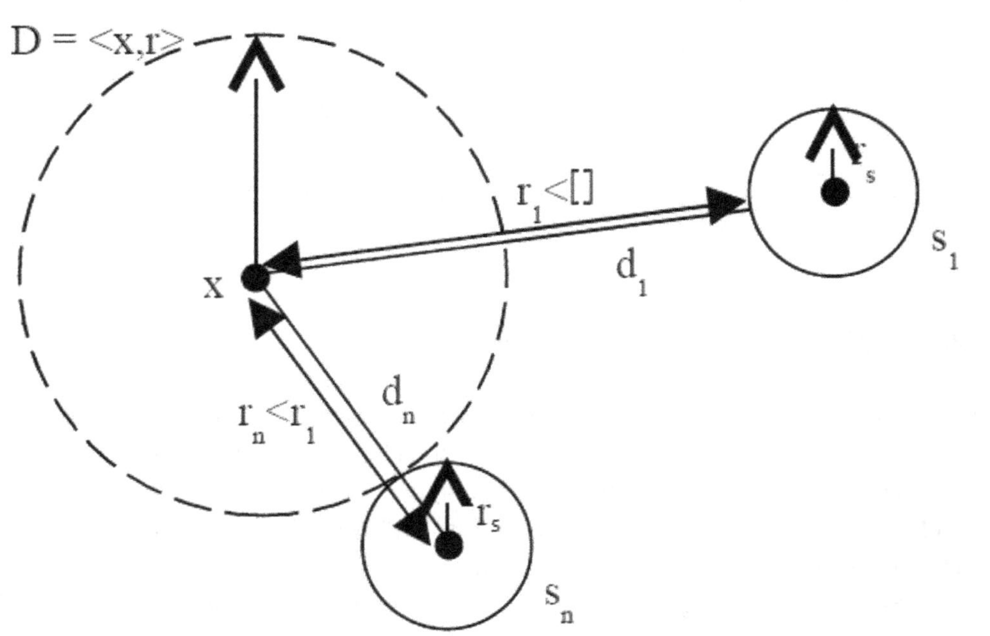

*Figure 8. Flowchart of negative selection algorithm with variable detectors*

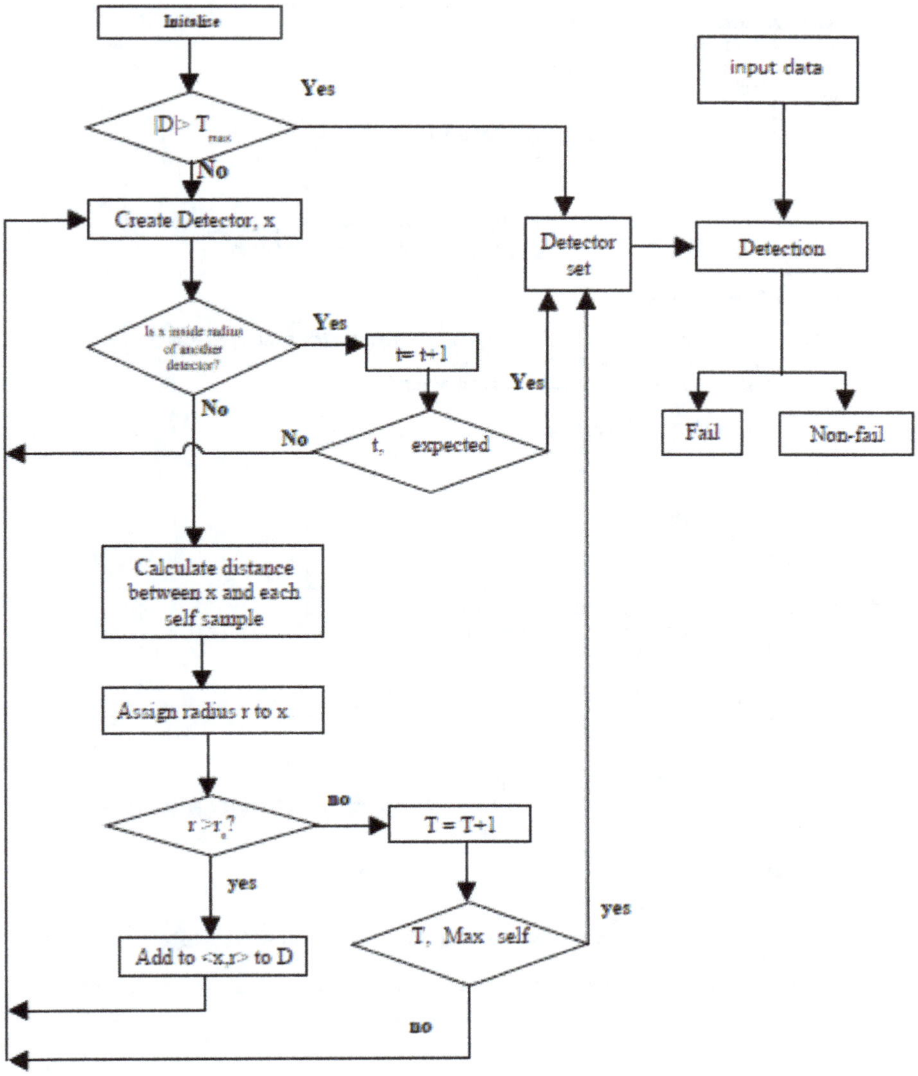


## CHARACTERIZATION OF NEGATIVE SELECTION ALGORITHM

There are three main elements determining the accuracy in negative selection algorithm. These are; detector, training and test tolerance. Training and test tolerance are determined after the trial. The set of detectors; on the other hand, is determined by randomly selecting at the beginning and then by evaluating affinity relationships with the training set. Again, in this detector set test set, self and nonself discrimination is performed considering affinity relationship. As can be seen here, random selection of a detector set is a disadvantage because it might not represent the desired space and this might affect accuracy.

*Figure 9. Detector generation algorithm of V-detector*

```
Require: {Generate detector set of V-detector}
S      : set of self samples
Tmax   : maximum number of detectors
Rs     : self radius
c0     : estimated coverage
cmax   : maximum proportion of self area that is allowed

1: D ← 0
2: repeat
3: t ← 0                                   {counter of "already covered" candidates}
4: T ← 0                                   {counter of "self matched" candidates}
5: r←∞
6: x ← random sample from [0, 1]n

7: for all every di in D = {di, i = 1, 2, · · · } do          {censored by existing detectors}
8:        dd ← Euclidean distance between di and x
9:        if dd < r(di) then {r(di) is the radius of detector di}
10:                t ← t + 1
11:                if t ≥ 1/(1 − c0) then
12:                        return D
13:                else
14:                        go to 4
15:                end if
16:        end if
17: end for

18: for all si in S do      {censored by self samples}
19:        d ← Euclidean distance between si and x
20:        if d − rs < r then
21:                r ← d − rs
22:        end if
23: end for

24: if r > 0 then {not matched by self samples}
25:        D ← D ∪ {< x,r >}               {< x,r > denotes a detector with center x and radius r}
26: else
27:        T ← T + 1
28: end if

29: if T > 1/(1 − cmax) then
30:        exit            {Exception case: too much area are matched by self samples}
31: end if
32: until |D| = Tmax    {Exception case: too many detectors to handle}
```

## DATA SET

In the current study, three data sets were used. These are data obtained from Parkinson, Carotid Doppler and Epilepsy patients. There are two classes in this data set: Patient and Normal. Normals are defined as "Self" and patients are defined as "Nonself". The class defined as Self in the dataset was taken as the training set. Of this dataset, 75% were randomly selected and trained. The whole Nonself class and 25% of the Self class were taken as the test set.

### Parkinson

The dataset was created by Max Little of the University of Oxford, in collaboration with the National Centre for Voice and Speech, Denver, Colorado, who recorded the speech signals. The original study published the feature extraction methods for general voice disorders.

1. **Title:** Parkinsons Disease Data Set
2. **Sources:** The dataset was created by Max Little of the University of Oxford, in collaboration with the National Centre for Voice and Speech, Denver, Colorado, who recorded the speech signals. The original study published the feature extraction methods for general voice disorders. Date received: 2008
3. **Number of Instances:** 195
4. **Number of Attributes:** 22 plus class
5. **Class Distribution:** (class value 0 is interpreted as "tested positive for Parkinson") (Box 1).

## Carotid Artery Doppler

The data were obtained by recording the related signals from the patients in the Radiology Clinic of Fırat University Medical Center.

1. **Title:** Carotid Artery Doppler Disease Data Set
2. **Sources:** The data were obtained by recording the related signals from the patients in the Radiology Clinic of Fırat University Medical Center. Date received: 2006
3. **Number of Instances:** 191
4. **Number of Attributes:** 4 plus class
5. **Class Distribution:** (class value 0 is interpreted as "tested positive for carotid doppler") (Box 2).

## Epilepsy

The data were obtained through the analysis of brain images in DICOM format taken from MR images archive in Aksaray University Aksaray Training and Research Hospital Neurology Clinic.

1. **Title:** Epilepsy Disease Data Set

*Box 1.*

| Class Value | Number of Instances |
|---|---|
| 0 | 147 |
| 1 | 48 |

*Box 2.*

| Class Value | Number of Instances |
|---|---|
| 2 | 136 |
| 3 | 55 |

2. **Sources:** The data were obtained through the analysis of brain images in DICOM format taken from MR images archive in Aksaray University Aksaray Training and Research Hospital Neurology Clinic. Date received: 2013
3. **Number of Instances:** 196
4. **Number of Attributes:** 50 plus class
5. **Class Distribution:** (class value 0 is interpreted as "tested positive for Epilepsy") (Box 3).

## RESEARCH AND EXPERIMENTATION

Throughout the research, two algorithms were used. These are Negative Selection (NS) and Variable Detector Negative Selection (V-DET NS) algorithms. These real valued algorithms were used to best predict the classification and detection (Gonzalez and Dasgupta, 2002). In the data set used in these algorithms that can distinguish Self and Nonself, self / nonself correspond to [0,1].

1. **The Negative Selection Algorithm (NSA):** The negative selection algorithm can be described as a mathematical representation of the maturation of T cells in the thymus gland. It uses self / nonself discrimination principles to distinguish between two normal and abnormal system states. In this application, normal represents healthy and abnormal represents unhealthy individuals. Unhealthy individuals are identified using detectors that have been trained in the set of sample healthy individuals.
2. **The Variable Detector Algorithm (V-Det NSA):** The algorithm is another modification of the basic negative selection algorithm. In the basic NS algorithm, the detectors are in the form of hyper-sphere and have a fixed detection radius. In variable detector algorithm, the detectors have a changing feature. Each detector's detection radius is changeable and therefore, the self / nonself discrimination is done with fewer detectors. Thus, the algorithm is made more efficient.

## EXPERIMENTS AND RESULTS

The characteristics of the data sets used for detection and classification are given in Table 1.

The usage status of the data set during training and testing phases is given in Table 2. During the training phase, 75% of the data received from normal subjects were used by randomly selecting. In the testing phase, all of the patients and 25% of the normals were used.

The parameters and values used in the negative selection algorithm are given in Table 3. *Detector* refers to the number of detectors, *TrainTol* refers to training tolerance and *TestTol* refers to test tolerance.

*Box 3.*

| Class Value | Number of Instances |
|---|---|
| 0 | 75 |
| 1 | 121 |

*Table 1. Distribution of the data*

| Data Sets | Number of Instance | Number of Attributes | Number of Patients | Number of Non Patients |
|---|---|---|---|---|
| Parkinson | 195 | 22 | 147 | 48 |
| Carotid Doppler | 191 | 4 | 136 | 55 |
| Epilepsi | 196 | 50 | 75 | 121 |

*Table 2. Data distributions used in the training and testing of the data sets*

| Data Sets | | Training Set | Test Set | Total | Partition of Dataset (%) |
|---|---|---|---|---|---|
| 1 | Parkinson | 0 | 147 | 147 | %75 (random selection) |
| | Normal | 36 | 12 | 48 | |
| 2 | Carotid | 0 | 136 | 136 | %75 (random selection) |
| | Normal | 45 | 10 | 55 | |
| 3 | Epilepsy | 0 | 75 | 75 | %75 (random selection) |
| | Normal | 90 | 31 | 121 | |

*Table 3. Parameters and values used in NSA*

| Data Sets | Parameters | Values |
|---|---|---|
| Parkinson | Detector | 300 |
| | TrainTol | 1.11 |
| | TestTol | 1.09 |
| | Iterasyon | 1000 |
| Carotid | Detector | 300 |
| | TrainTol | 0.94 |
| | TestTol | 0.92 |
| | Iterasyon | 1000 |
| Epilepsy | Detector | 300 |
| | TrainTol | 0.034 |
| | TestTol | 0.039 |
| | Iterasyon | 1000 |

Parameters and values used in V-DET Negative Selection algorithm are given in Table 4. Here, *Tmax* represents the maximum detector number and *Rs* represented self Radius value.

The evaluation of the power of any screening test to determine the health problem is made by comparing it with a reference test that gives a definite result for this health problem. This can be made through the comparison of the results of a screening test with the results of the reference test.

The performance of a classifier is measured by using the accuracy, sensitivity, specificity and precision. By using these variables, epidemiological criteria related to strength and adequacy of the screening test are found. These parameters are computed as formulated (1), (2), (3) and (4).

- **Sensitivity:** It is defined as the ability of a screening test to find those with the health problem among individuals who actually have this health problem. It can be briefly defined as a screening test's patient detection ability.

Sensitivity = TP/(TP+FN) (1)

- **Specificity:** It is defined as the ability of a screening test to find healthy individuals among the individuals who are actually healthy. It can be briefly defined as a screening test's healthy individual detection ability.

Specificity = TN/(TN+FP) (2)

- **Precision:** It is the percentage of individuals who have the health problem within the individuals that the screening test finds to have the health problem.

Precision = TP/(TP+FP) (3)

- **Accuracy:** It is defined as the ability of the screening test to make a correct diagnosis. The strength or validity of a test is defined as the percentage of the number of sick or healthy persons diagnosed correctly by the screening test in all outcomes.

*Table 4. Parameters and values used in V-DET NSA*

| Data Sets | Parameters | Values |
|---|---|---|
| Parkinson | Tmax | 500 |
| | Rs | 0.03 |
| | Iterasyon | 1000 |
| Carotid | Tmax | 500 |
| | Rs | 0.01 |
| | Iterasyon | 1000 |
| Epilepsy | Tmax | 500 |
| | Rs | 0.01 |
| | Iterasyon | 1000 |

$$\text{Accuracy} = (TP+TN)/(TP+FP+FN+TN) \quad (4)$$

TP,TN,FN, and FP are determined as follows:

- **True Positive (TP):** By counting the number of patients with patients correctly identified as patients.
- **True Negative (TN):** By counting the number of non-patients correctly identified as non-patients.
- **False Negative (FN):** By counting the number of patients with patients incorrectly identified as non-patients.
- **False Positive (FP):** By counting the number of non-patients incorrectly identified as patients.

We evaluate the performance of algorithms in terms of sensitivity (1), specificity (2), precision (3) and accuracy (4).

Performance values of the evaluated negative selection algorithm are given in Table 5.

Performance values of the evaluated V-DET negative selection algorithm are given in Table 6.

In the first algorithm, on the test data, 92.46%, 93.40% and 90.57% classification accuracy and 92.45%, 91.46% and 92.21% detection accuracy were achieved. In the second algorithm, on the test data, 94.34%, 94.52% and 91.51% classification accuracy and 94.23%, 94.40% and 89.29% detection accuracy were achieved.

Considering the values given above, the patient detection and classification accuracy of negative selection algorithm is over 90% in general in different medical data sets and different conditions and this is a highly promising result.

## CONCLUSION

The basic purpose of this study was to investigate the suitability of negative selection algorithms for the resolution of medical problems. This section presents a summary of the findings, summarizes the results

*Table 5. Classification and detection accuracy of NSA on the test data*

| Data Sets | Accuracy | Sensitivity | Specificity | Precision |
|---|---|---|---|---|
| Parkinson | 0.9246 | 1.0000 | 0.0000 | 0.9245 |
| Carotid Doppler | 0.9340 | 1.0000 | 0.7742 | 0.9146 |
| Epilepsy | 0.9057 | 0.9467 | 0.8065 | 0.9221 |

*Table 6. Classification and detection accuracy of V-DET NSA on the test data*

| Data Sets | Accuracy | Sensitivity | Specificity | Precision |
|---|---|---|---|---|
| Parkinson | 0.9434 | 1.0000 | 0.2500 | 0.9423 |
| Carotid Doppler | 0.9452 | 1.0000 | 0.2000 | 0.9440 |
| Epilepsy | 0.9151 | 1.0000 | 0.7097 | 0.8929 |

of these findings, and provides suggestions for further research. Two AIS algorithms were used for the classification of medical data and the problem of patient detection: Negative Selection (NS) and Variable Detector Negative Selection (V-Det NSA). The proposed approach is built on supervised learning, which aims to classify the data as normal (self) and patient (nonself). The operation is performed in two stages, learning and classification.

In the present study, the classification performance of AIS algorithms on the medical data sets was tested. Besides patient detection, high classification performance is also an important criterion. It was found that detection and classification performance of AIS changed depending on the data set. The number of the detectors used, the number of trained data and tolerance value were found to be the determinants of the performance.

In the Negative Selection Algorithm, the set of detectors that determine the classification performance is very important. Therefore, different hybrid methods and optimization techniques should be used for the detector set to represent the desired space. Moreover, by changing affinity criterion, we can affect the performance.

As a result, the Negative Selection Algorithm has shown significant success in terms of classification performance over medical data.

## REFERENCES


Akbal, E., & Ergen, B. (2007). A Performance Comparison of User and Access Point based Artificial Immune Systems for intrusion detection on a Wireless Local Area Network. *International Journal of Computer Science and Network Security*, *7*, 119–124.

Aydın, İ., Karaköse, M., & Akın, E. (2009). a fault diagnosis work using fuzzy logic and negative selection algorithm. *Journal of the Faculty of Engineering and Architecture of Gazi University*, *24*(4), 745–753.

Aydın, İ., Karaköse, M., & Akın, E. (2009). Artificial Immune System Based Diagnostic Model Using Genetic Algorithm. *Dokuz Eylul Unıversıty Faculty Of Engıneerıng Journal Of Scıence And Engıneerıng*, *11*(31), 58–72.

Ba-Karait, N. O. S., Shamsuddin, S. M., & Sudirman, R. (2009). Swarm Negative Selection Algorithm for Electroencephalogram Signals Classification. *Journal of Computational Science*, *5*(12), 998–1005.

Bedbiab, E., & Kholladi, M. (2010). The Negative Selection Algortihm A Supervised Learning Approach For Skin Detection And Clasification. *IJCSNS International Journal of Computer Science and Network Security*, *10*(11), 48–54.

Boskovic, D., & Mehra, R. (2000). Intelligent Adaptive Control of a Tailless Advanced Fighter Aircraft under Wing Damage. *Journal of Guidance Control, and Dynamics*, *23*(5), 876–884. doi:10.2514/2.4617

Castro, N., & Timmis, J. (2002). *Artificial Immune Systems: A Novel Paradigm to Pattern Recognition*. University of Paisley.

Dasgupta, D. (1996). *Using Immunological Principles in Anomaly Detection. In Proceedings of Artificial Neural Networks in Engineering*. ANNIE.

Dasgupta, D. (1998). *Artificial immune systems and their applications. Springer-Verlag.*

Dasgupta, D., & Attoh-Okine, N. 1997). Immunity-Based Systems: A Survey. *Proceedings of the IEEE International Conference on Systems, Man and Cybernetics*. doi:10.1109/ICSMC.1997.625778

Dasgupta, D., & Forrest, S. (1995). *Tool Breakage Detection in Milling Operations using a Negative-Selection Algorithm.* Technical Report CS95-5. Computer Science, University of New Mexico.

Dasgupta, D., & Forrest, S. (1996). Novelty Detection in Time Series Data Using Ideas From Immunology. *Proceedings of the ISCA'96.*

Dasgupta, D., & Forrest, S. (1996). Novelty Detection in Time Series Data Using Ideas from Immunology. *Proceedings of The 5th International Conference on Intelligent Systems.*

Dasgupta, D., & Forrest, S. (1999). Artificial Immune systems in Industrial Applications. In *Proceedings of the Second International Conference on Intelligent Processing and Manufacturing of materials (IPMM)*. doi:10.1109/IPMM.1999.792486

de Castro, L. N., & Timmis, J. (2002). *Artificial Immune Systems: A New Computational Intelligence Approach*. London: Springer.

De Castro, L. N., & Von Zuben, F. J. (2000). Artificial immune systems: Part I- Basic theory and applications. DCA-RT 02/00, Brasil, 23-28.

Engin, O., & Döyen, A. (2004). Artificial Immune Systems And Applications In Industrial Problems. *Journal of Science of Gazi University*, *17*(1), 71–84.

Forrest, S. (1994). Self-Non self Discrimination in a Computer. *Proceedings of the IEEE Symposium on Security and Privacy*, 202.

Forrest, S., Perelson, A., & Allen, L. (1994). Self-Nonself Discrimination in a Computer. *Proceedings of the IEEE Symposium on Research in Security and Privacy*, 202-212.

Gonzalez, F. A., & Dasgupta, D. (2002). Anomaly Detection Using Real-Valued Negative Selection. *Genetic Programming and Evolvable Machines*.

Igawa, K., & Ohashi, H. (2009). A negative selection algorithm for classification and reduction of the noise effect. *Applied Soft Computing*, *9*(1), 431–438. doi:10.1016/j.asoc.2008.05.003

Ji, Z. (2006). *Negative Selection Algorithms: From The Thymus To V-Detector* (PhD dissertation). The Universty of Memphis.

Ji, Z., & Dasgupta, D. (2007). Revisiting Negative Selection Algorithms. *Evolutionary Computation, Massachusetts Institute of Technology*, *15*(2), 223–251. doi:10.1162/evco.2007.15.2.223 PMID:17535140

Kaymaz, E. D. (2007). *Predicting The Cellular Localization Sites Of Protein Patterns With K-nn Classification Algorithm Based Artificial Immune System* (Master Thesis). Institute of Science of Fırat University, Elazığ.

Kodaz, H. (2007). *Information Gain Based Artificial Immunity Recognition System* (PhD thesis). Institute of Science of Selcuk University, Konya.

Kodaz, H., Özşen, S., Arslan, A., & Güneş, S. (2009). Medical application of information gain based artificial immune recognition system (AIRS): Diagnosis of thyroid disease. *Expert Systems with Applications*, *36*(2), 3086–3092. doi:10.1016/j.eswa.2008.01.026

Latifoğlu, F., Kodaz, H., Kara, S., & Güneş, S. (2007). Medical application of Artificial Immune Recognition System (AIRS): Diagnosis of atherosclerosis from carotid artery Doppler signals. *Computers in Biology and Medicine*, *37*(8), 1092–1099. doi:10.1016/j.compbiomed.2006.09.009 PMID:17156772

Li, D., Liu, S., & Zhang, H. (2016). A boundary-fixed negative selection algorithm with online adaptive learning under small samples for anomaly detection. *Engineering Applications of Artificial Intelligence*, *50*, 93–105. doi:10.1016/j.engappai.2015.12.014

Magna, G., Casti, P., Jayaraman, S. V., Salmeri, M., Mencattini, A., Martinelli, E., & Di Natale, C. (2016). Identification of mammography anomalies for breast cancer detection by an ensemble of classification models based on artificial immune system. *Knowledge-Based Systems*, *101*, 60–70. doi:10.1016/j.knosys.2016.02.019

Mori, K., Tsukiyama, M., & Fukuda, T. (1997). Artificial Immunity Based Management System for a Semiconducter Production Line. *IEEE International Conference on Systems, Man, and Cybernetics*, *1*, 852-856.

Özşen, Ş. (2008). *Development Of A Problem-Based Artificial Immune System For Bıomedical Classification Problems And Application On Biomedical Classification Problems* (PhD thesis). Institute of Science of Selcuk University, Konya.

Perelson, A. S., & Oster, G. F. (1979). Theoretical studies of clonal selection: Minimal antibody repertoire size and reliability of self-non-self discrimination. *Journal of Theoretical Biology*, *81*(4), 645–670. doi:10.1016/0022-5193(79)90275-3 PMID:94141

Polat, K. (2004). *Going To Medical Diagnosis By Using Artificial Immune Recognition Systems With Feature Selection* (Master thesis). Institute of Science of Selcuk University, Konya.

Sertkaya, C. (2009). *Use of Artificial Immune System in Computer Security* (Master thesis). Institute of Science of Sakarya University, Sakarya.

Singh, S. (2002). Anomaly Detection Using Negative Selection Based on the R-Contiguous Matching Rule. In *Proceedings of ICARIS 2002*. University of Kent at Canterbury Printing Unit.

Taylor, D., & Corne, D. (2003). An Investigation of the Negative Selection Algorithm for Fault Detection in efrigeration Systems. In *Proceeding of Second International Conference on Artificial Immune Systems (ICARIS)*. Napier University.

Timmis, J. (2001). aiVIS - Artificial Immune Network Visualisation. *EuroGraphics UK 2001 Conference Proceedings.*

Timmis, J. (2007). Artificial immune systems-Today and tomorrow. *Natural Computing: An International Journal*, *6*(1), 1-18.

## ADDITIONAL READING

Afshinmanesh, F., Marandi, A., & Rahimi-Kian, A. (2005, November). A novel binary particle swarm optimization method using artificial immune system. In *Computer as a Tool, 2005. EUROCON 2005. The International Conference on*(Vol. 1, pp. 217-220). IEEE. doi:10.1109/EURCON.2005.1629899

Armitage, G. C. (1999). Development of a classification system for periodontal diseases and conditions. *Annals of Periodontology*, *4*(1), 1–6. doi:10.1902/annals.1999.4.1.1 PMID:10863370

Ashcraft, M. L., Fries, B. E., Nerenz, D. R., Falcon, S. P., Srivastava, S. V., Lee, C. Z., & Errera, P. et al. (1989). A psychiatric patient classification system: An alternative to diagnosis-related groups. *Medical Care*, *27*(5), 543–557. doi:10.1097/00005650-198905000-00009 PMID:2498585

Balthrop, J., Esponda, F., Forrest, S., & Glickman, M. (2002, July). Coverage and generalization in an artificial immune system. In *Proceedings of the 4th Annual Conference on Genetic and Evolutionary Computation* (pp. 3-10). Morgan Kaufmann Publishers Inc.

Banerjee, S. (2017). An artificial immune system approach to automated program verification: Towards a theory of undecidability in biological computing. *PeerJ Preprints, 5*, e2690v1.

Botter, M. L. (2000). The use of information generated by a patient classification system. *The Journal of Nursing Administration*, *30*(11), 544–551. doi:10.1097/00005110-200011000-00010 PMID:11098254

Cayirli, T., Veral, E., & Rosen, H. (2008). Assessment of patient classification in appointment system design. *Production and Operations Management*, *17*(3), 338–353. doi:10.3401/poms.1080.0031

Chabot, G., & Fox, M. (2005). The creation of a patient-classification system in an outpatient infusion center setting. *Oncology Nursing Forum*, *32*(3), 535–538. doi:10.1188/05.ONF.535-538 PMID:15902758

Chen, P. Y., Gensollen, T., & Hero, A. O. (2017, March). AMOS: An automated model order selection algorithm for spectral graph clustering. In *Acoustics, Speech and Signal Processing (ICASSP), 2017 IEEE International Conference on* (pp. 6448-6452). IEEE.

Dasgupta, D., Ji, Z., & Gonzalez, F. (2003). Artificial immune system (AIS) research in the last five years. In Evolutionary Computation, 2003. CEC'03. The 2003 Congress on (Vol. 1, pp. 123-130). IEEE.

de Abreu, C. C. E., Duarte, M. A. Q., & Villarreal, F. (2017). An immunological approach based on the negative selection algorithm for real noise classification in speech signals. *AEÜ. International Journal of Electronics and Communications*, *72*, 125–133. doi:10.1016/j.aeue.2016.12.004

Dindo, D., Demartines, N., & Clavien, P. A. (2004). Classification of surgical complications: A new proposal with evaluation in a cohort of 6336 patients and results of a survey. *Annals of Surgery*, *240*(2), 205–213. doi:10.1097/01.sla.0000133083.54934.ae PMID:15273542

Dudek, G. (2017). Artificial immune system with local feature selection for short-term load forecasting. *IEEE Transactions on Evolutionary Computation*, *21*(1), 116–130. doi:10.1109/TEVC.2016.2586049

Engin, O., & Döyen, A. (2004). A new approach to solve hybrid flow shop scheduling problems by artificial immune system. *Future Generation Computer Systems*, *20*(6), 1083–1095. doi:10.1016/j.future.2004.03.014

Fagerström, L., & Rauhala, A. (2007). Benchmarking in nursing care by the RAFAELA patient classification system–a possibility for nurse managers. *Journal of Nursing Management, 15*(7), 683–692. doi:10.1111/j.1365-2934.2006.00728.x PMID:17897144

Forthman, M. T., Dove, H. G., & Wooster, L. D. (2000). Episode Treatment Groups (ETGs): A patient classification system for measuring outcomes performance by episode of illness. *Topics in Health Information Management, 21*(2), 51–61. PMID:11143280

Fries, B. E., & Cooney, L. M. Jr. (1985). Resource utilization groups: A patient classification system for long-term care. *Medical Care, 23*(2), 110–122. doi:10.1097/00005650-198502000-00002 PMID:3919222

Giovannetti, P., & Johnson, J. M. (1990). A new generation patient classification system. *The Journal of Nursing Administration, 20*(5), 33–40. doi:10.1097/00005110-199005000-00009 PMID:2335788

Harmer, P. K., Williams, P. D., Gunsch, G. H., & Lamont, G. B. (2002). An artificial immune system architecture for computer security applications. *IEEE Transactions on Evolutionary Computation, 6*(3), 252–280. doi:10.1109/TEVC.2002.1011540

Hlusko, D. L., & Nichols, B. S. (1996). Can you depend on your patient classification system? *The Journal of Nursing Administration, 26*(4), 39–44. doi:10.1097/00005110-199604000-00012 PMID:8774471

Hofmeyr, S. A., & Forrest, S. (2000). Architecture for an artificial immune system. *Evolutionary Computation, 8*(4), 443–473. doi:10.1162/106365600568257 PMID:11130924

Hu, B. (2016). An Improved Detector Set Generation Mechanism in Negative Selection Algorithm. *DEStech Transactions on Computer Science and Engineering*, (icte).

Hunt, J. E., & Cooke, D. E. (1996). Learning using an artificial immune system. *Journal of Network and Computer Applications, 19*(2), 189–212. doi:10.1006/jnca.1996.0014

Kim, J., & Bentley, P. (1999). Negative selection and niching by an artificial immune system for network intrusion detection. In *Late Breaking Papers at the 1999 Genetic and Evolutionary Computation Conference* (pp. 149-158).

Kim, J., & Bentley, P. J. (2002). Towards an artificial immune system for network intrusion detection: an investigation of dynamic clonal selection. In *Evolutionary Computation, 2002. CEC'02. Proceedings of the 2002 Congress on* (Vol. 2, pp. 1015-1020). IEEE.

Le Boudec, J. Y., & Sarafijanovic, S. (2004, January). An artificial immune system approach to misbehavior detection in mobile ad hoc networks. In BioADIT (pp. 396-411). doi:10.1007/978-3-540-27835-1_29

Letournel, E. (2007). Acetabulum fractures: Classification and management. *Orthopedic Trauma Directions, 5*(05), 27–33. doi:10.1055/s-2007-980136

Liu, Z., Li, T., Yang, J., & Yang, T. (2017). An improved negative selection algorithm based on subspace density seeking. *IEEE Access: Practical Innovations, Open Solutions.*

Lu, H., Chen, J., Yan, K., Jin, Q., Xue, Y., & Gao, Z. (2017). A hybrid feature selection algorithm for gene expression data classification. *Neurocomputing.*

Maluf, K. S., Sahrmann, S. A., & Van Dillen, L. R. (2000). Use of a classification system to guide nonsurgical management of a patient with chronic low back pain. *Physical Therapy*, *80*(11), 1097–1111. PMID:11046197

Mohi-Aldeen, S. M., Mohamad, R., & Deris, S. (2016). Application of Negative Selection Algorithm (NSA) for test data generation of path testing. *Applied Soft Computing*, *49*, 1118–1128. doi:10.1016/j.asoc.2016.09.044

Nasaroui, O., Gonzalez, F., & Dasgupta, D. (2002). The fuzzy artificial immune system: Motivations, basic concepts, and application to clustering and web profiling. In *Fuzzy Systems, 2002. FUZZ-IEEE'02. Proceedings of the 2002 IEEE International Conference on* (Vol. 1, pp. 711-716). IEEE.

Padmanabhan, S., Chandrasekaran, M., Ganesan, S., & Navakanth, P. (2017, March). Optimal Solution for an Engineering Applications Using Modified Artificial Immune System. [). IOP Publishing.]. *IOP Conference Series. Materials Science and Engineering*, *183*(1), 012025. doi:10.1088/1757-899X/183/1/012025

Parlak, B., & Uysal, A. K. (2018). On Feature Weighting and Selection for Medical Document Classification. In *Developments and Advances in Intelligent Systems and Applications* (pp. 269–282). Cham: Springer. doi:10.1007/978-3-319-58965-7_19

Powell, B. M., Kalsy, E., Goswami, G., Vatsa, M., Singh, R., & Noore, A. (2017, May). Attack-Resistant aiCAPTCHA using a Negative Selection Artificial Immune System. In *Proc. 38th S&P, 2nd Workshop on BioSTAR*.

Rainio, A. K., & Ohinmaa, A. E. (2005). Assessment of nursing management and utilization of nursing resources with the RAFAELA patient classification system–case study from the general wards of one central hospital. *Journal of Clinical Nursing*, *14*(6), 674–684. doi:10.1111/j.1365-2702.2005.01139.x PMID:15946275

Sahan, S., Polat, K., Kodaz, H., & Gunes, S. (2007). A new hybrid method based on fuzzy-artificial immune system and k-nn algorithm for breast cancer diagnosis. *Computers in Biology and Medicine*, *37*(3), 415–423. doi:10.1016/j.compbiomed.2006.05.003 PMID:16904096

Sotiropoulos, D. N., & Tsihrintzis, G. A. (2017). Artificial Immune Systems. In *Machine Learning Paradigms* (pp. 159–235). Springer International Publishing. doi:10.1007/978-3-319-47194-5_7

Su, Q., Wang, Y., Jiang, X., Chen, F., & Lu, W. C. (2017). A Cancer Gene Selection Algorithm Based on the KS Test and CFS. *BioMed Research International*, 2017.

Tan, K. C., Goh, C. K., Mamun, A. A., & Ei, E. Z. (2008). An evolutionary artificial immune system for multi-objective optimization. *European Journal of Operational Research*, *187*(2), 371–392. doi:10.1016/j.ejor.2007.02.047

Tarakanov, A., & Dasgupta, D. (2000). A formal model of an artificial immune system. *Bio Systems*, *55*(1), 151–158. doi:10.1016/S0303-2647(99)00093-3 PMID:10745119

Timmis, J., Neal, M., & Hunt, J. (2000). An artificial immune system for data analysis. *Bio Systems*, *55*(1), 143–150. doi:10.1016/S0303-2647(99)00092-1 PMID:10745118

Van Slyck, A. (1991). A Systems Approach to the Management of Nursing Services-Part II: Patient Classification System. *Nursing Management*, *22*(4), 23–25. doi:10.1097/00006247-199104000-00009 PMID:1901400

Walts, L. M., & Kapadia, A. S. (1995). Patient classification system: An optimization approach. *Health Care Management Review*, *21*(4), 75–82. doi:10.1097/00004010-199623000-00009 PMID:8922967

Wen, T., Xu, A., & Tang, J. (2016). Study on extension negative selection algorithm. *International Journal of High Performance Computing and Networking*, *9*(1-2), 1–7. doi:10.1504/IJHPCN.2016.074653

Zhang, F. S., Li, S. W., Hu, Z. G., & Du, Z. (2017). Fish swarm window selection algorithm based on cell microscopic automatic focus. *Cluster Computing*, *20*(1), 485–495. doi:10.1007/s10586-017-0752-4

## KEY TERMS AND DEFINITIONS

**Artificial Immune System:** A general research approach – method of artificial intelligence as inspired from mechanisms of biological immune systems.

**Classification:** As a problem-solving way in artificial intelligence, classification is the operation on finding different groups of objects intelligently in order to reach to the desired solution findings.

**Negative Selection Algorithm:** A type of algorithm included within artificial immune system, inspired by the mechanism of self-nonself discrimination behavior in biological immune systems.

**Patient Classification:** A classification operation determining separate groups of patients having different features according to diseases or any other information.

# Chapter 5
# Differential Diagnosis of Erythematous Squamous Diseases With Feature Selection and Classification Algorithms

**Aydın Çetin**
*Gazi University, Turkey*

**Tuba Gökhan**
*Gazi University, Turkey*

## ABSTRACT

*In this chapter, the differential diagnosis of erythematous diseases was determined using data mining and machine learning algorithms. In this chapter, data mining and its application to differential diagnosis of erythematous squamous diseases were discussed. A dermatology dataset from UCI Machine Learning Repository was used for the study. The dataset consists of 366 data items with 34 attributes. Initially, feature selection was made, and then classification was performed by using various algorithms. The number of attributes has been reduced from 34 to 19 as a result of the integration of the correlation-based filter methods and various heuristic search methods. The evaluation results show that Naive Bayes has 100% success rate in classification of psoriasis, seborrheic dermatitis, lichen planus, rose disease, chronic dermatitis, and pityriasis rubra pilaris diseases with 19 attributes selected with feature extraction algorithms.*

## INTRODUCTION

In primary health care institutions where dermatologic diseases are frequent, there are frequent cases of inaccurate diagnosis or no diagnosis. Incorrect diagnosis causes wrong treatment. Inaccuracies in diagnosing diseases also lead to problems such as difficulties in disease management and inadequate information of patients, unnecessary referrals and laboratory examinations. In the field of medicine, it is important to examine the theoretical knowledge and disease management in order to be able to obtain

DOI: 10.4018/978-1-5225-4769-3.ch005

a successful diagnosis in the diagnosis and treatment of common dermatological diseases (Yilmaz, 2016). The differential diagnosis of erythematous squamous disease is a real problem in dermatology. Erythematous squamous disease, psoriasis, seborrheic dermatitis, lichen planus, Pityriasis Rosea, chronic dermatitis and pityriasis rubra pilaris have similar clinical features. Biopsy is usually carried out for diagnosis of the disease. However, the histopathological features of these diseases are similar and even while these diseases show the characteristics of another disease at earlier stages, they can show their own characteristics when disease progresses (Güvenir, Demiröz, & İlter, 1998).

Today, as the amount of information on the world increases day by day with the data being stored in the digital environment, the number of databases is also increasing at a similar or even higher rate. This increase makes it difficult to retrieve meaningful information. Therefor there is a good reason to believe that smart data analysis will be more pervasive in medicine. Data mining is a multidisciplinary field that serves as a bridge between many technical fields such as database technology, statistics, artificial intelligence, machine learning, pattern recognition and data visualization. Data mining is applied in many fields such as astronomy, biology, finance, marketing, insurance, medicine (Ozekes, 2003).

In this chapter, Data mining and its application to differential diagnosis of erythematous squamous diseases were discussed. A dermatology dataset from UCI Machine Learning Repository which was shared by Prof. Dr. Nilsel İlter from Gazi University and Altay Güvenir from Bilkent University, Turkey, was used for the study (Güvenir et al, 1998). The dataset consists of 366 data items with 34 attributes. Initially, feature selection was made and then, classification was performed by using various algorithms. The number of attributes has been reduced from 34 to 19 as a result of the integration of the correlation based filter methods and various heuristic search methods.

## DATA MINING

Data Mining is the process of obtaining previously unknown, valid and applicable information from large data sources. With data mining relationships between the variables can be deduced and future predictions can be made (Kaya & Özel, 2014). The process of obtaining raw data to knowledge using data mining is shown in Figure 1.

As shown in Figure 1, in the data mining process, the process begins with obtaining the data of the probing that is desired to be examined first. Noisy and inconsistent data is removed from the data set by data cleansing. After selecting and extracting the attributes on the data warehouse, the appropriate models for the data and probing are created, and the suitability and adequacy of these models are evaluated. Intelligent methods are applied to capture data patterns while models are being created. In the pattern evaluation step, interesting patterns representing information obtained according to predetermined criteria are defined. In the information presentation step, the obtained information is presented to the user (Kaya & Özel, 2014).

## FEATURE SELECTION

The complexity and size of databases in the real world has prompted researchers to reduce data sizes in the data mining. Size reduction can basically be done in two ways. First, to create a random sample space from existing data, and second, to reduce the amount of available attributes. There are two main

*Figure 1. Data mining processes*

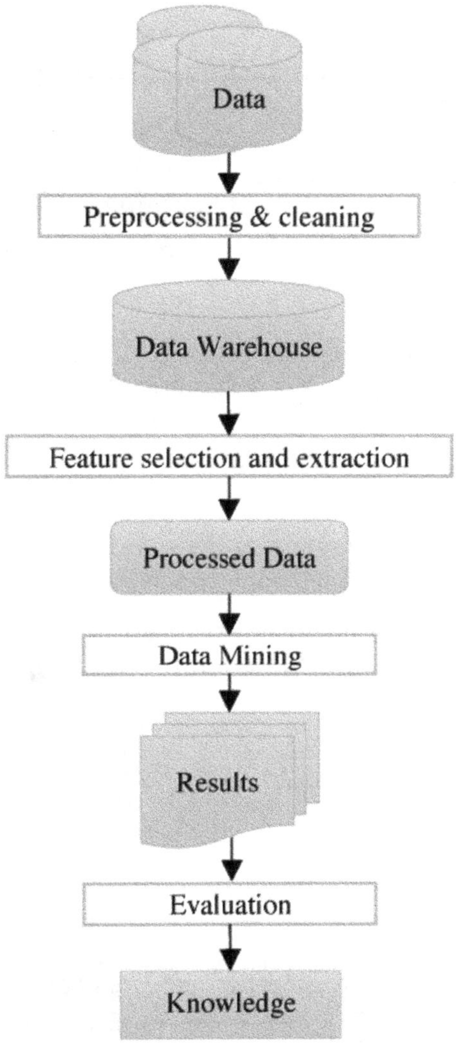


purposes of data reduction: to reduce data complexity and to reduce the amount of error in the task (Classification, clustering, association analysis) to be performed (Freitas & Kaestner, 2002).

Attribute selection is the basic precondition for the implementation of data mining algorithms. The overall objective of the attribute selection is to determine the subset of attributes that are most relevant for the intended purpose among the available attributes. The selection of the attribute allows the calculation time of the data mining algorithm to be used to be reduced, thereby increasing performance. For this reason, attribute selection is an important area of research in data mining. As with many data mining problems, choosing an attribute involves simultaneous optimization of multiple targets (Freitas & Kaestner, 2002). Figure 2 shows the general feature selection scheme used for classification.

Attribute selection can be defined as an optimization problem based on finding the best solution set among the solutions (Freitas & Kaestner, 2002). Although there are many attribute selection methods, these methods can be examined in three basic groups: filter, wrapper and embedded models. Figures 3-5 show schematics of these three models.

*Figure 2. A general framework of feature selection for classification*
*Tang, Alelyani, & Liu, 2015.*

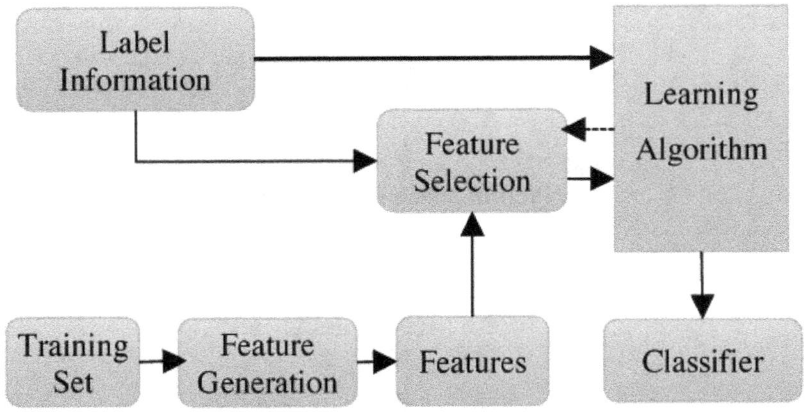


*Figure 3. Filter methods*
*Naqvi, 2012.*

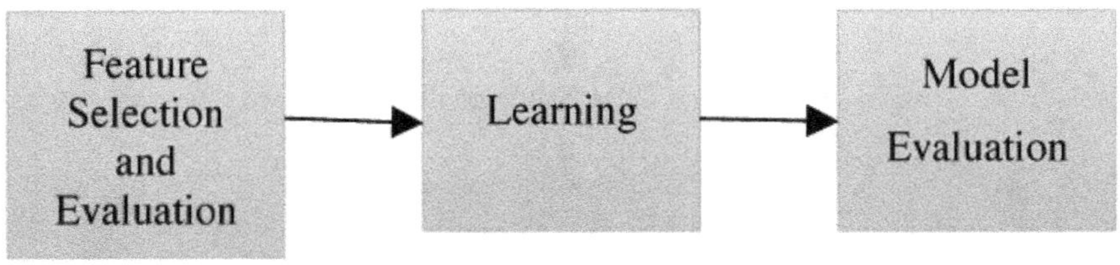


*Figure 4. Wrapper methods*
*Naqvi, 2012.*

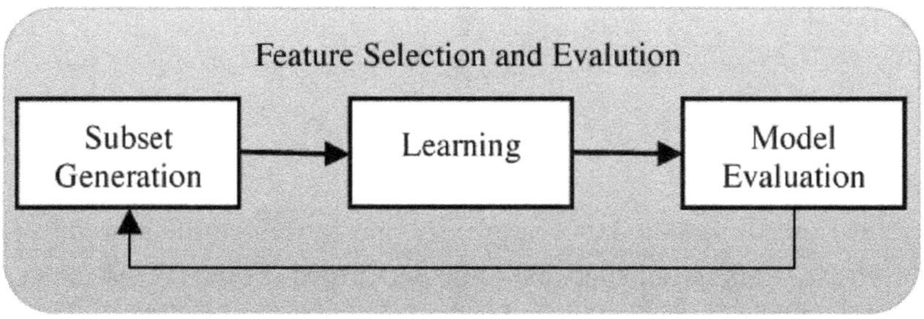


*Figure 5. Embedded methods*
*Naqvi, 2012.*

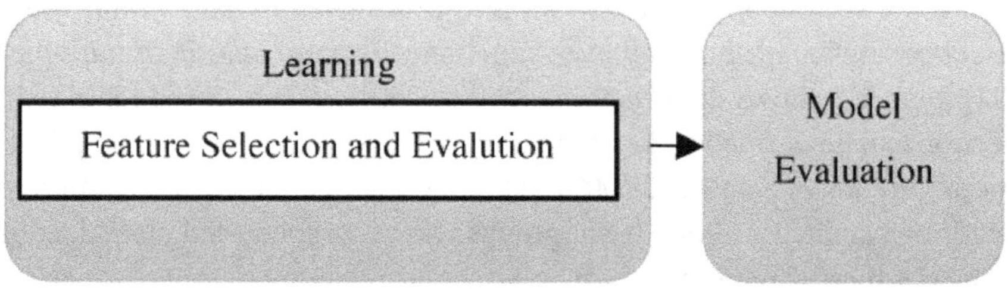

The greatest advantage of the feature selection algorithms using the Filter model is that it is fast and independent of the classification algorithm. Without any learning algorithm, attributes are selected using metrics such as information, dependency, and consistency. Because these methods are less complex and faster than wrapper models, they are more suited to large datasets. However, the disadvantage is that the classification rate is low (Gümüşçü, Aydilek, & Taşaltın, 2016).

Feature selection algorithms with a wrapper model are calculated by classifying success rate by modifying a selected set of selected features. Classification success rate is tried to be increased by elimination or selection according to the calculated classification success rate. This process is performed iteratively until the most ideal set of attributes is found. The greatest advantage is that the wrapper model feature selection algorithms have a high classification success rate (Gümüşçü, Aydilek, & Taşaltın, 2016).

Embedded feature selection algorithms are feature selection algorithms embedded in the classification algorithm. Attribute selection process and learning method parameter selection are performed simultaneously. In this method, the computational complexity that occurs in wrapper methods can be reduced (Gümüşçü, Aydilek, & Taşaltın, 2016). The feature selection algorithms in the literature are summarized in Table 1.

Filter modeled feature selection algorithms use statistical assumptions to select attributes by scoring and thresholding attributes by calculating the contribution of the attributes to the result. The most commonly used filter-model feature selection algorithms in the literature are ordered by correlation-based feature selection (CFS), information gain (IG), common knowledge, Chi-square, ReliefF, and F-score (Gümüşçü, Aydilek, & Taşaltın, 2016).

In this chapter, Correlation-based Feature Selection (CFS) has been used as an Attribute Evaluation tool in Filter-type feature selection algorithms.

## Correlation-Based Feature Selection

Correlation-based Feature Selection(CFS) is a simple filter algorithm that sorts feature subsets according to a heuristic evaluation function based on correlation. CFS is a method that identifies and removes unnecessary, not-needed and noise-related features very quickly.

Generally, it provides more than half of the properties and thus improves the success performance of learning methods. It evaluates the value of a subset of qualifications taking into account the individual estimation ability of each feature and the degree of redundancy between them. With low inter-correlation, subclasses of features that are highly correlated with the class are preferred. (Machine Learning Group at the University of Waikato, 2017). CFS generally specifies the best subset of features and generally matches search strategies such as forward selection, backward elimination, two-sided search, best first search, and genetic search. The CFS defined by equation (1).

*Table 1. Advantages and disadvantages of feature selection methods*

| Feature Selection Methods | Advantages | Disadvantages |
|---|---|---|
| Filter Model | Fast, Independent | Low success rate |
| Wrapper Model | High success rate, Easy to implement | Slow, higher processing load |
| Embedded Model | Low processing load, High success rate | Integrated work with Decision Trees |

Eskidere, 2012.

*Figure 6. The process of CFS feature selection*

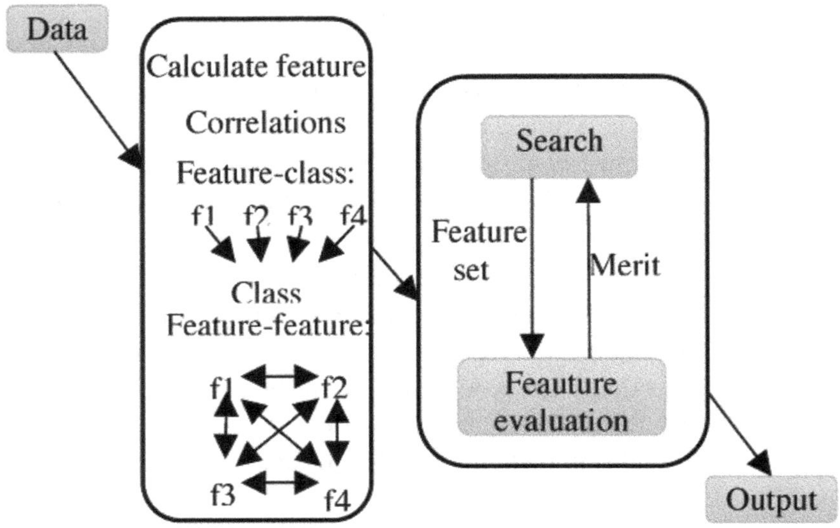


$$Merit_s = \frac{k\overline{r}_{cf}}{\sqrt{k + k(k-1)\overline{r}_{ff}}} \tag{1}$$

where $Merit_s$ is the heuristic merit of a feature subset S containing k features, $k_{cf}$ is the average feature-class correlation, and $r_{ff}$ is the average feature-feature inter-correlation. Equation 1 is, in fact, Pearson's correlation, where all variables have been standardized (Peng, Liu, Deng, Feng, & Liao, 2012).

## Heuristic Search Methods

Filter modeled attribute selection methods are evaluated for attribute subset. However, the examination of all possible subset of features in the search field is costly. For this reason, heuristic search algorithms are generally used in feature selection. In this chapter, CFS method, were examined along with 5 heuristic search algorithms namely, particle swarm optimization (PSO) search, genetic search, rank search, forward/backward greedy algorithm and best first search algorithm.

### PSO Search

Particle swarm optimization (PSO) proposed by Kennedy and Eberhart in 1995, is an optimization technique inspired by the behavior of birds. (Kennedy & Eberhart, 1995). The PSO is based on the knowledge sharing of the candidate particles, reaching the optimal solution. Each particle changes its position using information such as the current position and velocity, the distance between the current position and the best position of a particle, the distance between the current position of a particle and the best position of the particle. The mathematical models used to update the speed and location information of particles are given in equations (2) and (3).

$$v_{i,d}^{k+1} = w^k v_{i,d}^k + c_1 r_{1,d} \left( K_{i,d} best - K_{i,d}^k \right) + c_2 r_{2,d} \cdot \left( K_{gd} best - K_{i,d}^k \right) \tag{2}$$

$$K_{i,d}^{k+1} = K_{i,d}^k + v_{i,d}^{k+1} \tag{3}$$

where $V_{i,d}$ is $i^{th}$ particle velocity in the $d^{th}$ dimension, $w$ is inertia weight factor, $c_1$, $c_2$ are acceleration constants, $r_{1,d}$, $r_{2,d}$ are random numbers between 0 and 1 in the $d^{th}$ dimension, $K_{i,d}$, is $i^{th}$ particle position in the $d^{th}$ dimension, $K_{gd} best$, is best position of the $i^{th}$ particle, and $K_{gd} best$ is best position of the group in the $d^{th}$ dimension. $c_1$ and $c_2$ are constants that affect the velocity of a particle depending on experience in itself and swarm, respectively (Erdogmuş & Yalçın, 2015).

## Genetic Search

Genetic Algorithm (GA) is a heuristic search technique based on parameter coding that tries to find solutions using random search techniques. Genetic algorithms are algorithms that simulate the evolutionary process of competing individuals and creating competitive endeavors based on Darwin's "survival of the fittest" principle (Horat, 2014).

Some concepts defined in genetic science are modeled and used in genetic algorithm. These concepts are mentioned below:

- **Gene:** It is the smallest unit carrying meaningful genetic information. Genetic algorithm is defined as a bit, the solution refers to a feature.
- **Chromosome (Individual):** It is formed by the association of genes. Expresses one of the possible solutions of the problem. In Genetic algorithm Chromosome is called an array.
- **Population (Generation):** Population is a cluster of chromosomes. It expresses a certain part of the possible solutions of the problem. The number of individuals in the population is determined at the beginning of the algorithm and remains constant.
- **Selection:** Genetic algorithm is an iterative solution. The selection process defines how individuals will be selected when a population of a given number of solutions goes to the next iteration.
- **Crossing:** The formation of two new individuals by the displacement of a part of the chromosomes of two selected individuals in the population. The crossover results in new individuals that are not in the population.
- **Mutation:** One or more genes of an individual are changed into different individuals. Mutation is one of the operators that allows diversity to occur in the population.
- **Fitness Function:** The objective of the problem is function. It is calculated according to the bit values of the individuals. The relevance function value indicates the quality of the individual's solution. The individuals to be transferred to the next generation are determined according to their fitness function values.

Steps for a genetic algorithm defined for an optimization approach are shown below.

**Step 1:** Chromosome design determines how the possible solutions of the problem will be expressed. Generally, the initial population is randomized.
**Step 2:** The fitness function of the probing is determined.
**Step 3:** The size of the initial population, the probability of crossing, the probability of mutation, and the stop criterion, which are the parameters of the genetic algorithm, are determined.
**Step 4:** The fitness function value is calculated for each individual in the initial population.
**Step 5:** By reproductive or selection, individuals of the new population are created.
**Step 6:** New individuals are subjected to crossover and mutation operators.
**Step 7:** The fitness values of the individuals in the new population are calculated.
**Step 8:** If the stopping criterion of the algorithm is met, it stops; if not, returns to the 5th step for the next iteration.
**Step 9:** The solution with the best fitness value of the last population is the solution that the algorithm finds.

### Rank Search

M. Gall and G. Holmes proposed a sorting search method in 2003 to compare feature selection algorithms (Hall & Holmes, 2003). The Rank Search technique sorts all the genes and evaluates the subset in the ascending order from the sorted list (I.e., first property, two first group, etc.). The best feature set is reported (Ruiz, Riquelme, & Aguilar-Ruiz, 2005). The main disadvantage of the Rank Search algorithm is that it usually selects relatively large sub-sets (Bermejo, Ossa, Gamez, & Puerta, 2012).

### Forward/Backward Greedy Algorithm

Greedy search takes care of changes to the existing sub-cluster by adding or removing features. A greedy search for a particular set of 'parents' reviews all possible 'children' subsets through the addition or removal of features. The subset of the best-measuring child then takes the place of the upper-lower cochlea and the process is repeated. The process ends when the process cannot be further improved (Williams, Zander, & Armitage, 2006).

### Best-First Search

The Best First search (BFS) is similar to a greedy search in terms of creating a new subset based on adding or removing features from the current subset. However, when the current path is no longer improving, it has the ability to go back to the lower set of choices to explore different possibilities. To avoid tracing the search back through all possibilities in the feature area, a limit is set on the number of unimpeded improper subgroups (Williams, Zander, & Armitage, 2006).

## CLASSIFICATION

Classification in the data mining is the process of classifying the present data according to the specified properties and predicting the class of this data when a new data is added. Classification techniques generate models that include classes, groups, relationships, or correlations. Estimates can be made according to these models. The accuracy percentage of the predictions made shows the percentage of success of the model over the data. In a data mining application, the success of algorithms is crucial in making accurate estimates for new data (Pala, 2013).

In this study, 7 different techniques were applied to create data models.

## Naive Bayes Algorithm

Naive Bayes is an augmented classification algorithm used to label data. This algorithm, which is often preferred in classifying the data, is very easy to use. In general, the Naive Bayesian classifier is based on the calculation of the likelihood values of the effects of each criterion. Naive Bayes computes the conditional probabilities of a given class of words or data to predict the likelihood of a class of documents or a class of data. Bayes' theorem can be expressed by the equation (4).

$$P(A/B) = P((B/A) \cdot P(A)) / P(B) \tag{4}$$

where P(A), the independent probability of A phenomenon (premise probability), P(B), the independent probability of B, P(B|A), the probability of event B when the event A is known (conditional probability), P(A|B), The probability of event A is known when event B is known. Thus, the class of the new incoming sample can be estimated by finding the cases that maximize P (A | B) (Çalış, Gazdağı, & Yıldız, 2013).

## Multilayer Perceptron

Multilayer perceptron (MLP) is the most commonly used and best-learned multi-layer artificial neural network model. It is known that the multi-layered sensor is a very powerful function in classification estimation problems. The goal of this model is to minimize the error difference between the intended result (output) of the network and the resultant produced. This model is also referred to as backpropagation network because it uses error propagation algorithm or backpropagation learning algorithm in the training of multi-layer sensor. (Pala, 2013).

## K-Nearest Neighbors Algorithm

K-Nearest Neighbors Algorithm (k-NN) is a learning-based algorithm. k-NN assigns the class label using the model that it learned from the training data in the test data. Every newcomer is assigned a class label according to this model, whose class label is obvious. The basic structure and operation of the algorithm is for each new incoming instance, looking at a class label looking at the nearest k neighbors. The following path is followed for the data of any class tag of the k-NN algorithm. The k value is first determined in the classifier of the kNN. According to the determined k values, the distances of the new data are calculated according to all the data in the learning set and a class tag for the new sample is predicted by looking at the class tags where the data with the minimum distance exists (Çalış, Gazdağı, & Yıldız, 2013)

## C4.5 Decision Tree

Decision tree is a method of the data mining classification model. Many decision tree algorithms exist in the literature. It is a decision tree algorithm that uses the C4.5 entropy concept presented by J. Ross Quinlan in 1993 (Quinlan, 1993). C4.5 operates first according to the depth principle and continuously sequences the data at each node to achieve the best branching separation criterion. The pseudocode of the C4.5 algorithm is as follows:

```
C4.5 (D, Target, Atts)                    (Mitchell,1997)
returns: a decision tree that correctly classifies the given examples
variables
D: Training set of examples
Target: Attribute whose value is to be predicted by the tree
Atts: List of other attributes that may be tested by the learned decision tree
create a Root node for the tree
if D are all positive then Root ← +
else if D are all negative then Root ← -
else if Attrs = ∅ then Root ← most common value of target in D
else
        A← the best decision attribute from Atts
        root ← A
        for each possible value vi of A
        add a new tree branch with A=vi
        Dvi ← subset of D that have value vi for A
        if Dvi = ∅ add then leaf ← most common value of Target in D
        else add the subtree C4.5(Dvi, Target, Atts-{A})
```

The C4.5 algorithm makes an entropy account and divides the tree branches through the most suitable variable. Entropy can be described mathematically as:

$$\sum_{i=1}^{n} p_i = 1 \tag{5}$$

If <p1, p2, … pn> expresses probabilities, then the sum of all these probabilities must be 1 (one). In this case the entropy will be as follows:

$$H\left(p_1, p_2, \cdots p_n\right) = \Sigma\left(p_i \log\left(\frac{1}{p_i}\right)\right) \tag{6}$$

The C4.5 algorithm uses the entropy value for the purpose of gain and performs the branch separation on the variable that provides the highest gain (Silahtaroğlu, 2016).

## Random Forest

Random Forest Algorithm is an algorithm that catches a high success rate in classification using many decision trees. In the random forest algorithm, it is important to select branching criteria and an appropriate pruning method in a similar way to other decision trees. Gini coefficient method is used in determining branching criteria. In order to construct the tree structure in the algorithm, the number of samples to be used in each node and the number of trees to be created must be determined. During classification, the decision tree is made up of K trees determined by the user. When a new object is to be classified, it is processed by these K decision trees and the highest one among the ratios obtained from each tree is selected to classify (Karakoyun & Hacıbeyoğlu, 2014).

## Decision Table

It was proposed by Kohavi as an alternative to decision trees. The decision table algorithm first chooses the property with the Wrapper method for all properties. A set of selected features starting with an empty set is added at each step to provide the best performance. The properties selected to create the decision table are divided into property-spacing pairs. As the labels of the kings formed by combining these two, the output values of the training samples that match these rules (the class which is the most for classification problems, the average of the outputs for regression problems) is used (Amasyalı, 2011).

## Zero Rule

The Zero Rule algorithm is used to measure the default error in a data set. The Zero Rule outputs the same value for all of the test samples. This value is the average of the outputs of all training samples. Accordingly, the square root of the mean square error of the Zero Rule (RMSE) is calculated by the following equation.

$$RMSE_{Zerorule} = \sqrt{\frac{1}{N}\sum_{i=1}^{N}\left(y^m - y^i_{real}\right)^2} \quad (7)$$

$$y^m = \frac{1}{T}\sum_{j=1}^{T} y^i_{real} \quad (8)$$

where $y^i_{real}$ j. the output value of the training sample, T, the number of training samples, N shows the number of test samples (Amasyalı, 2011).

## MATERIALS AND METHODS

In order to apply the data mining, it is necessary to establish the appropriate probing models. Models used in data mining are examined under two main headings as Predictive and Descriptive. Predictive models, a model developed by the movement of the result known and established data resulting from utilizing this model is intended to estimate the value of the results for the unknown data sets. In descriptive models, it is possible to identify patterns in existing statements that can be used to guide decision making. Data mining models can be analyzed under three main headings:

1. Classification and Regression
2. Clustering
3. Association Rules

Classification and regression models are predictive while clustering, association rules are descriptive models (Ozekes, 2003).

This study mainly consists of two phases. First, the feature selection was applied on the data set which was determined as the material. After that, appropriate models were created by using classification methods and erythematous squamous disease differential diagnosis was determined from common dermatological diseases in the field of medicine.

## Problem Definition

Among the dermatological diseases, a large group of diseases with frequent erythema and sclerosis is common. Erythema is called deep redness as a result of congestion (blood collection) in the capillaries. The erythematic result is that when the finger is pressed with the finger, the melted part fades. It can be caused by various reasons, mainly infection and allergy (Mosby 2013; Anderson, 2009).

The cells that make up the stratum corneum (the outermost part of the epidermis) are filled with a protein called keratin and have lost their nuclei. As a result of differences in the formation of the upper skin cells (Parakeratosis), cells in the stratum corneum layer do not fill with keratin and do not lose their nuclei. As a result, the underlying lesions are called squamous. Squamous are separated from the skin

*Figure 7. The approach used in the study*

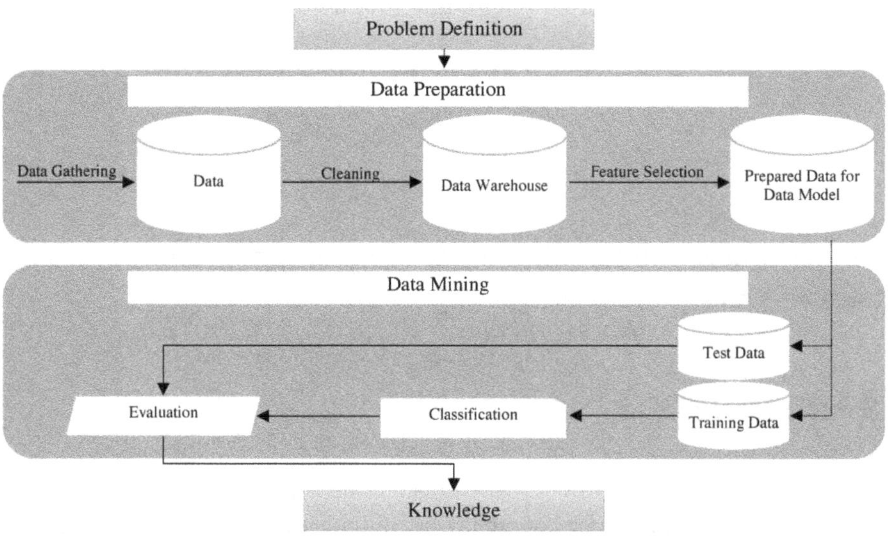

in the form of flake-bran. Lesions are strictly limited, erythematous and squamous papules or plaques. Unlike eczematous diseases that show scouring, there is not a lot of crushing, exorcism and irrigation. Distinction and distribution of lesions are of great importance in differential diagnosis (Kundakcı, 2011) .

## Psoriasis

Psoriasis is a chronic disease that can be widespread throughout the body as it can be seen in a limited area with a few plaques. The most common age is between 10 to 30. The likelihood of being seen equals men and women. It forms between 1% and 3% of dermatoses. It is a disease that has not yet been diagnosed. The disease is caused by environmental factors in people with genetic predisposition. Factors that trigger or exacerbate the disease include trauma, acute streptococcal infections, drugs and emotional stress. The differential diagnosis of psoriasis diseases should be made with diseases with seborrheic dermatitis, lichen planus, pityriasis rose, erythema-scarred lesions (Karaduman, 2000).

## Seborrheic Dermatitis

Red itchy rash with raised scales in hair; Psoriasis is a common chronic disease that resembles eczema or allergic reactions. It is thought that the disease is caused by fungi because 70% of disease cases have detected a kind of fungus. It is more common in newborns and adults aged 30-60 years. It is more common in men and in people with oily skin. It is seen in the body as well as in the hair. The involvement in the subarachnoid, eyelids, eyebrows and periphery, nasal edges, navel pouch, under the chest, back region is quite tight (Oakley, 1997).

The cause of the disease is not clearly known. Stress, genetic makeup, yeast that normally lives on the bottom, drugs, and cold, dry air are the causes of the disease. Bright red plaques form on the bottom of the disease. Itching and burning may occur. The peeled scales are white or yellowish and the deep image is moist and oily. Differential diagnosis is made with psoriasis, eczema, pityriasis rosea diseases (Dessinioti & Katsambas, 2013).

## Lichen Planus

It is a common, itchy, common inflammatory disease involving the skin, mucosa, nails and hair follicles. It is a little more common in middle-aged women and women. The cause is not clearly known. Infections, psychological, genetic, enzymes, immunologic factors, drugs can cause the onset or exacerbation of the disease. The main lesion is a flat, polygonal-shaped, bright papule over purple-red. The papules group together to form plaques 1-2 cm in diameter. White-laced Wickham lines on the surface of bright-faced papules are typical manifestations. Lesions can be found in any part of the body. It's a very itchy disease. Exercise may not develop due to severe itching. It is a chronic illness. When old lesions disappear, the outbreak of the new is observed. Differential diagnosis includes special placement areas, bright violet, typical papules, Wickham lines (Başkan & Tunalı, 2004).

## Pityriasis Rosea

It is a symmetrically distributed, mild inflammatory exanthema, especially seen in the body. It is more common in young adults. The cause is a disease that is not known but it is exacerbated by stress (Rapini,

Bolognia, & Jorizzo, 2007). Autumn and winter months are more common. The disease starts with a bright red plate 2-7 cm in diameter called "medallion" on the upper part of the body. Then it is in pale pink color. There is desquamation from the center, which is slightly sediment, towards the periphery. The plaques that come out from Madras are smaller in diameter. It has no local or systemic symptoms. Half of the patients may have itching. The differential diagnosis should be made with seborrheic dermatitis, second cycle syphilis, drug reactions, psoriasis and lichen planus (Chuh, 2003).

### Chronic Dermatitis

Chronic dermatitis is an inflammatory and allergic skin disease that often comes into contact with many substances in the skin, resulting in susceptibility or by the effects of various genetic and environmental factors. Also known as eczema (Nedorost, 2012). Eczema often occurs at 1 year of age. This can take up to puberty or disappear after 2-2.5 years. In some patients, it can take years to heal. It is seen on average 2% of children. There are many types of eczema. The most common cause is the allergic. The eczema seen in this way is called atopic eczema. It is common in patients with asthma and hay fever. Emotional, psychological problems also increase the risk of this disease (McAleer, Flohr, & Irvine, 2012).

As a symptom of dermatitis, there are itching, skin rash, redness in hands, feet and widespread. It is very visible on the elbows. It can be seen in oily skin and hair. Differential diagnosis is made with psoriasis. Physical examination and skin examination (McAleer, Flohr, & Irvine, 2012).

### Pityriasis Rubra Pilaris

Pityriasis Rubra Pilaris (PRP) is a genodermatosis characterized by dandruff on a red ground in the form of widespread plaques, keratoderma in the palms and soles, plugging and redness around the hair follicles (Rapini, Bolognia, & Jorizzo, 2007).

The cause of the disease is not clearly known. It is common in early childhood and in 50's. PRP often starts on the outer surfaces of arms and legs, especially in the knees and elbows, shoulders and hips. Sometimes the face and hair can start at the bottom too. The skin is fried and deeply dandruff has formed on it. The papules are observed in 1mm deep. As the PRP progresses, thickening of the hands and feet starts and these areas turn orange. At the same time cracks and fissures develop in the hands and feet. In some patients fever and shivering can be seen. PRP that starts in early childhood begins after trauma or a streptococcal infection. 75% of those starting between the ages of 3 and 5 are spontaneously disappearing. It is the remainder of the disease afterwards. Skin biopsy should be performed for differential diagnosis. Biopsy should be differentiated from other diseases (Ngan & Writer, 2003).

## Data Preparation

When the data set was constructed, 12 clinical features of the patients were examined first. Afterwards, the skin samples of the patients were taken and 22 histopathologic features were evaluated. Histopathological features were analyzed and determined by examining skin samples under a microscope. Thus, a dermatology dataset created and shared by Prof. Dr. Nilsel İlter from Gazi University and H.Altay Güvenir from Bilkent University, Turkey (Güvenir et al. 1998). The dataset consists of 366 data items with 34 attributes. 33 of these attributes are linear values and 1 is nominal value. Features in the dataset are shown in Table 2.

*Table 2. Feature information in dataset*

| Clinical Features | |
|---|---|
| Erythema | 1 |
| Scaling | 2 |
| Definite borders | 3 |
| İtching | 4 |
| Koebner phenomenon | 5 |
| Polygonal papules | 6 |
| Follicular papules | 7 |
| Oral mucosal involvement | 8 |
| Knee and elbow involvement | 9 |
| Scalp involvement | 10 |
| Family history,(0 veya 1) | 11 |
| Age(linear) | 34 |
| **Histopathological Features** | |
| Melanin incontinence | 12 |
| Eosinophils in the infiltrate | 13 |
| PNL infiltrate | 14 |
| Fibrosis of the papillary dermis | 15 |
| Exocytosis | 16 |
| Acanthosis | 17 |
| Hyperkeratosis | 18 |
| Parakeratosis | 19 |
| Clubbing of the rete ridges | 20 |
| Elongation of the rete ridges | 21 |
| Thinning of the suprapapillary epidermis | 22 |
| Spongiform pustule | 23 |
| Munro micro abcess | 24 |
| Focal hypergranulosis | 25 |
| Disappearance of the granular layer | 26 |
| Vacuolisation and damage of the basal layer | 27 |
| Spongisis | 28 |
| Saw-tooth appearance of retes | 29 |
| Follicular horn plug | 30 |
| Perifollicular parakeratosis | 31 |
| İnflammatory monoluclear infiltrate | 32 |
| Band-like infiltrate | 33 |

In the family history attribute shown in Table 2, if the disease was observed in the family history, it was marked as "1", otherwise it was marked as "0". The age attribute indicates the age of the patient. Other features are evaluated according to a degree in the range of 0-3. 0, indicates that no sign of the property exists. 3, the largest possible amount. 1 and 2 represent relative intermediate values.

In the dataset, the 8 missing data were removed and data reduction is done by the owner of the data set. The distribution of disease classes in the data warehouse obtained is shown in Table 3.

## Feature Selection

After the data gathering and cleaning steps, feature selection and extraction are performed to improve both performance and to avoid noise data by being able to identify the subset of data sets with the most relevant attributes.

In this step, the data set is encoded in the .arff file format. The description of the file structure with the .arff extension and the dataset example section are shown in Figure 8.

In the study, Correlation-based Feature Selection (CFS) has been used as an Attribute Evaluation tool in Filter-type feature selection algorithms. The generated data warehouse was examined by CFS method with 5 different search methods. Experimental results are shown in Table 4. As seen in Table 4, the Best First Method and the Greedy Stepwise method suggest the same attributes, while the other three algorithms have different results.

*Table 3. Class distribution*

| Code | Class | Number of Samples |
|---|---|---|
| 1 | Psoriasis | 111 |
| 2 | Seboreic Dermatitis | 60 |
| 3 | Lichen Planus | 71 |
| 4 | Pityriasis Rosea | 48 |
| 5 | Cronic Dermatitis | 48 |
| 6 | Pityriasis Rubra Pilaris | 20 |

*Table 4. Feature selection results*

| Attribute Evaluator | Search Method | Selection Mode | Selected Attributes |
|---|---|---|---|
| CfsSubsetEval | Best First | Full Training | 2,3,4,5,7,9,13,14,15,16,20,21,22,25,26,28,29,31,33 |
| | Greedy Stepwise | Full Training | 2,3,4,5,7,9,13,14,15,16,20,21,22,25,26,28,29,31,33 |
| | GeneticSearch | Full Training | 2,3,4,5,7,9,13,14,15,16,20,21,22,26,27,28,29,31,33 |
| | RankSearch | Full Training | 2,3,4,5,6,7,8,9,10,11,12,13,14,15,16, 19,20,21,22,23,24,25,26,27,28,29,30,31,33,34 |
| | PSO Search | Full Training | 2,3,4,5,6,7,9,13,14,15,16,20,21,22,, 26,27,28,29,30 |

In data mining, different algorithms may have different parameters. These parameters may differ depending on the algorithm selected or the data mining software used. Therefore, the choice of parameters will affect the performance of the model to be formed. For this reason, it has been decided to create 4 different models in line with the recommendation of CFS based search methods. Thus, each approach has been evaluated and compared.

## Classification

When creating the model, determination of the learning and test clusters that are used also has an effect on the performance of the model. Different methods can be used to separate the current data set as a learning set and a test set.

In Cross validation method, all the data number k is divided into subsets of approximately equal size. This classification algorithm is trained and tested k times. As a test data, one of the folds is taken every time, and the remaining folds are taken to form the training data. As a result, k different test results are found for each training-test configuration (Coşkun & Baykal, 2011).

In the Percentage Split method, the data is used as a random % k training data and the remaining data are used for the test.

The basic concepts used to evaluate model performance are error rate, precision, Recall and F-criterion. The success of the model is related to the number of samples that are correctly categorized and the number of samples that are categorized incorrectly.

The performance information of the results achieved in the test result can be expressed by the complexity matrix. In the confusion matrix, the rows represent the actual numbers of the samples in the test set, and the columns represent the estimate of the model.

The most popular and simple method used to measure the model success is the accuracy of the model. The number of correctly classified samples (TP + TN) is the ratio of the total sample counts (TP + TN + FP + FN). Error rate is the one's complement of that value. In other words, the number of misclassified samples (FP + FN) is the ratio of the total number of samples (TP + TN + FP + FN).

$$Accuracy = \frac{TP + TN}{TP + FP + FN + TN} \tag{9}$$

$$Error\ Rate = \frac{FP + FN}{TP + FP + FN + TN} \tag{10}$$

$$Precision = \frac{TP}{TP + FP} \tag{11}$$

$$Recall = \frac{TP}{TP + FN} \tag{12}$$

*Figure 8. Sample of .arff file*

```
@relation dermatoloji

@attribute erythema {0,1,2,3}
@attribute scaling {0,1,2,3}
@attribute definiteBorders {0,1,2,3}
@attribute itching {0,1,2,3}
@attribute koebnerPhenomenon {0,1,2,3}
@attribute polygonalPapules {0,1,2,3}
@attribute follicularPapules {0,1,2,3}
@attribute oralMucosalInvolment {0,1,2,3}
@attribute kneeandElbowInvolment {0,1,2,3}
@attribute scalpInvolment {0,1,2,3}
@attribute familyHistory {0,1}
@attribute melaninIncontinence {0,1,2,3}
@attribute eosinophilsintheinfiltrate {0,1,2,3}
@attribute PNLinfiltrate {0,1,2,3}
@attribute fibrosisofthePapillaryDermis {0,1,2,3}
@attribute exocytosis {0,1,2,3}
@attribute acanthosis {0,1,2,3}
@attribute hyperkeratosis {0,1,2,3}
@attribute parakeratosis {0,1,2,3}
@attribute clubbingoftheReteRidges {0,1,2,3}
@attribute elongationoftheReteRidges {0,1,2,3}
@attribute thinkingoftheSuprapapillaryEpidermis {0,1,2,3}
@attribute spongiformPustule {0,1,2,3}
@attribute munroMicroabcess {0,1,2,3}
@attribute focalHypergranulosis {0,1,2,3}
@attribute disappearanceoftheGranularLayer {0,1,2,3}
@attribute vacuolisationandDamageofBasalLayer {0,1,2,3}
@attribute spongiosis {0,1,2,3}
@attribute saw_ToothAppearanceofRetes {0,1,2,3}
@attribute follicularHornPlug {0,1,2,3}
@attribute perifollicularParakeratosis {0,1,2,3}
@attribute inflammatoryMonoluclearInflitrate {0,1,2,3}
@attribute band_LikeInfiltrate {0,1,2,3}
@attribute age numeric
@attribute class {1,2,3,4,5,6}

@data
2,2,0,3,0,0,0,0,1,0,0,0,0,0,0,3,2,0,0,0,0,0,0,0,0,0,0,3,0,0,0,1,
0,55,2
3,3,3,2,1,0,0,0,1,1,1,0,0,1,0,1,2,0,2,2,2,2,2,1,0,0,0,0,0,0,0,1,
0,8,1
```

*Table 5. Confusion matrix*

| | | Predicted | |
|---|---|---|---|
| | | **Positive** | **Negative** |
| **True** | **Positive** | True Positives (TP) | False Negatives (FN) |
| | **Negative** | False Positives (FP) | True Negatives (TN) |

$$F\text{-}Measure = \frac{2 \times Recall \times Precision}{Recall + Precision} \tag{13}$$

Precision and Recall criteria alone are not enough to make a meaningful comparison. Evaluating both criteria together gives more accurate results. For this reason the F-criterion, harmonic mean of precision and recall, criterion is defined (Coşkun & Baykal, 2011).

In this study, the obtained models were evaluated on 7 different classification algorithms by applying cross validation and percentage split at different values. The accuracy ratios arising in the classification algorithms are shown in Table 6.

As shown in Table 6, the Naive Bayes algorithm, as a result of the applied feature selection methods, performed 100% correct classification in 3 models. The Zero Rule algorithm is not considered sufficient for any model with about 30% correct classification. Best First / Greedy Stepwise search method based model showed the same results in large scale when compared to the Genetic Search based model. Despite the fact that the PSO Search algorithm generally makes a high-level correct classification, 100% correct classification is not observed.

## Evaluation

When the results of the feature selection process are evaluated in the direction of the experimental results, it is determined from the Search Methods performed together with the CfsSubsetEval method:

- Best First Search and Greedy Stepwise algorithms have chosen the same attributes. Of the 34 qualifications listed in the dataset, 19 were deemed sufficient for classification.
- Genetic search showed different behavior than Best First and Greedy Stepwise in only 1 attribute.
- It has been observed that the Rank Search algorithm suffers from very few attributes.
- The PSO algorithm has a completely different distribution from the others.
- When we examine the attributes that Best First and Greedy Stepwise algorithms recommend to remove from the dataset:
- Attributes 1, 17, and 32 cannot be used as differential diagnosis because they have a very large proportion of the disease.
- 6, 8 and 12 were eliminated because it attributes largely to create a single disease diagnosis does not make sense for the other 5 of them.
- The presence of attributes 18 and 19 does not change the disease state.
- 10, 11, 23, 24, 27, and 30, the number of attributes, and 0 (that is, if not) can have 6. If not 0, it may be indicative for more than one disease.

Given these conclusions, Best First and Greedy Stepwise algorithms show correct results.

There are different results observed in the classification process. The accuracy of the classification process is determined as the basic performance criterion. The error rate, precision, recall and F-criterion results are also considered. Table 7 shows the error rate, precision, recall and F-criterion results of the Naive Bayes algorithm. Table 8 shows the confusion matrix obtained when classification is applied by the model Naive Bayes algorithm which is formed by 66% training data.

*Table 6. Classification result with accuracy rate*

| Classification Algorithms | Training Data Set | Feature Selection Search Methods | | | |
|---|---|---|---|---|---|
| | | Best First/Greedy Stepwise | Genetic Search | Rank Search | PSO Search |
| Naive Bayes | Percentage Split: 66% | **100** | **100** | **100** | 99,1803 |
| | Percentage Split: 60% | **100** | **100** | 98,6014 | 99 |
| | Percentage Split: 50% | **100** | **100** | 98,8827 | 99,4413 |
| | 2 Fold Cross Validation | 98,324 | 98,324 | 97,2067 | 98,0447 |
| | 3 Fold Cross Validation | 98,6034 | 98,6034 | 98,6034 | 98,324 |
| Zero Rule | Percentage Split: 66% | 35,2459 | 35,2459 | 35,2459 | 33,5664 |
| | Percentage Split: 60% | 33,5664 | 33,5664 | 33,5664 | 33,5664 |
| | Percentage Split: 50% | 32,9609 | 32,9609 | 32,9609 | 32,9609 |
| | 2 Fold Cross Validation | 31,0056 | 31,0056 | 31,0056 | 31 |
| | 3 Fold Cross Validation | 31,0056 | 31,0056 | 31,0056 | 31 |
| C4.5 Decision Tree J48 | Percentage Split: 66% | 94,2623 | 94,2623 | 94,2623 | 95,9016 |
| | Percentage Split: 60% | 95,1049 | 95,1049 | 94,4056 | 96,5035 |
| | Percentage Split: 50% | 96,0894 | 96,0894 | 95,5307 | 97,2067 |
| | 2 Fold Cross Validation | 92,1788 | 91,8994 | 93,2961 | 93,2961 |
| | 3 Fold Cross Validation | 94,1341 | 94,1341 | 93,2961 | 94,9712 |
| Multilayer Perceptron | Percentage Split: 66% | 98,3607 | 98,3607 | 98,3607 | 99,1803 |
| | Percentage Split: 60% | 98,6014 | 98,6014 | 98,6014 | 99,3007 |
| | Percentage Split: 50% | 98,8827 | 98,8827 | 98,8827 | 98,8827 |
| | 2 Fold Cross Validation | 97,486 | 97,486 | 97,7654 | 96,9274 |
| | 3 Fold Cross Validation | 97,2067 | 97,2067 | 96,648 | 97,486 |
| K-Nearest Neighbors with value 1 | Percentage Split: 66% | 99,1803 | 99,1803 | 94,2623 | 99,1803 |
| | Percentage Split: 60% | 97,9021 | 97,9021 | 94,4056 | 98,6014 |
| | Percentage Split: 50% | 97,7654 | 97,7654 | 94,9721 | 98,324 |
| | 2 Fold Cross Validation | 96,0894 | 96,0894 | 94,6927 | 95,5307 |
| | 3 Fold Cross Validation | 96,0894 | 96,0894 | 94,9721 | 96,3687 |
| Decision Table | Percentage Split: 66% | 90,9836 | 90,9836 | 90,9836 | 90,9836 |
| | Percentage Split: 60% | 81,1189 | 79,7203 | 79,7203 | 79,7203 |
| | Percentage Split: 50% | 91,6201 | 91,6201 | 91,6201 | 77,6536 |
| | 2 Fold Cross Validation | 87,4302 | 87,7095 | 87,4302 | 85,1955 |
| | 3 Fold Cross Validation | 86,3128 | 85,7542 | 85,7542 | 83,7989 |
| Random Forest | Percentage Split: 66% | 96,7213 | 96,7213 | 97,541 | 98,3607 |
| | Percentage Split: 60% | 97,9021 | 97,9021 | 97,9021 | 97,9021 |
| | Percentage Split: 50% | 98,8827 | 98,324 | 98,324 | 98,324 |
| | 2 Fold Cross Validation | 96,0894 | 96,9274 | 97,2067 | 96,3687 |
| | 3 Fold Cross Validation | 95,8101 | 96,3687 | 96,3687 | 96,648 |

*Table 7. Naive Bayes detailed accuracy by class*

| | Tp Rate | Fp Rate | Precision | Recall | F-Measure | Class |
|---|---|---|---|---|---|---|
| | 1,000 | 0,000 | 1,000 | 1,000 | 1,000 | 1 |
| | 1,000 | 0,000 | 1,000 | 1,000 | 1,000 | 2 |
| | 1,000 | 0,000 | 1,000 | 1,000 | 1,000 | 3 |
| | 1,000 | 0,000 | 1,000 | 1,000 | 1,000 | 4 |
| | 1,000 | 0,000 | 1,000 | 1,000 | 1,000 | 5 |
| | 1,000 | 0,000 | 1,000 | 1,000 | 1,000 | 6 |
| Weighted Avg. | 1,000 | 0,000 | 1,000 | 1,000 | 1,000 | |

*Table 8. Confusion matrix*

| a | b | c | d | e | f | |
|---|---|---|---|---|---|---|
| 43 | 0 | 0 | 0 | 0 | 0 | a=1 |
| 0 | 20 | 0 | 0 | 0 | 0 | b=2 |
| 0 | 0 | 25 | 0 | 0 | 0 | c=3 |
| 0 | 0 | 0 | 14 | 0 | 0 | d=4 |
| 0 | 0 | 0 | 0 | 16 | 0 | e=5 |
| 0 | 0 | 0 | 0 | 0 | 4 | f=6 |

As seen in Table 8, with Naive Bayes, when 66% of the 358 data, i.e. 236 data, are used for training, the remaining 122 data are fully classified.

The Naive Bayes algorithm has 100% correct classification in 3 different models created as a result of feature selection. These results have also been tested with different test sets.

As a result of the obtained results, the PSO algorithm as Feature Search and Search method was insufficient in determining the differential diagnosis of erythema-scarring diseases. The Rank Search method has not shown the expected increase in system performance due to very few attributes. For this reason, the model based on the Rank search method is not suggested for solving this problem.

As a result of this study, the recommended approach for the classification of malignant-squamous diseases is shown in Table 9.

*Table 9. Best solution*

| Feature Selection Model | Attribute Evaluator | Search Method | Training Data Set | Classification Method | Accuracy Rate |
|---|---|---|---|---|---|
| Filter Method | CFSSubsetEval | Best First && Greedy Stepwise&& Genetic Search | %66 | Naive Bayes | %100 |
| | | | %60 | | |
| | | | %50 | | |

Based on the filter model in the proposed method, the selection of correlation-based attributes can be integrated with any of Best First or Greedy Stepwise or Genetic Search search algorithms. The resulting dataset is modeled to generate a training set of 66% or 60% or 50%. The generated model is classified by the Naive Bayes algorithm. Thus, 100% success is achieved.

## CONCLUSION

In this study, the differential diagnosis of erythematous diseases was determined using data mining and machine learning algorithms. In the study, dermatology dataset which was shared in UCI Machine Learning Repository was used. The dataset consists of 366 data items with 34 attributes. Initially, feature selection was made and then, classification was performed by using various algorithms. The number of attributes has been reduced from 34 to 19 as a result of the integration of the correlation based filter methods and various heuristic search methods. The evaluation results show that Naive Bayes has 100% success rate in differential diagnosis of Erythematous Squamous Diseases by classification of Psoriasis, Seborrheic Dermatitis, Lichen Planus, Pityriasis Rosea, Chronic Dermatitis and Pityriasis Rubra Pilaris with feature extraction algorithms.

## REFERENCES

Amasyalı, S. (2011). *Operasyon İşlemlerinin Hızlandırılmasında Veri Madenciliğinin Kullanılması*. İstanbul: Yıldız Teknik Üniversitesi Fen Bilimleri Enstitüsü.

Anderson, D. M. (2009). *Mosby's Dictionary Of Medicine, Nursing & Health Professions (4th ed.)*. Elsevier.

Aybek Yılmaz, A. (2016). *Aile Hekimlerinin Birinci Basamakta Sik Görülen Dermatolojik Lezyonlar Hakkindaki Bilgi Düzeylerinin İncelenmesi*. İzmir: İzmir Katip Çelebi Üniversitesi Atatürk Eğitim Ve Araştirma Hastanesi.

Başkan, E. B., & Tunalı, Ş. (2004). Sık Rastlanan Pediatrik Deri Hastalıkları - Derleme. *The Journal of Current Pediatrics*, 2-4.

Bermejo, P., Ossa, L. D., Gamez, J. A., & Puerta, J. M. (2012). Fast Wrapper Feature Subset Selection In High-Dimensional Datasets By Means Of Filter Re-Ranking. *Knowledge-Based Systems*, *25*(1), 35–44. doi:10.1016/j.knosys.2011.01.015

Çalış, K., Gazdağı, O., & Yıldız, O. (2013). Reklam İçerikli Epostaların Metin Madenciliği Yöntemleri Ile Otomatik Tespiti. *Bilişim Teknolojileri Dergisi*, 1-7.

Chuh, A. (2003). *Clinical Dermatology: A Clinical Guide To Diagnosis And Therapy* (4th ed.). Journal Of The European Academy Of Dermatology And Venereology.

Coşkun, C., & Baykal, A. (2011). Veri Madenciliğinde Sınıflandırma Algoritmalarının Bir Örnek Üzerinde Karşılaştırılması. *Akademik Bilişim*, *2011*, 1–8.

Dessinioti, C., & Katsambas, A. (2013). Seborrheic Dermatitis: Etiology, Risk Factors, And Treatments: Facts And Controversies. *Clinics in Dermatology*, *31*(4), 343–351. doi:10.1016/j.clindermatol.2013.01.001 PMID:23806151

Erdogmuş, P., & Yalçın, E. (2015). The Solutions Of Unconstrained Optimization Benchmark Problems With Particle Swarm Optimization. *Journal of Advanced Technology Sciences*, 14-22.

Eskidere, Ö. (2012). A Comparison Of Feature Selection Methods For Diagnosis Of Parkinson's Disease From Vocal Measurements. *Journal of Engineering and Natural Sciences*, 402-414.

Freitas, A. A., & Kaestner, C. A. (2002). Attribute Selection With A Multiobjective Genetic Algorithm. *16th Brazilian Symposium On Artificial Intelligence* (pp. 280-290). Porto De Galinhas/Recife: Springer-Verlag.

Gümüşçü, A., Aydilek, İ. B., & Taşaltın, R. (2016). Mikro-Dizilim Veri Sınıflandırmasında Öznitelik Seçme Algoritmalarının Karşılaştırılması. *Harran Üniversitesi Mühendislik Dergisi*, *1*(1), 1–7.

Güvenir, H. A., Demiröz, G., & İlter, N. (1998, July). Learning Differential Diagnosis Of Erythemato-Squamous Diseases Using Voting Feature Intervals. *Artificial Intelligence in Medicine*, *13*(3), 147–165. doi:10.1016/S0933-3657(98)00028-1 PMID:9698151

Hall, M., & Holmes, G. (2003). Benchmarking Attribute Selection Techniques For Discrete Class Data Mining. *IEEE Transactions on Knowledge and Data Engineering*, *15*(6), 1437–1447. doi:10.1109/TKDE.2003.1245283

Horat, B. (2014). *Genetik Algoritma İle Ankara İlinde Mobese (Kent Güvenlik Yönetim Sistemi) Yerleştirme Optimizasyonu*. Ankara: Gazi Üniversitesi Bilişim Enstitüsü.

Ilter, N., & Guvenir, H. A. (1998). *Dermatology Data Set.* Şubat 1, 2017 Tarihinde The Uci Machine Learning Repository: Https://Archive.Ics.Uci.Edu/Ml/Datasets/Dermatology

Kaliyeva, S. (2013). *Bilimsel Makalelerin Metin İşleme Yöntemleri İle Siniflandirilmasi*. Ankara, Türkiye: Gazi Üniversitesi Fen Bilimler Enstitüsü.

Karaduman, A. (2000). Psoriazis. *Sürekli Tıp Eğitimi Dergisi*, 12-19.

Karakoyun, M., & Hacıbeyoğlu, M. (2014). Biyomedikal Veri Kümeleri İle Makine Öğrenmesi Siniflandirma Algoritmalarinin İstatistiksel Olarak Karşilaştirilmasi. *Deü Mühendislik Fakültesi Mühendislik Bilimleri Dergisi*, 30-41.

Kaya, M., & Özel, S. A. (2014). *Açık Kaynak Kodlu Veri Madenciliği Yazılımlarının Karşılaştırılması.* Mersin: Akademik Bilişim.

Kennedy, J., & Eberhart, R. (1995). Particle Swarm Optimization. *Neural Networks, Proceedings. Ieee International Conference On*, 1942-1948.

Kundakcı, N. (2011). Eritemli Skuamlı Hastalıklar. Ankara: Academic Press.

Machine Learning Group At The University Of Waikato. (2017). *Weka.Attributeselection Class Cfssubseteval*. Retrieved from Http://Weka.Sourceforge.Net/Doc.Stable/: Http://Weka.Sourceforge.Net/Doc.Stable/Weka/Attributeselection/Cfssubseteval.Html

Mcaleer, M., Flohr, C., & Irvine, A. (2012). Management Of Difficult And Severe Eczema In Childhood. BMJ (Clinical Research Ed.), 35-41. doi:10.1136/bmj.e4770

Mosby. (2013). *Mosby's Medical Dictionary* (9th ed.). St. Louis, MO: Elsevier.

Naqvi, G. (2012). *A Hybrid Filter-Wrapper Approach For Feature Selection*. Örebro University.

Nedorost, S. T. (2012). Generalized Dermatitis In Clinical Practice. Springer Science & Business Media.

Ngan, V., & Writer, S. (2003). *Pityriasis Rubra Pilaris*. Dermnet New Zealand.

Oakley, A. (1997). *Seborrhoeic Dermatitis*. Dermnet New Zealand.

Ozekes, S. (2003). Veri Madenciliği Modelleri Ve Uygulama Alanlari. *İstanbul Ticaret Üniversitesi Dergisi, 2*(3), 65-82.

Pala, T. (2013). *Tibbi Karar Destek Sisteminin Veri Madenciliği Yöntemleriyle Gerçekleştirilmesi*. İstanbul: Marmara Üniversitesi Fen Bilimleri Enstitüsü.

Peng, X., Liu, P., Deng, Y., Feng, B., & Liao, B. (2012). A Novel Feature Selection Method Based On Cfs In Cancer Recognition. *2012 IEEE 6th International Conference on Systems Biology (Isb)* (pp. 226-231). Xi'an, China: IEEE.

Quinlan, J. R. (1993). C4.5: Programs For Machine Learning. *Machine Learning*, 235–240.

Rapini, R. P., Bolognia, J. L., & Jorizzo, J. L. (2007). *Dermatology: 2-Volume Set*. St. Louis, MO: Mosby.

Ruiz, R., Riquelme, J. C., & Aguilar-Ruiz, J. S. (2005). Incremental Wrapper-Based Gene Selection From Microarray Data For Cancer Classification. *Pattern Recognition*, 2383–2392.

Silahtaroğlu, G. (2016). *Veri Madenciliği Kavram Ve Algoritmaları*. İstanbul: Papatya Bilim.

Tang, J., Alelyani, S., & Liu, H. (2015). *Feature Selection For Classification: A Review. In Data Classification Algorithms And Applications* (pp. 37–64). Chapman And Hall/CRC.

Williams, N., Zander, S., & Armitage, G. (2006). A Preliminary Performance Comparison Of Five Machine. *ACM Sigcomm Computer Communication Review Learning Algorithms For Practical Ip Traffic Flow Classification*, 7-15.

## ADDITIONAL READING

Abdia, M. J., & Giveki, D. (2013). Automatic detection of erythemato-squamous diseases using PSO–SVM based on association rules. *Engineering Applications of Artificial Intelligence, 26*(1), 603–608. doi:10.1016/j.engappai.2012.01.017

Aggarwal, C. C. (2015). *Data Classification: Algorithms and Applications (Chapman & Hall/CRC Data Mining and Knowledge Discovery Series)* (1st ed.). U.S.A: Chapman & Hall. doi:10.1007/978-3-319-14142-8_10

Aggrwal, C. C. (2015). *Data Mining: The Textbook*. USA: Springer. doi:10.1007/978-3-319-14142-8

Ansari, N., Hou, E., (1997). Heuristic Search Methods. *Computational Intelligence for Optimization,* 11-25.

Antonio Giusti, G. R. (2013). *Classification and Data Mining (Studies in Classification, Data Analysis, and Knowledge Organization) 2013th Edition*. Berlin: Springer.

Athitsos, V., & Sclaroff, S. (2005). Boosting nearest neighbor classifiers for multiclass recognition. *Computer Vision and Pattern Recognition - Workshops, 2005. CVPR Workshops. IEEE Computer Society Conference on.* San Diego, CA, USA.

Azar, A. T., El-Said, S. A., Balas, V. E., & Olariu, T. (2013). Linguistic Hedges Fuzzy Feature Selection for Differential Diagnosis of Erythemato-Squamous Diseases. *Soft Computing Applications, 195*, 487–500. doi:10.1007/978-3-642-33941-7_43

Badrinath, N., Gopinath, G., Ravichandran, K. S., & Soundhar, R. G. (2016). Estimation of automatic detection of erythemato-squamous diseases through AdaBoost and its hybrid classifiers. *Artificial Intelligence Review, 45*(4), 471–488. doi:10.1007/s10462-015-9436-8

Basu, A., Roy, S. S., & Abraham, A. (2015). A Novel Diagnostic Approach Based on Support Vector Machine with Linear Kernel for Classifying the Erythemato-Squamous Disease. *Computing Communication Control and Automation (ICCUBEA), 2015 International Conference on.* Pune, India.

Dash, M., & Liu, H. (1997). Feature selection for classification. *Intelligent Data Analysis, 1*(1-4), 131–156. doi:10.1016/S1088-467X(97)00008-5

Fidan, U., Ozkan, N., & Calikusu, I. (2016). Clustering and classification of dermatologic data with Self Organization Map (SOM) method. *Medical Technologies National Congress (TIPTEKNO)*, 2016. Antalya, Turkey. doi:10.1109/TIPTEKNO.2016.7863075

Frank, E., Hall, M., Trigg, L., Holmes, G., & Witten, I. H. (2004). Data mining in bioinformatics using Weka. *Bioinformatics (Oxford, England), 20*(15), 2479–2481. doi:10.1093/bioinformatics/bth261 PMID:15073010

Han, J., Kamber, M., & Pei, J. (2012). Data Mining: Concepts and Techniques, Third Edition (The Morgan Kaufmann Series in Data Management Systems) 3rd Edition. United State of America: Morgan Kaufmann.

Holmes, G., Donkin, A., & Witten, I. (1994). WEKA: a machine learning workbench. *Intelligent Information Systems,1994. Proceedings of the 1994 Second Australian and New Zealand Conference on* (s. 357-361). Brisbane, Qld., Australia: IEEE.

Hornik, K., Buchta, C., & Zeileis, A. (2009). Open-source machine learning: R meets Weka. *Computational Statistics, 24*(2), 225–232. doi:10.1007/s00180-008-0119-7

Janert, P. K. (2011). *Data Analysis with Open Source Tools: A Hands-On Guide for Programmers and Data Scientists* (1st ed.). USA: O'Reilly Media Inc.

Kelleher, J. D., Namee, B. M., & D'Arcy, A. (2015). *Fundamentals of Machine Learning for Predictive Data Analytics: Algorithms, Worked Examples, and Case Studies (MIT Press)* (1st ed.). Massachusetts Institue of Technology.

Lahijanian, B., Farahani, F. V., & Zarandi, M. F. (2016). A new multiple classifier system for diagnosis of erythemato-squamous diseases based on rough set feature selection. *Fuzzy Systems (FUZZ-IEEE), 2016 IEEE International Conference on.* Vancouver, BC, Canada.

Liu, T., Hu, L., Ma, C., Wang, Z.-Y., & Chen, H.-L. (2015). A fast approach for detection of erythemato-squamous diseases based on extreme learning machine with maximum relevance minimum redundancy feature selection. *International Journal of Systems Science*, *46*(5), 919–931. doi:10.1080/00207721.2013.801096

Menai, M. E., & Altayash, N. (2014). Differential Diagnosis of Erythemato-Squamous Diseases Using Ensemble of Decision Trees. *International Conference on Industrial, Engineering and Other Applications of Applied Intelligent Systems*, 369-377.

Mola, F., Conversano, C., & Vichi, M. (2017). Classification, (Big) Data Analysis and Statistical Learning (Studies in Classification, Data Analysis, and Knowledge Organization) 1st ed. 2017 Edition. Switzerland: Springer.

Olatunji, S. O., & Arif, H. (2013). Identification of Erythemato-squamous Skin Diseases Using Extreme Learning Machine and Artificial Neural Network. *ICTACT Journal On Soft Computing*, *4*(1), 2229–6959. doi:10.21917/ijsc.2013.0090

Özçift, A., & Gülten, A. (2013). Genetic algorithm wrapped Bayesian network feature selection applied to differential diagnosis of erythemato-squamous diseases. *Digital Signal Processing*, *23*(1), 230–237. doi:10.1016/j.dsp.2012.07.008

Pappa, G. L., Freitas, A. A., & Kaestner, C. A. (2002). *Attribute Selection with a Multi-objective Genetic Algorithm*. Berlin, Heidelberg: Advances in Artificial Intelligence. doi:10.1007/3-540-36127-8_27

Parpinelli, R. S., Lopes, H. S., Freitas, A. A., & Pr, C. (2001). An Ant Colony Based System for Data Mining:Applications to Medical Data.

Patil, T. R., & Sherekar, M. S. (2013). Performance Analysis of Naive Bayes and J48 Classification Algorithm for Data Classification. *International Journal Of Computer Science And Applications*, 256-261.

Pearl, J. (1984). *Heuristics: Intelligent search strategies for computer problem solving. United States: Addison-Wesley Pub.* Reading, MA: Co., Inc.

Ravichandran, K. S., Narayanamurthy, B., Ganapathy, G., Ravalli, S., & Sindhura, J. (2014). An efficient approach to an automatic detection of erythemato-squamous diseases. *Neural Computing & Applications*, *25*(1), 105–114. doi:10.1007/s00521-013-1452-5

Spiliopoulou, M., Schmidt-Thieme, L., & Janning, R. (2014). *Data Analysis, Machine Learning and Knowledge Discovery (Studies in Classification, Data Analysis, and Knowledge Organization) 2014th Edition*. Switzerland: Springer.

Witten, I. H., Frank, E., Hall, M. A., & Pal, C. J. (2016). *Data Mining: Practical Machine Learning Tools and Techniques*. Morgan Kaufmann.

Xie, J., & Wang, C. (2010). Using support vector machines with a novel hybrid feature selection method for diagnosis of erythemato-squamous diseases. *Expert Systems with Applications*, *38*(5), 5809–5815. doi:10.1016/j.eswa.2010.10.050

## KEY TERMS AND DEFINITIONS

**Best First Search:** Best first search creates new subsets based on the addition or removal of features to the current subset. However, it has the ability to backtrack along the subset selection path to explore different possibilities when the current path no longer shows improvement. To prevent the search from backtracking through all possibilities in the feature space, a limit is placed on the number of non-improving subsets that are considered.

**Classification:** Classification in data mining is the process of classifying the present data according to the specified properties and predicting the class of this data when new data is added. Classification techniques generate models that include classes, groups, relationships, or correlations. Estimates can be made according to these models.

**Data Mining:** Data mining is the process of obtaining previously unknown, valid, and applicable information from large data sources. With data mining relationships, the variables can be deduced and future predictions can be made.

**Feature Selection:** Feature selection is the basic precondition for the implementation of data mining algorithms. The overall objective of the attribute selection is to determine the subset of features that are most relevant for the intended purpose among the available features. The selection of the feature allows the calculation time of the data mining algorithm to be used to be reduced, thereby increasing performance.

**Genetic Search:** Genetic algorithm (GA) is a heuristic search technique based on parameter coding that tries to find solutions using random search techniques. Genetic algorithms are algorithms that simulate the evolutionary process of competing individuals and creating competitive endeavors based on Darwin's "survival of the fittest" principle.

**Heuristic Search Methods:** Heuristic search methods are search procedures that rely on task-independent information to facilitate the search. In many instances, the act of problem solving can be regarded as a systematic process of constructing or searching for a solution.

**Naïve Bayes Algorithm:** Naive Bayes is an augmented classification algorithm used to label data. The Naive Bayesian classifier is based on the calculation of the likelihood values of the effects of each criterion.

# Chapter 6
# A General Medical Diagnosis System Formed by Artificial Neural Networks and Swarm Intelligence Techniques

**Pandian Vasant**
*Universiti Teknologi Petronas, Malaysia*

## ABSTRACT

*One of the most popular applications of artificial intelligence within the medical field is developing medical diagnosis systems. Because artificial-intelligence-based techniques are able to use pre-data and instant data flow for making predictions, it is an easy task to design intelligent systems that can give advice to people or perform diagnosis-based decision making. So, it has been an important research interest to design and develop intelligent systems, which are able to make diagnoses for medical purposes. In this sense, the objective of this chapter is to introduce a general medical diagnosis system that can be used for detecting diseases. In detail, the system employs artificial neural networks and swarm-intelligence-based techniques to form a general framework of intelligent diagnosis. The chapter briefly focuses on the infrastructure of the system and discusses its diagnosis potential.*

## INTRODUCTION

The field of Biomedical Engineering is rapidly improving with unstoppable innovations in different technologies. It is clear that innovations in computer and electronics technologies have a remarkable role on the improvement of the Biomedical Engineering. Of course, discoveries and inventions within the research works of medical – health is the main triggering factor of all these improvements. On the other hand, Artificial Intelligence has a good role in changing the way of solutions on Biomedical Engineering problems and it seems that this scientific field of the future will always have a great potential in the people's future. Not only Biomedical Engineering but also almost all other fields in the life are now associated with intelligent solution approaches, methods, and techniques introduced – suggested by the Artificial Intelligence.

DOI: 10.4018/978-1-5225-4769-3.ch006

If we examine the research works performed within the field of Biomedical Engineering, it can be seen that there is a big diversity of different works (Bonato, 2003; Bronzino, 1999; Enderle & Bronzino, 2012; Guimard et al., 2007; Kutz, 2003; Melchels et al., 2010). In time, there have been a great influence by the Artificial Intelligence into Biomedical Engineering and finally, the era of intelligent Biomedical Engineering applications has begun. Since first applications, there have been many different improvements and developments that can be associated with billions of research works. Among all of these works, the research interest on designing and developing intelligent medical diagnosis systems has a good popularity because the expert support mechanism of Artificial Intelligence based solutions can be employed in an easy and practical way. Because there are many different diseases to be considered, it is important for the researchers to design an effective enough system to deal with that diagnosis problems. There are different kinds of intelligent medical diagnosis oriented studies in the associated literatures (Amato et al., 2013; Bennett & Hauser, 2013; Choi et al., 2017; Kononenko, 2001; Londhe, 2017; Miller, 2012; Papaioannou et al., 2016; Patel et al., 2009; Polat et al., 2006; Rajan, 2017; Waal, 2017). One of widely chosen way to perform intelligent diagnosis is based on running Machine Learning techniques to learn previously something about pre-data and apply the trained system over newly encountered problem instances. Here, it is important to focus on which instances affect the objective disease and which kind of information do researchers need as something like an expert knowledge towards diagnosis.

In this chapter, it is aimed to introduce a general medical diagnosis system that can be used for detecting diseases. In detail, the system employs Artificial Neural Networks (ANN) and Swarm Intelligence (SI) based techniques to form a general framework of intelligent diagnosis. As default, the designed system employs five different SI techniques: Artificial Bee Colony (ABC), Cuckoo Search (CS), Particle Swarm Optimization (PSO), Genetic Algorithm (GA), and Differential Evaluation Algorithm (DE). As it can be understood, training processes of the formed ANN model is done with these algorithms. Except from the main structure of the system for diagnosis operations, it also employs a coding framework called as Intelli-Diagnose Partial Programming (IDP2), which is the key element of the developed system. The chapter briefly focuses on infrastructure of the system and discusses about its diagnosis potentials. The authors think that the work is an alternative contribution to the associated literatures, as based on a traditional but effectively supported with the IDP2.

Based on the topic and objective of the chapter, the remaining content is organized as follows: The next, second section provides explanations regarding to infrastructure of the system. In this context, it employs some essential information for the ANN and the SI related optimization algorithms for the training phase of ANN models. Following that, the third section is devoted to the coding framework of the developed system. In detail, some essential codes and their functions within the IDP2 are expressed briefly. After that section, the chapter is ended by discussing about conclusions and some future work under the fourth section.

## INFRASTRUCTURE OF THE DIAGNOSIS SYTEM

The designed and developed system along this study employs a simple form of the ANN technique that can be trained by SI optimization techniques. In order to understand better about the employed techniques, this section provides some essential information about them as provided under the following sub-titles.

## Artificial Neural Networks

Artificial Neural Networks (ANN) is a widely used Artificial Intelligence technique and it is based on the idea of simulating the brain and dynamics shown by biological neurons. A typical ANN structure is based on the following working mechanism:

- Typical models of ANN are often distributed, parallel data processing systems to achieve the mechanism of learning from the environment and improving the solution process via learning (Deperlioglu & Kose, 2011).
- ANN models include some processing objects connected to each other over some weight values. These objects are called as artificial neurons and each artificial neuron employ input(s), weight value(s), a summation function, a transfer function, and also an output (Deperlioglu & Kose, 2011; Elmas, 2003; Yegnanarayana, 2009).
- In standard cases, all inputs received by the artificial neurons are summed with the summation function and fed through the chosen transfer function. If the output obtained with the transfer function is input for another artificial neuron, the new value is then become an input for the next artificial neuron(s) (Anderson & McNail, 1992; Deperlioglu & Kose, 2011; Yegnanarayana, 2009).

Majority of problems in which ANN models are used are as follows (Carpenter, 1989; Cochocki & Unbehauen, 1993; Elmas, 2003; Miller et al., 1995; Ripley, 1994):

- Prediction,
- Classification,
- Pattern recognition,
- Control,
- Associative memory.

In the literature, there are many different kinds of known ANN models (i.e. Multi-Layer Perceptron, Hopfield, Elman) - (Deperlioglu & Kose, 2011; Elmas, 2003; Yegnanarayana, 2009). Within the study considered in this chapter, two types of ANN models: Multi-Layer Feed-Forward and Cascade Feed-Forward ANN has been chosen to be included in the developed system. Figure 1, and Figure 2 shows default Multi-Layer Feed-Forward and Cascade Feed-Forward ANN models respectively (Tengeleng & Armand, 2014).

In a traditional manner, there are some algorithms like Back-Propagation Algorithm (Oztemel, 2003; Phansalkar & Sastry, 1994; Yegnanarayana, 2009) or Marquardt Algorithm (Hagan & Menhaj, 1994) to train ANN models. But today, it is a popular research interest to use Swarm Intelligence (SI) based optimization algorithms for the same purpose because of their advantages in speed and efficiency (Gudise & Venayagamoorthy, 2003; Ilonen et al., 2003; Karaboga et al., 2007; Kose & Arslan, 2017; Montana & Davis, 1989; Valian et al., 2011).

*Figure 1. A Multi-Layer Feed-Forward Artificial Neural Network model*
*Tengeleng & Armand, 2014.*

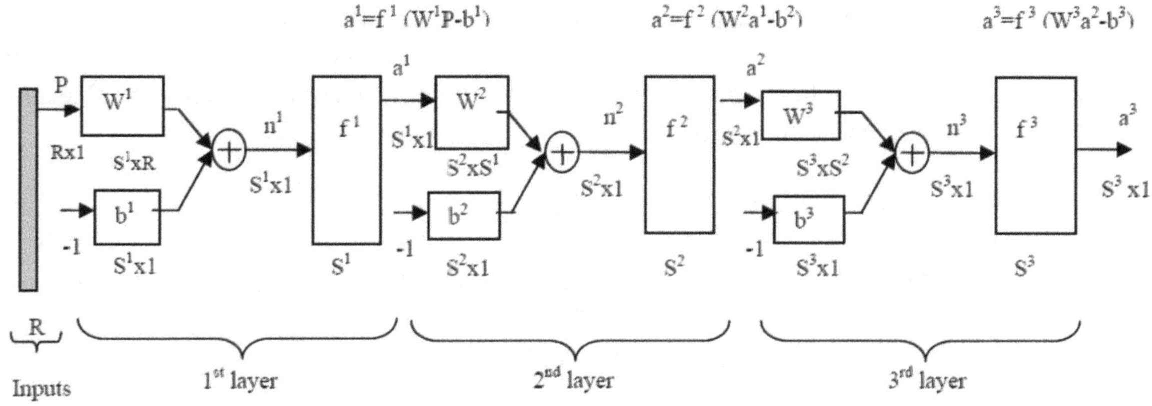


*Figure 2. A Cascade Feed-Forward Artificial Neural Network model*
*Tengeleng & Armand, 2014.*

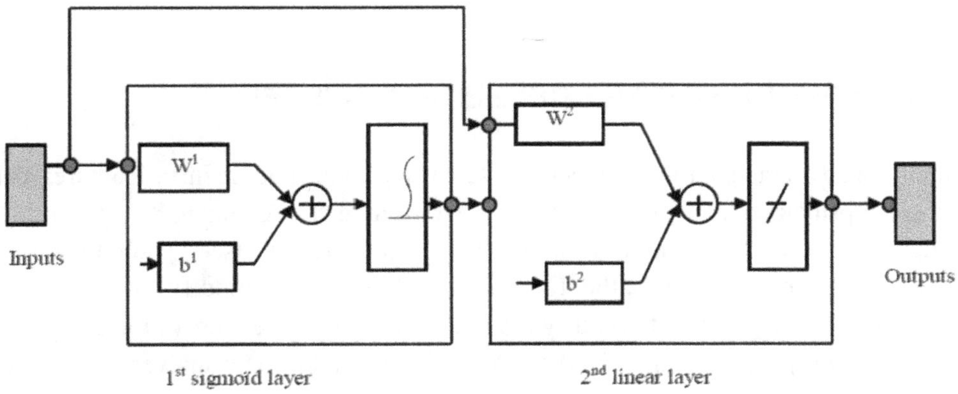


## Optimization Algorithms for Training

The default structure of the medical diagnosis system consists of different Swarm Intelligence (SI) based optimization algorithms for training the ANN models created by users. Even there are two types of ANN as Multi-Layer Feed-Forward and Cascade Feed-Forward, any optimization algorithm can be used for training phases. At this point, the concept of SI should be explained briefly to understand more about working features and functions of the used algorithms.

The concept of SI is associated with the groups of particles, which can behave intelligently and in a collaboration with other particles, without control of a central authority (Blum & Li, 2008; Engelbrecht, 2006; Garnier et al., 2007; Kose, 2015). Because of the advantage of collaborative behaviors, SI has become an important research approach to deal with real world problems. Even it is related with robotics and examined in a swarm engineering interest (Brambilla et al., 2013). From a general perspective, SI is focused on the problems that can be modeled in the form of optimization. Here, the nature has become the essential source of inspiration for improving the literature of SI with newer algorithms – techniques.

In the associated literature, there are too many SI algorithms introduced by researchers (Brownlee, 2011; Yang, 2010). But only some of them are strong enough to overcome wide number of optimization problems. In the developed medical diagnosis system in this study, a total of five different SI algorithms listed as below have been provided to use:

- Artificial Bee Colony (ABC) (Karaboga, 2005,2010),
- Cuckoo Search (CS) (Yang, & Deb, 2009),
- Particle Swarm Optimization (PSO) (Eberhart & Kennedy, 1995; Shi, & Eberhart, 1998),
- Genetic Algorithm (GA) (Holland, 1992, 2012),
- Differential Evaluation Algorithm (DE) (Storn & Price, 1997).

In the formed intelligent medical diagnosis system, the related SI algorithms have gone through the ANN models. Figure 3 represents a general schema of the medical diagnosis system.

Main advantage of the system is related to its unique coding framework. The next section is devoted to some information about this framework.

## CODING FRAMEWORK FOR THE DEVELOPED SYSTEM

The most important advantage of the introduced system is its unique coding infrastructure. As developed by considering the principles of Object Oriented Programming (Langtangen, 2016; Smith, 2015), the system provides a flexible, fast running programming framework for users. Called briefly as Intelli-Diagnose Partial Programming (IDP2), the framework enables users to modify some parts of the developed system. In detail, it is possible to modify the system by adding i.e. new training algorithms to the infrastructure, defining specific features for ANN or performing additional analyzes, thanks to the IDP2.

*Figure 3. General schema of the medical diagnosis system*

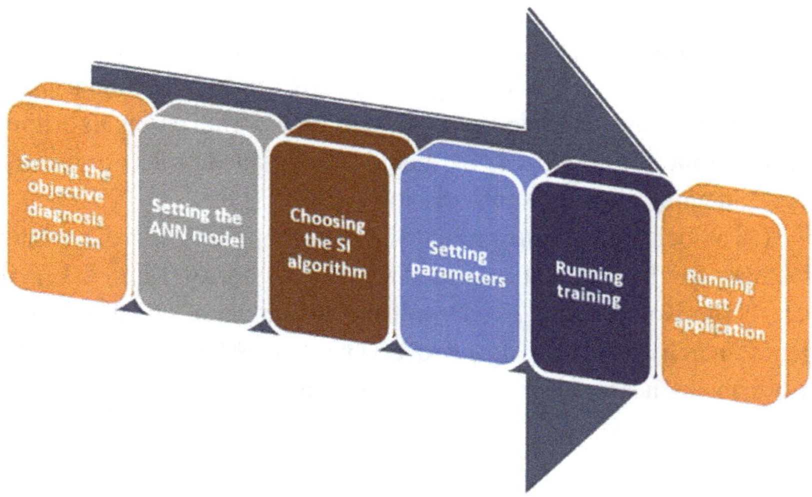

In the current form of the system, a small console interface has been viewed while using it. Users can write some codes to direct the system according to their needs. In order to improve clarity of research outputs, there are some visual tools to support the system outputs as early forms of the visual features – functions employed under the system infrastructure. In addition to the console interface of the system, there is an ongoing development regarding to a graphical user interface (GUI).

The following sub-titles introduces some codes that can be used over the system and via IDP2.

## Essential Using and Building Codes

Essential using and building codes of the system are for creating ANN models, choosing a SI algorithm for training, determining the objective diagnosis data, adjusting that data, and many other operations regarding to medical diagnosis operations. Codes that can be examined under this title are generally connected with the main class structure of the IDP2, so they require some advanced knowledge on computer programming and Object Oriented Programming to build a revised version of the IDP2. On the other hand, there are some types of codes included here to enable users for adding new SI algorithms, defining new parameters and perform some additions to the default solution infrastructure of the system.

The system comes with an .exe file, which is used to run its console interface. After running the console interface – system, the user can type essential using and building codes explained in Table 1, and Table 2 respectively, in order to perform diagnosis experiments.

As it was mentioned before the IDP2 allows also adding a new algorithm or adjusting some features of the default system. Some codes regarding to these operations are shown in Table 3.

*Table 1. Essential using and building codes: 1*

| Code | Function |
|---|---|
| *open;* | Opens the open file dialog to call back a saved project. |
| *open();* | Opens the saved project mentioned under the path as parameter. |
| *new;* | Creates a new project (wants the user to mention a new and press Enter). |
| *network.ff;* | Creates a Feed-Forward ANN model with default structure. |
| *network.cff;* | Creates a Cascade Feed-Forward ANN model with default structure. |
| *add.layer(,,,);* | Adds a layer to the current ANN with four parameters: layer name, type of layer (1:input; 2:hidden; 3:output), number of neurons, and type of activation function (function name) if it is desired to have all neurons with the same function. |
| *remove.layer();* | Removes the layer whose name is mentioned as the parameter. |
| *layer_name.add(,);* | Enables the user to add a specific neuron with two parameters: neuron name, and its activation function to the related layer. |
| *layer_name.remove();* | Removes the neuron whose name is mentioned as the parameter, from the related layer. |
| *data.use();* | Defines the data to be used for the diagnosis process. A name for the data is mentioned as the parameter (Data file that can be used by the system should be in .xls or .csv forms. |
| *data_name.check();* | Checks the data if it has a structural error (i.e. number of rows, columns, data types). |
| *data_name.use_as(,);* | Defines the data to be used for training, test, or application process (the first parameter: 1: training, 2: test, 3: application, and the second parameter is the total number of rows). |

*Table 2. Essential using and building codes: 2*

| Code | Function |
|---|---|
| *chose.algorithm();* | Determines which algorithm to use for training purpose (the abbreviation of the algorithm is mentioned as the parameter). |
| *list.algorithms();* | Lists all the algorithms with their name and abbreviations. |
| *list.active_projects();* | Lists all active projects currently running on the background. |
| *pass.project();* | Pass to the active project mentioned as the parameter. |
| *close.project();* | Closes the project mentioned as the parameter (The system asks for saving in case any unsaved changes in the project). |
| *save.project();* | Saves the project mentioned as the parameter. |
| *list.parameters();* | Lists all parameters regarding to active ANN model and the chosen algorithm to inform the user about which parameters he / she should determine. |
| *set.parameter_name();* | Set the value of the parameter (the value is the parameter of the code – function). |
| *run.training();* | Runs the training process of the project mentioned as the parameter. |
| *diagnose(,);* | Perform a diagnosis over the project mentioned as the first parameter, by using the data mentioned as the second parameter. |

*Table 3. Essential using and building codes: 3*

| Code | Function |
|---|---|
| *system.add.algorithm(,);* | Starts adding the algorithm mentioned as in the first parameter: full name, and the second parameter: abbreviation. |
| *algorithm_name.step(,);* | Adds a step with the name as the first parameter and short definition as the second parameter. Some of the definitions are as below:<br>• allparticles.do(parameter_name, calculation);<br>• bestparticle.do(parameter_name, calculation);<br>• worstparticle.do(parameter_name, calculation);<br>• nonbestparticle.do(parameter_name, calculation);<br>• evaluation.do(evaluation_mechanism_name);<br>• elimination.do(parameter_name, calculation);<br>• set.do(parameter_name, calculation);<br>• set_zero.do(parameter_name);<br>• set_start();<br>• set_end(); |
| *algorithm_name.stepremove();* | Removes the step, whose name is mentioned as the parameter. |
| *algorithm.add.loop();* | Adds a loop with the stopping criteria as the parameter. |
| *algorithm.set_type();* | Sets the type of the algorithm with the parameter; 1: continuous, 2: combinatorial. |
| *algorithm.name.finish();* | Finishes the algorithm adding. |
| *algorithm_name.test();* | Tests the algorithm for its consistency on optimization. |
| *algorithm.add_xml();* | Adds an algorithm, whose details are indicated in the form of XML. |
| *system.set_ann(,);* | Sets a parameter of the default ANN model types stored in the system. |
| *system.set_own();* | Sets a default parameter – feature of the system (i.e. default name, default activation functions, default available algorithms, default iteration numbers, default ANN parameters like learning rate, or momentum). |
| *system.reset();* | Resets all made changes in the system to the default. |

### Codes for Additional Features: Functions

In addition to the mentioned codes here, the user can also run some additional features – functions (i.e. showing graphics, viewing additional statistics) by using some specific codes. These codes are presented in Table 4 briefly.

## CONCLUSION

In this chapter, a general Artificial Intelligence based system for medical diagnosis has been introduced. Infrastructure of the system consists of a flexible Artificial Neural Networks (ANN) model and five different Swarm Intelligence techniques: Artificial Bee Colony (ABC), Cuckoo Search (CS), Particle Swarm Optimization (PSO), Genetic Algorithm (GA), and Differential Evaluation Algorithm (DE) respectively. By using the coding framework of the system, it is possible to create – modify the model of Artificial Neural Network and determine which algorithm to train the model. Also, it is possible to examine training and diagnosis processes with different output data that can be exported into files or viewed over specific graphics. The system also has its own programming language called as Intelli-Diagnose Partial Programming (IDP2) to enable users for modifying some features and functions of the system by adding i.e. new training algorithms to the infrastructure, defining specific features for ANN or performing additional analyzes if they have enough knowledge about the IDP2. Generally, the system can be modified to diagnose any type of disease after training the system with appropriate data. In order to improve training performances, well-known optimization algorithms have been coded rather than traditional algorithms like i.e. Backpropagation Algorithm. The chapter has been written to introduce programmatic aspects of the diagnosis system and there will be future research efforts as expressed follows:

- General diagnosis performance of the system will be evaluated with different datasets of different diseases.

*Table 4. Codes for additional features: Functions*

| Code | Function |
|---|---|
| *draw_g(,);* | Draws a graphic for the objective data in the first parameter and with the graphic type in the second parameter (i.e. 1: plot, 2: pie) |
| *analyze();* | Performs a general analyze over a project (project name in the parameter) that was run several times. |
| *export(,);* | Export the mentioned type of data (the first parameter is project name; the second parameter: 1: for all information about the project, 2: for the obtained training data, 3: for the obtained testing data, 4: for the obtained application data, 5: weight values of the ANN model, 6: error values, 7: testing accuracy data) |
| *draw_s();* | Draws a general schema of the performed diagnosis operation – project. |
| *draw_instant();* | Draws instant statistical information during the training, test, or application processes (the parameter is for defining which parameter to be shown. i.e. weights for the weight values of the ANN model, error for instantly changing error value during training). |
| *statistics();* | Shows a specific statistic related to the system, according to the mentioned parameter value (i.e. 1: total accuracy performance of the system so far, 2: total number of considered diagnosis problems, 0: all statistical information about the system). |

- As it can be understood the system has currently a console based using structure. Further developments on a flexible graphical user interface (GUI) is currently developed and will be ready within next possible reports.
- In addition to the current features like data export or visual graphical functions, there will be more tools in order to improve using experience in new versions.
- Coding framework will be continued to improve with new updates according to further studies and needs.
- Future works will include also comparative studies done with alternative diagnosis approaches taking place in the associated literature.

## REFERENCES

Amato, F., López, A., Peña-Méndez, E. M., Vaňhara, P., Hampl, A., & Havel, J. (2013). Artificial neural networks in medical diagnosis. *Journal of Applied Biomedicine*, *11*(2), 47–58. doi:10.2478/v10136-012-0031-x

Anderson, D., & McNeill, G. (1992). Artificial neural networks technology. *Kaman Sciences Corporation*, *258*(6), 1–83.

Bennett, C. C., & Hauser, K. (2013). Artificial intelligence framework for simulating clinical decision-making: A Markov decision process approach. *Artificial Intelligence in Medicine*, *57*(1), 9–19. doi:10.1016/j.artmed.2012.12.003 PMID:23287490

Blum, C., & Li, X. (2008). Swarm intelligence in optimization. In *Swarm Intelligence* (pp. 43–85). Springer Berlin Heidelberg. doi:10.1007/978-3-540-74089-6_2

Bonato, P. (2003). Wearable sensors/systems and their impact on biomedical engineering. *IEEE Engineering in Medicine and Biology Magazine*, *22*(3), 18–20. doi:10.1109/MEMB.2003.1213622 PMID:12845812

Brambilla, M., Ferrante, E., Birattari, M., & Dorigo, M. (2013). Swarm robotics: A review from the swarm engineering perspective. *Swarm Intelligence*, *7*(1), 1–41. doi:10.1007/s11721-012-0075-2

Bronzino, J. D. (1999). *Biomedical engineering handbook* (Vol. 2). CRC Press.

Brownlee, J. (2011). *Clever algorithms: nature-inspired programming recipes*. Jason Brownlee.

Carpenter, G. A. (1989). Neural network models for pattern recognition and associative memory. *Neural Networks*, *2*(4), 243–257. doi:10.1016/0893-6080(89)90035-X

Choi, Y. J., Baek, J. H., Park, H. S., Shim, W. H., Kim, T. Y., Shong, Y. K., & Lee, J. H. (2017). A computer-aided diagnosis system using artificial intelligence for the diagnosis and characterization of thyroid nodules on ultrasound: Initial clinical assessment. *Thyroid*, *27*(4), 546–552. doi:10.1089/thy.2016.0372 PMID:28071987

Cochocki, A., & Unbehauen, R. (1993). *Neural networks for optimization and signal processing*. John Wiley & Sons, Inc.

Deperlioglu, O., & Kose, U. (2011). An educational tool for artificial neural networks. *Computers & Electrical Engineering*, *37*(3), 392–402. doi:10.1016/j.compeleceng.2011.03.010

Eberhart, R., & Kennedy, J. (1995). A new optimizer using particle swarm theory. In *Micro Machine and Human Science, 1995. MHS'95., Proceedings of the Sixth International Symposium on* (pp. 39-43). IEEE. doi:10.1109/MHS.1995.494215

Elmas, C. (2003). *Artificial neural network – Theory, architecture, training, implementation*. Ankara, Turkey: Seçkin Press. (In Turkish)

Enderle, J. D., & Bronzino, J. D. (2012). *Introduction to biomedical engineering*. Academic press.

Engelbrecht, A. P. (2006). *Fundamentals of computational swarm intelligence*. John Wiley & Sons.

Garnier, S., Gautrais, J., & Theraulaz, G. (2007). The biological principles of swarm intelligence. *Swarm Intelligence*, *1*(1), 3–31. doi:10.1007/s11721-007-0004-y

Gudise, V. G., & Venayagamoorthy, G. K. (2003). Comparison of particle swarm optimization and backpropagation as training algorithms for neural networks. In *Swarm Intelligence Symposium, 2003. SIS'03. Proceedings of the 2003 IEEE* (pp. 110-117). IEEE. doi:10.1109/SIS.2003.1202255

Guimard, N. K., Gomez, N., & Schmidt, C. E. (2007). Conducting polymers in biomedical engineering. *Progress in Polymer Science*, *32*(8), 876–921. doi:10.1016/j.progpolymsci.2007.05.012

Holland, J. (1995). *Hidden order: How adaptation builds complexity*. Basic Books.

Holland, J. H. (2012). Genetic algorithms. *Scholarpedia*, *7*(12), 1482. doi:10.4249/scholarpedia.1482

Ilonen, J., Kamarainen, J. K., & Lampinen, J. (2003). Differential evolution training algorithm for feed-forward neural networks. *Neural Processing Letters*, *17*(1), 93–105. doi:10.1023/A:1022995128597

Karaboga, D. (2005). *An idea based on honey bee swarm for numerical optimization* (Vol. 200). Technical report-tr06, Erciyes University, Engineering Faculty, Computer Engineering Department.

Karaboga, D. (2010). Artificial bee colony algorithm. *Scholarpedia*, *5*(3), 6915. doi:10.4249/scholarpedia.6915

Karaboga, D., Akay, B., & Ozturk, C. (2007). Artificial bee colony (ABC) optimization algorithm for training feed-forward neural networks. In *International Conference on Modeling Decisions for Artificial Intelligence* (pp. 318-329). Springer Berlin Heidelberg. doi:10.1007/978-3-540-73729-2_30

Kononenko, I. (2001). Machine learning for medical diagnosis: History, state of the art and perspective. *Artificial Intelligence in Medicine*, *23*(1), 89–109. doi:10.1016/S0933-3657(01)00077-X PMID:11470218

Kose, U. (2015). Present state of swarm intelligence and future directions. In *Encyclopedia of Information Science and Technology* (3rd ed.; pp. 239–252). IGI Global. doi:10.4018/978-1-4666-5888-2.ch023

Kose, U., & Arslan, A. (2017). Optimization of self-learning in Computer Engineering courses: An intelligent software system supported by Artificial Neural Network and Vortex Optimization Algorithm. *Computer Applications in Engineering Education*, *25*(1), 142–156. doi:10.1002/cae.21787

Kutz, M. (2003). *Standard handbook of biomedical engineering and design*. McGraw-Hill.

Langtangen, H. P. (2016). Object-oriented programming. In *A Primer on Scientific Programming with Python* (pp. 567–643). Berlin: Springer. doi:10.1007/978-3-662-49887-3_9

Londhe, V. (2017). Brain MR Image Segmentation for Tumor Detection using Artificial Neural. *Brain, 6*(1).

Melchels, F. P., Feijen, J., & Grijpma, D. W. (2010). A review on stereolithography and its applications in biomedical engineering. *Biomaterials, 31*(24), 6121–6130. doi:10.1016/j.biomaterials.2010.04.050 PMID:20478613

Miller, P. L. (Ed.). (2012). *Selected topics in medical artificial intelligence.* Springer Science & Business Media.

Miller, W. T., Werbos, P. J., & Sutton, R. S. (Eds.). (1995). *Neural networks for control.* MIT Press. doi:10.1007/978-3-642-57760-4

Montana, D. J., & Davis, L. (1989). Training Feedforward Neural Networks Using Genetic Algorithms. *IJCAI.*

Oztemel, E. (2003). *Artificial neural networks.* Istanbul: Papatya Press. (In Turkish)

Papaioannou, V., Economou, G. P. K., & Tsakalidis, A. (2016). A robust smart device app assisting medical diagnosis. *Artificial Intelligence Review*, *5*(2), 82.

Patel, V. L., Shortliffe, E. H., Stefanelli, M., Szolovits, P., Berthold, M. R., Bellazzi, R., & Abu-Hanna, A. (2009). The coming of age of artificial intelligence in medicine. *Artificial Intelligence in Medicine, 46*(1), 5–17. doi:10.1016/j.artmed.2008.07.017 PMID:18790621

Phansalkar, V. V., & Sastry, P. S. (1994). Analysis of the back-propagation algorithm with momentum. *IEEE Transactions on Neural Networks, 5*(3), 505–506. doi:10.1109/72.286925 PMID:18267819

Polat, K., Şahan, S., & Güneş, S. (2006). A new method to medical diagnosis: Artificial immune recognition system (AIRS) with fuzzy weighted pre-processing and application to ECG arrhythmia. *Expert Systems with Applications, 31*(2), 264–269. doi:10.1016/j.eswa.2005.09.019

Rajan, J. R. (2017). Prognostic system for early diagnosis of pediatric lung disease using artificial intelligence. *Current Pediatric Research, 21*(1).

Ripley, B. D. (1994). Neural networks and related methods for classification. *Journal of the Royal Statistical Society. Series B. Methodological*, 409–456.

Shi, Y., & Eberhart, R. (1998). A modified particle swarm optimizer. In *Evolutionary Computation Proceedings, 1998. IEEE World Congress on Computational Intelligence., The 1998 IEEE International Conference on* (pp. 69-73). IEEE. doi:10.1109/ICEC.1998.699146

Smith, B. (2015). Object-oriented programming. In *Advanced ActionScript 3* (pp. 1–23). Apress.

Storn, R., & Price, K. (1997). Differential evolution–a simple and efficient heuristic for global optimization over continuous spaces. *Journal of Global Optimization, 11*(4), 341–359. doi:10.1023/A:1008202821328

Tengeleng, S., & Armand, N. (2014). Performance of using cascade forward back propagation neural networks for estimating rain parameters with rain drop size distribution. *Atmosphere*, *5*(2), 454–472. doi:10.3390/atmos5020454

Valian, E., Mohanna, S., & Tavakoli, S. (2011). Improved cuckoo search algorithm for feedforward neural network training. *International Journal of Artificial Intelligence & Applications*, *2*(3), 36–43. doi:10.5121/ijaia.2011.2304

Waal, I. (2017). Skin cancer diagnosed using artificial intelligence on clinical images. *Oral Diseases*. PMID:28326656

Yang, X. S. (2010). *Nature-inspired metaheuristic algorithms*. Luniver press.

Yang, X. S., & Deb, S. (2009). Cuckoo search via Lévy flights. In *Nature & Biologically Inspired Computing, 2009. NaBIC 2009. World Congress on* (pp. 210-214). IEEE.

Yegnanarayana, B. (2009). *Artificial neural networks*. PHI Learning Pvt. Ltd.

## ADDITIONAL READING

Akay, M. (2000). *Nonlinear Biomedical Signal Processing: Fuzzy logic, neural networks, and new algorithms*. Wiley-IEEE Press. doi:10.1109/9780470545362

Alonso, F., Caraça-Valente, J. P., González, A. L., & Montes, C. (2002). Combining expert knowledge and data mining in a medical diagnosis domain. *Expert Systems with Applications*, *23*(4), 367–375. doi:10.1016/S0957-4174(02)00072-6

Chan, F. T., & Tiwari, M. K. (2007). Preface: Swarm Intelligence, Focus on Ant and Particle Swarm Optimization. In Swarm Intelligence, Focus on Ant and Particle Swarm Optimization. InTech.

Cohen, P. R. (1995). *Empirical methods for artificial intelligence* (Vol. 139). Cambridge, MA: MIT press.

Cohen, P. R., & Feigenbaum, E. A. (Eds.). (2014). *The handbook of artificial intelligence* (Vol. 3). Butterworth-Heinemann.

De, S. K., Biswas, R., & Roy, A. R. (2001). An application of intuitionistic fuzzy sets in medical diagnosis. *Fuzzy Sets and Systems*, *117*(2), 209–213. doi:10.1016/S0165-0114(98)00235-8

Deepa, S. N., & Devi, B. A. (2011). A survey on artificial intelligence approaches for medical image classification. *Indian Journal of Science and Technology*, *4*(11), 1583–1595.

Doi, K. (2007). Computer-aided diagnosis in medical imaging: Historical review, current status and future potential. *Computerized Medical Imaging and Graphics*, *31*(4), 198–211. doi:10.1016/j.compmedimag.2007.02.002 PMID:17349778

Dorigo, M., & Birattari, M. (2007). Swarm intelligence. *Scholarpedia*, *2*(9), 1462. doi:10.4249/scholarpedia.1462

Fox, J., & Das, S. K. (2000). *Safe and sound: artificial intelligence in hazardous applications* (Vol. 1). Menlo Park, CA/Cambridge: AAAI Press/MIT Press.

Fritz, P., Kleinhans, A., Kuisle, F., Albu, P., Fritz-Kuisle, C., & Alscher, M. D. (2017). Medical diagnosis as a linguistic game. *BMC Medical Informatics and Decision Making*, *17*(1), 103. doi:10.1186/s12911-017-0488-3 PMID:28693559

Garnier, S., Gautrais, J., & Theraulaz, G. (2007). The biological principles of swarm intelligence. *Swarm Intelligence*, *1*(1), 3–31. doi:10.1007/s11721-007-0004-y

Ginsberg, M. L. (1988). Multivalued logics: A uniform approach to reasoning in artificial intelligence. *Computational Intelligence*, *4*(3), 265–316. doi:10.1111/j.1467-8640.1988.tb00280.x

Guzmán, J. C., Melin, P., & Prado-Arechiga, G. (2017). Neuro-Fuzzy Hybrid Model for the Diagnosis of Blood Pressure. In *Nature-Inspired Design of Hybrid Intelligent Systems* (pp. 573–582). Springer International Publishing. doi:10.1007/978-3-319-47054-2_37

Handl, J., & Meyer, B. (2007). Ant-based and swarm-based clustering. *Swarm Intelligence*, *1*(2), 95–113. doi:10.1007/s11721-007-0008-7

Hayashi, J., Kunieda, T., Cole, J., Soga, R., Hatanaka, Y., Lu, M., . . . Fujita, H. (2001). A development of computer-aided diagnosis system using fundus images. In *Virtual Systems and Multimedia, 2001. Proceedings. Seventh International Conference on* (pp. 429-438). IEEE. doi:10.1109/VSMM.2001.969697

Hinchey, M. G., Sterritt, R., & Rouff, C. (2007). Swarms and swarm intelligence. *Computer*, *40*(4), 111–113. doi:10.1109/MC.2007.144

Horvitz, E. J., Breese, J. S., & Henrion, M. (1988). Decision theory in expert systems and artificial intelligence. *International Journal of Approximate Reasoning*, *2*(3), 247–302. doi:10.1016/0888-613X(88)90120-X

Hsu, C. C., & Ho, C. S. (2004). A new hybrid case-based architecture for medical diagnosis. *Information Sciences*, *166*(1), 231–247. doi:10.1016/j.ins.2003.11.009

Hudson, D. L., & Cohen, M. E. (2000). *Neural networks and artificial intelligence for biomedical engineering* (pp. I–XXIII). New York: IEEE.

Hussain, W., Ishak, W., & Siraj, F. (2002). Artificial intelligence in medical application: an exploration.

Karaboga, D., & Akay, B. (2009). A survey: Algorithms simulating bee swarm intelligence. *Artificial Intelligence Review*, *31*(1-4), 61–85. doi:10.1007/s10462-009-9127-4

Kononenko, I. (1993). Inductive and Bayesian learning in medical diagnosis. *Applied Artificial Intelligence an International Journal*, *7*(4), 317–337. doi:10.1080/08839519308949993

Krause, J., Ruxton, G. D., & Krause, S. (2010). Swarm intelligence in animals and humans. *Trends in Ecology & Evolution*, *25*(1), 28–34. doi:10.1016/j.tree.2009.06.016 PMID:19735961

Kulikowski, C. A. (1980). Artificial intelligence methods and systems for medical consultation. *IEEE Transactions on Pattern Analysis and Machine Intelligence*, *PAMI-2*(5), 464–476. doi:10.1109/TPAMI.1980.6592368

Liao, S. H. (2005). Expert system methodologies and applications—a decade review from 1995 to 2004. *Expert Systems with Applications*, *28*(1), 93–103. doi:10.1016/j.eswa.2004.08.003

López, B., & Plaza, E. (1993). Case-based planning for medical diagnosis. *Methodologies for intelligent systems*, 96-105.

Majeed, H., Sridharan, S., Mir, M., Ma, L., Min, E., Jung, W., & Popescu, G. (2016). Quantitative phase imaging for medical diagnosis. *Journal of Biophotonics*. PMID:27539534

Mangasarian, O. L., Setiono, R., & Wolberg, W. H. (1990). Pattern recognition via linear programming: Theory and application to medical diagnosis. *Large-scale numerical optimization*, 22-31.

Matousek, P., & Stone, N. (2016). Development of deep subsurface Raman spectroscopy for medical diagnosis and disease monitoring. *Chemical Society Reviews*, *45*(7), 1794–1802. doi:10.1039/C5CS00466G PMID:26455315

McPhee, S. J., Papadakis, M. A., & Tierney, L. M. (Eds.). (2010). *Current medical diagnosis & treatment 2010*. New York: McGraw-Hill Medical.

McPherson, R. A., & Pincus, M. R. (2017). *Henry's Clinical Diagnosis and Management by Laboratory Methods E-Book*. Elsevier Health Sciences.

Meesad, P., & Yen, G. G. (2003). Combined numerical and linguistic knowledge representation and its application to medical diagnosis. *IEEE Transactions on Systems, Man, and Cybernetics. Part A, Systems and Humans*, *33*(2), 206–222. doi:10.1109/TSMCA.2003.811290

Mornet, S., Vasseur, S., Grasset, F., & Duguet, E. (2004). Magnetic nanoparticle design for medical diagnosis and therapy. *Journal of Materials Chemistry*, *14*(14), 2161–2175. doi:10.1039/b402025a

Nikovski, D. (2000). Constructing Bayesian networks for medical diagnosis from incomplete and partially correct statistics. *IEEE Transactions on Knowledge and Data Engineering*, *12*(4), 509–516. doi:10.1109/69.868904

Ogiela, M. R., & Tadeusiewicz, R. (2003). Artificial intelligence structural imaging techniques in visual pattern analysis and medical data understanding. *Pattern Recognition*, *36*(10), 2441–2452. doi:10.1016/S0031-3203(03)00089-X

Park, S. H., & Lee, S. P. (1998). EMG pattern recognition based on artificial intelligence techniques. *IEEE Transactions on Rehabilitation Engineering*, *6*(4), 400–405. doi:10.1109/86.736154 PMID:9865887

Park, Y. J., Kim, B. C., & Chun, S. H. (2006). New knowledge extraction technique using probability for case-based reasoning: Application to medical diagnosis. *Expert Systems: International Journal of Knowledge Engineering and Neural Networks*, *23*(1), 2–20. doi:10.1111/j.1468-0394.2006.00321.x

Parpinelli, R. S., & Lopes, H. S. (2011). New inspirations in swarm intelligence: A survey. *International Journal of Bio-inspired Computation*, *3*(1), 1–16. doi:10.1504/IJBIC.2011.038700

Poli, R., Kennedy, J., & Blackwell, T. (2007). Particle swarm optimization. *Swarm Intelligence*, *1*(1), 33–57. doi:10.1007/s11721-007-0002-0

Poole, D., Goebel, R., & Aleliunas, R. (1987). Theorist: A logical reasoning system for defaults and diagnosis. In The Knowledge Frontier (pp. 331-352). Springer New York. doi:10.1007/978-1-4612-4792-0_13

Rahwan, I., & Simari, G. R. (Eds.). (2009). *Argumentation in artificial intelligence* (Vol. 47). Heidelberg: Springer.

Schmid-Saugeona, P., Guillodb, J., & Thirana, J. P. (2003). Towards a computer-aided diagnosis system for pigmented skin lesions. *Computerized Medical Imaging and Graphics*, *27*(1), 65–78. doi:10.1016/S0895-6111(02)00048-4 PMID:12573891

Schwartz, W. B., Patil, R. S., & Szolovits, P. (1987). Artificial Intelligence in Medicine Where Do We Stand? *Jurimetrics*, *27*(4), 362–369. PMID:3821801

Sen, K., & Ghosh, K. (2017). Developing Effective Crowdsourcing Systems for Medical Diagnosis: Challenges and Recommendations.

Sterckx, S., Cockbain, J., & Pennings, G. (2017). Patenting medical diagnosis methods in Europe: Stanford University and time-lapse microscopy. *Reproductive Biomedicine Online*, *34*(2), 166–168. doi:10.1016/j.rbmo.2016.10.014 PMID:27908603

Stoitsis, J., Valavanis, I., Mougiakakou, S. G., Golemati, S., Nikita, A., & Nikita, K. S. (2006). Computer aided diagnosis based on medical image processing and artificial intelligence methods. *Nuclear Instruments & Methods in Physics Research. Section A, Accelerators, Spectrometers, Detectors and Associated Equipment*, *569*(2), 591–595. doi:10.1016/j.nima.2006.08.134

Szolovits, P., & Pauker, S. G. (1978). Categorical and probabilistic reasoning in medical diagnosis. *Artificial Intelligence*, *11*(1-2), 115–144. doi:10.1016/0004-3702(78)90014-0

Woolley, A., Kostopoulou, O., & Delaney, B. C. (2016). Can medical diagnosis benefit from "unconscious thought"? *Medical Decision Making*, *36*(4), 541–549. doi:10.1177/0272989X15581352 PMID:25852079

Zhang, H. Z., Wang, K. Q., Zhang, D., Pang, B., & Huang, B. (2005). Computer aided tongue diagnosis system. In *Engineering in Medicine and Biology Society, 2005. IEEE-EMBS 2005. 27th Annual International Conference of the* (pp. 6754-6757). IEEE. doi:10.1109/IEMBS.2005.1616055

Zhao, S. (2017). Mining Medical Causality for Diagnosis Assistance. In *Proceedings of the Tenth ACM International Conference on Web Search and Data Mining* (pp. 841-841). ACM. doi:10.1145/3018661.3022752

## KEY TERMS AND DEFINITIONS

**Artificial Neural Networks:** Artificial neural networks are techniques of artificial intelligence inspired by brain and natural neurons to form models of parallel processing units called artificial neurons in order to deal with problems.

**Intelligent Diagnosis System:** Intelligent diagnosis system is a software- (and sometimes hardware-) based system that employs artificial-intelligence-based approaches, methods, or techniques to deal with diagnosis problems.

**Nature-Inspired Techniques:** Nature-inspired techniques are typical intelligently designed problem-solving techniques included within the field of artificial intelligence inspired by the natural dynamics including the ones from humans, other living organisms, and lifeless objects.

**Swarm Intelligence:** Swarm intelligence is a sub-research interest under the field of artificial intelligence and focused on using particle-based approaches to deal with objective optimization problems.

# Chapter 7
# Artificial Intelligence Applications on Classification of Heart Sounds

**Huseyin Coskun**
*Usak University, Turkey*

**Tuncay Yigit**
*Suleyman Demirel University, Turkey*

## ABSTRACT

*The aim of this chapter is to classify normal and extra systole heart sounds using artificial intelligence methods. Initially, both heart sounds have been passed from Butterworth, Chebyshev, Elliptic digital filter in specific frequency values to remove noise. Afterwards, features of heart sounds have been obtained for classification. For this process, wavelet transform and Mel-frequency cepstral coefficients (MFCC) methods have been applied. Training and test data have been created for classifier by taking means and standard deviation of gained feature. Support vector machine (SVM) and artificial neural network (ANN) methods have been used for classification of these heart sounds. Using wavelet and MFCC features, classification success of SVM has been obtained as 93.33% and 100%, respectively. Using wavelet and MFCC features, classification success of ANN has been obtained as 83.33% and 90%, respectively.*

## INTRODUCTION

Biological signal is defined as the signal taken from living body through transducer. These signals have provided lots of information about the living body. These signals are split in two, in that some of its are based on electricity origin or some of its are physical origin. Heart sounds (HS) is defined as physical biomedical signals. HS are occurred from result of mechanical movement of heart. These sounds can be occurred with different properties because of different heart disease.

According to WHO (World Health Organization), heart diseases is the first cause of death around the world. Every year, most people lose their life because of heart diseases more than other reasons. In 2004 nearly 17.1 million people have been died due to heart diseases. This number points a rate of 29%

DOI: 10.4018/978-1-5225-4769-3.ch007

of all deaths in that year. But unfortunately this number is expected to be 23.4 million in 2030 (WHO, 2004). In Europe, heart diseases form 42% of all deaths and expenditure for treatment of the diseases is almost 192 billion € in total (Allender et al., 2008).

Early diagnosis of heart diseases is crucial importance in treatment. Early discern and short time diagnosis of heart disorder is vital importance for preventing serious heart disease. Furthermore, it plays effective role in decrease of total treatment expenditure. Doppler ultrasonography and ECG (Electrocardiography) are the systems that require time and expert knowledge. HS are gained by result of auscultation and these sounds are one of the related basic medical techniques used by doctors in diagnosis for heart disease. Moreover, auscultation is a method which is harmless, easier and faster for patient. In addition to this, it is vital importance to obtain medical information such as disease and other cardiac data of the patient with heart disease until patient is taken to hospital. In this study, nature inspired artificial intelligence methods that is used classification of normal and extra systole HS are discussed. The flowchart of learning model using in this study is given in Figure 1. Besides, all MATLAB code belong to each step is explained throughout the study. Readers can use all of these codes in one m-file by combining and they can implement them in its own research.

## BACKGROUND

As far as there are different studies in signal processing and classification of different heart sounds (HS). The methods in used in signal processing are wavelet transform for feature extractions in these studies. To classify, researcher use different methods of SVM and ANN. The main ones can be listed as follows. Zheng et al. (2015) have used least square support vector machine (LS-SVM) for the implementation of intelligent diagnosis of chronic heart failure. Patidar et. al. (2014) have classified cardiac sound signals using LS-SVM with various kernel functions for detection and identification of heart valve disorders. Lubaib et al. (2016) have classified normal and abnormal HS samples using SVM, K Nearest Neighbour (KNN), Bayesian and Gaussian Mixture Model (GMM) Classifiers. Güraksın et al. (2010), have classified HS using ANN (artificial neural network) on pocket computer which can run Windows OS. In their study, they have used Discrete Fourier Transform (DFT) to frequency analyzing of HS. It has

*Figure 1. The learning model of heart sound classification systems*

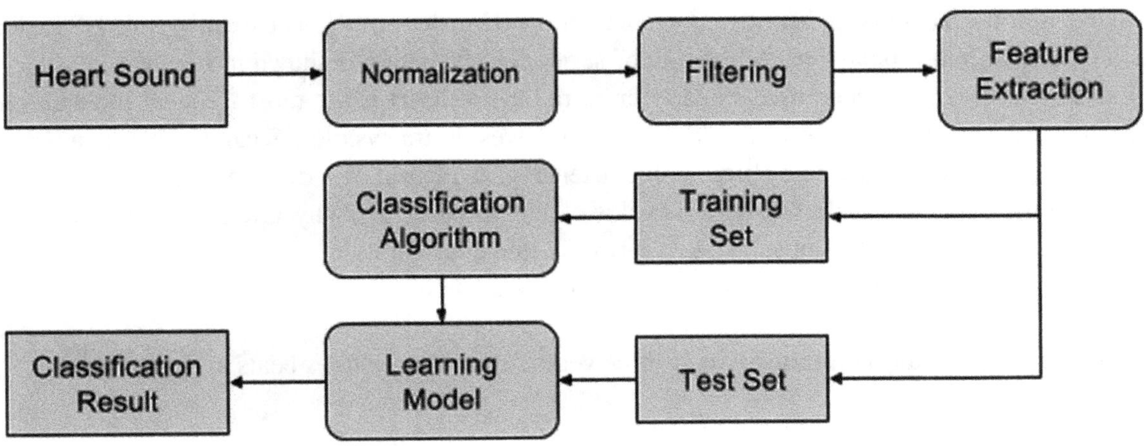

been noticed that the test set has been predicted correctly in the ratio of 91.6% by the neural network in 3 class label as Normal, Pulmonary Stenosis, Mitral Stenosis thanks to this mobile application. Gupta et al. (2007) have used DWT to get the features of HS and ANN to classify these features in 3 class label as Normal, Systolic murmur and Diastolic murmur. Azmy (2015) have used new mother wavelet and SVM to classify for normal and abnormal. Maglogiannis et al. (2009) have develop a diagnosis system for the identification of heart valve diseases based on the SVM classifying classification of HS. Sun et al. (2014) have used SVM to detection of ventricular septal defects. Ari et al. (2010) have used Least Mean Square (LMS) based SVM to detection cardiac abnormality from HS signal. Guermoui et al. (2013) have used different wavelet type to classify HS with SVM into 5 classes: normal HS (NHS), aortic stenosis (AS), aortic insufficiency (Al) mitral stenosis (MS), and mitral insufficiency (MI). Sh-Hussain et al. (2012) have used Mel-frequency cepstral coefficients (MFCC) to feature extraction of HS and then have used Hidden Markov Model (HMM) to classify for different auscultation area of HS. Das et al. (2009) have used neural network ensembles to diagnosis of valvular heart disease. In this study, developed ensemble-based methods creates new models by combining the posterior probabilities or the predicted values from multiple predecessor models to create more effective models. Das et. al (2009) et al. obtained 97.4% classification accuracy from the experiments made on data set containing 215 samples. Turkoglu et al. (2003) have develop an intelligent system for diagnosis of the heart valve disease with wavelet packet neural network. The model developed by Turkoglu et al. (2003) consists of two layers: wavelet and multi-layer perceptron. Wavelet layer is used for feature extraction of Doppler signals and the multi-layer perceptron is used is a feed-forward neural network to classification. Sengur (2008) have developed an expert system based on principal component analysis, artificial immune system and fuzzy k-NN for diagnosis of valvular heart diseases. In Sengur‘s study, classification results are obtained as 95.9% sensitivity and 96% specificity rate.

The principle area of interest in this chapter is to classification of extra systole HS and normal HS. Normal HS result from the vibrations of heart valves between auricle and ventricle. A normal HS is like the sounds “lub dub, lub dub” sounds. In medicine, “lub” is named as S1 sound while “dub” sound is called as the S2 sound. Every “lub dub” sound is one beat. Normal heart can have 60 - 100 beats in a minute. In Figure 2, normal HS, S1, S2 parts and “lub dub” sound pattern is shown.

Range between inception of S1 and S2 sounds are called as systole. Inception of S2 sound and starting of the S1 sound again is called as diastole. Systole range between S1 and S2 sounds in a normal heart follow the range of diastole between every beat. This situation is shown in Figure-1 presenting HS diagram. The extra systole range, which is one of the HS in an abnormal heart, is the event of hearing the S1 sound after the S1 sound or hearing S2 again after the S2 instead of S1. The extra systole HS include sound design of “lub-lub dub” or “lub dub-dub” series. This situation is shown in Figure 3.

Extra systole which is identified as early or extra beat of heart is the most frequent incident seen in irregular beats resulting from ventricles and heart valves. Extra systole HS can be seen in a healthy person. Extra systole HS are associated with existence of symptom of a disease in cases that it often happens. If the frequency is above a definite value of age, it speeds up tachycardia heart disease (Israel, 2014). Thus, extra systole HS must be taken seriously into consideration.

Incidents in which extra systole HS pose a risk are listed below

1. In EKG and Holter monitoring two or three ventricular extra systoles beats in series is risky.

*Figure 2. Normal heart sound S1, S2 sections, frequency values*

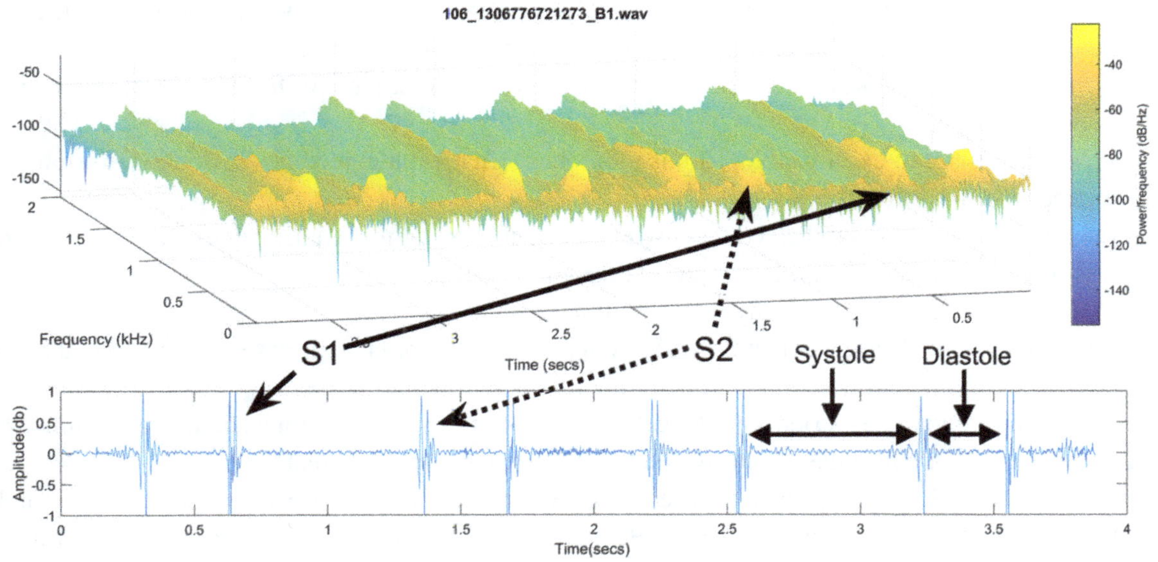


*Figure 3. Heart sound with extra systole pattern and frequency values*

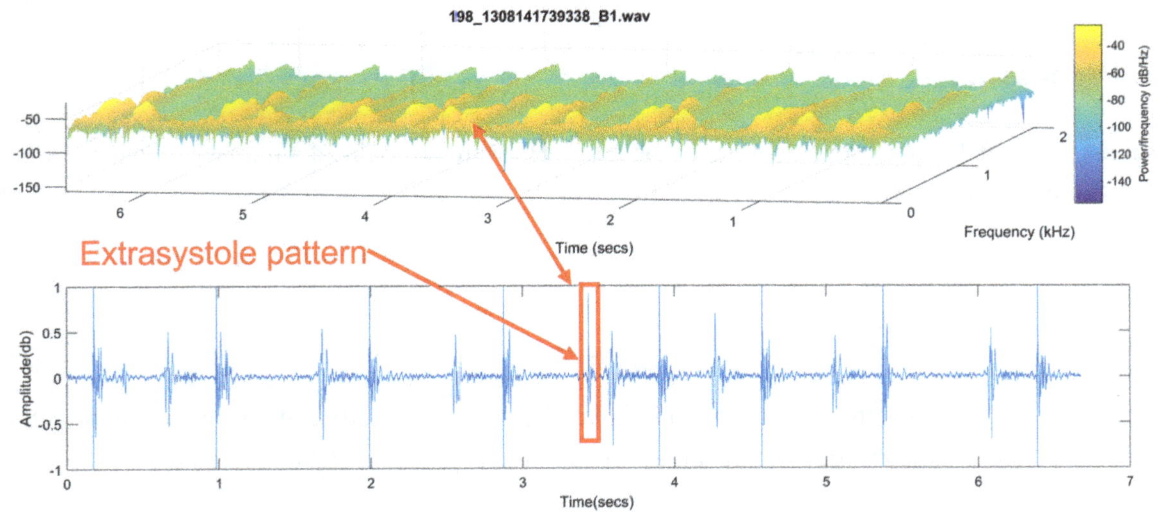


2. If people with cardiovascular disease experience the following cases or operations, occasion of ventricular extra systoles are risky. Researchers point out that the death rate is high in these people, as compared to the ones having no extra systole.
   a. Heart attack,
   b. By-pass operation,
   c. Stent operation
   d. Angiography
   e. Embolism even it is rare,

3. It was witnessed that heart deaths in ventricular extra systoles in every ten beat or more than it occurs with a chance of 2.5 times more than for people, those who haven't experienced it. If the number of extra systole beats is less than that number, it doesn't matter.
4. When extra systoles are multi formed in the sense of different parts of heart, it increases the risk of death.
5. If rise in number of extra systoles, if it is more than 7 in a minute, it comes in doublet or triplet, so it raises the risk of death.
6. In another study, it was indicated that high number of extra systoles during ephor is not important. It is more important when it takes place at the end of ephor.
7. Continuous bigeminal beats may have important results.
8. It is an unexpected case when extra systoles in patients with organic disease have bigeminal beats.

People who frequently experience such beats or has symptoms related with it should be examined by a cardiologist. They are harmless in many people and rarely treatment is required.

In this chapter, HS database which we used have there are four types of HS including 320 normal, 95 murmured, 46 with extra systole design, and 195 unlabeled ones. These sounds were recorded by using Digiscope numbered stethoscope (Bentley et al., 2011).

## METHOD

### Pre-Processing

HS must be normalized before extraction of the noise. HS were normalized with the statement in Equation 1 (Sung et al., 2015).

$$HS_{norm}[n] = \frac{HS[n]}{\max\left(\left|HS[n]\right|\right)} \tag{1}$$

where $HS_{norm}[n]$ denotes the normalized HS signal. There is MATLAB code of reading and plotting normalized extra systole HS from heart sound database (Bentley et al., 2011) in Figure 4.

### Filtering

Cardiac information in HS includes low frequency values. A normal S1 sound is in frequency of 70 -150 milliseconds and 25-45 hertz. A Normal S2 sound is nearly between 60 - 120 milliseconds and with 50 hertz (Phua et al., 2008). Sounds such as lung sound, body movement, and even environmental sounds get mix to HS and these sounds make noise during auscultation with an electronic stethoscope (Zhao et al., 2013). If these noises aren't extracted from heart sounds, accuracy of features belonging to HS can be affected (Zhao et al., 2013). In this case, it is not expected to have correct results from classification. It was noticed by Butterworth and Chebyshev are the best, fastest, and simple filters for extraction of noises from heart sounds (Cherif et al., 2014; Coskun et al.,2016; Smith, 1999). In Figure 5, With the same level ad values, Butterworth, Chebyshev and Elliptic filters within MATLAB™ program were applied to some extra systole heart sounds in heart sounds data base.

*Figure 4. Reading, normalizing, and plotting extra systole heart sound*

```
% Reading Heart Sound from WAV file and plotting
[HeartSound,HeartSoundFrq] = audioread('198_1308141739338_B1.wav');
SamplingNumber = (1/ HeartSoundFrq)*length(HeartSound); % Calculating Samplerate
HeartSoundVal = linspace(0, SamplingNumber,length(HeartSound));
HeartSoundVal = HeartSoundVal/max(abs(HeartSoundVal)); % Normalization of Heart Sound Values
subplot(2,1,1)
spectrogram(HeartSound,100,80,100, HeartSoundFrq,'xaxis')
title('198_1308141739338_B1.wav');
view(-77,72)
shading interp
subplot(2,1,2)
plot(HeartSoundVal, HeartSound);
xlabel('Time(secs) ')
ylabel('Amplitude(db)');
```

*Figure 5. Different filter type design and applying on heart sound*

```
%Design Butterworth bandpass filter that passes  between 20 and 140 Hz
StartFrq = 20 / (HeartSoundFrq/2); % Calculate Start Frequency
EndFrq = 140 / (HeartSoundFrq/2); % Calculate End Frequency
[bwb,bwa] = butter(FilterLevel, StartFrq/EndFrq,'bandpass'); % Create band pass butterworth filter
FilteredHeartSoundButter = filtfilt(bwb,bwa,HeartSound);% Apply band pass butterworth filter
%% Parameter of Chebyshev and Elliptic bandpass filter
Wp = [20 140] / (HeartSoundFrq/2); % passband range of band pass filter
Ws = [15 145] / (HeartSoundFrq/2); % stopband range of band pass filter
Rp = 0.1; % with less than 0.1 dB of ripple
Rs = 40; % at least 40 dB of ripple
[nc,Wp] = cheb1ord(Wp,Ws,Rp,Rs);  % the lowest order nc of the Chebyshev Type I
[cb,ca] = cheby1(nc,Rp,Wp,'bandpass'); % returns the coefficients of an ncth-order bandpass digital Chebyshev Type I
FilteredHeartSoundChebyshev = filtfilt(cb,ca,HeartSound);
[nc,Wp] = ellipord(Wp,Ws,Rp,Rs); %the lowest order nc of the Elliptic
[eb,ea] = ellip(nc,Rp,Rs,Wp); %returns the coefficients of an ncth-order bandpass digital Elliptic
FilteredHeartSoundElliptic = filtfilt(eb,ea,HeartSound);
```

In Figure 6, it is observed that Chebyshev and Elliptic filters are more successful in denosing. and noise extraction was shown with red arrow mark. Taking into consideration all of the data, Elliptic filter was selected to be used in the classification.

## Feature Extraction

Distinctive values that special to only its own class must be used for every class in classification application so that elements in class are collected on its own area. These values are called as feature. The purpose of the feature extraction is data transform to distinctive data arrays that can represent themselves with smaller size to reduce calculation process (Phua et al., 2008). There must be feature extraction process which has minimum diameter class and maximum distance among classes for classification method. Feature array which belong to time and frequency characteristic of signal is used to find high rate accuracy in digital signal classification process (Liang et al., 2005). HS signals are nonstationary signal, features which belong to both time domain and frequency domain has better representation capability for classification. To achieve this features in time domain and frequency domain, mathematical operations are used.

*Figure 6. Noisy heart sound (left), noiseless heart sound (middle), spectrum of noiseless heart sound (right)*

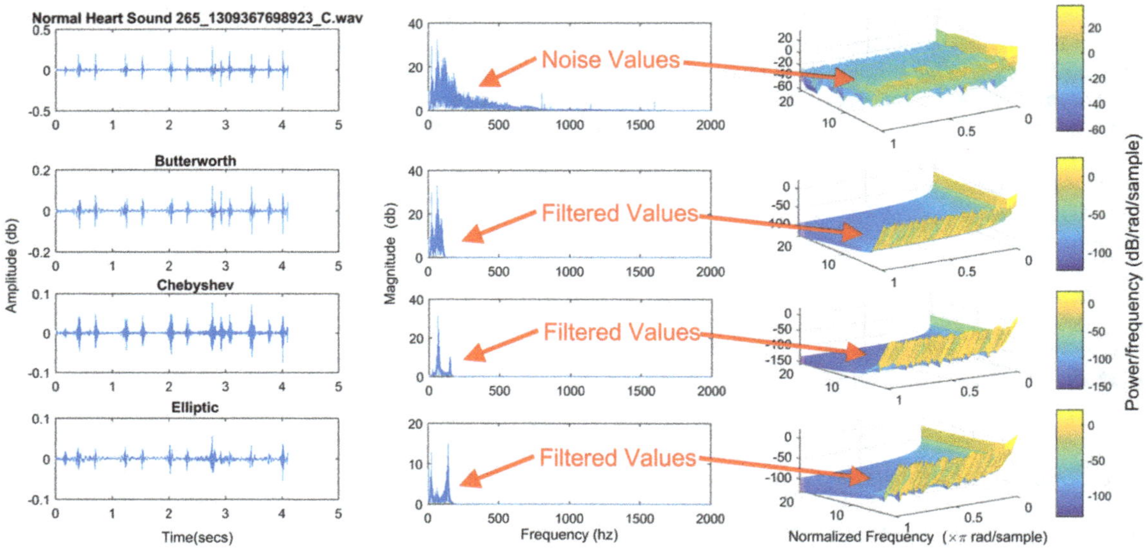


## Discrete Wavelet Transform

Wavelet transform (WT) is transform technique that can separate sound signal to different frequency information and represent each information with resolution where is in scale (Zhang et al., 2001). WT of signal that is function of time is depend on time and frequency variables. Wavelets present more sensitive transform technique for time and frequency analysis. WT is more outstanding than other time-frequency analysis methods now that it can perform window sizing that can be change to wide in low frequency and to narrow in high frequency for time-frequency analysis of continuous signal (Daubechies, 1990).

Continuous Wavelet Transform (CWT) calculate integral of value that was occurred result of multiplication with value which was obtained by mother wavelet shifts and scales during entire time domain. Also CWT can be defined as transform that shift and scale variables are continuous value. At the end of this calculation, wavelet coefficients are obtained by time position and scale value. There is very much working load in CWT thereby calculation performance reduce and working time extend. Mathematical equation of CWT is in Equation 2;

$$\Psi_{a,b}(t) = \frac{1}{\sqrt{a}}\Psi\left(\frac{t-b}{a}\right) \tag{2}$$

In this equation, $a > 0$, a,b ∈ $\Re$ and $a$ is scale variable, $b$ is transform variable, $x(t)$ is value of signal in $t$ time, $\Psi$ is wavelet function (mother wavelet), $1/\sqrt{a}$ is normalization factor, $\Psi(a,b)$ is wavelet transform of signal (Pérez-Rendón et al., 2004). Shifting and scaling variables take discrete value in Discrete Wavelet Transform (DWT). Signal is examined at specific scale thereby better calculating performance is obtained by CWT and correct results are obtained as CWT. Shifting and scaling is done by taking multiples of 2 in DWT and dyadic wavelet transform is obtained. Mathematical equation of DWT in Equation 3.

$$\Psi_{j,k}(t) = 2^{-\frac{j}{2}} \Psi\left(2^{-j} \cdot t - k\right) \tag{3}$$

In this equation

$$a = 2^j, b = k \cdot 2^j = k \cdot a, j, k \in Z^2,$$

*j* and *k* are scaling and transform number respectively. DWT functions create an orthogonal cluster. DWT coefficients are found with mathematical equation in Equation 4 (Mahmoodabadi et al., 2005).

$$c_{j,k} = 2^{j/2} \int_{-\infty}^{\infty} f(x) \cdot \Psi\left(2^j x - k\right) dx \tag{4}$$

Signal in DWT must be separated to high and low frequencies fractionally (Mallat, 2008). Detail coefficient of signal is calculated by using high pass filters to calculate variance with high frequency in sound signal. Approximation coefficient of signal is calculated by using low pass filters to calculate changing with low frequency in sound signal. These filters are applied by using convolution method.

Convolution method is a mathematical operation on two functions, producing a third function that is generally viewed as a modified version of any of original functions, giving the area overlap between the original functions as a function of the amount that one of the original functions are translated. In other words, it can be defined as one signal passing through other signal. As such, Convolution method is similar to cross-correlation (Ahn, 2015).

$$y[n] = x[n] * h[n] = \sum_{k=-\infty}^{\infty} x[k] \cdot h[n-k] \tag{5}$$

Convolution method is defined in Equation 5. *x[n], h[n], y[n]* are defined as input signal, impulse response and output respectively. The * (star) operator is symbolized as convolution method. For decomposition of heart sound signal to different frequency segments, k level approximation and detail coefficients are obtained at k step by applying convolution method to time signal *x[n]* with high pass and low pass filters in Equation 6 and Equation 7 respectively.

$$cD_1 = y_{low}[k] = \sum_{n=-\infty}^{\infty} x[n] \cdot h[2k-n] \tag{6}$$

$$cA_1 = y_{high}[k] = \sum_{n=-\infty}^{\infty} x[n] \cdot g[2k-n] \tag{7}$$

First level approximation coefficients ($cA_1$) represent high scale and low frequency component of signal for $k = 1$. First level detail coefficients ($cD_1$) represent low scale and high frequency component of signal. After first level decomposition, the highest frequency value in signal became $\pi$ instead $\pi / 2$, therefore half of signal can be eliminated by Nyquist Criterion. By this means, signal has been applied down sampling with 2 as shown in Figure 7.

Approximation coefficients are smaller than original signal, they have value that can be represent original signal nearly because they are produced with low pass filter (Mallat, 2008).

If two filter output signal at same decomposition level are passed through integration filters and are collected with up sampling, original $x(n)$ signal can be achieved. The signal which applied DWT can be created by wavelet coefficients again therefore it is accepted more effective, reliable and prevalent method in data compression in image and sound processing etc., noise removal, feature extraction problems.

The db6 type and 6th level wavelet approximation coefficients ($a_6$) attained with the with MATLAB™ program. Original heart sound signal is represented as ($s$) and wavelet coefficients are represented as ($a6$) in Figure 8.

There is extraction of wavelet features of heart sound in in Figure 9.

It is necessary that only sixth level wavelet approximation coefficients are used as feature set as *WaveletFeatures*. In Figure 9 *CoefficientsLengt* variable have all coefficients length. The value of sixth level wavelet approximation coefficients length is first element of *CoefficientsLengt* variable. Because of this, first 81 values of *Coefficients* variable are used as feature set.

*Figure 7. Wavelet tree, level 1 and level 2 coefficients*

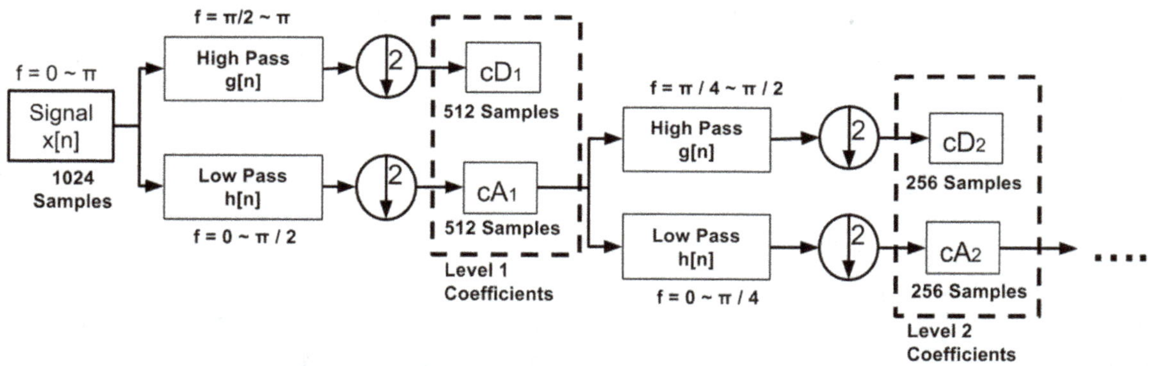


*Figure 8. Normal heart sound (top) and approximation coefficients at sixth level in db6 type via MATLAB™ (bottom)*

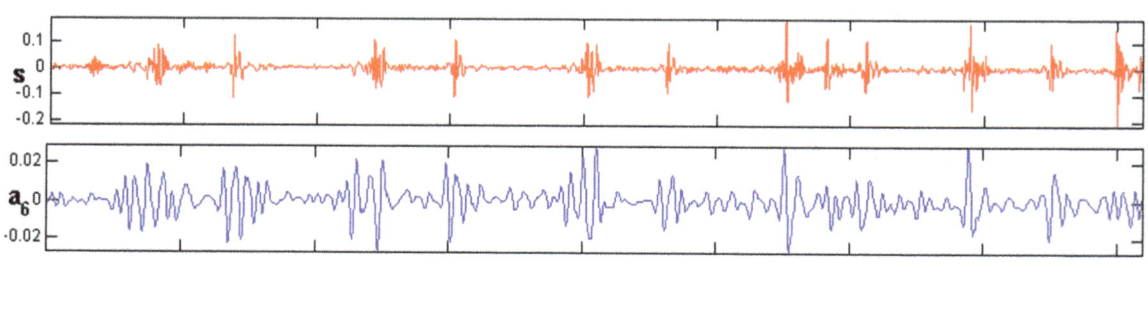

*Figure 9. MATLAB code of wavelet features extraction*

```
%% Wavelet Transform
WaveletLevel = 6;
WaveletName = 'db6';
[Coefficients,CoefficientsLength] = wavedec(FilteredHeartSoundElliptic,WaveletLevel,WaveletName);
%WaveletFeatures = Coefficients(1:CoefficientsLength(1));
WaveletFeatures = appcoef(Coefficients,CoefficientsLength,WaveletName,WaveletLevel);
```

## Mel-Frequency Cepstrum

Mel-Frequency Cepstrum (MFC) is a transform of a sound signal using short-term power spectrum based on linear cosine transform of a log power spectrum. Log power spectrum of sound signal is on a nonlinear mel-scale of frequency and mimics the human hearing (Li, 2000). The MFCC is widely used in automatic speech recognition, music and speech classification problem. MFCC is based on human hearing sense which cannot receive frequencies over 1 Khz (Vergin et al., 1996). Also MFCC based on variation of human ear's limit bandwidth with frequency (Dao et al, 2017; Openshaw et al., 1993). After MFC transform of sound signal, Mel-Frequency Cepstral Coefficients (MFCCs) are gained using Fast Fourier Transform (FFT). This transform consists of these steps: Pre-Emphasis, Framing, Windowing, Mel-Filter Bank, Log energy computation and Discrete Cosine Transform (DCT). These steps are shown in Figure 10.

In pre-emphasis step, a high pass filter is applied on HS. Calculation belong to this step is given in Equation 8.

$$HS(n) = hs(n) - a * hs(n-1) \tag{8}$$

In Equation 8 *HS(n)* is defined as output signal and $a$ value is used between 0,9 and 1,0 in general. Purpose of this step is to balancing of high frequency values which are suppressed during human hearing. In this way, importance of sections with high frequency can be increased. HS have become more keen and have smaller volume after pre-emphasis step.

The MATLAB code of this step is given in Figure 11. In framing step, HS signal is split as frame length of 20-30 ms. Generally, frame size is equal to square of 2 to easily using of FFT. For example, FTT size can be 512, 1024 or 2048. Each frames of HS are multiplied by a Hamming Window to protect continuity of first and last point in windowing step. If the signal in a frame is defined as *s(n), n = 0,1,.....*

*Figure 10. Mel-frequency cepstrum calculation steps*

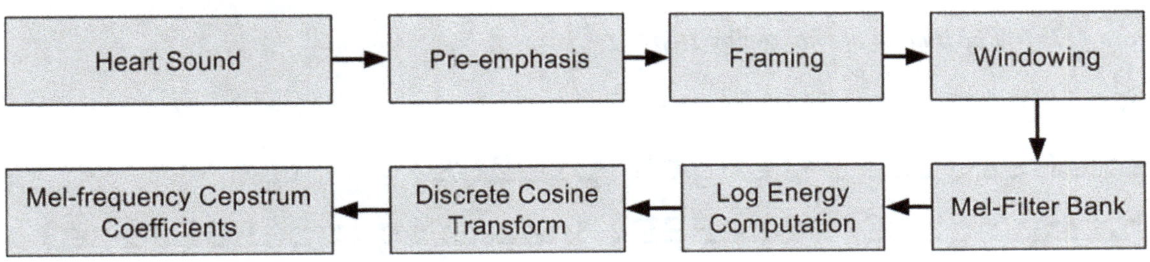

*Figure 11. MATLAB code of pre-emphasis step*

```
filterNum = 20; % Filter number
mfccNum = 20; % mfcc Number
%% Pre-emphasis Step
a=0.95;
preHSE = filter([1, -a], 1, FilteredHeartSoundElliptic);
```

*N-1*, it becomes as *s(n) * w(n)* after windowing step and *w(n)* is Hamming Window. Hamming window is explained in Equation 9 and value of *a* is taken as 0,46.

$$w(n) = \begin{cases} (1-a) - a\cos\left(\dfrac{2\pi n}{N-1}\right), 0 \le n \le (N-1) \\ 0, otherwise \end{cases} \quad (9)$$

The MATLAB code of windowing step is given in Figure 12.

Mel filter step is applied in frequency domain. Therefore, each frame is transported to frequency domain from time domain using FFT. This step is carried out to obtain frequency response of each frame. Discrete Fourier Transform (DFT) is used to obtaining mel filter. DFT is a linear transform which takes as input a complex signal $x$ of length *N* and gives as output a complex signal *X* of length $N, X = W \cdot x$. The *W* is a complex *NxN* matrix. DFT is given in Equation 10.

$$X(k) = \sum_{n=0}^{N-1} x(n) e^{\frac{-j2\pi kn}{N}}, k = 0,1,2,..,(N-1) \quad (10)$$

The MATLAB code of Discrete Fourier Transform using in mel-filter bank step is given in Figure 13 (Matworks, 2016).

Some group of triangular filter are used to calculate weighted summation filter spectrum component in Mel filter (melody filter) step of MFC transform. These filters have band pass frequency response and they are found at even intervals in mel-scale. The MATLAB code of obtaining mil-filter bank parameter is given in Figure 14.

Triangular filters are used as the amount of the obtained coefficients. The MATLAB code of applying triangular filter is given in Figure 15.

*Figure 12. MATLAB code of windowing step*

```
%% Hamming Window Step
preHSESize =length(preHSE);
hammingHSE =preHSE.*hamming(preHSESize);
```

*Figure 13. MATLAB code of discrete fourier transform*

```
% Appliying FFT to move frequency domain
[frameSize, frameCount]=size(hammingHSE);
freqStep = HeartSoundFrq /frameSize;
% Spectrum
z = fft(hammingHSE);
% Frequency vector
fftFreq = freqStep*(0:frameSize/2)';
% taking One side
z = z(1:length(fftFreq), :);
% Assuming N is even, symmetric data is multiplied by 2 for one side
z(2:end-1, :)=2*z(2:end-1, :);
% Magnitude spectrum
magSpec=abs(z);
% Power in db
powerSpecDb=20*log(magSpec);
```

*Figure 14. The MATLAB code of mel-filter bank parameter*

```
%% Get parameters for triangular filter bank
fLow=0;
fHigh=HeartSoundFrq /2;
% Calculate the frequencies of the triangular filters
for i=1:filterNum+2
    fLowLinFreq= 700*(exp(fLow/1125)-1);
    fHighLinFreq = (700*(exp(fHigh/1125)-1));
    linFreq = fLowLinFreq + (i-1)*(fHighLinFreq - fLowLinFreq)/(filterNum+1);
    f(i) = 700*(exp(linFreq/1125)-1);
end
triFilterBank=[];
for i=1:filterNum
    triFilterBank=[[f(i); f(i+1); f(i+2)], triFilterBank];
end
```

Obtained sample filter using in mel-filter bank step is given in Figure 16 (Jang, 2010).

Original frequency scale is transformed to mel-scale in log energy computation step. This transform equation is given Equation 11. $a$ value can be 2095 or 1125, $f$ denotes the real frequency, and *Mel (f)* denotes the perceived frequency in Equation 11. Mel (melody) is a unit of pitch, In other words, Mel is

*Figure 15. The MATLAB code of triangular filter*

```
%% Appliying triangular filter.
for i=1:filterNum
    triFilterCoef(i)=dot(powerSpecDb, trimf(fftFreq, triFilterBank(:,i)));
end
```

*Figure 16. Frequency response of triangular filter bank with mel intervals*

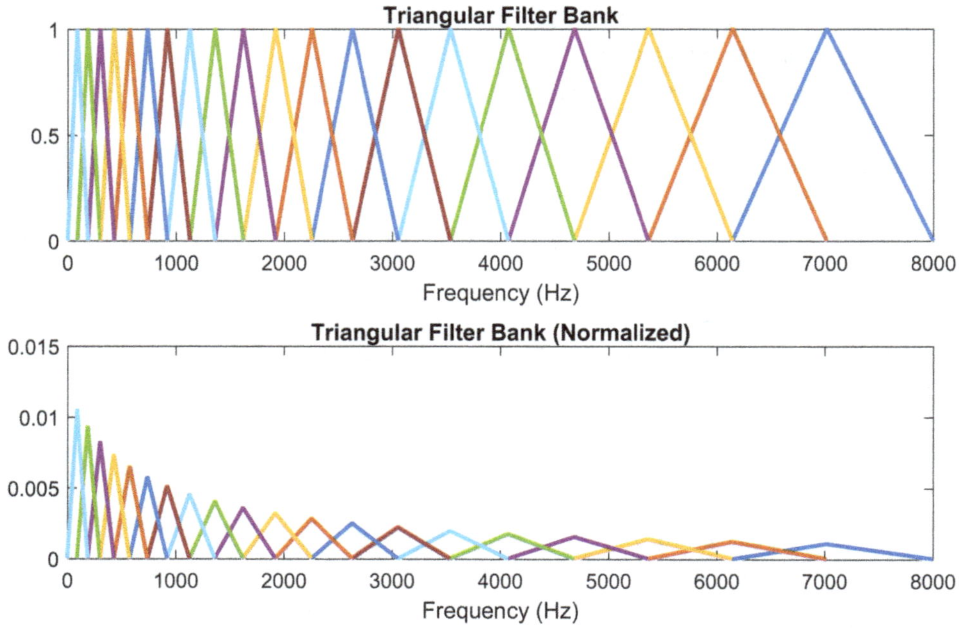


a unit of frequency measurement of sound or tone which is perceived (Proakis et al., 2007). As reference 1 kHz has been selected and it has been accepted that equals to 1000 Mel.

$$Mel(f) = a \cdot \log_{10}\left(1 + \frac{f}{700}\right) \tag{11}$$

The MATLAB code of log energy computation step is given in Figure 17.

In Mel-filter bank process each filter output is the sum of its filtered spectral components (Hai et al., 2015). Mel filter bank process is given Figure 18.

DCT is applied for transporting of coefficients from frequency domain to time domain in last step. DCT is given in Equation 12.

$$C_j = \sum_{n=1}^{M} L_n \cdot \cos\left(\frac{(n-0,5)\cdot j \cdot \pi}{M}\right), j = 0,1,2,\ldots,J \tag{12}$$

*Figure 17. The MATLAB code of log energy computation*

```
%% Log Energy Computation Step
logEnergy=10*log10(sum(FilteredHeartSoundElliptic.*FilteredHeartSoundElliptic));
mfcc=[logEnergy; mfcc];
```

*Figure 18. Mel-filter bank process*

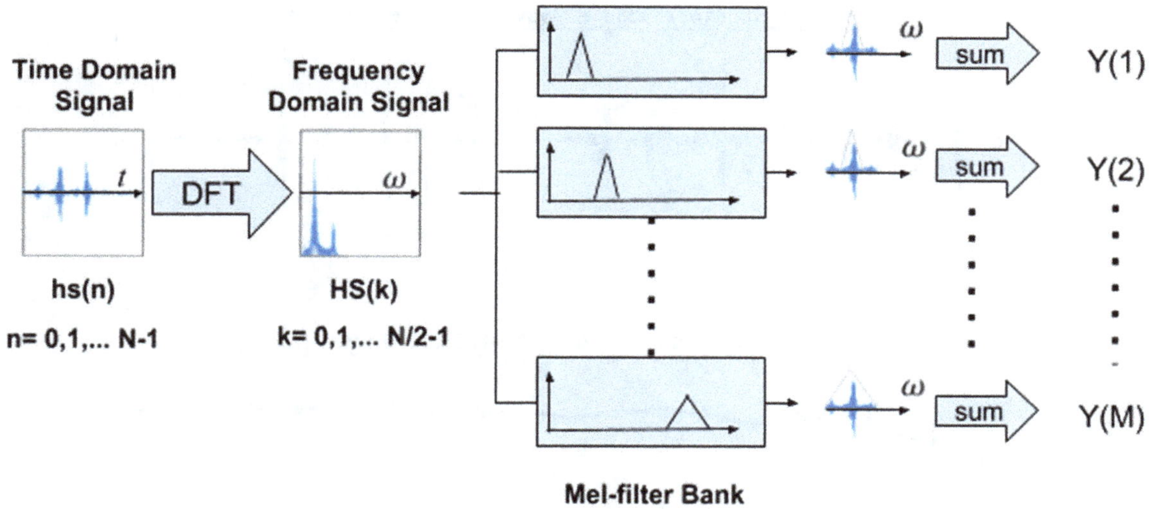


In Equation 12, *M* denotes number of filter in filter bank, *J* denotes number of cepstrum coefficients, $L_n$ denotes calculated log energy output of *i*th filter in previous step of DCT and *C* denotes MFCC. The MATLAB code of DCT step is given in Figure 19.

MFCC vector of 198_1308141739338_B1.wav extra systole HS in database is given in Figure 20.

In this study, 20 triangular filters were used and 20 MFCC coefficients were gained for each HS in database.

Once these features of the HS are obtained, it can be passed to the classification process. Averages (AVG) and standard deviations (SD) of Wavelet and MFFC features of normal and extra systole HS are taken to use as training and test data for input of classification algorithm. The MATLAB codes of AVG and SD of Wavelet and MFCC features are given Figure 21.

Means and standard deviation values of five of normal and extra systoles are given in Table 1.

The codes in Figure 21 need to be applied on both normal and extra systole heart sound. Accordingly, there need to be two MFCC and Wavelet features vector belong to normal and extra systole respectively at the end of this step.

*Figure 19. The MATLAB code of discrete cosine transform*

```
%% Discrete Cosine Transform Step
mfcc=zeros(mfccNum, 1); % mffc array
mfcc=[logEnergy; mfcc];
for i=1:mfccNum
    coef = cos((pi/filterNum)*i*((1:filterNum)-0.5))';
    mfcc(i) = sum(coef.*triFilterCoef');
end
```

*Figure 20. MFCC vector of extra systole HS*

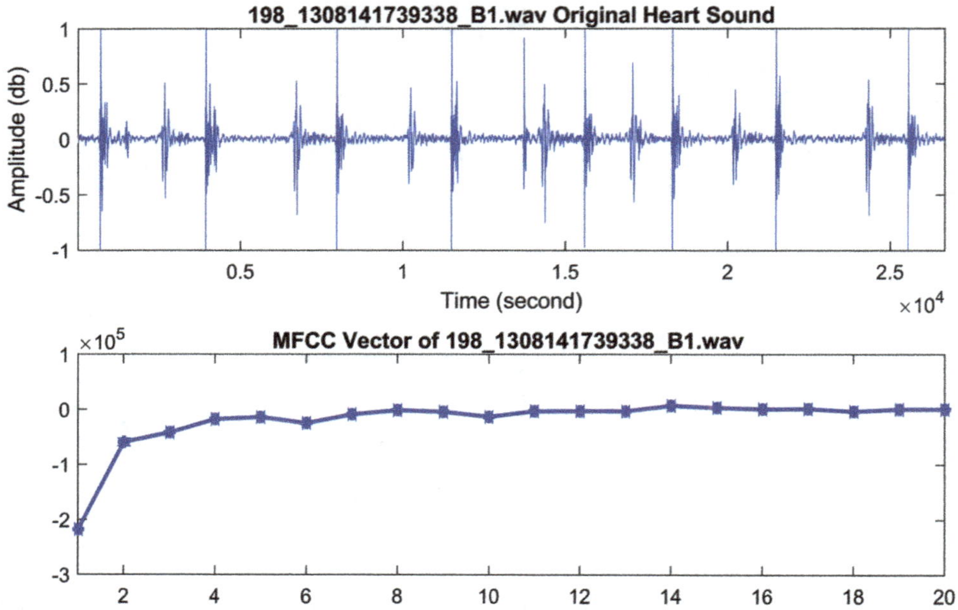


*Figure 21. The MATLAB code of averages and standard deviations of wavelet MFCC features*

```
%% Averages and Standard Deviation of Wavelet Feature Set
AVGWaveletFeatures =  mean(WaveletFeatures);
SDWaveletFeatures  =  std(WaveletFeatures);
%% Averages and Standard Deviation of MFCC Features Set
AVGMFCCFeatures =  mean(mfcc);
SDMFCCFeatures  =  std(mfcc);
```

*Table 1. Means and standard deviation values of HS*

| No. | Heart Sound Name | Class | MFCC Features | | Wavelet Features | |
|---|---|---|---|---|---|---|
| | | | Mean | Standard Deviation | Mean | Standard Deviation |
| 1 | 127_1306764300147_C2 | E | -7,758768 | 2,9663109 | -0,00197 | 0,1029 |
| 2 | 128_1306344005749_A | E | -7,878848 | 2,9293624 | 0,00102 | 0,0848 |
| 3 | 130_1306347376079_D | E | -7,674664 | 2,8264248 | 0,00734 | 0,0518 |
| 4 | 134_1306428161797_C1 | E | -8,332766 | 3,0741382 | -0,00407 | 0,0614 |
| 5 | 138_1306762146980_B | E | -7,287901 | 2,7722949 | -0,00387 | 0,1268 |
| 6 | 103_1305031931979_B | N | -4,34994 | 1,616149 | -0,000856 | 0,0087727 |
| 7 | 103_1305031931979_D1 | N | -4,45345 | 1,641038 | 0,0007087 | 0,0115639 |
| 8 | 103_1305031931979_D2 | N | -3,52541 | 1,311176 | -0,0003998 | 0,0095613 |
| 9 | 103_1305031931979_D3 | N | -6,44657 | 2,343937 | -4,45766E-05 | 0,0056838 |
| 10 | 106_1306776721273_B1 | N | -5,79628 | 2,217364 | -0,0003562 | 0,0110406 |

E: Extra systole N: Normal

## Classification

In machine learning and statistics, classification can be defined as a problem that research a data set belongs to which in category or class. This category or class represents a learning model that is determined earlier with classification algorithm using training data similar this data set. It is necessary to know that this training data belongs to which of class in creating learning model. Along the study, 45 extra systoles patterned heart sounds were used from HS database. At this point 30 of them were used as training data for classification while remaining 15 ones were for the test. Also Support Vector Machine (SVM) and Artificial Neural Network (ANN) classification methods are used to create different classification model. In this section, these two classification algorithm has been discussed in detail. Firstly, general information about classification methods have been given. Afterwards, mathematical definitions of each method has been explained in step by step. After these explanations, MATLAB applications of HS classification using each method have been given to more understandable.

If sample class label and class data was found positive after the calculation, it is defined as true positive (TP) and when sample class label is positive but class data is found as negative it is called as false- negative (FN). When sample class label is negative and class data was found negative it is called as true – negative. Finally, if when sample class label is negative but class data was found positive, it is then called as false positive (FP). The explanations in Table 2 follows as: A = Line Total, B = Classification accuracy of every class in percentage, C = Number of heart sound which classified in X class while in Y class.

Also to measure success of classification, certainty ($c$), sensitiveness ($s$), accuracy ($a$) values and confusion matrix need to be calculate via Equation 13, 14, 15 with the use of definitions provided in Table 2.

$$c = TN \,/\, (TN + FP) \tag{13}$$

$$s = TP \,/\, (TP + FN) \tag{14}$$

$$a = (TP + TN) \,/\, N \tag{15}$$

*Table 2. Calculation of performance analysis*

| | | Real Status | | | | |
|---|---|---|---|---|---|---|
| | | **Positive** | **Negative** | **A** | **B** | **C** |
| **Predicted Status** | **Positive** | True Positive (TP) | False Positive (FP) | TP + FP | TP*100/A | FP |
| | **Negative** | False Negative (FN) | True Negative (TN) | FN + TN | TN*100/A | FN |
| **Sum** | | TP + FN | FP + TN | | | |

## Preparing Data to Classification

All steps which are up to classification have been applied on one heart sound with MATLAB. Whereas, vector data of means and standard deviations which is used in classification step involves all heart sounds values. For this reason, the reader needs to apply the steps which are up to classification step in a loop to obtain these vectors up to the number of heart sounds.

The means and standard deviation values of normal and extra systole HS features need to be transformed to a specific format to classification. Therefore, primarily these values have been transformed to column vector while they're row vector in MATLAB. Transforming to column vector of MFCC and Wavelet features are given in Figure 22.

Afterwards, first 30 columns and last 15 columns of normal and extra systole heart sound features vector have been selected as training data and test data respectively in Figure 23.

In Figure 23, mean variables starting with AVG and standard deviation variables starting with SD represent MFCC features that respectively belong to extra systole and normal heart sound which mentioned in last paragraph of Feature Extraction section. In this study, definition of MFCC features of each sound is given and explained. However, if Wavelet features of heart sounds are wanted to use for classification, mean and standard deviation variables starting with AVG and SD of Wavelet features in Figure 22 need to be created and to be used in creating of Training and Test Data of each heart sound type in Figure 23.

*Figure 22. Column vector MFCC and wavelet features*

```
%% Row vector of Mean and SD transform to column vector
AVGMFCCFeatures = AVGMFCCFeatures';
SDMFCCFeatures = SDMFCCFeatures';
```

```
%% Row vector of Mean and SD transform to column vector
AVGWaveletFeatures = AVGWaveletFeatures';
SDWaveletFeatures = SDWaveletFeatures';
```

*Figure 23. Creating training and test data vectors of normal and extra systole HS*

```
%Creating Training and Test Data
% of Extra systole and Normal Heart Sound
TraniningDataExtra = [AVGMFCCFeaturesExtra(1:30)
                      SDMFCCFeaturesExtra(1:30)];
TestDataExtra = [AVGMFCCFeaturesExtra(31:45)
                      SDMFCCFeaturesExtra(31:45)];

TraniningDataNormal = [AVGMFCCFeaturesNormal(1:30)
                      SDMFCCFeaturesNormal(1:30)];
TestDataNormal = [AVGMFCCFeaturesNormal(31:45)
                      SDMFCCFeaturesNormal(31:45)];
```

The training vectors of normal and extra systole heart sound have been merged to transform to just one training data in Figure 24.

A column vector named *Class* with 60 values has been created to define output values of classifier in Figure 25. First 30 values of *Class* vector have been assigned as +1 to represent extra systole heart sounds and last 30 values of its have been assigned as -1 to represent normal heart sounds.

## Support Vector Machines

Support vector machines (SVM), developed by Vladimir Vapnik and Alexey Yakovlevich Chervonenkis in 1960s, a is a machine learning function, which is based on statistical learning theory and used in pattern recognition, curve fitting, decision structures and classification. Briefly, SVM can be defined as separation of data on the same plane belonging to different classes thanks to bound line formed by linear equation (Cortes,et al., 1995). SVM is a core based learning approach giving fast and correct results in such issues; linear and nonlinear classification, regression analysis, identification of outlier, function and density estimation. It is especially used in classification problems when patterns between variants of data to be classified in data mining are not known.

Advantages and Disadvantages of SVM can be listed as follows (Nitze, 2012; Çomak,et al., 2007; Auria, 2008; Cawley et al., 2010):

- It doesn't require rudiment assumption about distribution.
- It gives high accuracy values.
- It can model complex decision boundaries.
- It can work with many independent variables.
- It has a regularisation parameter, which makes avoiding over-fitting.
- SVM is defined by a convex optimisation problem which has not local minima.
- It can work with data which cannot be classified as linear or nonlinear.
- In SVM, overfitting problem is less than other classification methods

*Figure 24. Creating a training data*

```
%Creating a training vector by merging two training vector
TrainingData = [TrainingDataNormal;TrainingDataExtra];
```

*Figure 25. Creating class variable*

```
% Value of 1 for Extra systole Heart Sound
Class = ones(60,1);
% Value of -1 for Normal Heart Sound
Class(1:30) = -1;
```

- It produces deterministic results instead of probabilistic results. (Like existent- non-existent, A class B class).
- In the context of core functions, it needs positive and constant symmetrical functions (Mercer Condition).

SVM can be analyzed under two main titles: linear and nonlinear. In continuation, only Linear SVM has been explained because of it is used HS classification.

## Linear Support Vector Machines

It is the first basic form of SVM. It is used to classify dual classed and multi classed data which are separated from each other by a linear function in basically. Training data is as follows when **n** dual dimensioned, and classed data set is considered;

$$(x_1, y_1), (x_2, y_2), (x_3, y_1), (x_4, y_1), \ldots..(x_i, y_i), i = 1, 2, 3..., n \tag{16}$$

In Equation 16, $x \in \Re^d$ resembles d-dimensioned features vector, ***y*** vector symbolizes class labels. For dual dimensioned *n* elemented input data $x \in \Re^d$ and $y \in \{+1, -1\}$. Support vectors, in condition that x point is $w = \left[w_1, w_2, w_3, \ldots.., w_n\right]^T$ and $w \in \Re^N$ identified in Equation 17 as *d* differentiation or decision function. The *b* in Equation 17 is a constant number symbolizing bias value.

$$d(x, w, b) = w^T x + b = \sum_{i=1}^{n} w_i x_i + b \tag{17}$$

According to Equation 17 separative striped line presented in Figure 27 can be defined in Equation 18:

$$w^T x + b = 0 \tag{18}$$

Canonical upper hyper planes are described as;

$$D1 = w^T x + b = +1 \tag{19}$$

$$D2 = w^T x + b = -1 \tag{20}$$

Let $w$ symbolizes weight vector in Equation 19 and 20, distance of $x_1$ point to upper plane on D1 line $d_1$, and distance of $x_2$ on D2 line to upper lane $d_2$ in turn symbolizes in class 1 and 2 distance of samples closest to upper plane to upper plane M (margin) = $d_1 + d_2$ is accepted as differentiation value. The M can be defined geometrically in Equation 21.

$$M = \frac{2}{\left\| w \right\|} \tag{21}$$

SVM aims to find an upper plane to take differentiation value M on countless number of upper planes to the highest level (Kechman, 2001). It is called as Interval Maximization. In order to maximize M it is necessary to minimize $\left\| w \right\|$. Dual classes and linear support vectors are shown in Figure 26 (Kechman, 2001).

Many different upper planes separating these two classes are represented in Figure 27 (Kechman, 2001). The best appropriate one is selected according to Interval Maximization rule. In order to identify this plane, two upper planes parallel to these lines for defining class boundaries must be determined. Separative line in the larger interval formed by upper planes which determine boundary line, that is the one on the right side is accepted as better separator according to interval maximisation. Upper planes for determining class boundaries are called as support vectors.

Weight vector is obtained according to separative lines, after a successful training stage. A new unknown ***x*** data set to be classified, obtains ***o*** output according to display function given in the Equation 22 (Kechman, 2001).

$$i_f = o = sign\left(d(x,w,b)\right) \tag{22}$$

The ***o*** support vector in this equation is a standard output from learning machine. According to this output, decision rules can be written as follows;

If $d\left(x_p, w, b\right) > 0, x_p$ belong to class 1 $\left(o = y_1 = +1\right)$

*Figure 26. Linear support vector machine*

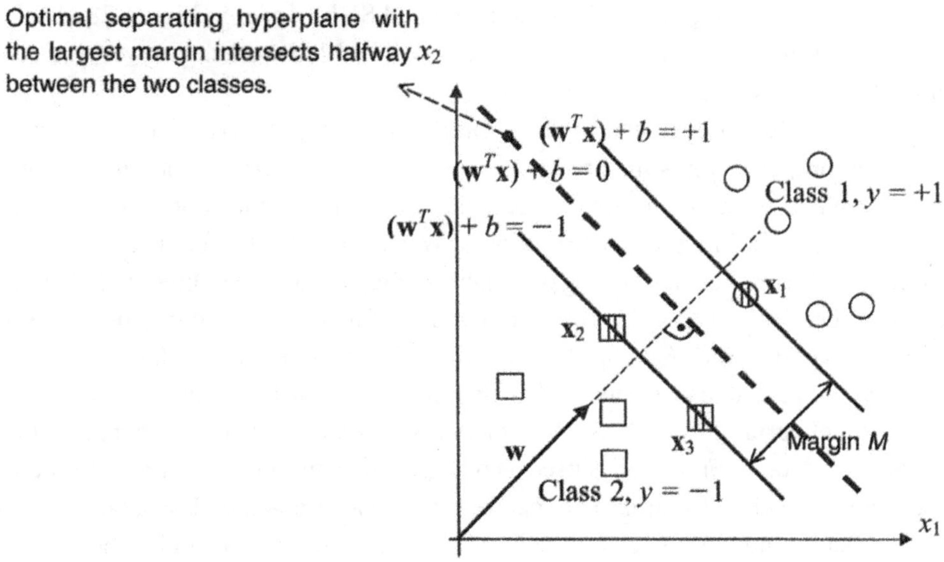

*Figure 27. Best separator plane selection*

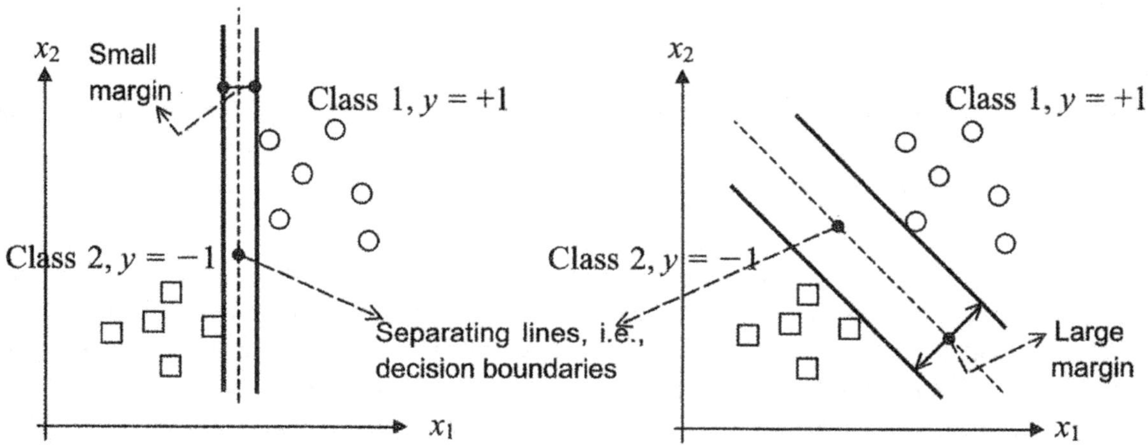


If $d\left(x_p, w, b\right) < 0, x_p$ belong to class 1 $\left(o = y_2 = -1\right)$

There are two functions as svmtrain and fitcsvm in MATLAB towards classification using SVM. svmtrain function is out of date since R2013b. But it can be still usable. fitcsvm function is introduced in R2014a. In this section, fitcsvm function has been used and explained to classification of HS. This function trains a SVM model for two-class (binary) classification on a low dimensional predictor data set and supports mapping the predictor data using kernel functions. fitcsvm function have many parameters but it has to use two parameters basically to classification. First parameter is sample data that used to train model, specified as a table. Each row and column of sample data match up with one sample record and feature respectively. This sample record represents a class element in each class. Other basic parameter of fitcsvm is response variable name. This parameter must be a class or character array, logical or numeric vector, or cell array of character vectors. In this study, TrainingData variable is used as sample data and class variable is used as response variable name in Figure 28. fitcsvm function using these parameters returned an output variable as a trained SVM model named SVMModel in Figure 28. The SVMModel has many properties about SVM classification of classes. The Mathworks.com web site can be visited or help command with fitcsvm can be used for this.

If training data and SVM are wanted to draw, *SupportVectors* property of *SVMModel* can be used. The *SupportVectors* property stores the predictor values for the support vectors, including the dummy variables. For example, if there are *m* support vectors and three predictors, one of which is a categorical variable with three levels. Training data distribution of Wavelet and MFFC features is given in Figure 29.

After creating SVM model using training data, labels of test data have to predict. SVM model is used to predict labels of test data. In MATLAB, there is predict function to predict labels of test data. This function has only two parameters. One of these parameter is SVM model and the other is predictor data to be classified. Predictor data can be numeric matrix or table in just the same way as training data. In this study, SVMModel variable is used as SVM model which created using fitcsvm function in Figure 28. TestData which created in Figure 23 is used as predictor data in Figure 30. predict function returned two output variable as label and score. Label can be defined as predicted class label. It can be class or character array, logical or numeric vector, or cell array of character vectors and is the same data type as

*Figure 28. Creating SVM model*

```
%Creating SVM Model using TrainingData and Class Label
SVMModel = fitcsvm(TrainingData,Class);
```

*Figure 29. Training data of wavelet (a) and MFCC (b) features*

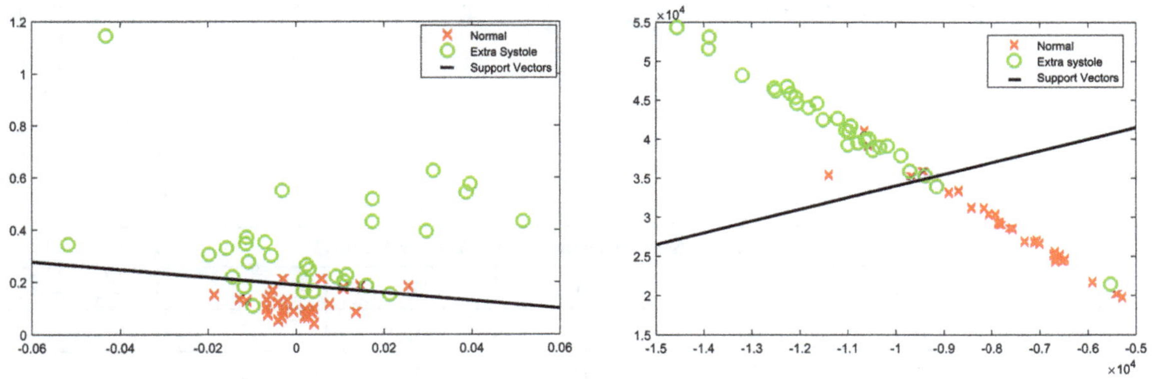


*Figure 30. Prediction of normal and extra systole HS test data*

```
%Predict Test Data of Normal and Extra systole Heart Sound
[LabelNormal, ScoreNormal] = predict(SVMModel ,TestDataNormal);
[LabelExtra, ScoreExtra] = predict(SVMModel ,TestDataExtra);
```

the observed class labels response variable name in fitcsvm function. Other output value named score is defined as predicted class scores or posterior probabilities returned as numeric column vector or matrix. For classification with two-class score is a two-column matrix with the same number of rows as test data.

Test data classification of Wavelet is given Figure 31 (a).

As shown in Figure 31 (a), normal HS data and data which classified in normal class while in extra systole class are shown in purple plus marker, also extra systole HS data and data which classified in extra systole class while in normal class are shown in blue star marker. Test data classification of MFCC features is given Figure 31 (b). Confusion matrix which calculated using Table 2 according to classification results is given in Table 3.

Performance analysis values calculated using values in Table 3 are given in Table 4.

## Artificial Neural Network

Artificial Neural Network (ANN) is modeling of the human brain layered and parallel structure which inspired from neural cell with all its functions. This modelling of human brain is realized in digital world which created by hardware and software. This model simulates learning process of human brain while

*Figure 31. Classification of test data of wavelet (a) and MFCC (b) features*

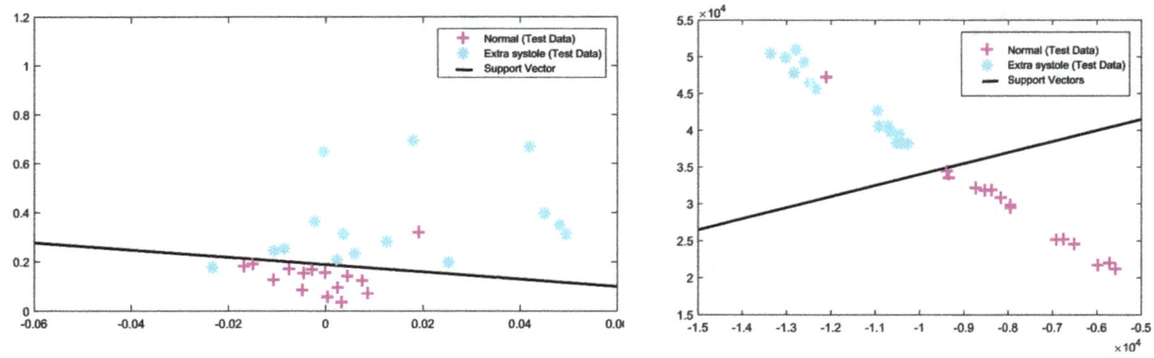


*Table 3. Confusion matrix*

| Number of Heart Sound Classified Using Wavelet Features | | | | | | |
|---|---|---|---|---|---|---|
| **Classes** | **Extra Systole** | **Normal** | **A** | **B** | **C** | |
| Extra systole | 14 | 1 | 15 | 93,33 | 2 | |
| Normal | 1 | 14 | 15 | 93,33 | 1 | |
| | 14 | 16 | 30 | | 3 | |
| **Number of Heart Sound Classified Using MFCC Features** | | | | | | |
| **Classes** | **Extra Systole** | **Normal** | **A** | **B** | **C** | |
| Extra systole | 15 | 0 | 15 | 100 | 0 | |
| Normal | 1 | 14 | 15 | 93,33 | 1 | |
| | 16 | 14 | 30 | | 1 | |

*Table 4. Values of performance analysis*

| Feature Set | Certainty | Sensitiveness | Accuracy |
|---|---|---|---|
| Wavelet Features | 0,93 | 0,9333 | 0,93 |
| MFCC Features | 1,0 | 0,9375 | 0,96 |

it working. Therefore, structure and working of biological neural cell should be analyzed to understand structure of this digital model and its working.

The structure of biological neural cell is given in Figure 32. A biological neuron consists of dendrite, cell body, axon and synapse. A dendrite transmits signals came from other neuron to cell body. The cell body has a nucleus of neuron and collect all signals which came from dendrites. The axons are responsible to transfer all signals which came from cell body to synapses. The synapses transfer all total information which came from axon to dendrite of another neuron cell after pre-processing. This pre-processing of information by synapse is highly important. Because, this pre-processing is made up of change of total signals according to specific threshold value. Thus, total signal is reduced to a specific

*Figure 32. Biological neuron cell*

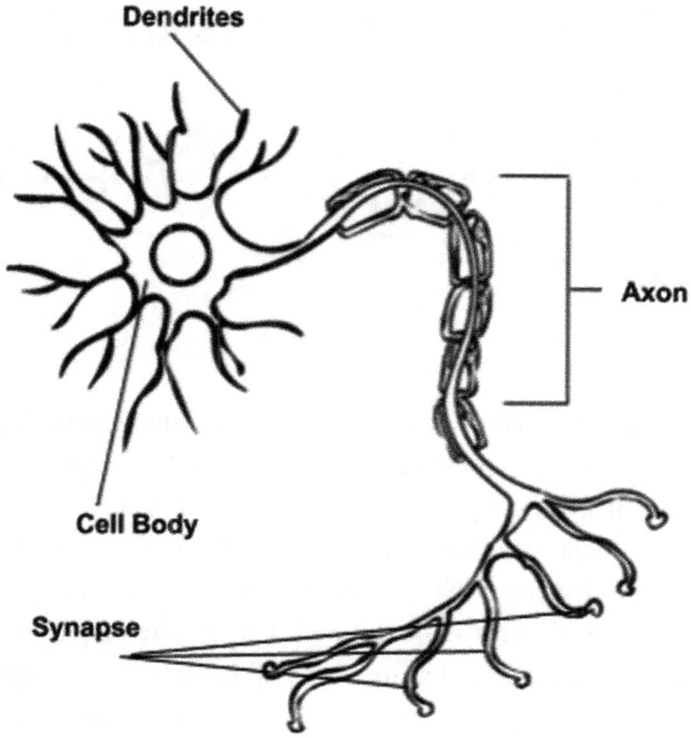


range and not transmitted as it is to another neuron. From this point of view, a correlation is occurred between every total signal that came from axon and transmitted to dendrite. Starting from this point, the hypothesis is emerged that learning process is occurred in synapses. On the basis of this hypothesis, the learning process of ANN model is perceived that updating weighting coefficients between dendrite and synapse. Herewith, this hypothesis can be converted to theory and artificial neuron is designed by inspiring by this working way of biological neuron (Fausett, 1994). The artificial neuron cell (ANC) is developed based on real biological neuron cell structure and its working way. The comparison of ANC and biological neuron cell is given in Table 5.

The basic structure of ANC that created using this table is given in Figure 33.

*Table 5. Comparison of artificial neural system and nervous system*

| Elements of Nervous System | Elements of Artificial Neural System |
|---|---|
| Neuron Cell | Artificial Neuron Cell |
| Synapse | Weights |
| Dendrite | Summing Function |
| Cell Body | Transfer Function |
| Axons | Outputs of Cell |

*Figure 33. Structure of artificial neural cell*

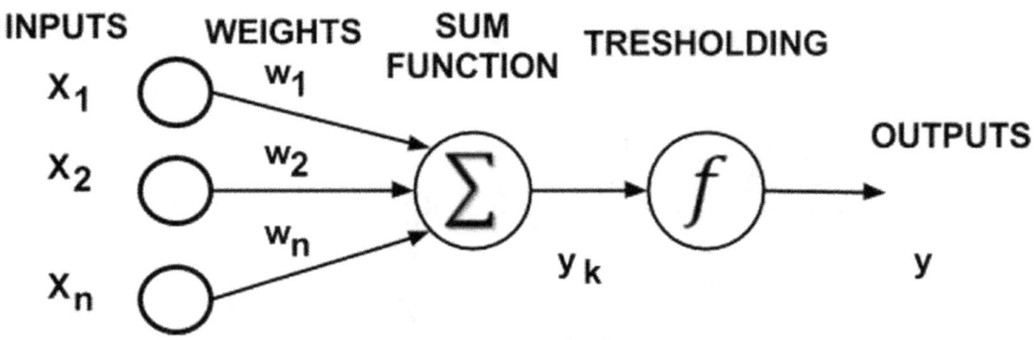


The dendrite of ANC has been indicated as $x_n$ and weights (importance degree) of each dendrite have been indicated as $w_n$. Thus $x_n$ represents inputs signals and $w_n$ represents the weights of these signals. The weighted sum of all input signals is obtained by sum function in cell body. This total signal is shown as $y_k$ and directed to threshold function in synapse. The result of threshold function is defined as $y$ and directed for feeding of another ANC. The purpose of ANC is to create output signal in response to input values and transmit these output values to another ANC. The weights $w_n$ which represent correlation between $x_n$ and $y$ rearrange according to every new inputs and output signals. This arrangement process is named the learning. The learning process continues until change of weights $w_n$ is to stable. ANN is created by interconnected ANC in layered structure. The ANN uses a mathematical model to learn by processing information with each ANC which created itself. This mathematical model is given Equation 23. The summing value of $y_k$ is sent by cell body to synapse over axon in Equation 23 (Fausett, 1994).

$$y_k = \sum_{i=1}^{h} x_i * w_i \tag{23}$$

Synapse create total output value by thresholding total signal values that came from axon. The thresholding function which defined as *f* in Equation 24 can be any mathematical function. Also this function can be named activation or transfer function.

$$y = f(y_k) \tag{24}$$

The importance of selecting of activation function can be explained that derivative of this function can be calculated by simply. There are three type of activation function basically.

- **Hard Limit Function:** This activation function is used for obtaining discrete result by input values. In other words, the outputs just become either +1 or 0. Thus, certain limit can be taken. The graphic and equation of hard limit function are given Figure 34 and Equation 25.

*Figure 34. Hard limit transfer function*

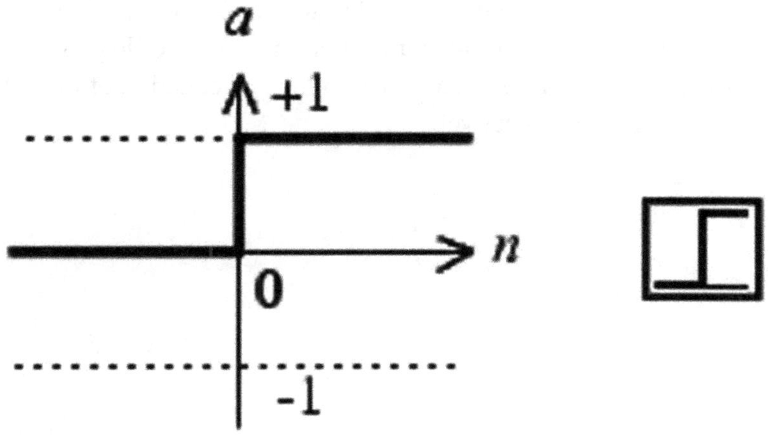


$$hard\lim(x) = \begin{cases} 1, x \geq 0 \\ 0, x < 0 \end{cases} \quad (25)$$

- **Saturating Linear Transfer Function:** This function is similar to Hard Limit Transfer Function. It responses linear increased value until certain value. When the response achieves until certain top limit, the response become discrete value don't increase. The graphic and equation of hard limit function are given Figure 35 and Equation 26.

$$sat\lim(x) = \begin{cases} 0, x \leq x_{\min} \\ mx + b, x_{\min} \leq x \leq x_{\max} \\ 1, x_{\max} \geq x \end{cases} \quad (26)$$

*Figure 35. Satlim transfer function*

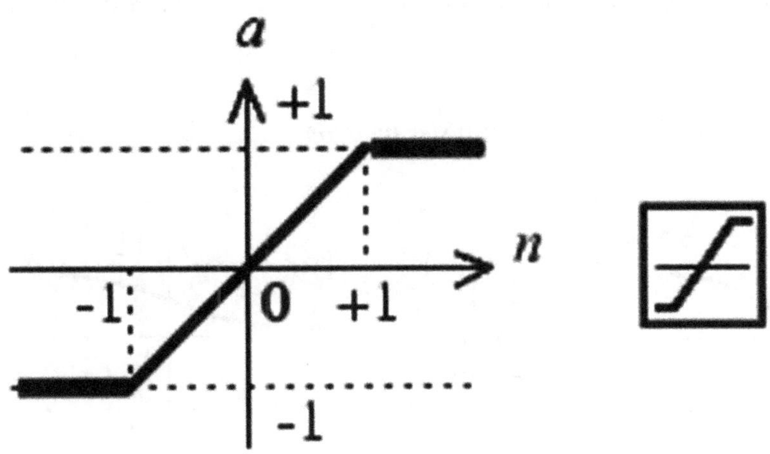

- **Sigmoid Transfer Function:** This function responses continuous value. The responses are not discrete absolutely. This function should be used to sensitive result is required. The sigmoid function consists of two functions, logistic and tangential. The values of logistic function range from 0 and 1 and -1 to +1 for tangential function. The graphic and equation of these functions are given Figure 36 and Equation 27-28 respectively.

$$\log sig(x) = \frac{1}{1+e^{-x}} \tag{27}$$

$$\tan sig(x) = \frac{2}{1+e^{-2x}} - 1 \tag{28}$$

There are different various of ANN that used different artificial intelligence application. The most commonly used ANN can be listed following;

- Feed-Forward Networks
- Radial Basis Function Networks
- Recurrent Networks
- Echo State Networks
- Hopfield Networks
- Self-Organizing Maps

In this study, Feed-forward neural networks (FFNNs) have been used to classification of heart sound. A Multilayer Perceptron (MLP) is an example of FFNNs. The MLP maps input array into appropriate output array. The MLP consists of multiple layers of nodes which connected to the next one (Baum, 1988). MLP is a supervised learning algorithm that learns a function $f(\bullet): R^m \rightarrow R^o$ by training on data set. In this function, $m$ is the number of dimensions for input array, $o$ is the number of dimension for output array (ScikitLearn, 2017). There is a input array $X = x_1, x_2, x_3, ..., x_m$ and a target y, and it can learn a non-linear function. Between the input and the output layer, there can be minimum one or more non-linear layers called hidden layers. Figure 37 shows hidden layer of MLP with scalar output.

*Figure 36. Log-sigmoid and Tan-sigmoid transfer function*

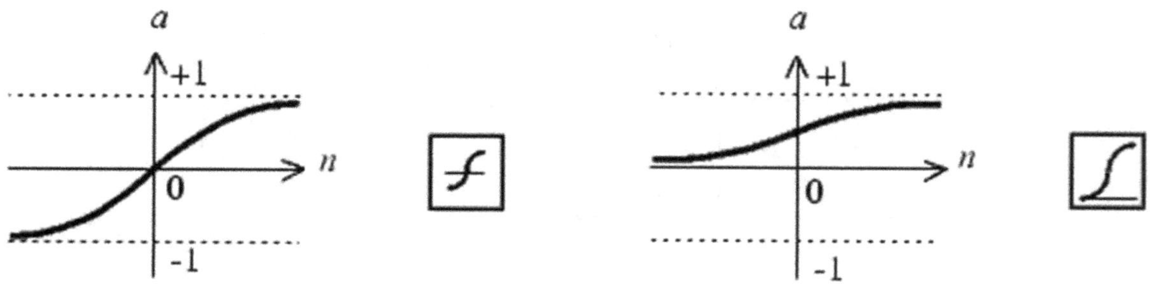

*Figure 37. Multilayer perceptron*

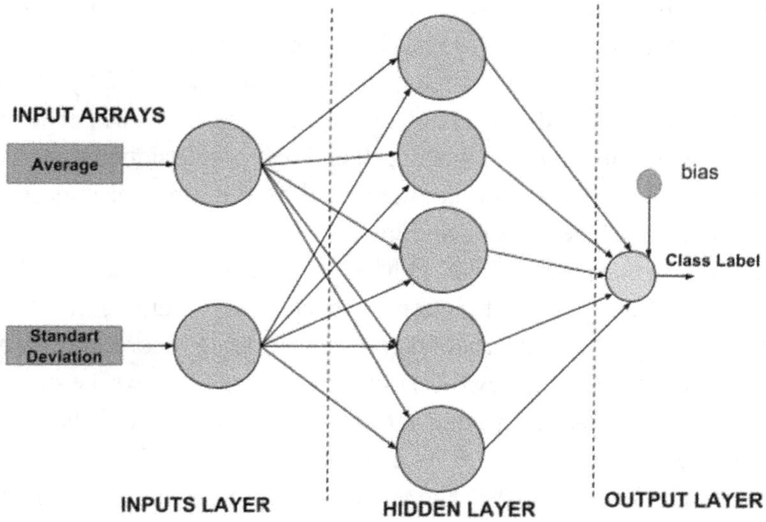


In Figure 20, leftmost layer is the input layer that consists of neurons representing the input features, average and standard deviation. In the hidden layer, each neuron transfers the values that came from previous layer with a weighted linear summation to next layer. The output layer at rightmost receives the values from the last hidden layer and transforms them into output values. The bias value can be used to shift activation function to the left or right by adding specific value. This can be critical for more successful learning for some situation. The advantages of Multi-Layer Perceptron are:

- An ability to learn how to do tasks based on the data given for training or initial experience.
- MLP do not make any assumption regarding the underlying probability density functions or other probabilistic information about the pattern classes under consideration in comparison to other probability based models (Su et al., 1996).
- MLP has capability to learn non-linear models and learng models in real-time (on-line learning)
- MLP produce the required decision function directly via training
- The stochastic nature of the learning process reduces the possibility of getting stuck in local minima in MLP
- They can be used for difficult to complex problems

The disadvantages of Multi-Layer Perceptron (MLP) include:

- MLP with hidden layers have a non-convex loss function where there exists more than one local minimum. Therefore different random weight initializations can lead to different validation accuracy (ScikitLearn, 2017).
- MLP requires tuning a number of hyperparameters such as the number of hidden neurons, layers, and iterations.
- MLP is sensitive to feature scaling.
- MLP can not be parallelized.

Firstly, the training and testing data set and class label of heart sounds needs to be prepared to classification using ANN. In this section only MFCC features of heart sound has been prepared to classification. The training data of each heart sound type has been converted to 2-by-30 matrix while is 30-by-20 matrix and they have been merged to transform to just one training data of 2-by-60 matrix in Figure 38.

The class label data of heart sounds have been defined as 2-by-60 matrix. The every column value to 30th column of first row of label data assigned as 0 and rest of it assigned as 1.

At second row, every column to 30th column assigned as 1 and rest of it assigned 0 in Figure 39. Figure 40 shows values of class label data of training and testing.

There is patternnet function to create ANN in MATLAB. The *net* variable with 5 hidden layer has been defined using this function and train function has been used to training *net* variable with Training data and class label in Figure 41. The *tr* variable represents training record of training data.

The *net* variable may be shown using view(net) function. Figure 42 shows the structure of net variable.

Finally, Test data needs to be classified using *net* variable. The *net* variable has been used as function and test data has been used as a parameter of its in Figure 43. The vec2ind function has been used to find index of misclassification value on test. The value of 2 represents as normal heart sound and value of 1 represents extra systole heart sound in *testIndicies* variable.

*Figure 38. Create training data of heart sounds for ANN*

```
% Training data of each type merge one training data
TrainingData = [TrainingDataExtra' TrainingDataNormal'];
% Test data of each type merge one one testing data
TestData = [TestDataExtra' TestDataNormal'];
```

*Figure 39. Creating class label data*

```
%Creating Class Label
TrainingClassLabel(1,1:30) = 1; % 1 for Extrasystole heart sound
TrainingClassLabel(2,1:30) = 0; % 0 for Normal heart sound
TrainingClassLabel(1,31:60)= 0;
TrainingClassLabel(2,31:60)= 1;
TestClassLabel(1,1:15) = 1;
TestClassLabel(2,1:15) = 0;
TestClassLabel(1,16:30)= 0;
TestClassLabel(2,16:30)= 1;
```

*Figure 40. The class label data values of training (top) and testing (bottom)*

|  | 1 | 2 | 3 | 4 | 5 | 6 | 7 | 8 | 9 | 10 | 11 | 12 | 13 | 14 | 15 | 16 | 17 | 18 | 19 | 20 | 21 | 22 | 23 | 24 | 25 | 26 | 27 | 28 | 29 | 30 |
|---|---|---|---|---|---|---|---|---|---|---|---|---|---|---|---|---|---|---|---|---|---|---|---|---|---|---|---|---|---|---|
| 1 | 1 | 1 | 1 | 1 | 1 | 1 | 1 | 1 | 1 | 1 | 1 | 1 | 1 | 1 | 1 | 1 | 1 | 1 | 1 | 1 | 1 | 1 | 1 | 1 | 1 | 1 | 1 | 1 | 1 | 1 |
| 2 | 0 | 0 | 0 | 0 | 0 | 0 | 0 | 0 | 0 | 0 | 0 | 0 | 0 | 0 | 0 | 0 | 0 | 0 | 0 | 0 | 0 | 0 | 0 | 0 | 0 | 0 | 0 | 0 | 0 | 0 |

|  | 31 | 32 | 33 | 34 | 35 | 36 | 37 | 38 | 39 | 40 | 41 | 42 | 43 | 44 | 45 | 46 | 47 | 48 | 49 | 50 | 51 | 52 | 53 | 54 | 55 | 56 | 57 | 58 | 59 | 60 |
|---|---|---|---|---|---|---|---|---|---|---|---|---|---|---|---|---|---|---|---|---|---|---|---|---|---|---|---|---|---|---|
| 1 | 0 | 0 | 0 | 0 | 0 | 0 | 0 | 0 | 0 | 0 | 0 | 0 | 0 | 0 | 0 | 0 | 0 | 0 | 0 | 0 | 0 | 0 | 0 | 0 | 0 | 0 | 0 | 0 | 0 | 0 |
| 2 | 1 | 1 | 1 | 1 | 1 | 1 | 1 | 1 | 1 | 1 | 1 | 1 | 1 | 1 | 1 | 1 | 1 | 1 | 1 | 1 | 1 | 1 | 1 | 1 | 1 | 1 | 1 | 1 | 1 | 1 |

|  | 1 | 2 | 3 | 4 | 5 | 6 | 7 | 8 | 9 | 10 | 11 | 12 | 13 | 14 | 15 | 16 | 17 | 18 | 19 | 20 | 21 | 22 | 23 | 24 | 25 | 26 | 27 | 28 | 29 | 30 |
|---|---|---|---|---|---|---|---|---|---|---|---|---|---|---|---|---|---|---|---|---|---|---|---|---|---|---|---|---|---|---|
| 1 | 1 | 1 | 1 | 1 | 1 | 1 | 1 | 1 | 1 | 1 | 1 | 1 | 1 | 1 | 1 | 0 | 0 | 0 | 0 | 0 | 0 | 0 | 0 | 0 | 0 | 0 | 0 | 0 | 0 | 0 |
| 2 | 0 | 0 | 0 | 0 | 0 | 0 | 0 | 0 | 0 | 0 | 0 | 0 | 0 | 0 | 0 | 1 | 1 | 1 | 1 | 1 | 1 | 1 | 1 | 1 | 1 | 1 | 1 | 1 | 1 | 1 |

*Figure 41. Creating training of net*

```
%% Create net with 5 hidden layer and train it.
net = patternnet(5);
[net,tr] = train(net,TrainingData,TrainingClassLabel);
view(net)
```

*Figure 42. The structure of net variable with 5 hidden layer*

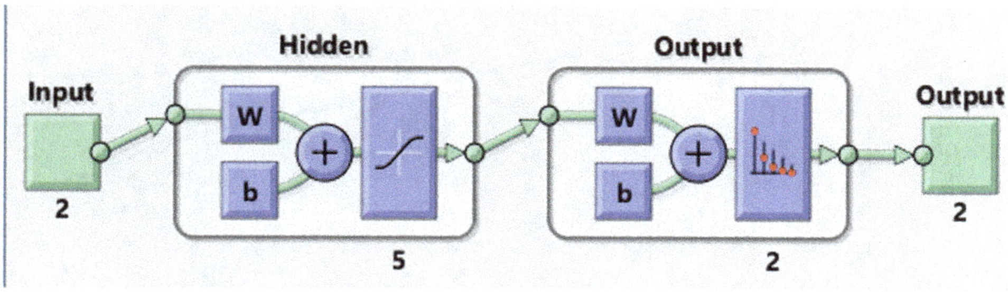


*Figure 43. Classification of test data*

```
TestResult = net(TestData);
testIndices = vec2ind(TestResult);
figure
plotperform(tr)
figure
plotconfusion(TestClassLabel,TestResult)
[c,cm] = confusion(TestClassLabel,TestResult);
fprintf('Percentage Correct Classification : %f%%\n', 100*(1-c));
fprintf('Percentage Incorrect Classification : %f%%\n', 100*c);
```

Also the plotperform function has been used with *tr* variable to draw classification performance. Figure 44 shows classification performance of heart sounds using ANN.

The plotconfusion function has been used with *TestClassLabel* and *TestResult* variable to draw confusion matrix of classification. The confusion matrix of classification is shown in Figure 45.

The confusion function has been used to calculate percentage of classification result in Figure 43. The *c* variable represents confusion value which is fraction of value misclassified. The *cm* variable represents n-by-n confusion matrix where *cm(m,n)* is the number of samples whose target is the *m*th class that was classified as *n*. The percentage correct classification has been found as 90% and percentage incorrect classification has been found as 10%.

Nowadays, artificial intelligence (A.I) and machine learning (ML) is used to assist people to solve their real-world complex problems in more effective and easier way with information which obtained from real-world. The A.I deals with real-world applications which is from health applications to autonomous vehicle or from smart houses to agricultural production or many fields like those. The A.I actualizes to

*Figure 44. The classification performance of ANN*

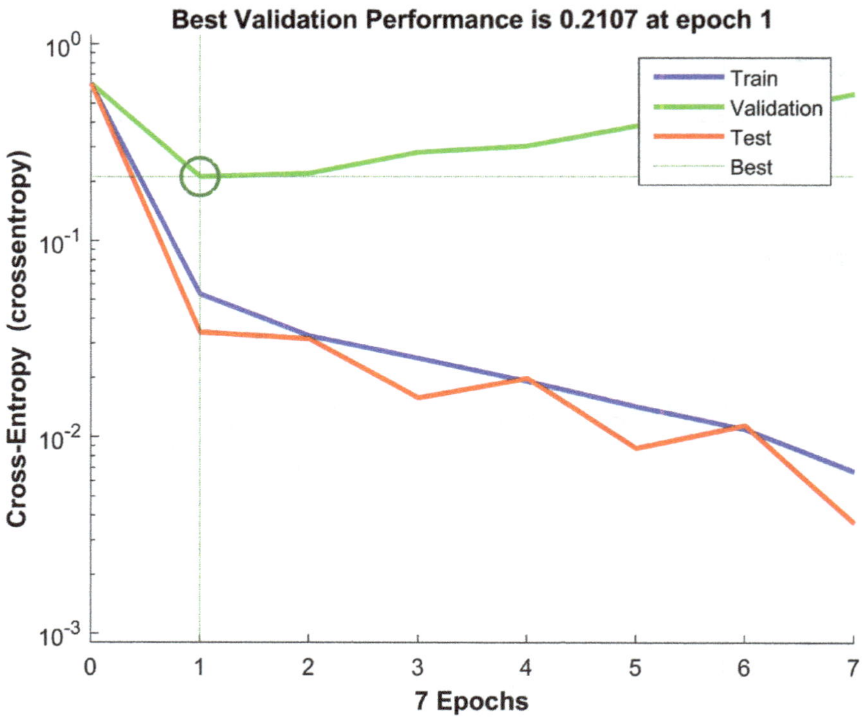


*Figure 45. The confusion matrix of classification*

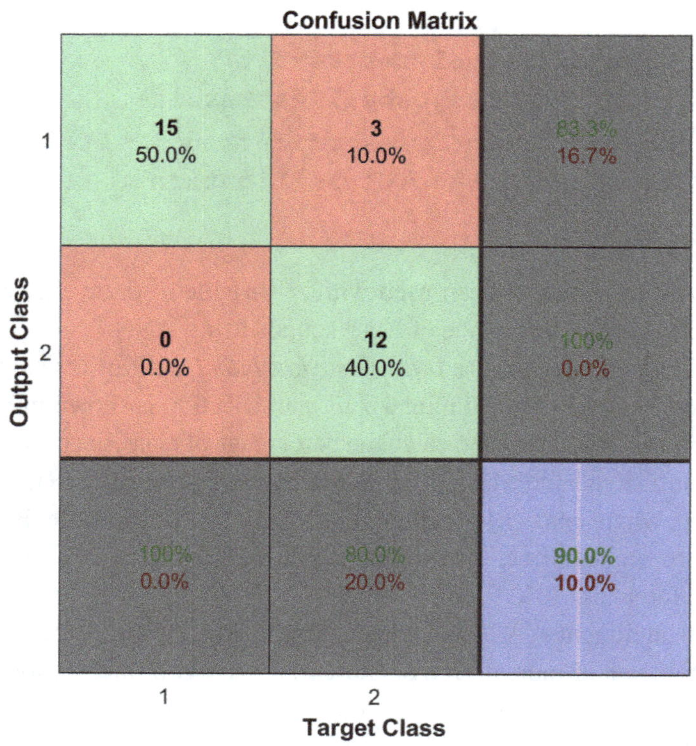

solve a problem by imitating intelligence or modes of living of some creatures which lives in group or single-handed in general. The A.I benefits from rules, definitions and process results of this creatures in problem field while solving a problem. These creatures consist of humans and animal mostly.

The health applications are one of the fields that artificial intelligence is used mostly. The many A.I methods are employed for diagnosing and treatment of the diseases, for assisting elderly or disability people by analyzing biomedical signal or image or by using robotic and to develop smart biomedical software or devices using expert system. These software or devices considerably benefit to physicians when decide about the diseases or assist to elderly or disability patient to maintain of its life. Some of fields which the A.I used in health; computer-aided interpretation of medical images, computer-aided detection (CADe) in radiology, heart sound analysis, probability of patients being cancer (IBM Watson Project), hospital attendant robots for the care of the elderly, mining medical records to provide more diagnosis information for clinical decision support system, design treatment plans, assist in repetitive jobs including medication management, drug creation. Therefore, these areas are considered, it is not known where the A.I will take human being.

In this sense, if we return to our issue, the developing, using and becoming prevalent of these devices or software is highly important to early diagnosis and fully treatment of the diseases. In the context of this section, it has been studied that heart diseases have been diagnosed with different heart sound signal.

## CONCLUSION

In this chapter, nature-inspired feature extraction and classification algorithm has been applied towards heart sound classification successfully. The classification rates of MFCC and Wavelet features of normal and extra systole heart sound with SVM and ANN are remarkable. As a result, the MFCC which nature-inspired feature extraction method is more successful than Wavelet method. Different MFCC characteristic and Wavelet type or level can be tried to get more successful classification result. The other nature-inspired feature extraction method like Tabu Research, Genetic Algorithm, Artificial Immune System, Artificial Bee Colony, Ant Colony Algorithms can be tried on heart sound classification. The experimental study in this chapter can be applied to the real-life experience on a mobile device application, an electronic stethoscope for a normal user or expert user in the hospital.

## REFERENCES

Ahn, S. H. (2015). *Digital Signal Processing - Convolution*. Retrieved from http://www.songho.ca/dsp/convolution/convolution.html

Allender, S., Scarborough, P., Peto, V., Rayner, M., Leal, J., Luengo-Fernandez, R., & Gray, A. (2008). *European cardiovascular disease statistics*. Brussels: European Heart Network.

Ari, S., Hembram, K., & Saha, G. (2010). Detection of cardiac abnormality from PCG signal using LMS based least square SVM classifier. *Expert Systems with Applications*, *37*(12), 8019–8026. doi:10.1016/j.eswa.2010.05.088

Auria, L., & Moro, R. A. (2008). *Support vector machines (SVM) as a technique for solvency analysis*. DIW Berlin Discussion Paper No. 811. 10.2139/ssrn.1424949

Azmy, M. M. (2015, December). Classification of normal and abnormal heart sounds using new mother wavelet and support vector machines. In *Electrical Engineering (ICEE), 2015 4th International Conference on* (pp. 1-3). IEEE. doi:10.1109/INTEE.2015.7416684

Baum, E. B. (1988). On the capabilities of multilayer perceptrons. *Journal of Complexity*, *4*(3), 193–215. doi:10.1016/0885-064X(88)90020-9

Bentley, P., Nordehn, G., Coimbra, M., & Mannor, S. (2011). *The pascal classifying heart sounds challenge 2011 (chsc2011) results*. Retrieved from http://www. peterjbentley. com/heartchallenge/index. html

Cawley, G. C., & Talbot, N. L. (2010). On over-fitting in model selection and subsequent selection bias in performance evaluation. *Journal of Machine Learning Research*, *11*(Jul), 2079–2107.

Cherif, L. H., Mostafi, M., & Debbal, S. M. (2014). Digital filters in Heart Sound Analysis. *International Journal of Clinical Medicine Research*, *1*(3), 97–108.

Çomak, E., Arslan, A., & Türkoğlu, İ. (2007). A decision support system based on support vector machines for diagnosis of the heart valve diseases. *Computers in Biology and Medicine*, *37*(1), 21–27. doi:10.1016/j.compbiomed.2005.11.002 PMID:16426598

Cortes, C., & Vapnik, V. (1995). Support-vector networks. *Machine Learning*, *20*(3), 273–297. doi:10.1007/BF00994018

Coskun, H., Yigit, T., & Deperlioglu, O. (2016, October). Effect of filter selection on classification of extrasystole heart sounds via mobile devices. In *Application of Information and Communication Technologies (AICT), 2016 IEEE 10th International Conference on* (pp. 1-5). IEEE. doi:10.1109/ICAICT.2016.7991820

Dao, V. L., Nguyen, V. D., Nguyen, H. D., & Hoang, V. P. (2016, December). Hardware Implementation of MFCC Feature Extraction for Speech Recognition on FPGA. In *International Conference on Advances in Information and Communication Technology* (pp. 248-254). Springer International Publishing.

Das, R., Turkoglu, I., & Sengur, A. (2009). Diagnosis of valvular heart disease through neural networks ensembles. *Computer Methods and Programs in Biomedicine*, *93*(2), 185–191. doi:10.1016/j.cmpb.2008.09.005 PMID:18951649

Daubechies, I. (1990). The wavelet transform, time-frequency localization and signal analysis. *IEEE Transactions on Information Theory*, *36*(5), 961–1005. doi:10.1109/18.57199

Fausett, L. (1994). *Fundamentals of neural networks: architectures, algorithms, and applications (No. 006.3)*. Prentice-Hall.

Guermoui, M., Mekhalfi, M. L., & Ferroudji, K. (2013, May). Heart sounds analysis using wavelets responses and support vector machines. In *Systems, Signal Processing and their Applications (WoSSPA), 2013 8th International Workshop on* (pp. 233-238). IEEE. doi:10.1109/WoSSPA.2013.6602368

Gupta, C. N., Palaniappan, R., Swaminathan, S., & Krishnan, S. M. (2007). Neural network classification of homomorphic segmented heart sounds. *Applied Soft Computing*, *7*(1), 286–297. doi:10.1016/j.asoc.2005.06.006

Guraksin, G. E., Ergun, U., & Deperlioglu, O. (2010). Classification of the heart sounds via artificial neural network. *International Journal of Reasoning-based Intelligent Systems*, *2*(3-4), 272–278. doi:10.1504/IJRIS.2010.036874

Hai, N. T., Van Thuyen, N., Mai, T. T., & Van Toi, V. (2015). MFCC-DTW Algorithm for Speech Recognition in an Intelligent Wheelchair. In *5th International Conference on Biomedical Engineering in Vietnam* (pp. 417-421). Springer.

Israel, C. W. (2014). Mechanisms of sudden cardiac death. *Indian Heart Journal*, *66*, S10–S17.

Jang, J. S. R. (2010). *Speech and audio processing toolbox*. Available from the link at the author's homepage at: http://www. cs. nthu. edu. tw/~ jang

Kechman, V. (2001). *Learning and soft computing*. Academic Press.

Li, S. Z. (2000). Content-based audio classification and retrieval using the nearest feature line method. *IEEE Transactions on Speech and Audio Processing*, *8*(5), 619–625. doi:10.1109/89.861383

Liang, B., Yaali, H., Songyang, L., Jianyun, C., & Lingda, W. (2005, October). Feature analysis and extraction for audio automatic classification. In *Systems, Man and Cybernetics, 2005 IEEE International Conference on* (Vol. 1, pp. 767-772). IEEE. doi:10.1109/ICSMC.2005.1571239

Lubaib, P., & Muneer, K. A. (2016). The heart defect analysis based on PCG signals using pattern recognition techniques. *Procedia Technology*, *24*, 1024–1031. doi:10.1016/j.protcy.2016.05.225

Maglogiannis, I., Loukis, E., Zafiropoulos, E., & Stasis, A. (2009). Support vectors machine-based identification of heart valve diseases using heart sounds. *Computer Methods and Programs in Biomedicine*, *95*(1), 47–61. doi:10.1016/j.cmpb.2009.01.003 PMID:19269056

Mahmoodabadi, S. Z., Ahmadian, A., Abolhasani, M. D., Eslami, M., & Bidgoli, J. H. (2006, January). ECG feature extraction based on multiresolution wavelet transform. In *Engineering in Medicine and Biology Society, 2005. IEEE-EMBS 2005. 27th Annual International Conference of the* (pp. 3902-3905). IEEE.

Mallat, S. (1999). *A wavelet tour of signal processing*. Academic Press.

Matworks. (2016). *Discrete Fourier Transform*. Retrieved from https://www.mathworks.com/help/signal/ug/discrete-fourier-transform.html

Nitze, I., Schulthess, U., & Asche, H. (2012). Comparison of machine learning algorithms random forest, artificial neural network and support vector machine to maximum likelihood for supervised crop type classification. *Proceedings of the 4th GEOBIA*, 7-9.

Openshaw, J. P., Sun, Z. P., & Mason, J. S. (1993, April). A comparison of composite features under degraded speech in speaker recognition. In *Acoustics, Speech, and Signal Processing, 1993. ICASSP-93., 1993 IEEE International Conference on* (Vol. 2, pp. 371-374). IEEE. doi:10.1109/ICASSP.1993.319316

Patidar, S., & Pachori, R. B. (2014). Classification of cardiac sound signals using constrained tunable-Q wavelet transform. *Expert Systems with Applications*, *41*(16), 7161–7170. doi:10.1016/j.eswa.2014.05.052

Pérez-Rendón, A. F., & Robles, R. (2004). The convolution theorem for the continuous wavelet tranform. *Signal Processing*, *84*(1), 55–67. doi:10.1016/j.sigpro.2003.07.014

Phua, K., Chen, J., Dat, T. H., & Shue, L. (2008). Heart sound as a biometric. *Pattern Recognition*, *41*(3), 906–919. doi:10.1016/j.patcog.2007.07.018

Proakis, J. G., & Manolakis, D. G. (2007). *Digital signal processing: principles, algorithms, and applications*. Pearson Prentice Hall.

ScikitLearn. (2017). *Neural network models (supervised)*. Retrieved from http://scikit-learn.org/stable/modules/neural_networks_supervised.html

Sengur, A. (2008). An expert system based on principal component analysis, artificial immune system and fuzzy k-NN for diagnosis of valvular heart diseases. *Computers in Biology and Medicine*, *38*(3), 329–338. doi:10.1016/j.compbiomed.2007.11.004 PMID:18177849

Sh-Hussain, H., Salleh, S. H., Ariff, A. K., Alhamdani, O., Tian-Swee, T., Noor, A. M., . . . Yusoff, K. (2012, July). Application of multipoint auscultation for heart sound diagnostic system (MAHDS). In *Information Science, Signal Processing and their Applications (ISSPA), 2012 11th International Conference on* (pp. 836-841). IEEE.

Smith, S. W. (1999). *The scientist and engineer's guide to digital signal processing*. California Technical Pub.

Su, M. C., Jean, W. F., & Chang, H. T. (1996, September). A static hand gesture recognition system using a composite neural network. In *Fuzzy Systems, 1996., Proceedings of the Fifth IEEE International Conference on* (*Vol. 2*, pp. 786-792). IEEE.

Sun, S., Wang, H., Jiang, Z., Fang, Y., & Tao, T. (2014). Segmentation-based heart sound feature extraction combined with classifier models for a VSD diagnosis system. *Expert Systems with Applications*, *41*(4), 1769–1780. doi:10.1016/j.eswa.2013.08.076

Sung, P. H., Thompson, W. R., Wang, J. N., Wang, J. F., & Jang, L. S. (2015). Computer-assisted auscultation: Patent ductus arteriosus detection based on auditory time–frequency analysis. *Journal of Medical and Biological Engineering*, *35*(1), 76–85. doi:10.1007/s40846-015-0008-9

Turkoglu, I., Arslan, A., & Ilkay, E. (2003). An intelligent system for diagnosis of the heart valve diseases with wavelet packet neural networks. *Computers in Biology and Medicine*, *33*(4), 319–331. doi:10.1016/S0010-4825(03)00002-7 PMID:12791405

Vergin, R., O'Shaughnessy, D., & Gupta, V. (1996, May). Compensated mel frequency cepstrum coefficients. In *Acoustics, Speech, and Signal Processing, 1996. ICASSP-96. Conference Proceedings., 1996 IEEE International Conference on* (Vol. 1, pp. 323-326). IEEE. doi:10.1109/ICASSP.1996.541097

WHO. (2004). *Global Burden Disease*. Retrieved from http://www.who.int/entity/healthinfo/global_burden_disease/GBD_report_2004update_full.pdf?ua=1

Zhang, Y., Wang, Y., Wang, W., & Liu, B. (2001). Doppler ultrasound signal denoising based on wavelet frames. *IEEE Transactions on Ultrasonics, Ferroelectrics, and Frequency Control*, *48*(3), 709–716. doi:10.1109/58.920698 PMID:11381694

Zhao, Z., Shen, Q., & Ren, F. (2013). Heart sound biometric system based on marginal spectrum analysis. *Sensors (Basel)*, *13*(2), 2530–2551. doi:10.3390/s130202530 PMID:23429515

Zheng, Y., Guo, X., Qin, J., & Xiao, S. (2015). Computer-assisted diagnosis for chronic heart failure by the analysis of their cardiac reserve and heart sound characteristics. *Computer Methods and Programs in Biomedicine*, *122*(3), 372–383. doi:10.1016/j.cmpb.2015.09.001 PMID:26387633

## ADDITIONAL READING

Akay, M., Medicine, I. E. i., & Society, B. (1998). *Time Frequency and Wavelets in Biomedical Signal Processing*. Wiley.

Akay, M., Medicine, I. E. i., & Society, B. (2000). *Nonlinear Biomedical Signal Processing, Dynamic Analysis and Modeling*. Wiley. doi:10.1109/9780470545379

Alpaydin, E. (2014). *Introduction to Machine Learning*. MIT Press.

Blanchet, G., & Charbit, M. (2015). Digital Signal and Image Processing using MATLAB, Volume 3: Advances and Applications, The Stochastic Case: Wiley.

Blinowska, K. J., & Zygierewicz, J. (2011). *Practical Biomedical Signal Analysis Using MATLAB*. Taylor & Francis.

Campbell, C., & Ying, Y. (2011). Learning with Support Vector Machines: Morgan & Claypool.

Cohen, A. (1986). *Biomedical Signal Processing: Time and frequency domains analysis*. CRC Press.

Deng, N., Tian, Y., & Zhang, C. (2012). *Support Vector Machines: Optimization Based Theory, Algorithms, and Extensions*. Taylor & Francis.

Giannakopoulos, T., & Pikrakis, A. (2014). *Introduction to Audio Analysis: A MATLAB® Approach*. Elsevier Science.

Giron-Sierra, J. M. (2016a). Digital Signal Processing with Matlab Examples, Volume 1: Signals and Data, Filtering, Non-stationary Signals, Modulation: Springer Singapore.

Giron-Sierra, J. M. (2016b). Digital Signal Processing with Matlab Examples, Volume 2: Decomposition, Recovery, Data-Based Actions: Springer Singapore.

Górriz, J. M., Lang, E. W., & Ramírez, J. (2011). *Recent Advances in Biomedical Signal Processing*. Bentham Science Publishers.

Guraksin, G. E., Ergun, U., & Deperlioglu, O. (2010). Classification of the heart sounds via artificial neural network. *International Journal of Reasoning-based Intelligent Systems*, *2*(3-4), 272–278. doi:10.1504/IJRIS.2010.036874

Ingle, V. K., & Proakis, J. G. (2011). Digital Signal Processing Using MATLAB: Cengage Learning.

Khandpur, R. (2004). Biomedical Instrumentation: Technology and Applications: McGraw-Hill Education.

Khanorkar, S. V. (2011). *Insights in Physiology: Jaypee Brothers*. Medical Publishers Pvt. Limited.

Kiasaleh, K. (2016). Biological Signals Classification and Analysis: Springer Berlin Heidelberg.

Lee, S. W., & Verri, A. (2003). Pattern Recognition with Support Vector Machines: First International Workshop, SVM 2002, Niagara Falls, Canada, August 10, 2002. Proceedings: Springer Berlin Heidelberg.

Lei, B., van der Heijden, F., Xu, G., Feng, M., Zou, Y., de Ridder, D., & Tax, D. M. J. (2017). *Classification, Parameter Estimation and State Estimation: An Engineering Approach Using MATLAB*. Wiley.

Lerch, A. (2012). An Introduction to Audio Content Analysis: Applications in Signal Processing and Music Informatics: Wiley-IEEE Press.

Madisetti, V. (2009). *Digital Signal Processing Fundamentals*. CRC Press. doi:10.1201/9781420046076

McLoughlin, I. V. (2016). *Speech and Audio Processing: A MATLAB-based Approach*. Cambridge University Press. doi:10.1017/CBO9781316084205

Najarian, K., & Splinter, R. (2012). *Biomedical Signal and Image Processing* (2nd ed.). Taylor & Francis.

Northrop, R. B. (2016). Signals and Systems Analysis. In *Biomedical Engineering* (2nd ed.). CRC Press.

Palaniappan, R. (2011). *Biological Signal Analysis*. Studentia Ventus Publishing.

Pearson, D. W., Steele, N. C., & Albrecht, R. F. (2012). Artificial Neural Nets and Genetic Algorithms: Proceedings of the International Conference in Alès, France, 1995: Springer Vienna.

Prasad, B., & Prasanna, S. R. M. (2008). Speech, Audio, Image and Biomedical Signal Processing using Neural Networks: Springer Berlin Heidelberg.

Quinquis, A. (2010). *Digital Signal Processing Using Matlab*. Wiley.

Ranganathan, N., Sivaciyan, V., & Saksena, F. B. (2007). *The Art and Science of Cardiac Physical Examination: With Heart Sounds and Pulse Wave Forms on CD*. Humana Press. doi:10.1007/978-1-59745-023-2

Rangayyan, R. M. (2015). *Biomedical Signal Analysis*. Wiley. doi:10.1002/9781119068129

Reddy, D. C. (2005). *Biomedical Signal Processing: Principles and Techniques: McGraw-Hill Education*. India: Pvt Limited.

Rocchesso, D. (2003). *Introduction to Sound Processing: Ass*. Culturale Mondo Estremo.

Semmlow, J., & Griffel, B. (2014). Biosignal and Medical Image Processing, Third Edition. doi:citeulike-article-id:13616543

Sharif, Z. (2001). Signal Analysis and Classification of Heart Sounds and Murmurs: Universiti Teknologi Malaysia.

*Signal Processing Theory and Methods I*. (2000). IEEE Operations Center.

Singh, B. (2016). *Computational Tools and Techniques for Biomedical Signal Processing*. IGI Global.

Smith, S. (2013). *Digital Signal Processing: A Practical Guide for Engineers and Scientists*. Elsevier Science.

Steeb, W. H. (2005). *Mathematical Tools in Signal Processing with C++ & Java Simulations*. World Scientific. doi:10.1142/5939

Stoean, C., & Stoean, R. (2016). *Support Vector Machines and Evolutionary Algorithms for Classification: Single or Together?* Springer International Publishing.

Vaseghi, S. V. (2008). *Advanced Digital Signal Processing and Noise Reduction*. Wiley. doi:10.1002/9780470740156

Walnut, D. F. (2002). An Introduction to Wavelet Analysis: Birkhäuser Boston.

Webster, J. (2009). Medical Instrumentation: Application And Design, 3Rd Ed: Wiley India Pvt. Limited.

Weeks, M. (2010). Digital Signal Processing Using MATLAB & Wavelets: Jones & Bartlett Learning.

Zölzer, U. (1997). *Digital Audio Signal Processing*. Wiley.

## KEY TERMS AND DEFINITIONS

**Artificial Neural Network:** Artificial neural network is a classification model based on the neural structure of the brain. It is employed to classify specific groups with some features. It has outputs which change according to input.

**Extra Systole:** A premature contraction of the heart that is independent of the normal rhythm of the heart and that arises in response to an impulse in some part of the heart other than the normal impulse from the sinoatrial (SA) node. The extra systole is followed by a pause, as the heart electrical system "resets" itself and the contraction following the pause is usually more forceful than normal. These more forceful contractions are frequently perceived as palpitations.

**Heart Sound:** A sound that are the noises generated by the beating heart and the resultant flow of blood through it.

**Mel-Frequency Cepstrum:** This term is a representation of the short-term power spectrum of a sound, based on a linear cosine transform of a log power spectrum on a nonlinear mel scale of frequency.

**Phonocardiography:** A term that is used to show the sound signal of the recording of all the sounds made by the heart during a cardiac cycle.

**Support Vector Machine:** A mathematical method used to classify specific group that has same properties.

**Wavelet Transform:** In mathematics, a wavelet series is a representation of a square-integrable function by a certain orthonormal series generated by a wavelet. In this study, it is applied to get feature of heart sounds.

# Chapter 8
# A Sample Application for Computer-Based Bacteria Counting by Using Watershed Transform

**Merve Nur Onal**
*Afyon Kocatepe University, Turkey*

**Nesibe Cekmez**
*Afyon Kocatepe University, Turkey*

**Gokhan Akarca**
*Afyon Kocatepe University, Turkey*

## ABSTRACT

*Bacteria counting process and bacterial counting by eye can be performed by manual counting in which a specialist counts the bacteria with naked eye or the automatic counting that utilizes computer-based techniques. The counting process becomes exhausting, long, and incorrect when the counting is performed by specialist. Therefore, there is a need for an efficient and more reliable method for the bacteria counting process performed automatically. In this chapter, a semi-automatic method has been designed for bacterial counting from the images of petri plates and photomicrograph. For this purpose, the image processing procedure has been applied to the images. In the image processing procedure, watershed transform was used to separate the adjacent bacteria and thus a more accurate counting was performed. The efficiency of the presented method has been shown for many different images, and the counting results of the proposed method have been compared with counting results of an expert. The method has been shown to be a successful and promising work for bacterial counting.*

DOI: 10.4018/978-1-5225-4769-3.ch008

## INTRODUCTION

Bacteria are a single-celled microscopic organism. They form a separate living category between plants and animals. Bacteria can be found in nature, in soil, in air, in water, in plants, in animals and even in humans. Bacteria are small enough not to be seen with the naked eye in sizes between 1 and 6 microns. Bacteria multiply by division. In addition to the harmful bacteria that make the disease, there are also types of bacteria that are useful. Generally, bacteria can have classified as disease causing bacteria, yeast bacteria, rotting bacteria, and nitrogen bacteria. They can live in very hot environments where other creatures cannot survive or even under very high pressure. They corrupt the structure of dead organisms and allows them to come to a state where organic materials are used by other organisms. These are useful aspects of bacteria. Bacteriologists report that there are more than 100 species of bacteria and classify them according to type, structure, growth, culture environment and staining type. Bacteriology is the field of researching bacteria in microbiology. Examination of microorganisms, production and staining in different culture environment are included in bacteriological studies. Experts working on this field are called bacteriologists.

Bacterial counting is one of the most basic and frequent applications. Not only to demonstrate efficacy of disinfectants, but also to medical examinations, as well as evaluations for food and drug safety. Bacteria counting with this much precaution is usually done by manual counting of well trained technicians. For this reason, manual numbering is very time-consuming and labor intensive in practice. In addition, manual counting procedure is an error-prone method especially when counting on large number of plates and taking into account the different results available to different technicians. One of the most important reasons for the different count results that can be obtained by different technicians is the indistinguishable bacterial overlap (Talas, 2017).

In recent years, computer vision and image processing techniques have been used in many areas. One of these is the colony counting and the analyzing. Over the past 10 years, high-tech image analysis techniques and software applications with studies on artificial intelligence have accelerated histologic cell analysis, accelerating with studies on artificial intelligence in general. Especially the development of histologic markers enables the same image and many different structures to be visible at the same time, enabling the groups to perform studies in the field of medical image processing. Analysis of medical images includes many methods such as image acquisition, image reconstruction, image derivation, image compression and storage, image analysis and image-based recognition. All of these methods can be used to simplify operations that are challenging in cell and bacterial counting stages.

In the earlier stages of studies on tissue fragments was usually performed using the thresholding method which is a very popular method in this area. Thresholding is basically performed at the pixel level on the gray level images. It is aimed to make it possible to distinguish between the background and the desired object to be seen (Otsu, 1979). In many studies done on this area, good results were obtained with a reasonable threshold value selected according to the gray level (Kittler & Illingworth, 1986). There are also empirical methods to choose the threshold value. Since these methods are able to perform cell counting purposes, experiments can be made to increase the number (Markiewicz, 2003). Semi-automatic systems provide interactive processing by allowing counting at the user selected threshold interval by examining pixels in the spatial plane (Colley et al., 1989).

In the literature there are several studies on computer based microorganism counting. One of these studies Gilchrist E. et.al. had given Spiral Plate Method for bacterial determination in 1973. In this study a method for determining the number of bacteria in a solution via a machine which deposits a known volume of sample on a rotating agar plate in an ever-decreasing amount in the form of an Archimedes spiral. After incubation, different colony densities become apparent on the surface of the plate. From this point of view, a modified counting grid is described related to area of the plate of the volume of sample. Thus the number of bacteria in the sample is estimated by counting an appropriate area of the plate (Gilchrist et al., 1973).

In another study Dubuisson and friends had given segmentation and classification of bacterial culture images in 1994. In this paper a procedure using simple image processing and pattern recognition techniques to automatically extract and count organisms using images of bacterial cultures is given (Dubuisson et al., 1994).

Mukherjee and friends had given water quality analysis which is based on a pattern recognition approach in 1995. In this method, authors described the development of a microcomputer-based system which is used for automatic counting of colonies of indicator organisms trapped on membrane filters from water samples. In this study, three different approaches based on mathematical morphology, distance transformation and fuzzy c-means clustering are applied respectively (Mukherjee et al., 1995).

In 1999, Kawai and friends have proposed a method for rapid enumeration of physiologically active bacteria in purified water for use in the pharmaceutical production process. According to this method, bacteria with potential for growth were enumerated using micro-colony technique and direct viable counting. As a result, the authors have said that this fast and useful technique would be useful for systematic monitoring of waterborne bacteria used for pharmaceutical production (Kawai et al., 1999).

In another study Kramer and friends had searched evaluation of the spiral plate and laser colony counting techniques for the enumeration of bacteria in foods in 2004. In this method, samples of food and dairy products, bacterial cultures and spore suspensions were examined by two operators using both the spiral plate and surface drop techniques for counting bacteria (Kramer et al., 2004).

Zhang and Chen had given an effective and robust method for automatic bacterial colony enumeration in 2007. In this study, they propose a fully automatic colony counter and compare its performance with Clono-Counter. These experimental results show that the proposed method outperforms Clono-Counter in terms of Precision, Recall, and F- measures (Zhang & Chen, 2007).

In another study with the watershed transform was performed by Beucher and Lantuejoul in 1979. They were applied the concept of watershed and divide lines to segmentation problems. They used it for segmenting the images of bubbles and SEM metallographic pictures (Beucher & Lantuejoul, 1979).

In this study, a computer application in which automatic bacterial counting can be performed has been developed. For this purpose, the image processing procedure has been applied to the images. In the image processing procedure, watershed transform was used for to separate the adjacent bacteria and thus a more accurate counting was performed. By the help of this application all the bacteria are labeled and counted. In addition, some properties like area of each bacterium are recorded by the help of performed application.

## BASIC PRINCIPLES OF MICROORGANISM COUNTING

The legal restrictions on the number of microorganisms in foodstuffs are influenced by a number of factors, including the type of microorganism, food types and the state of development of the country. These legal regulations, increasingly competitive conditions in the food trade are primarily forcing food manufacturers to produce more hygienic products. Therefore, the number of microorganisms in food should be determined correctly. Determination of the number of microorganisms in a food sample often requires a long analysis period. This period can range from 1 to 10 days depending on the type of microorganism counted and the method used in standard cultural methods. These long analysis alternatives to the high-speed methods are high analysis cost and / or faulty disadvantages such as the possibility of obtaining results and / or the difficulty of analysis. The main reason why standard culture analysis methods take a long time is microorganism to develop in the solid medium to form colonies or to develop in the liquid medium generation is needed.

Rapid methods, which are alternatives to the long analysis period, can lead to disadvantages such as high analysis cost and / or possibility of obtaining erroneous results and / or difficulty of analysis. These are:

- The employee's eyes are tough and exhausted.
- Exemplary particles can be mixed with cells.
- Microorganisms may not always be uniformly distributed on the preparation; cell clusters may be seen.
- When painting, some cells may not be fully dyed, so they cannot be counted.

However, automatic bacterial and cell counting both reduces disadvantages and reduces error rates. Dilution of microorganisms is usually performed by standard cultural methods. The results of the census are given as an "accepted value" according to statistical rules in a sense. Standard cultural methods with made microbiological analysis have how many disadvantages. Standard cultural analyzes are widely used and accepted both in analyzes conducted by legal control agencies today and in auto control analyzes in enterprises. However, methods that are defined as fast are increasingly accepted in food industry enterprises. In addition, some rapid analysis methods that are not fully adopted by international analysis institutions are accepted as standard cultural analysis today. (Halkman & Ayhan, 2017).

## Automatic Bacteria Counting

Automated cell counting is found in many important applications that are fundamental to the health of many inhabitants of wholesale food. For example, milk, cheese, etc. in the food field. It is used in the follow-up of fruit and vegetable development in order to ensure that useful or harmful bacteria in the products are counted and all necessary information about the product is collected. It is widely used especially in the field of medicine and diagnosis. Cell counting is important because it is used for control purposes in healthy individuals. Allows early detection of any disease. Early diagnosis of the disease means shortening the time of intervention and is of vital importance. The accuracy of measurement results is increased when too many samples can be examined. The automated methods are far superior to the methods performed by the expert because of the large number of bacteria counts and the low statistical error rate. Errors can generally be caused structure of the sample, techniques of specialist and faulty equipment (Çetinel & Kamanlı, 2016).

## IMAGE PROCESSING

Image analyzing systems consist of a video camera connected to an image-array microprocessor and host computer. The image is digitized and is converted into matrix parts called pixels. The camera can be coupled to a microscope (including electron microscope). The system stores, enhances and displays image on the computer screen. The use of computer allows presentation of the data in various forms, including plotting histograms of the original grey-level images, enhancing for low-contrast images, assigning colors to various grey-level values, enumerating cells and identifying special features, measuring parameters, determining relative densities and performing data analyses. (Erkmen &Bozoglu,2008)

### Image Contrast and Contrast Enhancement

Contrast, also referred to as contrast, is the degree of difference in distinguishing an object from a gray level or other color component in a digital image. Contrast refers to the difference between the brightest part and the darkest part of an image. In contrast enhanced images, objects and colors are more distinct. At low contrast, color tones or gray levels are closer to each other.

First in contrast stretching; minimum and maximum pixel values in the image. Later; the formula for each pixel in the image is applied to the individual equations 1 (Bayram, 2017).

$$\text{Gout} = (\text{gin} - \text{gmin}) \times ((\text{goutmax} - \text{goutmin}) / (\text{gmax} - \text{gmin})) + \text{goutmin} \quad (1)$$

In Equation 1, the gray value after gout contrast stretching, gin represents the gray value of the pixel in the image. gmax ve gmin refers to the maximum and minimum values in the image. goutmax refers to the maximum value of the gray value range to be stretched. goutmin refers to the minimum value of the gray value range to be stretched.

Another method of contrast enhancement is histogram equalization. If gray level an image, image of the g min and gmax approximately pixcel value to image detail selection can be more difficult. The histograms of images with low contrast show a narrow range of distribution. The histogram equalization method is often used to improve the contrast of such images. Histogram equalization is performed with equation 2. (Karakuş, 2006).

$$f'(x, y) =((f(x, y) - f(x,y)min) / (f(x,y)\text{max} - f(x,y)min)) \times 255 \quad (2)$$

Here; $f(x, y)$ real pixel value, $f(x, y)min$ and $f(x, y)max$ represent the lowest and highest pixel values that the image has. Equality $f'(x, y)$, the pixel value of the contrast enhancement (Şenol, 2014).

Histogram equalization (HE) is the most preferred contrast enhancement method in image processing applications because of its simplicity and high efficiency (Gonzalez and Woods, 2008). The HE is used for distributing the pixel values of the image to its dynamic intensity range by using the input image's linear cumulative histogram. HE based image enhancement is used in many applications such as in medical image processing (Sundaram et al., 2011), speech recognition (De la Torre et al., 2005), satellite image processing (Ganesan & Rajini, 2014), and others.

Image stretching is performed on bacterial image to adjust the intensity value of the image to a specific range by using imadjust function. The stretchlim function returns a pair of gray values that can be used by imadjust to increase the image contrast. Then, imadjust maps the intensity values of the bacterial image so that 1% of data on lower and higher values of the image is saturated respectively (Iza et al., 2017). This step will increase the contrast of the output image which is of the same size as the original bacterial image that contains mapped values.

In general, since the pixel intensities of a low-light image are confined to a small range of dark values, these contrast enhancement techniques attempt to boost the pixel intensities to enhance the contrast The performance of image processing and computer vision algorithms, including object detection, classification, segmentation, and tracking, is strongly affected by the quality of an input image. Therefore, low-light image enhancement, which is a process that improves the contrast of images obtained from environments with low light ratios and bring out obscured details, is necessary for the effectiveness of the subsequent image processing and analyses (Jaemoon et al.,2017).

Watershed Transform

Segmentation in image processing is one of the most common problems. However, there are many techniques available for many specific situations. The segmentation of gray-scale images is especially important among these techniques. These techniques may be divided into two groups as based on contour detection and those involving region growing (Beucher & Meyer, 1993). The watershed transformation can be included to the second group among these two groups.

The watershed transformation which is used in image segmentation is a powerful tool in image processing. The concept of watershed method was first used by Beucher and Lantuejoul in 1979. By the help of watershed method, they divide the lines to segment the images of bubbles and SEM metallographic pictures (Beucher & Lantuejoul, 1979).

The watershed transformation enables the boundary of the object to be determined by taking advantage of the topological structure of the object. In Figure 1 an example of the principle of watershed is given.

Assume that the lower points in the figure 2 are B1, B2, ..., Bz in the image with the coordi$_{n}$ate points I(i,j). The points of the catchment basins was indicated by CBm in relation with the minimum region Bz(x,y) represented by X(n) accordingly I(x,y)<n (Salman and Bahrani, 2010).

$$X[n] = \{(x,y) \mid I(x,y) < n\} \quad (3)$$

*Figure 1. The principle of watershed*
*Salman & Bahrani, 2010.*

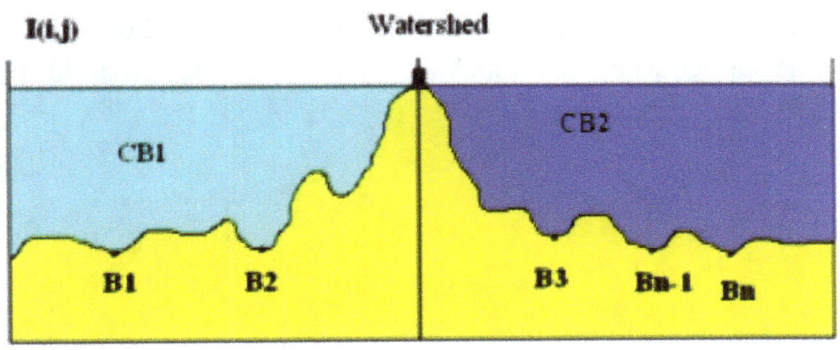

In equation 3 X[n] is the coordinate points of I(i,j) lying under the plain by geometrically with I(i,j)=n. The image filled with the water topographically with a filling increment value which begins from n=tmax-1 to n=tmin+1 (Salman and Bahrani, 2010). Two or more floods coming from different ground can be combined during the flood. In order to avoid this, a dam is built at this point. At the end of this process, the dams emerged. These dams define the basin (watershed) belonging to a function identified by f. They separate various catchment basins CBm, each containing only a minimum Bz (Beucher, 1992).

Under the level I(i,j)=n, the marker will used black color for the points in coordinates X[n] and for the other points, the marker used white color. Therefore, the number of dots below the water level is necessary (Salman and Bahrani, 2010).

$$C_nB_m = CB_m \cap X[n] \tag{4}$$

In equation 4, the coordinates in catchment basins which are water filling at the level n, related to Bm is represented by $CnB_m$. Then the union the filling water of the points of catchment basins of level n which is represented by C[n] (Salman & Bahrani, 2010):

$$C[n] = \bigcup_{m=1}^{z} C_nB_m \tag{5}$$

Finally the union of all the catchment basins which is represented by Ctmax+1 (Salman & Bahrani, 2010):

$$C_{t\max+1} = \bigcup_{m=1}^{z} CB_m \tag{6}$$

It can be understood that C[n] is a subgroup of X[n] form equation 4 and 5. So the watershed lines are prepared when C[tmin+1] = X[tmin+1]. Then the reconstruction of C[n-1] at level n procedure is performed. C[n] can be obtained from C[n-1] by assuming S as the set of the connected component in X[n], at s ∈ S[n] there are three assumption (Salman & Bahrani, 2010):

1. s is incorporated into C[n-1] to produce C[n] where s ∩ C[n-1] is empty and the assumption is verify at a new minimum is encountered.
2. s is incorporated in to C[n-1] to produce C[n] because of s ∩ C[n-1] contains more than one connected component of C[n-1].
3. s ∩ C[n-1] contains multiple connected components of C[n-1]. This indicates that whole or one part of the rim which is separating two or more catchment basins from each other is encumbered.

As we mentioned before, two or more floods during the flooding process can be merged with the other floods from different minima. It is desired to get rid of this situation. For this reason, a dam is built on the points of the surface where the floods would merge. At the end of this period only dams are emerging (Beucher, 1992).

## PROPOSED COMPUTER BASED BACTERIAL COUNTING SYSTEM

In this study, a computer software was designed for automatic bacterial counting as shown in Figure 2. The first stage of the method presented in this section describes how microscopic images are obtained and the steps to be applied to the image to improve the accuracy of the method. The basic steps of the proposed method are shown in Figure 3 as a block diagram.

In this study, an application for computer based bacterial counting was developed. In the first phase of this work, it is firstly time to get raw images to be used in our work. The bacterial images used in this study were taken from Afyon Kocatepe University Food Engineering Microbiology Laboratory.

In the second phase of our work, a user interactive software design was performed for selecting the optimal contrast settings. In this software, user can adjust the contrast by seeing the effects on the image instantly. In Figure 4, contrast adjustment part of the software was given. As seen in Figure 4, the contrast of the image could be done by interactively by the user. In Figure 5 an example of the contrast adjustment procedure was given.

In the third phase of our work, a threshold value is used to create binary images from a grayscale image. If the threshold value is small in an image, the object will suffer losses. Conversely, if the threshold value is selected to large, the object is detected incorrectly. In our software design selecting the threshold value is user interactive again. The threshold value can be set manually by setting in the slider as shown in Figure 6, to determine the threshold value for ambient light (darkness or lightness) or ambient conditions affecting the image. It has been decided that there will be more accurate results in this way. An example for this procedure was given in Figure 7. In Figure 7, grayscale image and its binary image with the appropriate threshold value was given.

In the designed software, when you click on the image processing button, separation with watershed is performed and the boundaries of each bacterium are identified and numbered. With the help of the watershed method, bacterial samples that are very close to each other or adjacent to each other are separated to ensure that the counting is more efficient as seen in fiGure 8.

*Figure 2. The interface of the software*

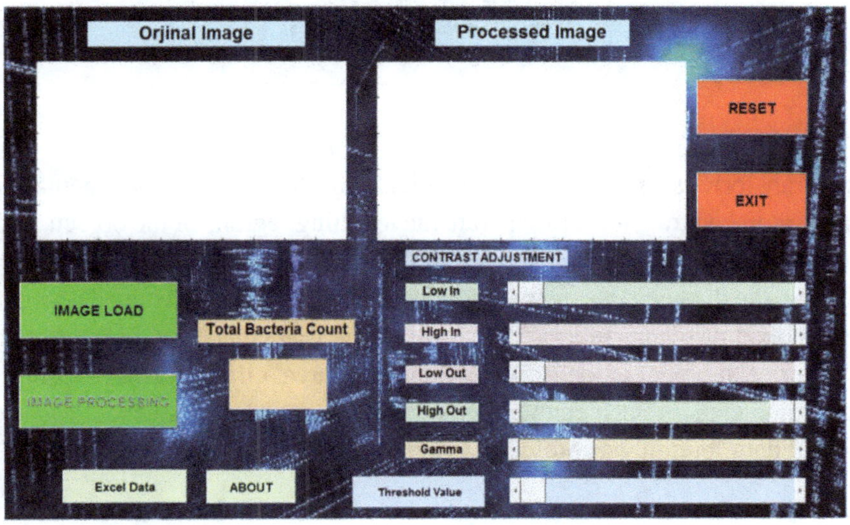

*Figure 3. The block diagram demonstration of the present method*

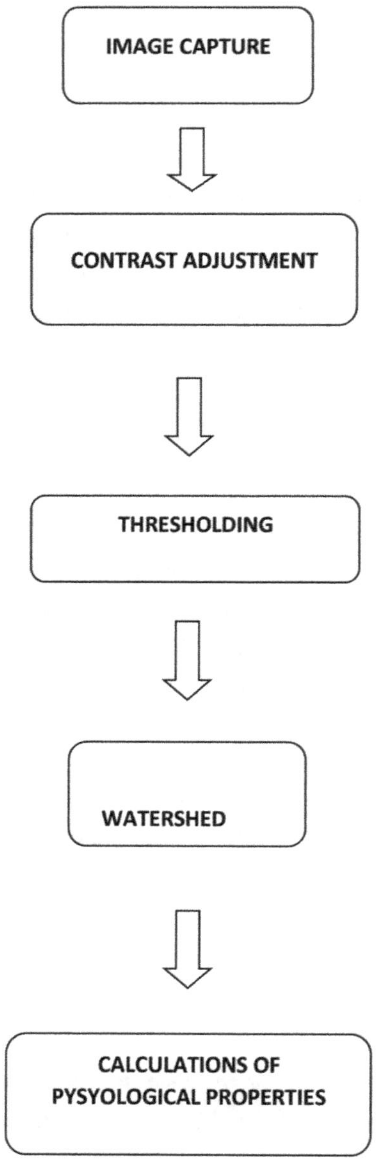


The watershed transform is a powerful mathematical morphology based segmentation algorithm. In this study we used watershed transform to separate the touching regions of the objects in the picture. The watershed method works very well when applied to the distance transform of a binary image. For each pixel in binary image, the distance transform assigns a number that is the distance between that pixel and the nearest nonzero pixel of binary image. The result of the watershed transform applied to the Picture is given in 8-d. After each object is identified and labeled, area, circumferences, center of gravity and equivalent circular diameters of each enumerated object are recorded in an excel file according to date and time. The 'excel data' button was provided to access this information.

*Figure 4. Contrast adjustment*

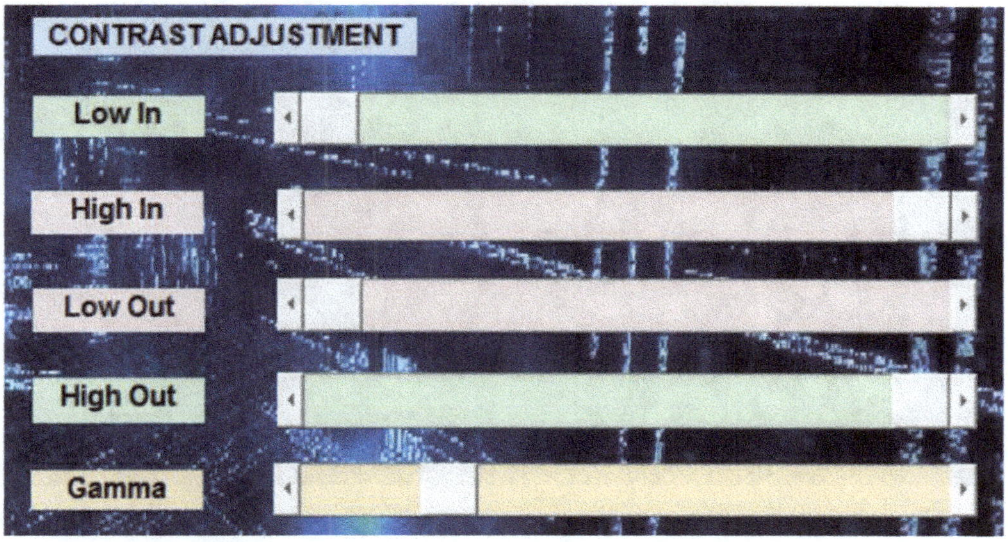


*Figure 5. Original image and contrast adjusted image*

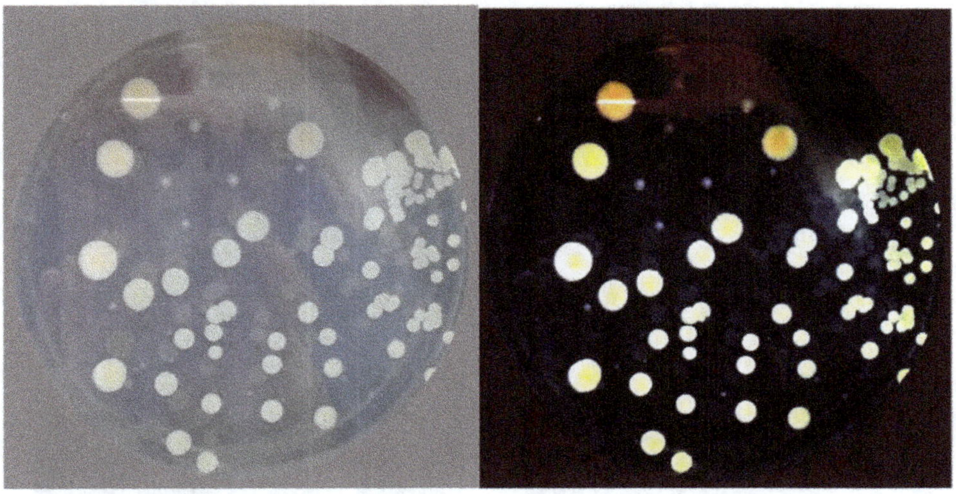

*Figure 6. The slider for threshold value*

*Figure 7. Gray scale image and binary image*

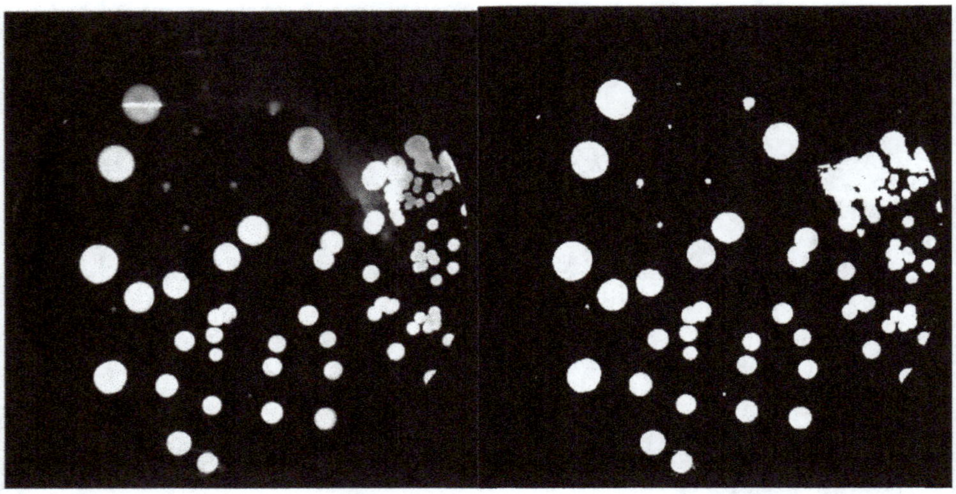

*Figure 8. Watershed stages*

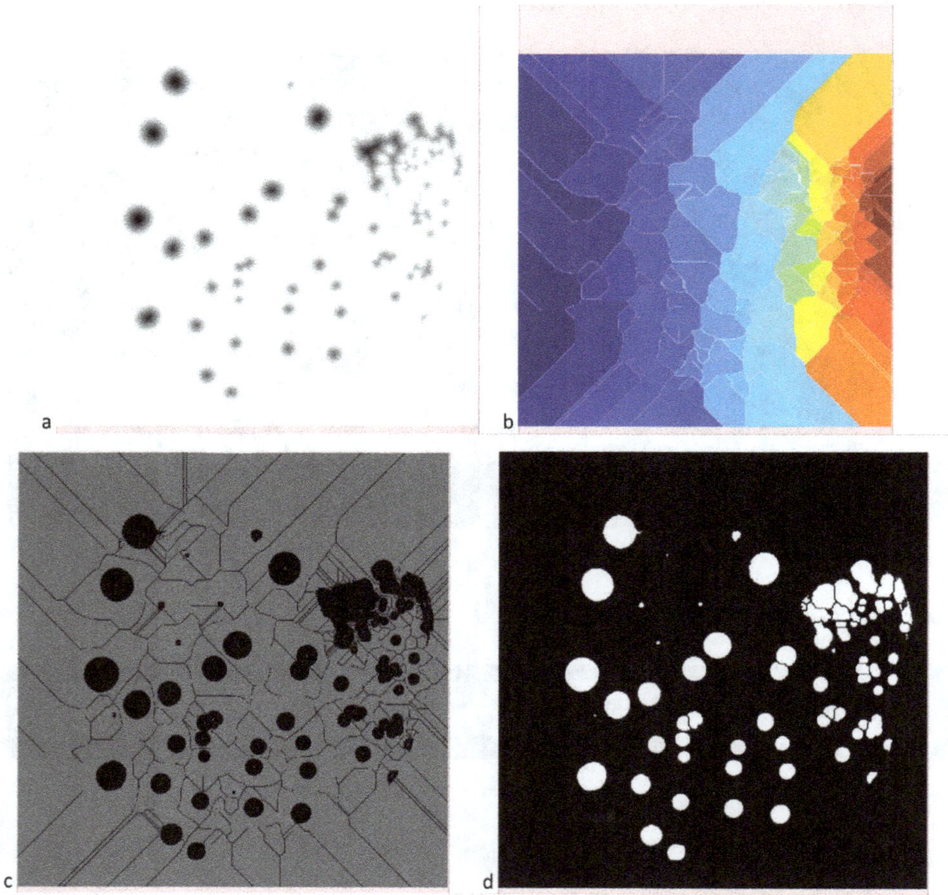

## RESULTS AND DISCUSSION

In this study, computer based automatic bacterial counting was performed. All the processing was performed by using MATLAB. In Figure 9 two sample raw images were given. Basin conversion done with watershed transform of the related images were given in Figure 10.

In Figure 11, pictures are given with the watershed transform applied. As you can see from the previous images some bacterial samples are connected to each other. As you can see from Figure 11, these connected objects are separated by the help of watershed transform. After that all the objects are labeled for the calculation of physical properties as mentioned before as seen in Figure 12.

The counting results for these two example images are given in Table 1. It is seen that these counts obtained from this study are within the acceptable limits. When the images used in the study are examined, it can be seen that the proposed method can count many objects in the image. If the user wishes to

*Figure 9. Original images*

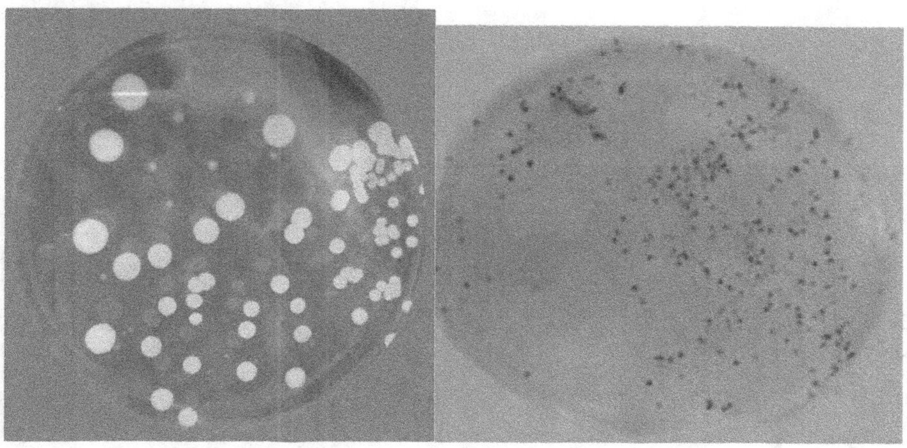

*Figure 10. Basin conversion done with watershed transform*

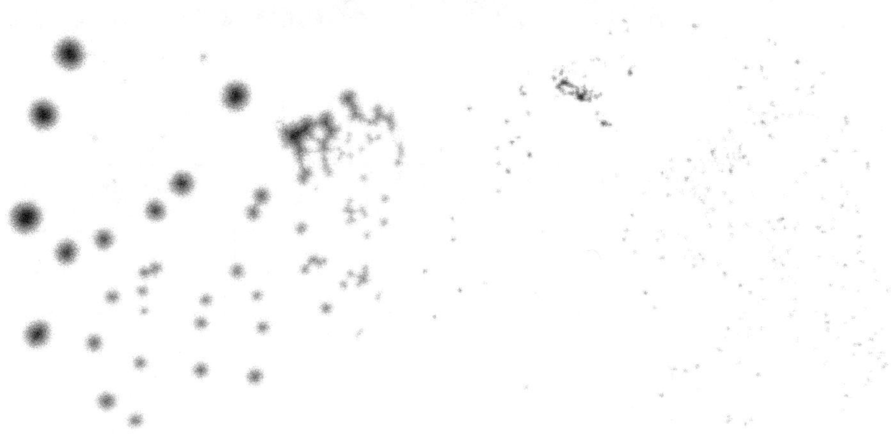

*Figure 11. Watershed applied state*

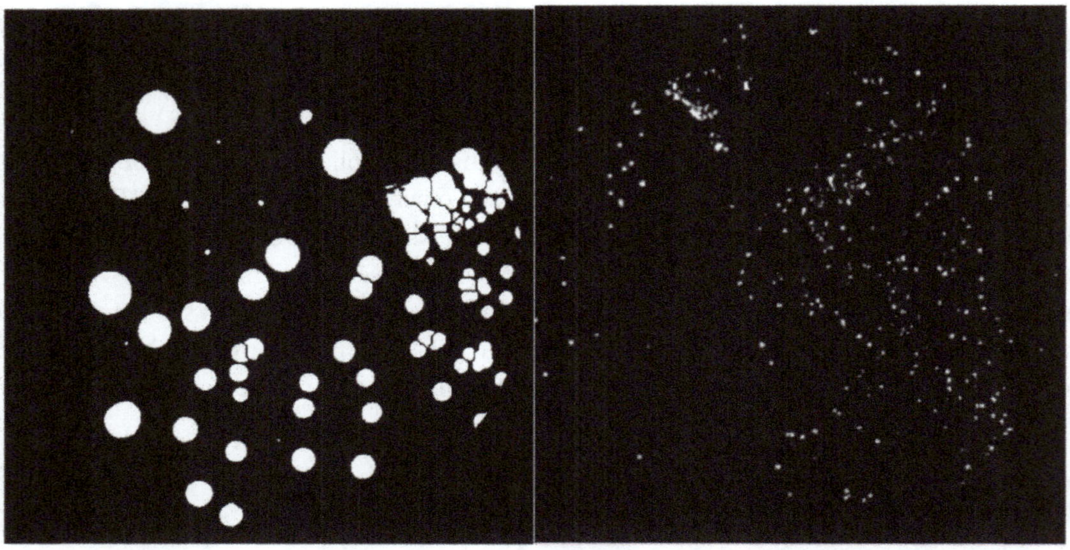

*Figure 12. Labeling process*

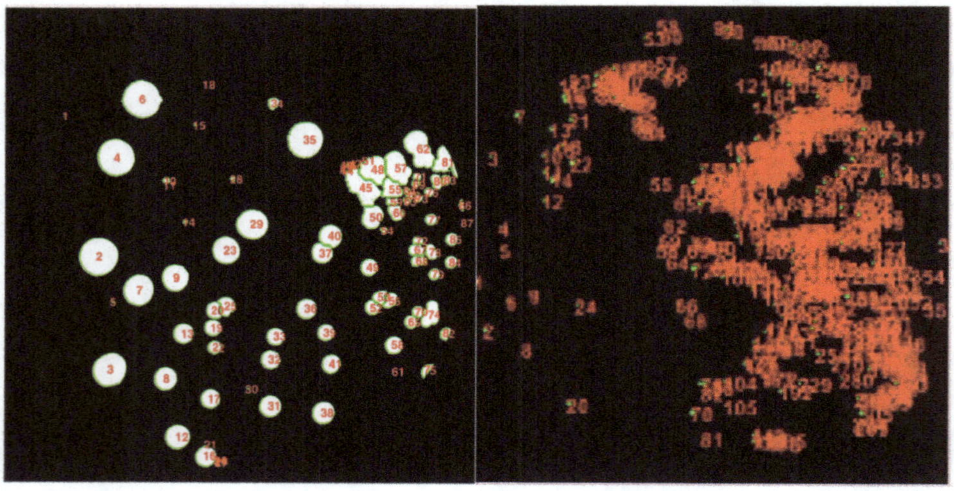

*Table 1. Counting results*

| Images | Expert Count | Count Results With Presented Method |
|---|---|---|
| 1 | 95 | 88 |
| 2 | 397 | 356 |

obtain detailed information for each bacterium after the labeling process, area, circumferences, center of gravity and equivalent circular diameters of each enumerated object are recorded in an excel file according to date and time.

## CONCLUSION

In today's food trade, one of the biggest priorities of food producers is to produce more hygienic products. In order to achieve this, the number of microorganisms in these food products needs to be determined correctly. Determination of the number of microorganisms in a food sample often requires a long analysis period. In this study, a computer-assisted bacterial counting method was introduced to reduce the time required for bacterial counting and to perform close counting. Therefore, some image processing techniques are applied on bacterial images such as watershed transformation. In the image processing procedure, watershed transform was used for to separate the adjacent bacteria and thus a more accurate counting was performed. The performance of the proposed method was tested on different samples and compared with the expert count on the same samples. According to the results obtained, it is seen that the proposed system achieves near-direct results in a shorter period of time. In this way, the results which are long period of time to reach have been obtained with high accuracy with a computer based algorithm. The obtained results show that the proposed method is successful and can be used for bacterial counting. Also, the physiological properties of each labeled bacterial obtained in this study could be used for determination of the bacterial species in the future studies. So that counting could be done according to the species.

## REFERENCES


Bayram, B. (2017). *Digital Image Processing*. Retrieved June 16, 2017, from http://www.yildiz.edu.tr/~bayram/sgi/saygi.htm

Beucher, S. (1992). The watershed transformation applied to image segmentation. Signal and image processing in microscopy and microanalysis. *Scanning Microscopy Supplement 6*, 299-314.

Beucher, S., & Lantuéjoul, C. (1979). Use of watersheds in contour detection. *International Workshop on Image Processing, Real-Time edge and motion detection/estimation*, 2.1-2.12.

Beucher, S., & Meyer, F. (1993). The morphological approach to segmentation: The watershed transform. In E. R. Dougherty (Ed.), *Mathematical Morphology in Image Processing* (pp. 433–481). Marcel Dekker, Inc.

Çetinel, G., & Kamanlı, A. F. (2016). An automatic embryonic stem cell counting method. *Sakarya University Journal of Science*, *20*(2), 83–97.

Colley, M., Kommoss, F., Bibbo, M., Dytch, H. E., Frnklin, W. A., Holt, J. A., & Wied, G. L. (1989). Assessment of hormone receptors in breast carcinoma by immunocytochemistry and image analysis. *Analytical and Quantitative Cytology and Histology*, *11*, 307–314. PMID:2803443

De la Torre, A., Peinado, A. M., Segura, J. C., Perez-Cordoba, J. L., Benitez, M. C., & Rubio, A. J. (2005). Histogram equalization of speech representation for robust speech recognition. *IEEE Transactions on Speech and Audio Processing*, *13*(3), 355–366. doi:10.1109/TSA.2005.845805

Dubuisson, M. P., Jain, A. K., & Jain, M. K. (1994). Segmentation and Classification of Bacterial Culture Images. *Journal of Microbiological Methods*, *19*(4), 279–295. doi:10.1016/0167-7012(94)90031-0

Erkmen, O., & Bozoğlu, T. F. (2008). *Food Microbiology-1* (1st ed.). Ankara: Gazi University Printing Depermant.

Ganesan, P., & Rajini, V. (2014). YIQ color space based satellite image segmentation using modified FCM clustering and histogram equalization. In ICAEE, Velore, India. doi:10.1109/ICAEE.2014.6838440

Gilchrist, J. E., Campbell, J. E., Donnelly, C. B., Peeler, J. T., & Delaney, J. M. (1973). Spiral Plate Method for Bacterial Determination. *Applied Microbiology*, *25*, 244–252. PMID:4632851

Gonzalez, R. C., & Woods, R. E. (2008). *Digital Image Processing* (3rd ed.). Upper Saddle River, NJ: Prentice Hall.

Halkman, A. K., & Ayhan, K. (2017). *Microorganism Counting*. Retrieved June 02, 2017, from http://www.mikrobiyoloji.org/TR/Genel/DosyaGoster.aspx?DIL=1&BELGEANAH=2927&DOSYAISIM=210010601.pdf

Iza, S. I., Siti, N. S., Muzaimi, M., & Noor, K. A. K. (2017). Automatic contrast enhancement of brain MR images using Average Intensity Replacement based on Adaptive Histogram Equalization (AIR-AHE). *Biocybernetics and Biomedical Engineering*, *37*, 24–34. doi:10.1016/j.bbe.2016.12.003

Jaemoon, L., Minhyeok, M., Chul, L., & Chang-Su, K. (2017). Contrast enhancement of noisy low-light images based on structure-texture-noise decomposition. *Journal of Visual Communication and Image Representation*, *45*, 107–121. doi:10.1016/j.jvcir.2017.02.016

Karakuş, D. (2006). *Determination of the structural properties of rocks with image analysing methods* (Doctoral Thesis). Dokuz Eylül University, Institute of Science, İzmir, Turkey.

Kawai, M., Yamaguchin, N., & Nasu, M. (1999). Rapid Enumeration of Physiologically Active Bacteria in purified water used in the Pharmaceutical Manufacturing Process. *Journal of Applied Microbiology*, *86*(3), 496–504. doi:10.1046/j.1365-2672.1999.00689.x PMID:10196754

Kittler, J., & Illingworth, J. (1986). Minimum error thresholding. *Pattern Recognition*, *19*(1), 41–47. doi:10.1016/0031-3203(86)90030-0

Kramer, J. M., Kendall, M., & Gilbert, R. J. (1978). Evaluation of the Spiral Plate and Laser Colony Counting techniques for the Enumeration of Bacteria in foods. *Journal of Applied Microbiology and Biotechnology*, *6*(3), 289–299. doi:10.1007/BF00508101

Markiewicz, T. (2003). Myelogenous leukemia cell image preprocessing for feature generation. *5th International Workshop on Computational Methods in Electrical Engineering*, 70–73.

Mukherjee, D., Pal, A., Sharma, S. E., & Majumder, D. D. (1995). Water Quality Analysis: A Pattern Recognition Approach. *Pattern Recognition*, *28*(2), 269–281. doi:10.1016/0031-3203(94)00089-5

Otsu, N. (1979). A threshold selection method from gray-level histograms. *IEEE Transactions on Systems, Man, and Cybernetics*, *9*(1), 62–66. doi:10.1109/TSMC.1979.4310076

Salman, S. D., & Bahrani, A. A. (2010). Segmentation of tumor tissue in gray medical images using watershed transformation method. *International Journal of Advancements in Computing Technology*, *2*(4), 123–127. doi:10.4156/ijact.vol2.issue4.13

Şenol, M. (2014). *Reading Optical Form From Web Camera With C#* (M.Sc. Thesis). Afyon Kocatepe University, Graduate School of Natural and Applied Sciences, Afyonkarahisar, Turkey.

Sundaram, M., Ramar, K., Arumugam, N., & Prabin, G. (2011). Histogram modified local contrast enhancement for mammogram images. *Applied Soft Computing*, *11*(8), 5809–5816. doi:10.1016/j.asoc.2011.05.003

Talas, D. (2017). *What is bacteria?* Retrieved June 08, 2017, from http://www.bilgiustam.com/bakteri-nedir/

Zhang, C., & Chen, W. B. (2007). An Effective and Robust Method for Automatic Bacterial Colony Enumeration. *Proceedings of the International Conference on Semantic Computing*, 581-588. doi:10.1109/ICSC.2007.20

## ADDITIONAL READING

Acharya, T., & Ray, A. K. (2005). *Image processing: principles and applications.* John Wiley & Sons. doi:10.1002/0471745790

Ao, J., Mitra, S., Long, R., Nutter, B., & Antani, S. (2012, April). A hybrid watershed method for cell image segmentation. In Image Analysis and Interpretation (SSIAI), 2012 IEEE Southwest Symposium on (pp. 29-32). IEEE. doi:10.1109/SSIAI.2012.6202445

Bovik, A. C. (2010). *Handbook of image and video processing*. Academic Press.

Chai, Y. H., Gao, L. Q., Lu, S., & Tian, L. (2006, June). Wavelet-based watershed for image segmentation algorithm. In Intelligent Control and Automation, 2006. WCICA 2006. The Sixth World Congress on (Vol. 2, pp. 9595-9599). IEEE.

Chuang, K. S., Tzeng, H. L., Chen, S., Wu, J., & Chen, T. J. (2006). Fuzzy c-means clustering with spatial information for image segmentation. computerized medical imaging and graphics, 30(1), 9-15.

Dhanachandra, N., & Chanu, Y. J. (2017). A Survey on Image Segmentation Methods using Clustering Techniques. *European Journal of Engineering Research and Science*, *2*(1), 15. doi:10.24018/ejers.2017.2.1.237

Duan, Y., Liu, F., Jiao, L., Zhao, P., & Zhang, L. (2017). SAR Image segmentation based on convolutional-wavelet neural network and markov random field. *Pattern Recognition*, *64*, 255–267. doi:10.1016/j.patcog.2016.11.015

Fischer, V., Kumar, M. C., Metzen, J. H., & Brox, T. (2017). Adversarial Examples for Semantic Image Segmentation. arXiv preprint arXiv:1703.01101.

Geissmann, Q. (2013). OpenCFU, a new free and open-source software to count cell colonies and other circular objects. *PLoS One*, *8*(2), e54072. doi:10.1371/journal.pone.0054072 PMID:23457446

Gowda, R. M., & Lingaraju, G. M. (2017). Texture-Based Watershed 3D Medical Image Segmentation Based on Fuzzy Region Growing Approach. In Advances in Computational Intelligence: Proceedings of International Conference on Computational Intelligence 2015 (pp. 233-243). Springer Singapore.

Hahn, H. K., & Peitgen, H. O. (2003, May). IWT-interactive watershed transform: a hierarchical method for efficient interactive and automated segmentation of multidimensional gray-scale images. In *Medical Imaging 2003* (pp. 643–653). International Society for Optics and Photonics. doi:10.1117/12.481097

Heimann, T., & Meinzer, H. P. (2009). Statistical shape models for 3D medical image segmentation: A review. *Medical Image Analysis*, *13*(4), 543–563. doi:10.1016/j.media.2009.05.004 PMID:19525140

Kashyap, R., & Gautam, P. (2017). Fast Medical Image Segmentation Using Energy-Based Method. Pattern and Data Analysis in Healthcare Settings, 35-60.

Khelifi, L., & Mignotte, M. (2017). EFA-BMFM: A multi-criteria framework for the fusion of colour image segmentation. *Information Fusion*, *38*, 104–121. doi:10.1016/j.inffus.2017.03.001

Lapeer, R. J., Tan, A. C., & Aldridge, R. (2002). A combined approach to 3D medical image segmentation using marker-based watersheds and active contours: the active watershed method. Medical Image Understanding-MIUA, 165-168.

Lempitsky, V., & Zisserman, A. (2010). Learning to count objects in images. In Advances in Neural Information Processing Systems (pp. 1324-1332).

Lim, J. S. (1990). *Two-dimensional signal and image processing*. Englewood Cliffs, NJ: Prentice Hall.

Londhe, V. (2017). Brain MR Image Segmentation for Tumor Detection using Artificial Neural. *Brain*, *6*(1).

Mangan, A. P., & Whitaker, R. T. (1999). Partitioning 3D surface meshes using watershed segmentation. *IEEE Transactions on Visualization and Computer Graphics*, *5*(4), 308–321. doi:10.1109/2945.817348

Maninis, K. K., Pont-Tuset, J., Arbelaez, P., & Van Gool, L. (2017). Convolutional Oriented Boundaries: From Image Segmentation to High-Level Tasks. *IEEE Transactions on Pattern Analysis and Machine Intelligence*, 1. doi:10.1109/TPAMI.2017.2700300 PMID:28475046

Maulik, U. (2009). Medical image segmentation using genetic algorithms. *IEEE Transactions on Information Technology in Biomedicine*, *13*(2), 166–173. doi:10.1109/TITB.2008.2007301 PMID:19272859

Mouelhi, A., Sayadi, M., & Fnaiech, F. (2013, May). Hybrid segmentation of breast cancer cell images using a new fuzzy active contour model and an enhanced watershed method. In Control, Decision and Information Technologies (CoDIT), 2013 International Conference on (pp. 382-387). IEEE. doi:10.1109/CoDIT.2013.6689575

Mukherjee, D. P., Pal, A., Sarma, S. E., & Majumder, D. D. (1995). Bacterial colony counting using distance transform. *International Journal of Bio-Medical Computing*, *38*(2), 131–140. doi:10.1016/0020-7101(94)01043-Z PMID:7729929

Na, S., & Heru, X. (2009, November). The segmentation of overlapping milk somatic cells based on improved watershed algorithm. In Artificial Intelligence and Computational Intelligence, 2009. AICI'09. International Conference on (Vol. 3, pp. 563-566). IEEE. doi:10.1109/AICI.2009.211

Nixon, M. S., & Aguado, A. S. (2012). *Feature extraction & image processing for computer vision.* Academic Press.

Parker, J. R. (2010). *Algorithms for image processing and computer vision.* John Wiley & Sons.

Pei, C., Wang, C., & Xu, S. (2010, December). Segmentation of the breast region in mammograms using marker-controlled watershed transform. In Information Science and Engineering (ICISE), 2010 2nd International Conference on (pp. 2371-2374). IEEE.

Pitas, I. (2000). *Digital image processing algorithms and applications.* John Wiley & Sons.

Pont-Tuset, J., Arbelaez, P., Barron, J. T., Marques, F., & Malik, J. (2017). Multiscale combinatorial grouping for image segmentation and object proposal generation. *IEEE Transactions on Pattern Analysis and Machine Intelligence*, *39*(1), 128–140. doi:10.1109/TPAMI.2016.2537320 PMID:26955014

Riddell, C., Brigger, P., Carson, R. E., & Bacharach, S. L. (1999). The watershed algorithm: A method to segment noisy PET transmission images. *IEEE Transactions on Nuclear Science*, *46*(3), 713–719. doi:10.1109/23.775604

Roy, P., Goswami, S., Chakraborty, S., Azar, A. T., & Dey, N. (2017). Image segmentation using rough set theory: a review. In Medical Imaging: Concepts, Methodologies, Tools, and Applications (pp. 1414-1426). IGI Global. doi:10.4018/978-1-5225-0571-6.ch059

Russ, J. C. (2016). *The image processing handbook*. CRC Press.

Salman, S. D., & Bahrani, A. A. (2010). Segmentation of tumor tissue in gray medical images using watershed transformation method. Intl. *Journal of Advancements in Computing Technology*, *2*(4).

Schlesinger, D., Jug, F., Myers, G., Rother, C., & Kainmüller, D. (2017). Crowd Sourcing Image Segmentation with iaSTAPLE. arXiv preprint arXiv:1702.06461.

Selinummi, J., Seppala, J., Yli-Harja, O., & Puhakka, J. A. (2005). Software for quantification of labeled bacteria from digital microscope images by automated image analysis. *BioTechniques*, *39*(6), 859–863. doi:10.2144/000112018 PMID:16382904

Sharma, N., & Aggarwal, L. M. (2010). Automated medical image segmentation techniques. *Journal of Medical Physics / Association of Medical Physicists of India*, *35*(1), 3. doi:10.4103/0971-6203.58777 PMID:20177565

Shen, D., Wu, G., & Suk, H. I. (2017). Deep learning in medical image analysis. Annual Review of Biomedical Engineering, (0).

Sonka, M., Hlavac, V., & Boyle, R. (2014). *Image processing, analysis, and machine vision*. Cengage Learning.

Torii, J., & Kouno, Y. (2017). Image Processing in X-Ray Fluoroscopy and Digital Tomography. In Respiratory Endoscopy (pp. 17-25). Springer Singapore.

van Heel, M., Harauz, G., Orlova, E. V., Schmidt, R., & Schatz, M. (1996). A new generation of the IMAGIC image processing system. *Journal of Structural Biology*, *116*(1), 17–24. doi:10.1006/jsbi.1996.0004 PMID:8742718

Yu, C. Y., & Li, Y. (2012, May). A watershed method for mr renography segmentation. In Biomedical Engineering and Biotechnology (iCBEB), 2012 International Conference on (pp. 700-703). IEEE. doi:10.1109/iCBEB.2012.45

Zhang, M., Zhang, L., & Cheng, H. D. (2010). A neutrosophic approach to image segmentation based on watershed method. *Signal Processing*, *90*(5), 1510–1517. doi:10.1016/j.sigpro.2009.10.021

Zhang, X., Jia, F., Luo, S., Liu, G., & Hu, Q. (2014). A marker-based watershed method for X-ray image segmentation. *Computer Methods and Programs in Biomedicine*, *113*(3), 894–903. doi:10.1016/j.cmpb.2013.12.025 PMID:24462387

Zhang, Y., Matuszewski, B. J., Shark, L. K., & Moore, C. J. (2008, July). Medical image segmentation using new hybrid level-set method. In BioMedical Visualization, 2008. MEDIVIS'08. Fifth International Conference (pp. 71-76). IEEE. doi:10.1109/MediVis.2008.12

## KEY TERMS AND DEFINITIONS

**Bacteria:** A type of organism, which is in the form of just a single cell.

**Computer-Based Bacterial Counting System:** A computer-based system that employs some techniques to count bacteria over a taken image.

**Contrast Adjustment:** The operation of modifying some specific contrast features/parameters of an image in order to obtain a more desired form of the image.

**Image Processing:** Modifying or analyzing images by using specific mathematical and logical techniques.

**Image Segmentation:** The operation of partitioning an image into some segments/regions.

**Microorganism:** An organism that can be seen well with a microscope and has a form of single or multi cell.

**Watershed Transform:** An image-processing technique that is used on gray-tone images and can benefit from a topographic surface-oriented approach to segment the image.

# Chapter 9
# Diabetes Diagnosis System Based on Support Vector Machines Trained by Vortex Optimization Algorithm

**Sadi Fuat Cankaya**
*Suleyman Demirel University, Turkey*

**Ibrahim Arda Cankaya**
*Suleyman Demirel University, Turkey*

**Tuncay Yigit**
*Suleyman Demirel University, Turkey*

**Arif Koyun**
*Suleyman Demirel University, Turkey*

## ABSTRACT

*Artificial intelligence is widely enrolled in different types of real-world problems. In this context, developing diagnosis-based systems is one of the most popular research interests. Considering medical service purposes, using such systems has enabled doctors and other individuals taking roles in medical services to take instant, efficient expert support from computers. One cannot deny that intelligent systems are able to make diagnosis over any type of disease. That just depends on decision-making infrastructure of the formed intelligent diagnosis system. In the context of the explanations, this chapter introduces a diagnosis system formed by support vector machines (SVM) trained by vortex optimization algorithm (VOA). As a continuation of previously done works, the research considered here aims to diagnose diabetes. The chapter briefly gives information about details of the system and findings reached after using the developed system.*

DOI: 10.4018/978-1-5225-4769-3.ch009

## INTRODUCTION

Artificial Intelligence is widely used in many different fields having remarkable role on increasing the amount of scientific developments. We cannot see any field that has not at least indirect support from intelligent approaches, methods, and techniques of Artificial Intelligence. It is possible to briefly just focus on how a different field is vital for us and how it can be improved efficiently thanks to the field of Artificial Intelligence. The research fields meeting under the framework of medical – health are known as such type of fields. When we consider the associated literature, we can see that there is a popular interest in i.e. designing diagnosis oriented systems to support doctors or any other workers in the fields of medical services (Choi et al., 2017; Jiang et al., 2017; Khubchandani et al., 2017; Kononenko, 2001; Stoitsis et al., 2006; Waal, 2017). When it comes which intelligent techniques to use for medical purposes, there is a great diversity of available Artificial Intelligence techniques.

Today, it can be observed that the diagnosis approach of Artificial Intelligence is generally based on using some past data – experiences to train an intelligent system and enabling that system to diagnose the objective disease(s) by using new data that may be new for the system. Because of the 'training mechanism' mentioned here, intelligent diagnosis systems are generally based on Machine Learning techniques. For example, Artificial Neural Networks or Support Vector Machines are some typical Machine Learning techniques used for diagnosis purposes (Abbass, 2002; Akay, 2009; Al-Shayea, 2011; Amato et al., 2013; Lee & Lee, 2003; Polat & Gunes, 2007; Wu et al., 1993). Along the systems formed over training of the chosen techniques, it is aimed to have as accurate as intelligent diagnosis infrastructure depending on the data sets of objective disease. It seems that there are many different types of medical and biomedical applications of Artificial Intelligence but diagnosis has a remarkable popularity because of the importance of 'early diagnosis of diseases', which is a vital factor for living organisms including humankind.

As associated with the explanations above, the objective of this chapter is to provide an alternative intelligent diagnosis system for detecting diabetes according to some given input data. In detail, the work is the continuation of a previously done work (Kose et al., 2015) and it focuses on some more aspects of the used Artificial Intelligence techniques. As different, some alternative parameter values and approaches have been used for the formed system, within this study. In detail, the diagnosis system is formed by a model of Support Vector Machines (SVM), which is trained by Vortex Optimization Algorithm (VOA). The system generally has a simple structure but employs the problem-solving approach of classification for reaching to its main objective. Along the chapter, background and details of the system and information about performed experiments are provided.

Considering the topic of the chapter, the remaining content is organized as follows: The next section provides essential information about the formed system, employed Artificial Intelligence techniques under it, in order to enable readers to have enough information about what has been done in the study. Following that, the third section explains some experimental operations and provides the obtained findings. Next, the fourth section focuses on some future research ideas and finally, the chapter ends with the last section on a brief discussion regarding to conclusions.

## A DIAGNOSIS SYSTEM FORMED BY SVM-VOA

The diagnosis system, which is used for diabetes here is a typical Machine Learning approach employing the mechanisms of training and test / application respectively. Although SVM is also a Data Mining technique (Deng & Tian, 2004), it is considered as a Machine Learning solution way by taking the objectives and application approaches followed in this study. Even Data Mining and Machine Learning often find common research focus on the real world based problems (Witten et al., 2016).

If we focus on the diagnosis system used in this study, we can see a general system of diagnosis. In other words, the system designed and developed here can be used for also general diagnosis purposes if appropriate adjustments are made for changing problems (new conditions, used input data…etc.). Even it is used for a specific disease or general diagnosing processes, the whole problem solution flow is as follows:

- Take the input data as solution objects of the Support Vector Machine (SVM),
- Adjust parameters of the Vortex Optimization Algorithm (VOA), which is the trainer of the SVM,
- Determine the kernel function of the SVM to be used for non-linear classification,
- Determine which parameter of the SVM to be optimized through the training,
- Perform the training process,
- Apply the trained SVM over objective input data of diagnosis.

In order to understand better about the background of the developed system, two essential techniques included in it should be explained briefly. Some essential information regarding to SVM and VOA are as follows:

*Figure 1. Solution flow in the formed system of SVM-VOA*

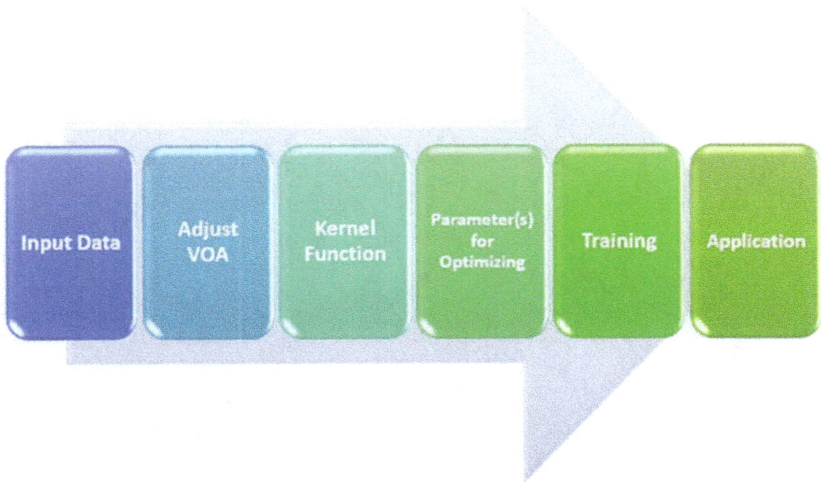

## Support Vector Machines

As introduced by Vapnik and Lerner (1997), Support Vector Machines (SVM) is a classification and regression technique, which can be examined under the subjects of both Data Mining and Machine Learning. The main approach of a typical SVM deals with the follows (Cortes & Vapnik, 1995; Vapnik & Lerner, 1997):

- SVM creates an optimally separating hyper-plane so that the related margin between objective groups is maximized.
- Data objects drawing the hyper-plane are called as support vectors (which gives the main name of the technique).
- According to amount and distribution of objective groups, different types of SVM can be employed. These two types of SVM are called as linear SVM and non-linear SVM respectively (Burges, 1998; Cristianini & Shawe-Taylor, 2000; Wang, 2005).
- As the simplest form of SVM, linear SVM uses a linear boundary to classify groups in a linear manner where the non-linear form of SVM employs non-linear boundaries (Burges, 1998; Cristianini & Shawe-Taylor, 2000; Smola & Schölkopf, 2004; Wang, 2005).

Figure 2 and Figure 3 respectively shows a typical SVM solution and a solved problem via non-linear SVM model.

*Figure 2. A typical SVM solution view*
*Kumar et al., 2014.*

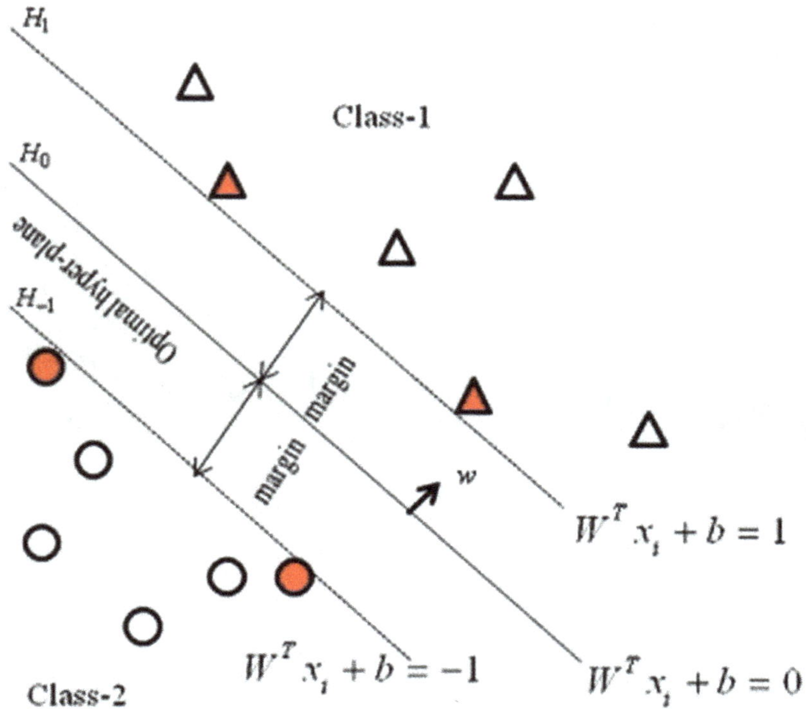

*Figure 3. An example for solved classification via non-linear SVM*
Welsh et al., 2013.

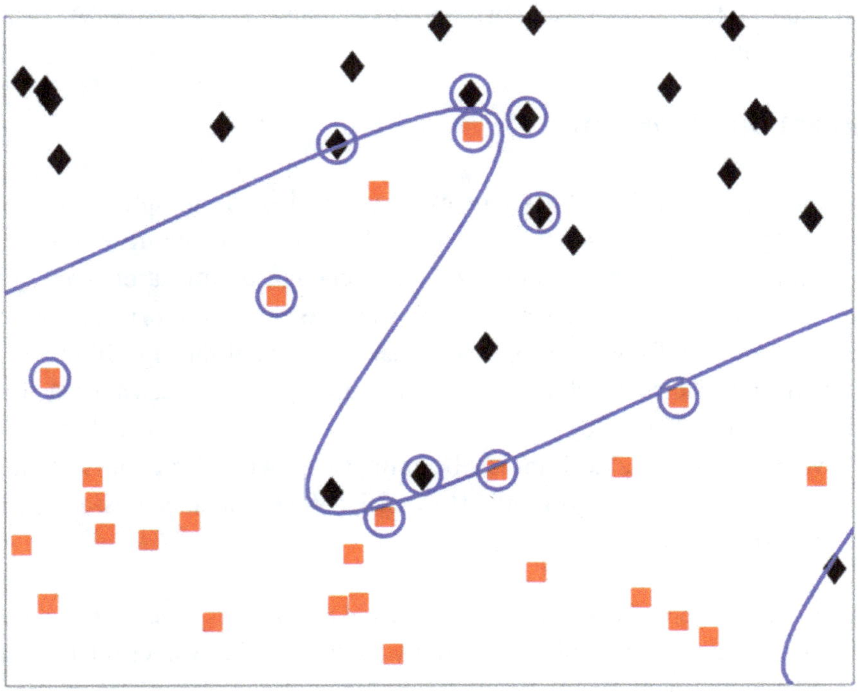

Except from linear SVM, non-linear types of SVM requires running special elements to reach to higher dimensions required for more complicated problems that cannot be solved via linear approaches. At this point, the model of SVM uses a kernel function to transform n-dimensional data set of m>n, including m-dimensional transformed into a new data set (Ozkaya, 2003). In the literature, there are three types of kernel functions as widely used within applications. These functions are given under the Equation 1, 2, and 3 as follows (Scholkopf et al., 1997; Scholkopf & Smola, 2001; Wang, 2005):

Polynomial kernel function:

$$K\left(x_i, x_j\right) = \left(1 + x_i^T x_j\right)^p \tag{1}$$

Linear kernel function:

$$K\left(x_i, x_j\right) = x_i^T x_j \tag{2}$$

Gaussian kernel function:

$$K\left(x_i, x_j\right) = e^{\frac{x_i - x_j^{\;2}}{2\sigma^2}} \tag{3}$$

In the studies regarding to training SVM, the approach of optimizing parameters of the related kernel functions is often followed. As it can be understood, this can be done by using an optimization algorithm – technique. In this study, this algorithm – technique is chosen as VOA, which is explained briefly under the next sub-title.

## Vortex Optimization Algorithm

Vortex Optimization Algorithm (VOA) is a recently developed Artificial Intelligence based optimization algorithm. As introduced by Kose and Arslan (2015), it has received many updates since its first introduction as aiming to provide practical but effective enough optimization way by inspiring from vortex dynamics in the nature. VOA generally employs some evaluation oriented operations to make it possible to improve solution flow by eliminating weak particles of the algorithm and supporting the solution space with fresh particles. In addition, it also uses a simple role-based approach among particles (Kose & Arslan, 2015, 2016; Kose, 2017).

After some developments, the algorithm has been improved with also an in-system approach dealing with bigger problems. A recent algorithmic flow of the VOA can be expressed as follows (Kose & Arslan, 2015, 2016; Kose, 2017):

**Step 1:** Determine initial parameters of the algorithm as $N$ for number of particles; vorticity ($v$) values of each particle; max. and min. limits for vorticity value and other values related to function, problem…etc.; and $e$ for elimination rate.

**Step 2:** Spread the particles randomly to the objective solution space and calculate fitness values for each of them. Update the v value of the particle with the best fitness value by using a random value. This particle is now a vortex and as storing its values as the best so far.

**Step 3:** Loop phase according to a stopping criteria:

**Step 3.1:** Accept each particle, whose fitness value is under the average fitness of all particles, as the 'vortex'. Remaining particles are in the status of 'normal' particle.

**Step 3.2:** Update $v$ value of each particle via Equation 4:

particlei_v_change = particlei_v + (random * (global_best_v / particlei_v))

*particlei_v = particlei_v_change* (4)

**Step 3.3:** Update the $v$ value of each vortex particle (except from the best particle) via random values.

**Step 3.4:** Update position of each particle (except from the best particle) via Equation 5:

*particlei_position+=(random*(particlei_v_change*(global_best_position–particlei_position)))* (5)

**Step 3.5:** Calculate fitness values with new positions of each particle. Accept the particle with the best value as a vortex (if it is not a vortex yet) and store its values as the best so far.

**Step 3.6:** If number of non-vortex particles is under the value of $e$, remove all non-particles from the solution space. Next, create new particles according to number of removed particles. Spread these new particles randomly to the solution space.

**Step 3.7**: Operate in-system optimization if it is necessary.

**Step 4:** Optimum value(s) at the end of the loop is the solution.

Although it is for early version of the VOA, Figure 4 explains the main solution flow mechanism designed in the algorithm (Kose & Arslan, 2015).

The next section is devoted to more information about the main research followed in this study over the SVM-VOA system.

## DIABETES DIAGNOSIS WITH SVM-VOA

The authors have focused on an approach of diabetes diagnosis by using the designed SVM-VOA system. Here, a widely used kernel function: Gaussian has been chosen to be used for the experimental process. In detail, VOA is employed to optimize the sigma (σ) parameter of the formed non-linear SVM model, in

*Figure 4. Main solution flow mechanism designed in the VOA*
*Kose & Arslan, 2015.*

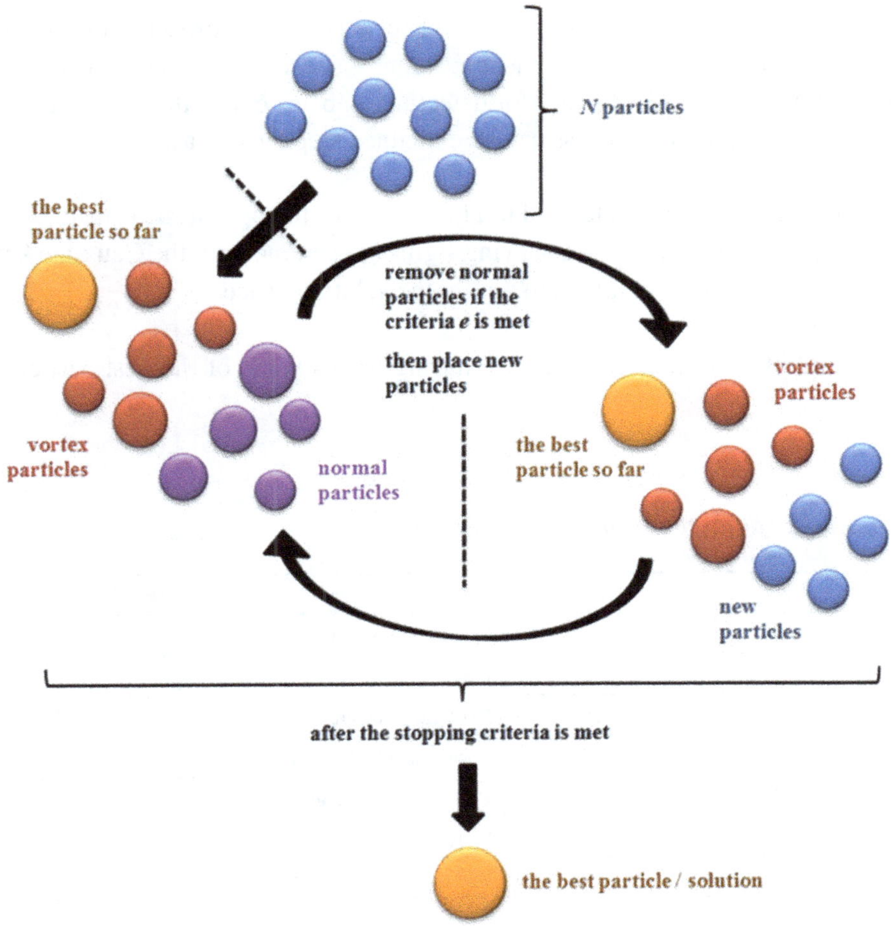

order to improve accuracy – success on detecting diabetes. The following sub-titles are for explaining the objective dataset, details of the solution process, and some findings had with several experimental studies.

## Diabetes Dataset

The diabetes dataset used in this study has been taken from the well-known repository source: UCI. Some essential information regarding to the dataset are as follows (Asuncion & Newman, 2007):

- Data is based on Pima Indians diabetes.
- There are a total of 8 different features included for each individual in the dataset (Details are given in Table 1).
- There are a total of 768 rows – data for each individual as including individuals with negative and positive diabetes.
- From 768 rows, data for 500 ones are related to individuals with negative diabetes whereas 268 ones are for the individuals with positive diabetes.

## Solution Process

The solution process followed here is very simple to apply. The whole process of training – optimizing the non-linear SVM is for just getting the most accurate – successful model. It is clear that the final model of trained SVM can be used directly for detecting diabetes out of the dataset used here. Details regarding to the designed solution process can be explained step by step as:

**Step 1:** Determine a number of particles and total iteration for the optimization task.
**Step 2:** Accept each particle of the VOA as having sigma (σ) parameter of the Gaussian kernel function, as different from their roles on position within the solution space.
**Step 3:** Run the VOA.
**Step 4:** In at the end of each new iteration, use the sigma (σ) value of the best particle in the kernel function.

*Table 1. Features in the dataset of Pima Indians diabetes*

| No. | Feature Detail |
|---|---|
| 1 | Number of pregnancy |
| 2 | Plasma glucose concentration |
| 3 | Diastolic blood pressure |
| 4 | Triceps skin fold thickness |
| 5 | 2-Hour serum insulin |
| 6 | Body mass index |
| 7 | Diabetes pedigree function |
| 8 | Age (years) |

**Step 5:** Run the SVM with the Gaussian kernel function employing the determined sigma (σ) value and evaluate the diabetes detection performance by finding the percentage of true classified individuals by dividing total number of true detection into total number of individuals – rows.
**Step 6:** Repeat the steps 4, and 5 until the stopping criteria of the VOA (iteration) is reached.
**Step 7:** The sigma (σ) parameter that achieved the globally best diagnosing SVM through iterations can then be used for diabetes diagnosis purposes.

As different from previous works done before, parameters of the VOA have been changed in this study to see effects of classification over the objective dataset. Some parameters used for four different experimental processes and the obtained diagnosis performances are shown in Table 2.

As it can be seen from the obtained findings, the SVM-VOA system can diagnose the diabetes with the objective data features accurately enough. Also, changing parameters of the VOA are very effective over the performance.

## FUTURE RESEARCH DIRECTIONS

In addition to the performed works of diagnosing diabetes over SVM-VOA system, the authors also think about some more further investigations related to the research interest considered. These can be mentioned briefly as follows:

- It is planned to use the system for different types of diseases and see if there is need for any improvements in the default structure designed system.
- The authors will also perform investigations on optimizing different parameters of the used kernel function. Additionally, same research focus for other kernel functions with different parameters will also be had for future studies.
- Obtained results with also in this study have enabled the authors to think about focusing on some kind of comparative works in which different intelligent optimization algorithms are used to train same models of SVM over same data sets.
- As different from the performed diagnosis experiments here, there are also alternative ones requiring some image processing steps in order to diagnose the objective disease (Deepa & Devi, 2011; El-Dahshan et al., 2010; Goldbaum et al., 1996; Hall et al., 1992; Stoitsis et al., 2006; Vyborny & Giger, 1994). So, there will be some future research efforts for also developing more advanced diagnosis system employing image processing oriented pre-operations.

*Table 2. Diabetes diagnosis: Detection processes with different settings*

| Process | VOA-*N* | VOA-*e* | Iteration | Best Sigma (σ) | Performance (True Detection %) |
|---|---|---|---|---|---|
| 1 | 75 | 25 | 3000 | 0,9256 | 75,72 |
| 2 | 100 | 50 | 4000 | 0,9182 | 78,63 |
| 3 | 150 | 50 | 5000 | 0,7731 | 88,47 |
| 4 | 200 | 75 | 5000 | 0,8614 | 82,19 |

- The authors will also focus on developing some new Artificial Intelligence techniques, which can also diagnose diseases. Investigations and applications in this manner are all future studies.
- The literature of medical diagnosis and Biomedical Engineering based problems in this manner is very wide. There are many different types of applications already introduced to the associated literature in each day. In this sense, the authors will be focused on employing newer studies for medical diagnosis and any other similar problems.

## CONCLUSION

In this chapter, a research process in which diabetes diagnosis has been tried to be made effectively via Support Vector Machine (SVM) technique was explained. The performed research is continuation on previously done works over diagnosing diabetes over newly developed Artificial Intelligence based algorithms to train other intelligent techniques and the authors have provided some more information about alternative efforts done in this manner. Considering the obtained results after applying the SVM model trained by Vortex Optimization Algorithm (VOA) for diagnosis purposes, it has been seen that it is possible to effectively determine diabetes thanks to the formed system with the training technique of VOA. It is also important in this study that the kernel function(s) of a SVM can be optimized by using appropriate algorithms with well-organized conditions. The diagnosis system considered here is an alternative for the associated literature as serving to the diagnosis phase, which is a vital factor along medical studies.

## REFERENCES

Abbass, H. A. (2002). An evolutionary artificial neural networks approach for breast cancer diagnosis. *Artificial Intelligence in Medicine*, *25*(3), 265–281. doi:10.1016/S0933-3657(02)00028-3 PMID:12069763

Akay, M. F. (2009). Support vector machines combined with feature selection for breast cancer diagnosis. *Expert Systems with Applications*, *36*(2), 3240–3247. doi:10.1016/j.eswa.2008.01.009

Al-Shayea, Q. K. (2011). Artificial neural networks in medical diagnosis. *International Journal of Computer Science Issues*, *8*(2), 150–154.

Amato, F., López, A., Peña-Méndez, E. M., Vaňhara, P., Hampl, A., & Havel, J. (2013). *Artificial neural networks in medical diagnosis*. Academic Press.

Asuncion, A., & Newman, D. (2007). *UCI machine learning repository*. Retrieved from http://mlearn.ics.uci.edu/MLRepository.html

Burges, C. J. (1998). A tutorial on support vector machines for pattern recognition. *Data Mining and Knowledge Discovery*, *2*(2), 121–167. doi:10.1023/A:1009715923555

Choi, Y. J., Baek, J. H., Park, H. S., Shim, W. H., Kim, T. Y., Shong, Y. K., & Lee, J. H. (2017). A computer-aided diagnosis system using artificial intelligence for the diagnosis and characterization of thyroid nodules on ultrasound: Initial clinical assessment. *Thyroid*, *27*(4), 546–552. doi:10.1089/thy.2016.0372 PMID:28071987

Cortes, C., & Vapnik, V. (1995). Support vector machine. *Machine Learning*, *20*(3), 273–297. doi:10.1007/BF00994018

Cristianini, N., & Shawe-Taylor, J. (2000). *An introduction to support vector machines and other kernel-based learning methods*. Cambridge University Press. doi:10.1017/CBO9780511801389

Deepa, S. N., & Devi, B. A. (2011). A survey on artificial intelligence approaches for medical image classification. *Indian Journal of Science and Technology*, *4*(11), 1583–1595.

Deng, N. Y., & Tian, Y. (2004). *A new method of data mining-support vector machine*. Science Publication.

El-Dahshan, E. S. A., Hosny, T., & Salem, A. B. M. (2010). Hybrid intelligent techniques for MRI brain images classification. *Digital Signal Processing*, *20*(2), 433–441. doi:10.1016/j.dsp.2009.07.002

Hall, L. O., Bensaid, A. M., Clarke, L. P., Velthuizen, R. P., Silbiger, M. S., & Bezdek, J. C. (1992). A comparison of neural network and fuzzy clustering techniques in segmenting magnetic resonance images of the brain. *IEEE Transactions on Neural Networks*, *3*(5), 672–682. doi:10.1109/72.159057 PMID:18276467

Jiang, F., Jiang, Y., Zhi, H., Dong, Y., Li, H., Ma, S., ... Wang, Y. (2017). Artificial intelligence in healthcare: past, present and future. *Stroke and Vascular Neurology*.

Khubchandani, P., Jha, K., Bijani, R., Lala, S., & Saindane, P. (2017). Medical Prediction using Artificial Intelligence. *International Journal of Engineering Science*, 4898.

Kononenko, I. (2001). Machine learning for medical diagnosis: History, state of the art and perspective. *Artificial Intelligence in Medicine*, *23*(1), 89–109. doi:10.1016/S0933-3657(01)00077-X PMID:11470218

Kose, U. (2017). *Development of Artificial Intelligence based Optimization Algorithms* (PhD. Thesis). Selcuk University, Institute of Natural Sciences, Dept. of Computer Engineering. (in Turkish)

Kose, U., & Arslan, A. (2015). On the Idea of a New Artificial Intelligence Based Optimization Algorithm Inspired From the Nature of Vortex. *BRAIN. Broad Research in Artificial Intelligence and Neuroscience*, *5*(1-4), 60–66.

Kose, U., & Arslan, A. (2016). Forecasting chaotic time series via anfis supported by vortex optimization algorithm: Applications on electroencephalogram time series. *Arabian Journal for Science and Engineering*, 1–12.

Kose, U., Guraksin, G. E., & Deperlioglu, O. (2015). Diabetes determination via vortex optimization algorithm based support vector machines. In *Medical Technologies National Conference (TIPTEKNO)* (pp. 1-4). IEEE. doi:10.1109/TIPTEKNO.2015.7374614

Kumar, Y., Dewal, M. L., & Anand, R. S. (2014). Epileptic seizure detection using DWT based fuzzy approximate entropy and support vector machine. *Neurocomputing*, *133*, 271–279. doi:10.1016/j.neucom.2013.11.009

Lee, Y., & Lee, C. K. (2003). Classification of multiple cancer types by multicategory support vector machines using gene expression data. *Bioinformatics (Oxford, England)*, *19*(9), 1132–1139. doi:10.1093/bioinformatics/btg102 PMID:12801874

Ozkaya, A. U. (2003). *Intelligent Arrhythmia Classification Based on Support Vector Machines* (MSc. Thesis). Istanbul Technical University, Institute of Natural Sciences, Dept. of Computer Engineering. (in Turkish)

Polat, K., & Güneş, S. (2007). Breast cancer diagnosis using least square support vector machine. *Digital Signal Processing*, *17*(4), 694–701. doi:10.1016/j.dsp.2006.10.008

Scholkopf, B., & Smola, A. J. (2001). *Learning with kernels: support vector machines, regularization, optimization, and beyond*. MIT Press.

Scholkopf, B., Sung, K. K., Burges, C. J., Girosi, F., Niyogi, P., Poggio, T., & Vapnik, V. (1997). Comparing support vector machines with Gaussian kernels to radial basis function classifiers. *IEEE Transactions on Signal Processing*, *45*(11), 2758–2765. doi:10.1109/78.650102

Smola, A. J., & Scholkopf, B. (2004). A tutorial on support vector regression. *Statistics and Computing*, *14*(3), 199–222. doi:10.1023/B:STCO.0000035301.49549.88

Stoitsis, J., Valavanis, I., Mougiakakou, S. G., Golemati, S., Nikita, A., & Nikita, K. S. (2006). Computer aided diagnosis based on medical image processing and artificial intelligence methods. *Nuclear Instruments & Methods in Physics Research. Section A, Accelerators, Spectrometers, Detectors and Associated Equipment*, *569*(2), 591–595. doi:10.1016/j.nima.2006.08.134

Vapnik, V., Golowich, S. E., & Smola, A. J. (1997). Support vector method for function approximation, regression estimation and signal processing. In Advances in neural information processing systems (pp. 281-287). Academic Press.

Vyborny, C. J., & Giger, M. L. (1994). Computer vision and artificial intelligence in mammography. *AJR. American Journal of Roentgenology*, *162*(3), 699–708. doi:10.2214/ajr.162.3.8109525 PMID:8109525

Waal, I. (2017). Skin cancer diagnosed using artificial intelligence on clinical images. *Oral Diseases*. PMID:28326656

Wang, L. (Ed.). (2005). *Support vector machines: theory and applications* (Vol. 177). Springer Science & Business Media. doi:10.1007/b95439

Welsh, R. C., Jelsone-Swain, L. M., & Foerster, B. R. (2013). The utility of independent component analysis and machine learning in the identification of the amyotrophic lateral sclerosis diseased brain. *Frontiers in Human Neuroscience*, 7. PMID:23772210

Witten, I. H., Frank, E., Hall, M. A., & Pal, C. J. (2016). *Data Mining: Practical machine learning tools and techniques*. Morgan Kaufmann.

Wu, Y., Giger, M. L., Doi, K., Vyborny, C. J., Schmidt, R. A., & Metz, C. E. (1993). Artificial neural networks in mammography: Application to decision making in the diagnosis of breast cancer. *Radiology*, *187*(1), 81–87. doi:10.1148/radiology.187.1.8451441 PMID:8451441

## ADDITIONAL READING


Adankon, M. M., & Cheriet, M. (2009). Support vector machine. In Encyclopedia of biometrics (pp. 1303-1308). Springer US.

Amari, S. I., & Wu, S. (1999). Improving support vector machine classifiers by modifying kernel functions. *Neural Networks*, *12*(6), 783–789. doi:10.1016/S0893-6080(99)00032-5 PMID:12662656

Baxt, W. G. (1990). Use of an artificial neural network for data analysis in clinical decision-making: The diagnosis of acute coronary occlusion. *Neural Computation*, *2*(4), 480–489. doi:10.1162/neco.1990.2.4.480

Baxt, W. G. (1991). Use of an artificial neural network for the diagnosis of myocardial infarction. *Annals of Internal Medicine*, *115*(11), 843–848. doi:10.7326/0003-4819-115-11-843 PMID:1952470

Bond, A. H., & Gasser, L. (Eds.). (2014). *Readings in distributed artificial intelligence*. Morgan Kaufmann.

Chi, M. T. (2014). *The nature of expertise*. Psychology Press.

Cohen, P. R., & Feigenbaum, E. A. (Eds.). (2014). *The handbook of artificial intelligence* (Vol. 3). Butterworth-Heinemann.

Console, L., & Torasso, P. (1991). On the co-operation between abductive and temporal reasoning in medical diagnosis. *Artificial Intelligence in Medicine*, *3*(6), 291–311. doi:10.1016/0933-3657(91)90002-S

Curran, K., Nichols, E., Xie, E., & Harper, R. (2010). An intensive insulinotherapy mobile phone application built on artificial intelligence techniques. *Journal of Diabetes Science and Technology*, *4*(1), 209–220. doi:10.1177/193229681000400126 PMID:20167186

Dilsizian, S. E., & Siegel, E. L. (2014). Artificial intelligence in medicine and cardiac imaging: Harnessing big data and advanced computing to provide personalized medical diagnosis and treatment. *Current Cardiology Reports*, *16*(1), 441. doi:10.1007/s11886-013-0441-8 PMID:24338557

Ferreira, C. (2006). *Gene expression programming: mathematical modeling by an artificial intelligence* (Vol. 21). Springer.

Freitag, D. (2000). Machine learning for information extraction in informal domains. *Machine Learning*, *39*(2), 169–202. doi:10.1023/A:1007601113994

Gandomi, A. H., Yang, X. S., Talatahari, S., & Alavi, A. H. (Eds.). (2013). *Metaheuristic applications in structures and infrastructures*. Newnes.

Goldbaum, M., Moezzi, S., Taylor, A., Chatterjee, S., Boyd, J., Hunter, E., & Jain, R. (1996). Automated diagnosis and image understanding with object extraction, object classification, and inferencing in retinal images. In *Image Processing, 1996. Proceedings., International Conference on* (Vol. 3, pp. 695-698). IEEE. doi:10.1109/ICIP.1996.560760

Goldberg, V., Manduca, A., Ewert, D. L., Gisvold, J. J., & Greenleaf, J. F. (1992). Improvement in specificity of ultrasonography for diagnosis of breast tumors by means of artificial intelligence. *Medical Physics*, *19*(6), 1475–1481. doi:10.1118/1.596804 PMID:1461212

Hanson, C. W. III, & Marshall, B. E. (2001). Artificial intelligence applications in the intensive care unit. *Critical Care Medicine*, *29*(2), 427–435. doi:10.1097/00003246-200102000-00038 PMID:11269246

Kahramanli, H., & Allahverdi, N. (2008). Design of a hybrid system for the diabetes and heart diseases. *Expert Systems with Applications*, *35*(1), 82–89. doi:10.1016/j.eswa.2007.06.004

Kalech, M., & Kaminka, G. A. (2007). On the design of coordination diagnosis algorithms for teams of situated agents. *Artificial Intelligence*, *171*(8), 491–513. doi:10.1016/j.artint.2007.03.005

Keedwell, E., & Narayanan, A. (2005). *Intelligent bioinformatics: The application of artificial intelligence techniques to bioinformatics problems*. John Wiley & Sons. doi:10.1002/0470015721

Kodratoff, Y., & Michalski, R. S. (2014). *Machine learning: an artificial intelligence approach* (Vol. 3). Morgan Kaufmann.

Kononenko, I. (2001). Machine learning for medical diagnosis: History, state of the art and perspective. *Artificial Intelligence in Medicine*, *23*(1), 89–109. doi:10.1016/S0933-3657(01)00077-X PMID:11470218

Mann, N. H. III, & Brown, M. D. (1991). Artificial intelligence in the diagnosis of low back pain. *The Orthopedic Clinics of North America*, *22*(2), 303–314. PMID:1826553

Michalski, R. S., Carbonell, J. G., & Mitchell, T. M. (Eds.). (2013). *Machine learning: An artificial intelligence approach*. Springer Science & Business Media.

Negnevitsky, M. (2005). *Artificial intelligence: a guide to intelligent systems*. Pearson Education.

Nilsson, N. J. (2014). *Principles of artificial intelligence*. Morgan Kaufmann.

Noble, W. S. (2006). What is a support vector machine? *Nature Biotechnology*, *24*(12), 1565–1567. doi:10.1038/nbt1206-1565 PMID:17160063

Polat, K., Güneş, S., & Arslan, A. (2008). A cascade learning system for classification of diabetes disease: Generalized discriminant analysis and least square support vector machine. *Expert Systems with Applications*, *34*(1), 482–487. doi:10.1016/j.eswa.2006.09.012

Pradhan, M., & Sahu, R. K. (2011). Predict the onset of diabetes disease using Artificial Neural Network (ANN). *International Journal of Computer Science & Emerging Technologies (E-ISSN: 2044-6004)*, *2*(2).

Ripley, P., Rose, I. G., Kelley, C. A., Bigio, I. J., Pickard, D., Briggs, G., & Bown, S. G. (2000, April). A comparison of Artificial Intelligence techniques for spectral classification in the diagnosis of human pathologies based upon optical biopsy. In *Biomedical Optical Spectroscopy and Diagnostics* (p. MC5). Optical Society of America. doi:10.1364/BOSD.2000.MC5

Shankaracharya, D. O., Samanta, S., & Vidyarthi, A. S. (2010). Computational intelligence in early diabetes diagnosis: A review. *The Review of Diabetic Studies; RDS*, *7*(4), 252–262. doi:10.1900/RDS.2010.7.252 PMID:21713313

Shiraishi, J., Li, Q., Appelbaum, D., & Doi, K. (2011, November). Computer-aided diagnosis and artificial intelligence in clinical imaging. [). WB Saunders.]. *Seminars in Nuclear Medicine*, *41*(6), 449–462. doi:10.1053/j.semnuclmed.2011.06.004 PMID:21978447

Steels, L. (1993). The artificial life roots of artificial intelligence. *Artificial life, 1*(1_2), 75-110.

Stefanelli, M. (2001). The socio-organizational age of artificial intelligence in medicine. *Artificial Intelligence in Medicine*, *23*(1), 25–47. doi:10.1016/S0933-3657(01)00074-4 PMID:11470215

Su, C. T., Yang, C. H., Hsu, K. H., & Chiu, W. K. (2006). Data mining for the diagnosis of type II diabetes from three-dimensional body surface anthropometrical scanning data. *Computers & Mathematics with Applications (Oxford, England)*, *51*(6), 1075–1092. doi:10.1016/j.camwa.2005.08.034

Temurtas, H., Yumusak, N., & Temurtas, F. (2009). A comparative study on diabetes disease diagnosis using neural networks. *Expert Systems with Applications*, *36*(4), 8610–8615. doi:10.1016/j.eswa.2008.10.032

Vas, P. (1999). *Artificial-intelligence-based electrical machines and drives: application offuzzy, neural, fuzzy-neural, and genetic-algorithm-based techniques* (Vol. 45). Oxford university press.

Wall, D. P., Dally, R., Luyster, R., Jung, J. Y., & DeLuca, T. F. (2012). Use of artificial intelligence to shorten the behavioral diagnosis of autism. *PLoS One*, *7*(8), e43855. doi:10.1371/journal.pone.0043855 PMID:22952789

Wang, L. (Ed.). (2005). *Support vector machines: theory and applications* (Vol. 177). Springer Science & Business Media. doi:10.1007/b95439

Weiss, S. M., & Kulikowski, C. A. (1979, August). EXPERT: A system for developing consultation models. In *Proceedings of the 6th international joint conference on Artificial intelligence-Volume 2* (pp. 942-947). Morgan Kaufmann Publishers Inc.

Witten, I. H., Frank, E., Hall, M. A., & Pal, C. J. (2016). *Data Mining: Practical machine learning tools and techniques*. Morgan Kaufmann.

Yang, X. S. (2010). *Nature-inspired metaheuristic algorithms*. Luniver press.

Yu, J., Wang, S., & Xi, L. (2008). Evolving artificial neural networks using an improved PSO and DPSO. *Neurocomputing*, *71*(4), 1054–1060. doi:10.1016/j.neucom.2007.10.013

Yudkowsky, E. (2008). Artificial intelligence as a positive and negative factor in global risk. *Global catastrophic risks, 1*(303), 184.

Zhang, L., Zhou, W., & Jiao, L. (2004). Wavelet support vector machine. *IEEE Transactions on Systems, Man, and Cybernetics. Part B, Cybernetics*, *34*(1), 34–39. doi:10.1109/TSMCB.2003.811113 PMID:15369048

## KEY TERMS AND DEFINITIONS

**Diagnosis System:** A diagnosis system is a general software-hardware supported technique that is able to diagnose objective diseases by analyzing input data with an artificial-intelligence-oriented solution way.

**Machine Learning:** An approach of artificial intelligence in which methods and techniques should go through some training/learning processes to be ready for solving objective problem.

**Optimization:** Optimization is the approach of finding the most appropriate value(s)/decision variable value(s) for an objective mathematical model of the real-world-based problem.

**Swarm Intelligence:** Swarm intelligence is an approach of artificial intelligence in which some groups of particles (artificial swarm) are used for searching for solutions within a solution space by considering generally an optimization framework.

**Support Vector Machine:** A support vector machine is a machine learning (artificial intelligence) and data mining technique that can perform classification and regression operations over the objective data by using an approach of support vectors determined over the data environment.

**Training:** The concept of training here is briefly the process of optimizing a machine-learning-based technique to solve newly encountered problems after learning enough about the problem thanks to some pre-data/knowledge/experience.

# Chapter 10
# Artificial Intelligence Ethics in Biomedical-Engineering-Oriented Problems

**Alice Pavaloiu**
*Paris School of Business, France*

## ABSTRACT

*The field of artificial intelligence has recently encountered some ethical questions associated with the future of humankind. Although it is a common question that has been asked for years, the existence of humankind against badly configured intelligent systems is more important nowadays. As a result of rapid developments in intelligent systems and their increasing role in our life, there is a remarkable anxiety about dangerous artificial intelligence. Because of that, some research interests gathered under some topics like machine ethics, future of artificial intelligence, and even existential risks are drawing researchers' interest. As associated with this state, the objective of this chapter is to examine ethical factors in using intelligent systems for biomedical-engineering-oriented purposes. The chapter firstly gives essential information about the background and then considers possible scenarios that may require ethical adjustments during design and development of artificial-intelligence-oriented systems for biomedical engineering problems.*

## INTRODUCTION

Nowadays, Artificial Intelligence (AI) has a remarkable place among scientific developments. Because it has a wide multidisciplinary scope and a flexible structure of mathematical – logical solution ways, it is widely employed in all fields of the life. Because successful results have been obtained with its applications, the trend on using it within every advanced, complex problems of different fields have become a traditional approach within scientific efforts. It is clear that AI has a great application scope (Farkas, 2003; Negnevitsky, 2005; McArthur et al., 2005; Pham & Pham, 1999; Ramesh et al., 2004;

DOI: 10.4018/978-1-5225-4769-3.ch010

Stahovich, 2001) and it is too effective nowadays on our works, studies, tasks, and generally problem solutions in the daily modern life (Kose & Pavaloiu, 2017). But as it has taken more place in people's life, the more anxieties regarding its possible dangerous behaviors have risen in time. Nowadays, there is a serious group of people, who believes that the AI will be effective on humankind's destiny and possibly cause dangerous and harmful situations while providing problem solutions. That's because sometimes a beneficial thing for somebody may not be beneficial at the same level to another one and even may cause dangerous for him / her. Furthermore, an intelligence more powerful than a human's intelligence can have better predictions for the future and cause dangerous states currently because of thinking for a better situation in the future, that a human cannot see – understand. It is also important that there are lots of moral dilemmas confusing our minds and it is an important issue to understand how an AI system can behave in such moral dilemmas. All of these anxieties and questions are trying to be solved – answered for all fields in which AI based systems – solutions take place widely. Biomedical Engineering is one of them and it has a remarkable importance because of its relation to the health and living organisms.

The applications of AI in Biomedical Engineering extend from Brain-Computer Interface and Neuroprosthetics, Sequence Analysis to Biomedical Imaging and Health-care robotics. The vast spectrum of applications generated from the AI – Biomedical Engineering symbiosis will be a major driver of a new technology which will reshape the personal tapestry of humanity and bring it one step closer to personalized medicine. Because of that this subject combination is a remarkable one, which should be examined in order to contribute to the associated literatures and also improve them. Furthermore, ethical anxieties appeared because of using intelligent systems should also be discussed widely as there is an improving anxiety in each time AI takes more place in our life. When we look at to the literature, we can see many examples of AI ethics oriented studies (Some of them to examine: Bostrom, 2003; Bostrom & Yudkowsky, 2014; Hawking et al., 2014; Moor, 2006; Russell et al., 2015; Vasant & Kose, 2017; Yudkowsky, 2008). But there is a need for dealing with ethical factors in specific application fields of AI. Here, Biomedical Engineering seems one of these specific application fields.

The objective of this chapter is to examine ethical factors in using intelligent systems for Biomedical Engineering oriented purposes. The chapter will firstly give essential information about the background and then consider possible scenarios that may require ethical adjustments during design and development of Artificial Intelligence oriented systems for Biomedical Engineering problems. From a general view, this chapter is both a reference for the interested readers and an additional approach in regard to the ethic-based discussions by taking the Biomedical Engineering as a target application field of intelligent systems.

Based on the objective and subject of the chapter, the remaining content is organized as follows: The second section is based on theoretical background and it provides about the related research subjects of Artificial Intelligence Ethics, Artificial Intelligence Safety, Machine Ethics, Future of Artificial Intelligence, and Existential Risks. Following to that section, the third section provides a general discussion about possible ethical – safety oriented issues that may appear while applying intelligent systems to problems of Biomedical Engineering. After that section, the fourth section provides some suggestions about future work in the context of the subject and finally the last section ends the paper with a brief discussion as the conclusion to this research work.

## ARTIFICIAL INTELLIGENCE ETHICS AND THE ASSOCIATED SUBJECTS

AI encountered an increasing research interest which was brought not only by the technological advancements during the past decades, but it also came with a varied palette of moral dilemmas, safety issues, machine ethics question marks and even existential risks.

### Artificial Intelligence Ethics

The controversy sparked around AI ethics was generated from multiple standpoints. Examples include the ethical decision-making process of AI which would lead to moral dilemmas if too rigidly programmed. Hidden bias or subjectiveness, which would affect AI moral conduct from learning or being taught and create major issues in an organization, for instance, in the process of hiring the best candidate. Social valence through entrainment - anthropomorphizing AI to the extent where marketers could take advantage of vulnerable clients who would agree to buy products in order to maintain the interaction with the AI system. Functional errors that could even lead to starting wars when autonomous weapons would come into play. Economic disparity, increasing the wealth gap, having a whole sector of unemployed people unable to work due to the lack of the new demanded set of skills would be another example.

Consequently, safe AI development and deployment is of extreme importance in the pursuit of technological advancements.

### Artificial Intelligence Safety

Creating safe AI and developing preventive methods for insuring an AI-friendly environment at all costs becomes crucial when AI connectedness, unpredictability rate and advancements increase in a fast-forward digital time, as previously stated by Nick Bostrom (2014). Robust and beneficial AI in the making is continuously drawing attention and action from the private sector, being developed closely with the academia. Solutions like instilling core, human values in AI are debated not only at a philosophical level, but also at an engineering level. In order to align people's set of values with machines', both philosophers and engineers agreed people must come back to what makes them human and get to an unbiased consensus before creating safe AI.

### Machine Ethics

M2M (machine to machine) communication, machine connectedness, especially when AI driven, became a concern when the issues of cyber-crime and the question of national security threat were brought up. Multiple accounts (Singularity Summit, 2017) regarding superintelligence and an unleashed AI consciousness unveil a plethora of scenarios focused around machine ethics.

### Future of Artificial Intelligence

Future implications include developing a culture of collective involvement, responsibility and accountability created by the industry and academia in conformity with the government, its regulatory framework and international organizations that would enable a safe transition to an era of smart machines. This would be the key to a thriving, abundant society with exponential technological advancements and benefits in all areas and domains. If AI is conceived and deployed from a standpoint where ethics and safety plays a major role in, while having instilled core, human values, it can have a positive impact upon the greater good without creating any ethical dilemmas.

## Existential Risks

Superintelligence through AI consciousness can be seen as an emerging phenomenon. If defined as an ensemble of the following traits: autonomy, self-awareness, environment-awareness while having the capacity of learning or self-training, the only parameter until the occurrence of AI consciousness will be only time.

As a result, AI ethics needs to be developed from the phase it is currently in, in order to prevent from happening any negative singularity scenario. Concerns about existential risks of humanity started shaping at a larger scale. Kurzweil (2005) and Bostrom (2002) underlined their concerns regarding existential risks if AI would become uncontrollable or being unable to align to people's core values. Solutions reside in the complex problematic of ethics and AI safety. In order to avoid any existential risk, it is humanity's duty to create a culture based on responsibility and accountability to insure not only safety, but also ethical deployment of AI in all fields.

*Figure 1. Artificial Intelligence ethics and the associated subjects*

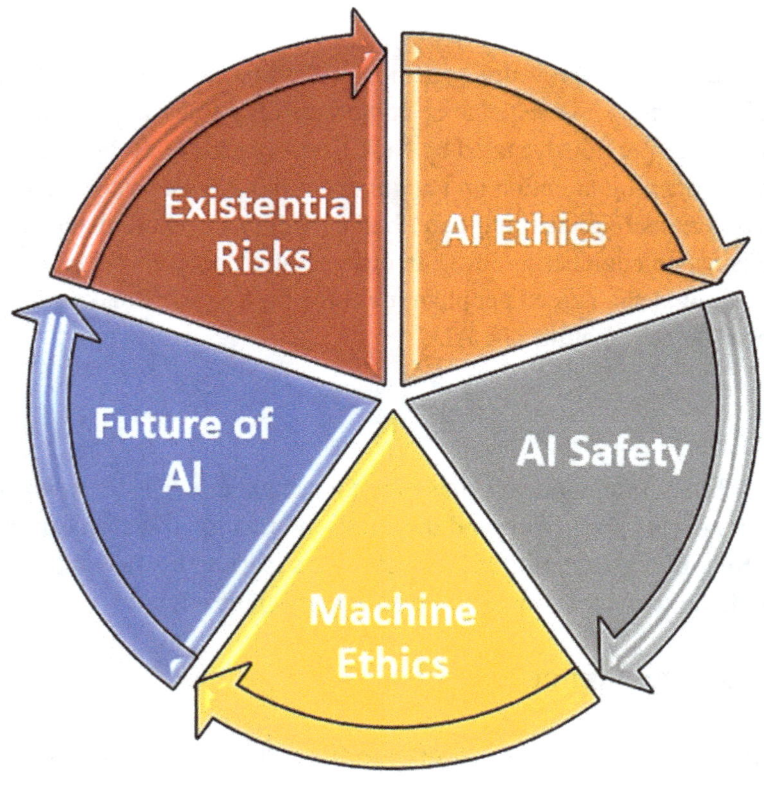

## ARTIFICIAL INTELLIGENCE ETHICS IN BIOMEDICAL ENGINEERING

Biomedical Engineering has relations with medical, health, and nature of living organisms. Because of that, research works realized in this field have more collaborative works of different fields (i.e. computer science, electronics, health) and because of that it meets with advantages of AI quite well. But because of anxieties regarding AI and current research subjects of AI Ethics, problems in the intersection of AI and Biomedical Engineering should be discussed. In this section, some possible scenarios have been taken into consideration, as follows:

### Discussion on Possible Scenarios

When we consider ethical issues on AI and Biomedical Engineering, these issues are often associated with moral dilemmas. As it was mentioned before, moral dilemmas – distresses have a remarkable place in AI Ethics and the associated subjects. In especially health, there is a remarkable research interest in such issues (Austin et al., 2005; Cohen & Erickson, 2006; Førde & Aasland, 2008; Hamric et al., 2006; Kälvemark et al., 2004). It can be understood, this issue could be improved greatly if we consider a future of AI taken active role in health and the associated fields. At this point, some possible scenarios can be discussed briefly as follows:

- **Value of the Patient:** Considering the medicine at hand and status of objective patients, it would be always a moral dilemma for AI systems to determine which patient to receive immediate treatment at the time of insufficient – low medicine or treatment opportunities. This dilemma could appear when the objective patients are all in very dangerous states and should receive immediate medical treatment. At this point, it is also important to see how collaborative AI based robots, which correspond to different type of Biomedical Engineering products can act collaboratively in such a moral dilemma. Is it necessary to use an hierarchical structure among them or such AI based robots should have the ability of creating exceptions in emergency states?
- **Keeping Alive or Leaving to Death:** For humans, it is important to see their friends or family of love to be alive at least a short time even they have cancer or any other mortal disease. But there could a moral dilemma at this time for an AI system, which is responsible for determining appropriate medicine for giving to the patient or preparing some mixtures, performing medical treatments…etc. Because AI system could think about saving resources because the objective patient will die soon. It could be not important for an AI system to think emotional and at least reduce patient's pain, because emotions are still not simulated in a real manner within AI based systems. This issue could be more dramatically if the AI system break up giving medicines and shortening the patient's life period towards death.
- **Medical Examinations of Criminals:** It could be a moral dilemma for AI based systems to give necessary importance on medical examinations and treatments for especially criminals. As humans, we even find it sometimes not appropriate to use all medical contributions for such people but because of Hippocratic Oath, it is not applicable practically. But what if an AI based system has more consciousness to decide on giving low priority for such patients because they are bad people, who has the potential to act similar if they become well?

- **Keeping the Balance of the Nature:** As humankind, we are all guilty for affecting the natural balance in a negative manner and not making too much thing to fix or at least tolerate that. At this point, it could be a moral dilemma for all kinds of AI systems having different roles in Biomedical Engineering products to have an unexpected consciousness to keep the balance of the nature and because of that, trying to control people's treatments or deaths in order to make the World a better place for all living organisms. At this point, a Biomedical examination device for detecting bacteria could find it useful to keep bacteria alive or not apply appropriate treatment procedures because of thinking widely for all living organisms, which are in a natural selection process and natural balance in this manner.
- **Laws for Biomedical Robots:** It is a widely discussed subject that the AI based systems of the future should have some laws like the ones by Asimov (MIT Technology Review, 2016). Although that is a wider subject, it is certainly associated with the behaviors that will be shown by AI systems dealing with Biomedical oriented problems. Running of similar laws like them may be a key point for ensuring safe AI systems as by resulting to elimination of many dilemmas like the ones mentioned under previous paragraphs. But it is also an important question that 'What if all these rules may cause confusion for the AI systems having better consciousness.
- **Global Sensitivity and Communication:** It could be another moral dilemma that how far an AI system should be sensitive to global medical issues appeared around the World. Could a wide collaboration – network of AI systems be beneficial to perform appropriate – ethical behaviors in Biomedical problems or is it necessary to create some limitations for AI based robots to deal with just their own specific works? This ethical issue gives a chance to discussion over communication among AI based robots. From a general perspective, it is a certain thing to have AI based systems in a network of communication with some other ones, but that does not mean not to have some borders – limitations in detail.

Many more possible scenarios can be given if we think more about the intersection of AI Ethics and Biomedical Engineering. Here, the most important thing to consider that there will be always some moral dilemmas – distresses and ethical issues until a certain model of AI systems of the future can be achieved. It is clear that ethical issues are still some open problems for AI because some thought problems can be examined clearly when levels of AI systems are improved with the support of some technologies like computers, electronics, communication. But although it is possible sometimes to discuss theoretically about ethical issues of AI, these discussions should be preparations for a safe future rising over AI.

## FUTURE RESEARCH DIRECTIONS

Considering the explanations – discussions made so far, some future research ideas regarding to ensuring Artificial Intelligence Ethics in Biomedical Engineering problems can be expressed briefly as follows:

- Ethical issues regarding to employment of Artificial Intelligence within Biomedical Engineering requires a wider discussion supported by experts from different fields including medical – health (as it is clear) and some other alternative fields like education, sociology, and psychology. So, it is required by the authors to achieve more research works in this manner by making brain storming

*Figure 2. Some possible scenarios of AI ethics in Biomedical Engineering*

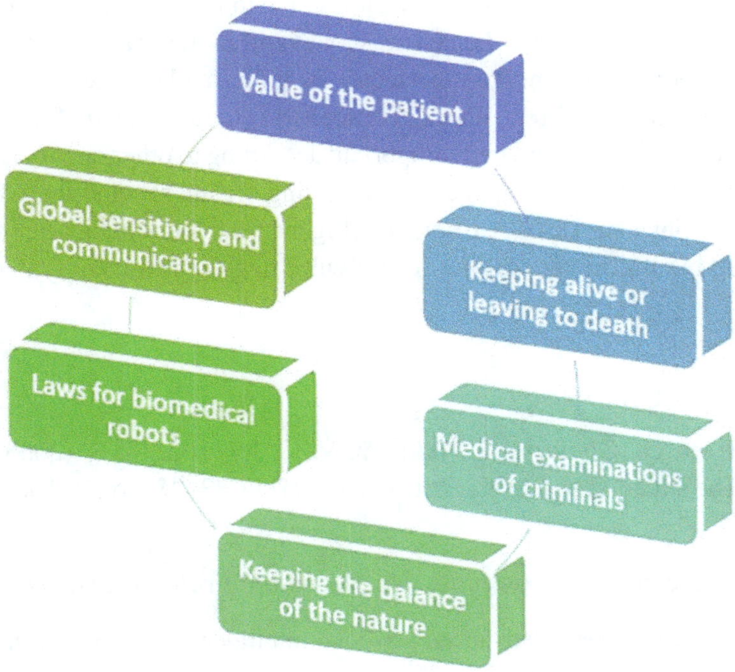


and collaborative research works to have idea about possible solution ways against any dangerous situations that intelligent systems as members of the future society may show.

- Because Biomedical Engineering problems include also design and development of machines – devices, it can be good way to discuss with electronics or mechanical engineers about any physical precautions that can be taken to eliminate dangerous behaviors that may rise from mind of an intelligent system but can be stopped in the physical world.
- In addition to the discussed scenarios here, there may be many different, alternative scenarios that can be discussed. That is another open door for the interested researchers.
- Although the content of this chapter is briefly focused on technical sides of the considered subjects, it is important for the interested researchers to examine the associated literature for having a good technical background – knowledge about what have been done so far.

## CONCLUSION

This chapter has provided a remarkable discussion on possible Artificial Intelligence Ethics issues when it comes to employ intelligent systems in Biomedical Engineering problems more and more. In detail, the chapter has also given some essential information about Artificial Intelligence Ethics and other research interests – topics associated with it. At this point, readers were enabled to have essential information about Machine Ethics, Artificial Intelligence Safety, Future of Artificial Intelligence, and Existential Risks. When we examine the literatures, it is possible to see same or similar research focuses with different topic titles. But it is clear that the anxieties about intelligent systems that may be dangerous for the humankind have taken many steps ahead by enabling the researchers to think about issues – problems and alternative solution ways in this manner. When it is come to Biomedical Engineering problems, the situation becomes more vital because of behaviors shown within medical – health operations are very important considering people's health. Possible scenarios indicated here are some ideas though by the authors and there may be many different, alternative scenarios to think about. Some fields like medical – health, natural sciences or military are more important according to other ones because they have great potential on determining the destiny of humans, other living organisms, and the World. Let's see what the technological developments, rise of Artificial Intelligence, and ongoing research efforts against any possible problems that intelligent systems may cause will bring to the humanity in time.

## REFERENCES

Austin, W., Lemermeyer, G., Goldberg, L., Bergum, V., & Johnson, M. S. (2005). Moral distress in healthcare practice: The situation of nurses. *HEC Forum*, *17*(1), 33–48. doi:10.1007/s10730-005-4949-1 PMID:15957267

Bostrom, N. (2002). The concept of existential risk. *Journal of Evolution and Technology / WTA*, 9.

Bostrom, N. (2003). Ethical issues in advanced artificial intelligence. *Science Fiction and Philosophy: From Time Travel to Superintelligence*, 277-284.

Bostrom, N. (2014). *Superintelligence: Paths, Dangers, Strategies*. OUP Oxford.

Bostrom, N., & Yudkowsky, E. (2014). The ethics of artificial intelligence. The Cambridge Handbook of Artificial Intelligence, 316-334.

Cohen, J. S., & Erickson, J. M. (2006). Ethical dilemmas and moral distress in oncology nursing practice. *Clinical Journal of Oncology Nursing*, *10*(6), 775–780. doi:10.1188/06.CJON.775-780 PMID:17193943

Farkas, I. (2003). Artificial intelligence in agriculture. *Computers and Electronics in Agriculture*, *1*(40), 1–3. doi:10.1016/S0168-1699(03)00007-3

Førde, R., & Aasland, O. G. (2008). Moral distress among Norwegian doctors. *Journal of Medical Ethics*, *34*(7), 521–525. doi:10.1136/jme.2007.021246 PMID:18591286

Hamric, A. B., Davis, W. S., & Childress, M. D. (2006). Moral distress in health care professionals. *Pharos*, *69*(1), 16–23. PMID:16544460

Hawking, S., Russell, S., Tegmark, M., & Wilczek, F. (2014). Stephen Hawking: Transcendence looks at the implications of artificial intelligence-but are we taking AI seriously enough? *The Independent*, *2014*(05-01), 9313474.

Kälvemark, S., Höglund, A. T., Hansson, M. G., Westerholm, P., & Arnetz, B. (2004). Living with conflicts-ethical dilemmas and moral distress in the health care system. *Social Science & Medicine*, *58*(6), 1075–1084. doi:10.1016/S0277-9536(03)00279-X PMID:14723903

Kose, U., & Pavaloiu, A. (2017). Dealing with Machine Ethics in Daily Life: A View with Examples. *International Virtual Conference on Advanced Scientific Results / Online Scientific Conference 2017 (ScieConf 2017)*, 200-205. DOI: doi:10.18638/scieconf.2017.5.1.454

Kurzweil, R. (2005). *The Singularity is Near: When Humans Transcend Biology*. Penguin.

McArthur, D., Lewis, M., & Bishary, M. (2005). The Roles of Artificial Intelligence in Education: Current Progress and Future Prospects. *Journal of Educational Technology*, *1*(4), 42–80.

MIT Technology Review. (2016). Do we need Asimov's laws? *MIT Technology Review*. Retrieved on July 29, 2017 from https://www.technologyreview.com/s/527336/do-we-need-asimovs-laws/

Moor, J. H. (2006). The nature, importance, and difficulty of machine ethics. *IEEE Intelligent Systems*, *21*(4), 18–21. doi:10.1109/MIS.2006.80

Negnevitsky, M. (2005). *Artificial intelligence: A guide to intelligent systems*. Pearson Education.

Pham, D. T., & Pham, P. T. N. (1999). Artificial intelligence in engineering. *International Journal of Machine Tools & Manufacture*, *39*(6), 937–949. doi:10.1016/S0890-6955(98)00076-5

Ramesh, A. N., Kambhampati, C., Monson, J. R. T., & Drew, P. J. (2004). Artificial intelligence in medicine. *Annals of the Royal College of Surgeons of England*, *86*(5), 334–338. doi:10.1308/147870804290 PMID:15333167

Russell, S., Dewey, D., & Tegmark, M. (2015). Research priorities for robust and beneficial artificial intelligence. *AI Magazine*, *36*(4), 105–114. doi:10.1609/aimag.v36i4.2577

Singularity Summit. (2012). *Ray Kurzweil on the singularity*. Retrieved on July 28, 2017 from https://www.youtube.com/watch?v=EXVrTCjetLg

Stahovich, T. F. (2001). Artificial intelligence for design. *Formal engineering design synthesis*, 228-269.

Vasant, P., & Kose, U. (2017). Moral Dilemma Scenarios Considering Possible Actions by Humans and Artificial Intelligence Based Systems. *Journal of Multidisciplinary Developments*, *2*(2), 39–49.

Yudkowsky, E. (2008). Artificial intelligence as a positive and negative factor in global risk. *Global Catastrophic Risks*, *1*(303), 184.

## ADDITIONAL READING

Allen, C., Wallach, W., & Smit, I. (2011). Why Machine Ethics?. Machine Ethics, 51-60.

Amigoni, F., & Schiaffonati, V. (2005). Machine ethics and human ethics: A critical view. In *Proceedings of the AAAI 2005 Fall Symposium on Machine Ethics* (pp. 103-104).

Anderson, M. (2005). Machine Ethics: *Papers from the 2004 AAAI Fall Symposium*;[November 4-6, 2005, Arlington, Virginia]. AAAI Press.

Anderson, M., & Anderson, S. L. (2006). Guest Editors' Introduction: Machine Ethics. *IEEE Intelligent Systems*, *21*(4), 10–11. doi:10.1109/MIS.2006.70

Anderson, M., Anderson, S. L., & Armen, C. (2004). Toward machine ethics. In *Proceedings of AAAI 2004 Workshop on Agent Organizations: Theory and Practice*.

Anderson, M., Anderson, S. L., & Armen, C. (2006). MedEthEx: a prototype medical ethics advisor. In Proceedings Of The National Conference On Artificial Intelligence (Vol. 21, No. 2, p. 1759). Menlo Park, CA; Cambridge, MA; London; AAAI Press; MIT Press; 1999.

Anderson, M., Anderson, S. L., & Armen, C. (2006). An approach to computing ethics. *IEEE Intelligent Systems*, *21*(4), 56–63. doi:10.1109/MIS.2006.64

Anderson, S. L. (2011). *Machine metaethics. Machine ethics* (pp. 21–27). Cambridge: Cambridge University Press. doi:10.1017/CBO9780511978036.004

Anderson, S. L., & Anderson, M. (2011). A Prima Facie Duty Approach to Machine Ethics and Its Application to Elder Care. In Human-Robot Interaction in Elder Care.

Beavers, A. F. (2011). 21 Moral Machines and the Threat of Ethical Nihilism. Robot ethics: The ethical and social implications of robotics, 333.

Bendel, O. (2015). Surgical, therapeutic, nursing and sex robots in machine and information ethics. In *Machine Medical Ethics* (pp. 17–32). Springer International Publishing. doi:10.1007/978-3-319-08108-3_2

Brundage, M. (2014). Limitations and risks of machine ethics. *Journal of Experimental & Theoretical Artificial Intelligence*, *26*(3), 355–372. doi:10.1080/0952813X.2014.895108

Casey, B. J. (2017). Amoral machines, or: How roboticists can learn to stop worrying and love the law.

Conitzer, V., Sinnott-Armstrong, W., Borg, J. S., Deng, Y., & Kramer, M. (2017). Moral Decision Making Frameworks for Artificial Intelligence. In AAAI (pp. 4831-4835).

Crnkovic, G. D., & Çürüklü, B. (2012). Robots: Ethical by design. *Ethics and Information Technology*, *14*(1), 61–71. doi:10.1007/s10676-011-9278-2

Danielson, P. (2011). Prototyping N-reasons: a computer mediated ethics machine. Machine ethics, 442-450.

Etzioni, A., & Etzioni, O. (2017). Incorporating Ethics into Artificial Intelligence. *The Journal of Ethics*, 1–16.

Goodall, N. J. (2014). Machine ethics and automated vehicles. In *Road vehicle automation* (pp. 93–102). Springer International Publishing. doi:10.1007/978-3-319-05990-7_9

Guarini, M. (2013). Introduction: Machine ethics and the ethics of building intelligent machines. *Topoi*, *32*(2), 213–215. doi:10.1007/s11245-013-9183-x

Gunkel, D. J. (2012). *The machine question: critical perspectives on AI, robots, and ethics*. MIT Press.

Jebari, K. (2015). Existential risks: Exploring a robust risk reduction strategy. *Science and Engineering Ethics*, *21*(3), 541–554. doi:10.1007/s11948-014-9559-3 PMID:24891130

Johnson, D. G. (2011). Computer Systems--Moral Entities but Not Moral Agents. Machine Ethics, 151-160.

Jonsen, A. R., Siegler, M., & Winslade, W. J. (2006). Clinical ethics: a practical approach to ethical decisions in clinical medicine. Univerza v Mariboru, Fakulteta za zdravstvene vede.

Komuda, R., Rzepka, R., & Araki, K. (2010). Social Factors in Kohlberg's Theory of Stages of Moral Development the Utility of (Web) Crowd Wisdom for Machine Ethics Research. In *Proceedings of the 5th International Conference on Applied Ethics* (p. 14).

LaGrandeur, K., & Hughes, J. J. (Eds.). (2017). *Surviving the Machine Age: Intelligent Technology and the Transformation of Human Work*. Springer. doi:10.1007/978-3-319-51165-8

Leben, D. (2017). A Rawlsian algorithm for autonomous vehicles. *Ethics and Information Technology*, *19*(2), 107–115. doi:10.1007/s10676-017-9419-3

Lim, H. C., Stocker, R., & Larkin, H. (2008). Review of trust and machine ethics research: Towards a bio-inspired computational model of ethical trust (CMET). In Proceedings of the 3rd International Conference on Bio-Inspired Models of Network, Information and Computing Sytems (p. 8). ICST (Institute for Computer Sciences, Social-Informatics and Telecommunications Engineering). doi:10.4108/ICST.BIONETICS2008.4728

Lindner, F., & Bentzen, M. M. (2017). The Hybrid Ethical Reasoning Agent IMMANUEL. In *Proceedings of the Companion of the 2017 ACM/IEEE International Conference on Human-Robot Interaction* (pp. 187-188). ACM. doi:10.1145/3029798.3038404

Mappes, T. A., & Zembaty, J. S. (1981). *Biomedical ethics* (p. 54). New York: McGraw-Hill.

McDermott, D. (2008). Why ethics is a high hurdle for AI. In *North American Conference on Computers and Philosophy*, Bloomington, IN.

McDermott, D. (2011). What matters to a machine. Machine ethics, 88-114.

McIntyre-Mills, J. (2012). Wellbeing and existential risk. In CASS research conference sustainable environmental policy and global governance, Beijing (pp. 18-19).

McLaren, B. (2005). Lessons in machine ethics from the perspective of two computational models of ethical reasoning. In *2005 AAAI Fall Symposium on Machine Ethics*.

McLaren, B. M. (2003). Extensionally defining principles and cases in ethics: An AI model. *Artificial Intelligence*, *150*(1-2), 145–181. doi:10.1016/S0004-3702(03)00135-8

Medsker, L. (2017). AI policy: Organizations, resources, and recent symposia. *AI Matters*, *3*(1), 9–11. doi:10.1145/3067682.3067684

Muehlhauser, L., & Helm, L. (2012). The singularity and machine ethics. In *Singularity Hypotheses* (pp. 101–126). Springer Berlin Heidelberg. doi:10.1007/978-3-642-32560-1_6

Pereira, L. M., & Saptawijaya, A. (2016). Bridging two realms of machine ethics. In *Programming Machine Ethics* (pp. 159–165). Springer International Publishing. doi:10.1007/978-3-319-29354-7_10

Pereira, L. M., & Saptawijaya, A. (2016). *Programming machine ethics* (Vol. 26). Springer. doi:10.1007/978-3-319-29354-7

Persson, D., & Erlandsson, L. K. (2002). Time to Reevaluate the Machine Society: Post-industrial Ethics from an Occupational Perspective. *Journal of Occupational Science*, *9*(2), 93–99. doi:10.1080/14427591.2002.9686497

Powers, T. M. (2011). Incremental machine ethics. *IEEE Robotics & Automation Magazine*, *18*(1), 51–58. doi:10.1109/MRA.2010.940152

Powers, T. M. (2011). One way to view the puzzle of machine ethics is to consider how. Machine Ethics, 464.

Price, H. (2013). Cambridge, cabs and Copenhagen: My route to existential risk. The New York Times, 27, 2013.

Shulman, C., Jonsson, H., & Tarleton, N. (2009). Machine ethics and superintelligence. Reynolds and Cassinelli, 95-97.

Sullins, J. P. (2006). When is a robot a moral agent. Machine Ethics, 151-160.

Tonkens, R. (2009). A challenge for machine ethics. *Minds and Machines*, *19*(3), 421–438. doi:10.1007/s11023-009-9159-1

Torrance, S. (2011). Machine ethics and the idea of a more-than-human moral world. Machine ethics, 115-137.

Veruggio, G., Solis, J., & Van der Loos, M. (2011). Roboethics: Ethics applied to robotics [from the guest editors]. *IEEE Robotics & Automation Magazine*, *18*(1), 21–22. doi:10.1109/MRA.2010.940149

Wallach, W. (2010). Robot minds and human ethics: The need for a comprehensive model of moral decision making. *Ethics and Information Technology*, *12*(3), 243–250. doi:10.1007/s10676-010-9232-8

Whitby, B. (2011). On Computable Morality An Examination of Machines. Machine ethics, 138.

Winter-Levy, S., & Trefethen, J. (2016). Safety First Entering the Age of Artificial Intelligence. *World Policy Journal*, *33*(1), 105–111. doi:10.1215/07402775-3545918

Yampolskiy, R. V. (2013). Artificial intelligence safety engineering: Why machine ethics is a wrong approach. Philosophy and theory of artificial intelligence, 389-396.

Yampolskiy, R. V. (2016). Taxonomy of Pathways to Dangerous Artificial Intelligence. In AAAI Workshop: AI, Ethics, and Society.

Yampolskiy, R. V., & Spellchecker, M. S. (2016). Artificial Intelligence Safety and Cybersecurity: a Timeline of AI Failures. arXiv preprint arXiv:1610.07997.

## KEY TERMS AND DEFINITIONS

**Artificial Intelligence Ethics/Ethical Artificial Intelligence:** Artificial intelligence ethics/ethical artificial intelligence is a concept or research interest associated with ensuring ethically acting problem-solving intelligent systems.

**Artificial Intelligence Safety:** Artificial intelligence safety is a research interest/topic focused on achieving safe intelligent systems and eliminating possible problems threatening the safety of such systems.

**Existential Risk:** Existential risk is a concept or research interest dealing with the possibility of something to be dangerous enough for threatening existence of humankind.

**Future of Artificial Intelligence:** Future of artificial intelligence is a research interest/topic that focused on the future states/issues regarding to the field of artificial intelligence and its applications.

**Machine Ethics:** Machine ethics is a research interest/topic dealing with developing intelligent machines acting ethically in the context of their environment and the objective problem.

**Moral Dilemma:** Moral dilemma is a typical conflict enabling one to decide among many actions including changing moral causes.

# Chapter 11
# Prediction of Electroencephalogram Time Series via Artificial Neuro-Fuzzy Inference System Trained by League Championship Algorithm

**Mehmet Mehdi Karakoc**
*Agri Ibrahim Cecen University, Turkey*

## ABSTRACT

*Time series analysis has a wide application interest in artificial-intelligence-oriented research studies. Because it is easy to run machine-learning-based solutions directly over time series, it has been a popular approach to use alternative types of intelligent systems to analyze time series. Regarding such works, time series prediction is known as a remarkable topic as followed by researchers from different fields. The objective of this chapter is to provide an alternative work by using artificial neuro-fuzzy inference system trained by the league championship algorithm, which is an optimization algorithm from the associated literature. As the application objective, electroencephalogram (EEG) time series have been tried to be predicted by using the designed ANFIS-LCA approach. The chapter briefly introduces details about the approach and reports findings from the performed prediction operations.*

## INTRODUCTION

It is certain that the humankind is currently experiencing a golden era in terms of technological developments and improvements. Although it is also often discussed that technological changes come with also some disadvantages, the big picture is known as full of practical and effective solutions for works – tasks – activities that we do. As a result of critical discoveries in both social and natural sciences, the

DOI: 10.4018/978-1-5225-4769-3.ch011

humankind had had a chance to focus more on inventions to form the future. In time, different kinds of inventions had led the humankind to deal with environmental factors and understand the nature better. Eventually, all these changes and remarkable additions to the human life have enabled us to live in fast transforming community. In this context, today's related transformations are associated with especially computer and communication technologies. The society of informatics in which we currently take active part is a result of innovative developments associated with computer and communication technologies. If we consider today's top trend research topics, Artificial Intelligence has a remarkable place among cause of these developments.

Artificial Intelligent is a flexible, effective and efficient scientific field, which is widely used for solving difficult problems. Today, there is almost no field in which Artificial Intelligence is not used. From natural sciences to social sciences, intelligent solution ways are often used in different kinds of problems and in that way, mathematically and logically effective systems often find their place in not only traditional difficult problem but also newly encountered problems (Akerkar, 2014; Flasiński, 2016). The literature of Artificial Intelligence is currently almost an infinite source of scientific approaches for difficult problems. Among all different difficult problem types, time series analysis has taken place in the context of research work introduced in this chapter. As the form of information – data points observed in a time period, time series are usable tools for understanding many facts and events related to the life. By just examining a graphical flow of changes regarding facts or events, it is possible for us to derive new information – explanations or understand easier about the objective fact or events (Box et al., 2015; Brown, 2004; Elliot et al., 2017; Weigend, 1994). It is possible to see use of time series in many remarkable fields including economy – business, education, health and even social media (Brooks et al., 2014; Costa et al., 200; Enders, 2008; Nguyen et al., 2012; Zeger et al., 2006). So, time series is a popular research topic attracting researchers' attention always.

Time series analysis has a wide application interest in Artificial Intelligence oriented research studies. Because it is easy to run Machine Learning based solutions directly over time series, it has been a popular approach to use alternative types of intelligent systems to analyze time series. Regarding to such works, time series prediction is known as a remarkable topic as followed by researchers from even different fields. In detail, the associated literature has many alternative examples of such research focuses benefiting from intelligent techniques for time series analysis (As some examples: Chatfield, 2016; Formisano et al., 2008; Längkvist et al., 2014; Prudêncio & Ludermir, 2004; Singh & Balasundaram, 2007; Tang et al., 2009).

When the research interest of time series analysis is examined, it can be seen that also time series prediction (forecasting) is one of popular problem approach under time series analysis done via Artificial Intelligence – Machine Learning. At this point, prediction capability of especially Machine Learning techniques within Artificial Intelligence is a remarkable factor on that and the literature is enough for having information about comprehensive applications of predictions (Alm et al., 2005; Chen & Asch, 2017; Mair et al., 2000; Shah, 2007; Shipp et al., 2002; Weiss & Kulikowski, 1991; Yuksel et al., 2017). Because it has been always important to have some idea about future states of an information – data flow even it is chaotic, intelligent techniques have also been applied in time series prediction. In that way, faster and effective prediction processes have been achieved easily. In the associated literature of intelligently time series analysis, it is possible to see also wide applications of intelligent time series prediction (Readers are referred to some recent examples as: Borovykh et al., 2017; Brockwell & Davis, 2016; Coelho et al., 2017; Elliot et al., 2017; Ince & Trafalis, 2017; Kose & Arslan, 2017; Qiu et al., 2017; Shi & Han, 2007; Toriyama et al., 2017; Xing et al., 2017; Yao et al., 2017; Zhang et al., 2017).

Considering the explanations above, objective of this chapter is to provide an alternative work of time series prediction by using Artificial Neuro-Fuzzy Inference System trained by the League Championship Algorithm, which is an optimization algorithm from the associated literature. As the application objective, Electroencephalogram (EEG) time series have been tried to be predicted by using the designed ANFIS-LCA approach. The chapter briefly introduces details about the approach and reports findings from the performed prediction operations.

The remaining content of the chapter is organized as follows: The second section provides some brief information about the employed Artificial Intelligence techniques: ANFIS and LCA for EEG time series prediction. Following that section, the third section explains essential features of the prediction-purpose ANFIS-LCA hybrid system the objective EEG time series, and focuses on the performed prediction applications along with also discussions over the findings. Finally, the chapter is ended with discussions about conclusions and future work, under the last – fourth section.

## ANFIS-LCA HYBRID SYSTEM

Before expressing about the exact application side of the research work forming the subject of this chapter, essential factors of the designed hybrid intelligent system: ANFIS-LCA should be taken into consideration. In that way, it is believed that the readers will have better idea about what was done along the related research processes. So, the following paragraphs are devoted to both Artificial Neuro-Fuzzy Inference System (ANFIS) and League Championship Algorithm (LCA).

### Artificial – Adaptive Neuro-Fuzzy Inference System

After some time regarding introduction of new Artificial Intelligence techniques, research works have been directed to combining some techniques to obtain more effective, hybrid intelligent solutions for the problems. In this context, especially nature-inspired techniques have been remarkable techniques to be employed for hybrid intelligent systems purposes (Melin et al., 2017). Here, Artificial Neuro-Fuzzy Inference System (which is also called as Adaptive Neuro-Fuzzy Inference System: ANFIS) has been a remarkable hybrid technique and it is employed often in problems. ANFIS briefly consists of two important Artificial Intelligence techniques: Artificial Neural Networks and Fuzzy Logic (Jang, 1993; Lin & Lee, 1996).

ANFIS has been first introduced by Jang (1992; 1993) in his research works. As default, an ANFIS comes with a Takagi-Sugeno model oriented Fuzzy Logic inference approach (Yen and Langari, 1999) employed over a Radial Basis Function Neural Network (RBFNN) including nodes having radial basis functions like Gaussian and Ellipsoidal ones (Jang, 1993; Lin and Lee, 1996). As general, ANFIS runs for the objective problem by benefiting from specific IF…THEN rules associated with first order Takagi-Sugeno model (Kose and Arslan, 2017). An ANFIS uses some parameters, which are called as premise and consequent and these parameters are optimized in order to make the ANFIS giving desired outputs to the training data (Jang, 1992; Jang, 1993). Traditionally, some learning algorithms like Gradient Decent and One Pass of Least Square Estimates, Gradient Decent Backpropagation and Gradient Decent and Least Square Estimates were being used for training (Guner, 2003). But today it is also possible to use an intelligent optimization algorithm for the same purpose.

*Figure 1. An example of ANFIS structure*
*Baseri et al., 2011.*

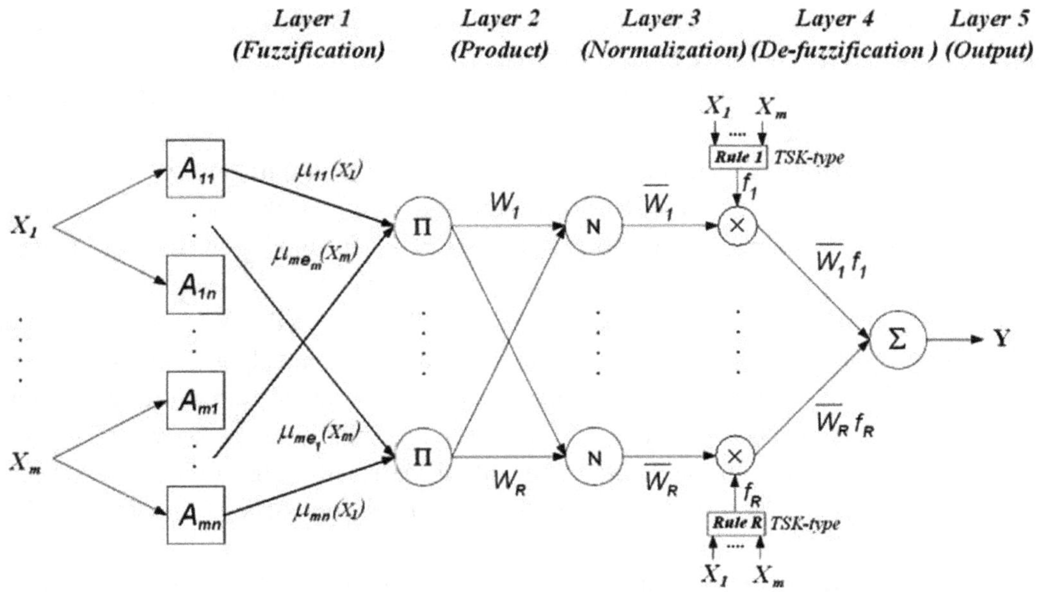


In the research work here, an ANFIS was trained by an alternative, optimization algorithm: League Championship Algorithm. Some essential information regarding this global optimization algorithm are given in the following paragraphs.

## League Championship Algorithm

Better solution ways than traditional ones and also intelligent techniques appeared in time have been a remarkable research interest in the associated literatures of optimization (Chong & Zak, 2013; Wang et al., 2007; Yourui, 2008; Yuksel et al., 2015). In this context, Artificial Intelligence based, intelligent optimization techniques have become widely used optimization tools. As introduced first by Kashan (2009), the League Championship Algorithm (LCA) in an optimization algorithm in the context of Artificial Intelligence oriented approaches. As an intelligent optimization technique, LCA inspires from championships performed for teams (i.e. soccer) and it uses some algorithmic steps to achieve an environment in which some teams (particles) make matches (in an iterative manner) to become champion (the most optimum solution) at the end (Kashan, 2009, 2014). With its details, LCA provides an evolutionary approach to deal with optimization problems considered. As a result of inspirations, some metaphors are used to define some mathematical and logical algorithmic steps forming the whole problem solution ways of optimization.

Briefly, the following definitions, acceptances and solution processes are used along the LCA (Kashan, 2009, 2014):

- In the algorithm, the term league is used for the population of solutions while, team is for defining each member within the population, team formation is for the solution, week is for the iteration and finally, playing strength is for the fitness value or function.

- The concept of playing strength is for if a team has enough ability to beat another one or not.
- Over a generated league schedule (which uses a simple, additional algorithm to make that) and along the weeks, each team makes matches with other ones and according to strengths determined, a team can win or lose a match (there is not tie).
- According to past matches and past best formation(s), the new formation of a team is determined before passing to the new week.
- New formations result to new determination of playing strength(s) in the new week(s).
- The whole league continues according to schedule along the weeks and the winner (the best – most optimum) team among all the past processes is the exact solution for the objective problem.

In detail, LCA employs different mathematical and logical steps for all processes regarding matches, updates over formations, determining winner or loser…etc. and a general algorithmic structure showing an exact championship environment of a population-based technique, which can be accepted as also a Swarm Intelligence technique (Blum & Li, 2008; Fister Jr et al., 2013; Kose & Vasant, 2018), is organized practically. In the context, of intelligent optimization studies, this algorithm can also be accepted as a metaheuristic technique with its essential problem solution approaches (Pardalos, 2005; Vasant, 2012; Vasant et al., 2017; Yang, 2011).

By taking also more details expressed by Kashan (2009; 2014) for LCA into consideration, default flow chart of the LCA can be drawn as in Figure 2 (Rezaei et al., 2018).

Based on the formed ANFIS-LCA, the next section explains the main prediction research approach done for EEG time series as considered in this chapter.

## PREDICTION APPLICATIONS OVER EEG TIME SERIES

The ANFIS-LCA system has been run over some EEG time series to see if it is effective to predict future states. In this manner, this section focuses on the performed applications and the obtained findings.

### Time Series Prediction With ANFIS-LCA

ANFIS-LCA system developed within the research work employs the following features and functions to deal with the time series prediction problem:

- The system consists of three inputs over Gaussian sets.
- There is a total of 9 rules to support the mechanism of ANFIS.
- In order to predict a time series, the ANFIS uses three past values of the time series. In detail, the system tries to predict the point at x(t + 3), which is output, by using the points over x(t), x(t - 3) and x(t – 6), which are inputs.
- As it is known, the training is done by the LCA, considered as the training technique – algorithm in the context of the introduced research.

*Figure 2. Flow chart of the LCA*
*Kashan, 2009, 2014; Rezaei et al., 2018.*

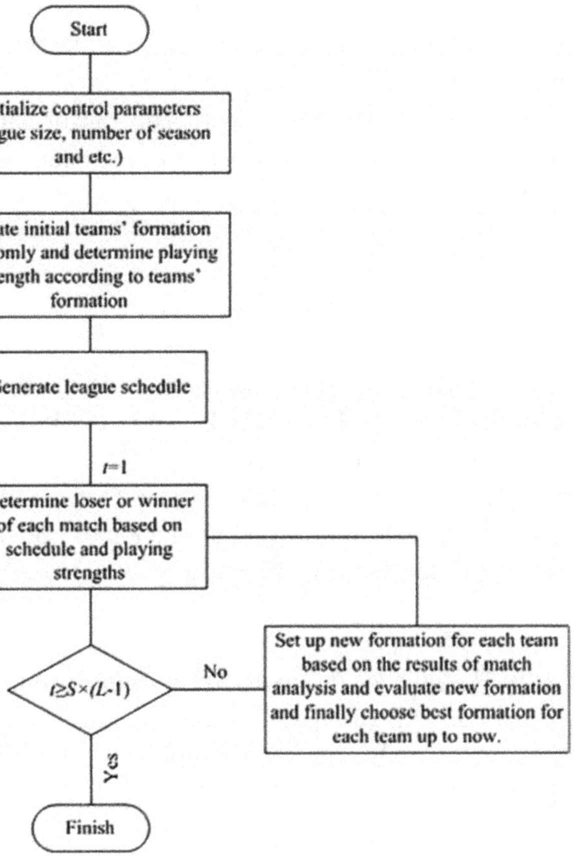


## Applications and Findings

For the applications of predicting EEG time series, the ANFIS-LCA has been applied over some data sets provided by Andrzejak et al. (2001). In this context, five different data sets were considered for the prediction applications. In detail, each time series data consists of 4000 rows over that output: x(t + 3) and the inputs: x(t), x(t - 3) and x(t - 6) respectively. For the prediction applications 50% of data were used for training and the remaining 50% were evaluated for understanding performance of the ANFIS-LCA. The related EEG time series are provided by Andrzejak et al. (2001) within five different sets called as A to E respectively, and the considered time series from each sets are:

- Z023 from Set A,
- O008 from Set B,
- N017 from Set C,
- F068 from Set D,
- S079 from Set E.

For evaluating the performance of the ANFIS-LCA, it has been compared over the Mean Average of Error (MAE) value, by considering also three system formations as the same ANFIS trained by Particle Swarm Optimization (PSO), Genetic Algorithms (GA) and Cuckoo Search (CS) separately (for more information about the alternative algorithms: Brownlee, 2011). The obtained findings are provided in Table 1 (best values are in bold).

As it is seen in the Table 1, ANFIS-LCA has provided better performances in three of five EEG time series, according to other systems. It is generally followed by the system of ANFIS-CS. For a visual representation of a prediction performance, Figure 3 provides the prediction of Z023 done by ANFIS-LCA.

## CONCLUSION

In this chapter, a time series prediction approach has been done by using an Artificial Neuro-Fuzzy Inference System (Adaptive Neuro-Fuzzy Inference System: ANFIS), which is trained by using an optimization algorithm: League Championship Algorithm (LCA). As an intelligent optimization technique, the algorithm has been used for training purposes rather than using traditional training processes done generally. As the objective application problem, prediction of Electroencephalogram (EEG) time series has been chosen as the input for the formed hybrid system. In detail, some prediction processes

*Table 1. Findings regarding MAE value for different ANFIS-training algorithm systems*

| EEG Time Series | ANFIS-LCA | ANFIS-PSO | ANFIS-GA | ANFIS-CS |
|---|---|---|---|---|
| Z023 | **12.4221** | 23.4012 | 23.8739 | 13.3560 |
| O008 | **15.1290** | 19.5109 | 18.4013 | 17.2906 |
| N017 | 10.5677 | 20.1119 | 21.8991 | **10.2983** |
| F068 | 16.4890 | 27.5320 | 26.2972 | **14.4791** |
| S079 | **20.9131** | 30.9311 | 36.7780 | 24.5192 |

*Figure 3. Prediction performance of ANFIS-LCA for Z023*

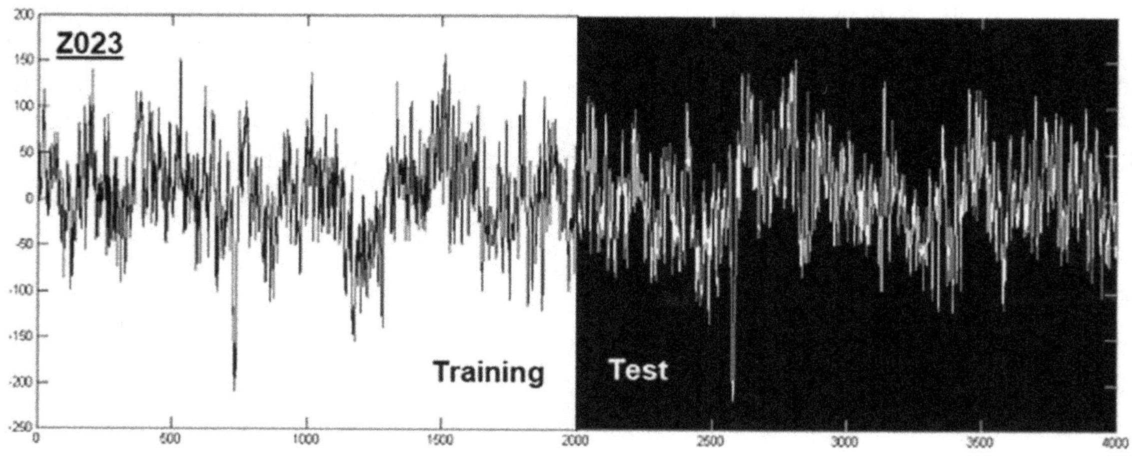

were done and according to the findings, it has been understood that the formed ANFIS-LCA system is effective enough to predict next steps of a typical EEG time series, which are key data in terms of health field, and Biomedical Engineering. In detail, it is important also that such a hybrid system of intelligent prediction has been successful over an example of chaotic time series. Moving from that, successful results obtained with this research have figured out an alternative system for predicting time series, which are important for health field, Biomedical Engineering and the associated fields dealing with especially chaotic time series.

## FUTURE RESEARCH DIRECTIONS

In addition to the performed research here, there are also some future work plans thought by the author. Some remarkable ones of them can be expressed as follows briefly:

- Future works will include adjustments regarding the formed ANFIS-LCA system and evaluating it after changed features – mechanisms and with also different EEG time series.
- In addition to the considered EEG time series, alternative medical data will also be used for prediction purposes.
- The hybrid system structure introduced in this chapter will be supported – updated with more alternative Artificial Intelligence techniques and different hybrid system structures will be evaluated in terms of time series prediction.
- The formed ANFIS model in this research will be trained with also some other intelligent optimization techniques and in this way, performances of different techniques on training ANFIS for the problem of time series prediction will be compared from different perspectives.

## REFERENCES

Akerkar, R. (2014). *Introduction to artificial intelligence*. PHI Learning Pvt. Ltd.

Alm, C. O., Roth, D., & Sproat, R. (2005). Emotions from text: machine learning for text-based emotion prediction. In *Proceedings of the conference on human language technology and empirical methods in natural language processing* (pp. 579-586). Association for Computational Linguistics. doi:10.3115/1220575.1220648

Andrzejak, R. G., Lehnertz, K., Mormann, F., Rieke, C., David, P., & Elger, C. E. (2001). Indications of nonlinear deterministic and finite-dimensional structures in time series of brain electrical activity: Dependence on recording region and brain state. *Physical Review. E*, *64*(6), 061907. doi:10.1103/PhysRevE.64.061907 PMID:11736210

Baseri, H., Rahmani, B., & Bakhshi-Jooybari, M. (2011). Selection of bending parameters for minimal spring-back using an ANFIS model and simulated annealing algorithm. *Journal of Manufacturing Science and Engineering*, *133*(3), 031010. doi:10.1115/1.4004139

Blum, C., & Li, X. (2008). Swarm intelligence in optimization. In *Swarm Intelligence* (pp. 43–85). Springer Berlin Heidelberg. doi:10.1007/978-3-540-74089-6_2

Borovykh, A., Bohte, S., & Oosterlee, C. W. (2017). *Conditional Time Series Forecasting with Convolutional Neural Networks.* arXiv preprint arXiv:1703.04691

Box, G. E., Jenkins, G. M., Reinsel, G. C., & Ljung, G. M. (2015). *Time series analysis: forecasting and control.* John Wiley & Sons.

Brockwell, P. J., & Davis, R. A. (2016). *Introduction to Time Series and Forecasting*. Springer. doi:10.1007/978-3-319-29854-2

Brooks, C. A., Thompson, C., & Teasley, S. D. (2014). Towards A General Method for Building Predictive Models of Learner Success using Educational Time Series Data. In LAK Workshops. Academic Press.

Brown, R. G. (2004). *Smoothing, forecasting and prediction of discrete time series*. Courier Corporation.

Brownlee, J. (2011). *Clever algorithms: nature-inspired programming recipes*. Jason Brownlee.

Chatfield, C. (2016). *The Analysis of Time Series: An Introduction*. CRC Press.

Chen, J. H., & Asch, S. M. (2017). Machine learning and prediction in medicine– beyond the peak of inflated expectations. *The New England Journal of Medicine*, *376*(26), 2507–2509. doi:10.1056/NEJMp1702071 PMID:28657867

Chong, E. K., & Zak, S. H. (2013). *An Introduction to Optimization* (Vol. 76). John Wiley & Sons.

Coelho, I. M., Coelho, V. N., da Luz, E. J. S., Ochi, L. S., Guimarães, F. G., & Rios, E. (2017). A GPU deep learning metaheuristic based model for time series forecasting. *Applied Energy*, *201*, 412–418. doi:10.1016/j.apenergy.2017.01.003

Costa, M., Goldberger, A. L., & Peng, C. K. (2002). Multiscale entropy analysis of complex physiologic time series. *Physical Review Letters*, *89*(6), 068102. doi:10.1103/PhysRevLett.89.068102 PMID:12190613

Elliot, A., Hsu, C. H., & Slodoba, J. (2017). *Time Series Prediction: Predicting Stock Price.* arXiv preprint arXiv:1710.05751

Enders, W. (2008). *Applied econometric time series.* John Wiley & Sons.

Fister, I., Jr., Yang, X. S., Fister, I., Brest, J., & Fister, D. (2013). *A brief review of nature-inspired algorithms for optimization.* arXiv preprint arXiv:1307.4186

Flasiński, M. (2016). *Introduction to Artificial Intelligence*. Springer. doi:10.1007/978-3-319-40022-8

Formisano, E., De Martino, F., & Valente, G. (2008). Multivariate analysis of fMRI time series: Classification and regression of brain responses using machine learning. *Magnetic Resonance Imaging*, *26*(7), 921–934. doi:10.1016/j.mri.2008.01.052 PMID:18508219

Guner, E. (2003). *Adaptive neuro fuzzy inference system applications in chemical processes* (Doctoral dissertation). METU.

Ince, H., & Trafalis, T. B. (2017). A hybrid forecasting model for stock market prediction. *Economic Computation and Economic Cybernetics Studies and Research*, *51*(3).

Jang, J. S. (1992). Self-learning fuzzy controllers based on temporal backpropagation. *IEEE Transactions on Neural Networks, 3*(5), 714–723. doi:10.1109/72.159060 PMID:18276470

Jang, J. S. (1993). ANFIS: Adaptive-network-based fuzzy inference system. *IEEE Transactions on Systems, Man, and Cybernetics, 23*(3), 665–685. doi:10.1109/21.256541

Kashan, A. H. (2009). League championship algorithm: a new algorithm for numerical function optimization. In *Soft Computing and Pattern Recognition, 2009. SOCPAR'09. International Conference of* (pp. 43-48). IEEE. doi:10.1109/SoCPaR.2009.21

Kashan, A. H. (2014). League Championship Algorithm (LCA): An algorithm for global optimization inspired by sport championships. *Applied Soft Computing, 16*, 171–200. doi:10.1016/j.asoc.2013.12.005

Kose, U., & Arslan, A. (2017). Forecasting chaotic time series via anfis supported by vortex optimization algorithm: Applications on electroencephalogram time series. *Arabian Journal for Science and Engineering, 42*(8), 3103–3114. doi:10.1007/s13369-016-2279-z

Kose, U., & Vasant, P. (2018). A Model of Swarm Intelligence Based Optimization Framework Adjustable According to Problems. In I. Zelinka, P. Vasant, V. Duy, & T. Dao (Eds.), *Innovative Computing, Optimization and Its Applications. Studies in Computational Intelligence* (Vol. 741). Cham: Springer. doi:10.1007/978-3-319-66984-7_2

Längkvist, M., Karlsson, L., & Loutfi, A. (2014). A review of unsupervised feature learning and deep learning for time-series modeling. *Pattern Recognition Letters, 42*, 11–24. doi:10.1016/j.patrec.2014.01.008

Lin, C. T., & Lee, C. G. (1996). *Neural fuzzy systems*. PTR Prentice Hall.

Mair, C., Kadoda, G., Lefley, M., Phalp, K., Schofield, C., Shepperd, M., & Webster, S. (2000). An investigation of machine learning based prediction systems. *Journal of Systems and Software, 53*(1), 23–29. doi:10.1016/S0164-1212(00)00005-4

Melin, P., Castillo, O., & Kacprzyk, J. (Eds.). (2017). *Nature-Inspired Design of Hybrid Intelligent Systems*. Springer. doi:10.1007/978-3-319-47054-2

Nguyen, L. T., Wu, P., Chan, W., Peng, W., & Zhang, Y. (2012, August). Predicting collective sentiment dynamics from time-series social media. In *Proceedings of the first international workshop on issues of sentiment discovery and opinion mining* (p. 6). ACM. doi:10.1145/2346676.2346682

Pardalos, P. M. (2005). Metaheuristics in Optimization. In *Workshop on the European Chapter on Metaheuristics and Large Scale Optimization, Vilnius, Lithuania May* (pp. 19-21). Academic Press.

Prudêncio, R. B., & Ludermir, T. B. (2004). Meta-learning approaches to selecting time series models. *Neurocomputing, 61*, 121–137. doi:10.1016/j.neucom.2004.03.008

Qiu, X., Ren, Y., Suganthan, P. N., & Amaratunga, G. A. (2017). Empirical mode decomposition based ensemble deep learning for load demand time series forecasting. *Applied Soft Computing, 54*, 246–255. doi:10.1016/j.asoc.2017.01.015

Rezaei, H., Bozorg-Haddad, O., & Chu, X. (2018). League Championship Algorithm (LCA). In *Advanced Optimization by Nature-Inspired Algorithms* (pp. 19–30). Singapore: Springer.

Shah, V. H. (2007, Spring). Machine learning techniques for stock prediction. *Foundations of Machine Learning*, 1-19.

Shi, Z., & Han, M. (2007). Support vector echo-state machine for chaotic time-series prediction. *IEEE Transactions on Neural Networks*, *18*(2), 359–372. doi:10.1109/TNN.2006.885113 PMID:17385625

Shipp, M. A., Ross, K. N., Tamayo, P., Weng, A. P., Kutok, J. L., Aguiar, R. C., & Ray, T. S. et al. (2002). Diffuse large B-cell lymphoma outcome prediction by gene-expression profiling and supervised machine learning. *Nature Medicine*, *8*(1), 68–74. doi:10.1038/nm0102-68 PMID:11786909

Singh, R., & Balasundaram, S. (2007). Application of extreme learning machine method for time series analysis. *International Journal of Intelligent Technology*, *2*(4), 256–262.

Tang, X., Yang, C., & Zhou, J. (2009). Stock price forecasting by combining news mining and time series analysis. In *Web Intelligence and Intelligent Agent Technologies, 2009. WI-IAT'09. IEEE/WIC/ ACM International Joint Conferences on* (Vol. 1, pp. 279-282). IEEE. doi:10.1109/WI-IAT.2009.48

Toriyama, N., Ono, K., & Orito, Y. (2017). Adaptive GA-based AR-hidden Markov model for time series forecasting. In *Evolutionary Computation (CEC), 2017 IEEE Congress on* (pp. 665-672). IEEE. doi:10.1109/CEC.2017.7969374

Vasant, P., Kose, U., & Watada, J. (2017). Metaheuristic Techniques in Enhancing the Efficiency and Performance of Thermo-Electric Cooling Devices. *Energies*, *10*(11), 1703. doi:10.3390/en10111703

Vasant, P. M. (Ed.). (2012). *Meta-heuristics optimization algorithms in engineering, business, economics, and finance*. IGI Global.

Wang, D. W., Wang, J. W., Wang, H., Zhang, R., & Guo, Z. (2007). *Intelligent optimization methods*. Higher Education Press.

Weigend, A. S. (1994). *Time series prediction: Forescasting the future and understanding the past* (No. 04; QA280, T5.). Academic Press.

Weiss, S. M., & Kulikowski, C. A. (1991). *Computer systems that learn: classification and prediction methods from statistics, neural nets, machine learning, and expert systems*. Morgan Kaufmann Publishers Inc.

Xing, F. Z., Cambria, E., & Zou, X. (2017). Predicting evolving chaotic time series with fuzzy neural networks. In *International Joint Conference on Neural Networks (IJCNN)*. Academic Press. doi:10.1109/ IJCNN.2017.7966252

Yang, X. S. (2011). Metaheuristic optimization: Algorithm analysis and open problems. *Experimental Algorithms*, 21-32.

Yao, B., Chen, C., Cao, Q., Jin, L., Zhang, M., Zhu, H., & Yu, B. (2017). Short-Term Traffic Speed Prediction for an Urban Corridor. *Computer-Aided Civil and Infrastructure Engineering*, *32*(2), 154–169. doi:10.1111/mice.12221

Yen, J., & Langari, R. (1999). *Fuzzy logic: Intelligence, Control, and Information* (Vol. 1). Upper Saddle River, NJ: Prentice Hall.

Yourui, H. (2008). *Intelligent optimization algorithm and application.* Beijing: National Defence Industry Press.

Yuksel, A. S., Cankaya, I. A., & Yuksel, M. E. (2015). An Analysis of RDF Storage Models and Query Optimization Techniques. *International Journal of Information Engineering and Electronic Business, 7*(2), 20–26. doi:10.5815/ijieeb.2015.02.04

Yuksel, A. S., Cankaya, S. F., & Uncu, I. S. (2017). Design of a Machine Learning Based Predictive Analytics System for Spam Problem. *Acta Physica Polonica A, 132*(3), 500–504. doi:10.12693/APhysPolA.132.500

Zeger, S. L., Irizarry, R., & Peng, R. D. (2006). On time series analysis of public health and biomedical data. *Annual Review of Public Health, 27*(1), 57–79. doi:10.1146/annurev.publhealth.26.021304.144517 PMID:16533109

Zhang, L., Liu, P., & Gulla, J. A. (2017). A Neural Time Series Forecasting Model for User Interests Prediction On Twitter. In *Proceedings of the 25th Conference on User Modeling, Adaptation and Personalization* (pp. 397-398). ACM. doi:10.1145/3079628.3079648

## ADDITIONAL READING

Abdulhamid, S. M., & Latiff, M. S. A. (2014). League Championship Algorithm based job scheduling scheme for infrastructure as a service cloud. *arXiv preprint arXiv:1410.2208.*

Abraham, A., & Jain, L. (2005). Evolutionary multiobjective optimization. *Evolutionary Multiobjective Optimization*, 1-6.

Akdeniz, E., Egrioglu, E., Bas, E., & Yolcu, U. (2018). An ARMA Type Pi-Sigma Artificial Neural Network for Nonlinear Time Series Forecasting. *Journal of Artificial Intelligence and Soft Computing Research, 8*(2), 121–132. doi:10.1515/jaiscr-2018-0009

Aminikhanghahi, S., & Cook, D. J. (2017). A survey of methods for time series change point detection. *Knowledge and Information Systems, 51*(2), 339–367. doi:10.1007/s10115-016-0987-z PMID:28603327

Apergis, N., Mervar, A., & Payne, J. E. (2017). Forecasting disaggregated tourist arrivals in Croatia: Evidence from seasonal univariate time series models. *Tourism Economics, 23*(1), 78–98. doi:10.5367/te.2015.0499

Bajpai, S., Jain, K., & Jain, N. (2011). Artificial neural networks. *International Journal of Soft Computing and Engineering (IJSCE), 1*(NCAI2011).

Billah, M., & Islam, N. (2016). An Early Diagnosis System for predicting Lung Cancer Risk Using adaptive Neuro Fuzzy Inference System and Linear Discriminant Analysis. *Journal of MPE Molecular Pathological Epidemiology.*

Blinowska, K. J., & Malinowski, M. (1991). Non-linear and linear forecasting of the EEG time series. *Biological Cybernetics, 66*(2), 159–165. doi:10.1007/BF00243291 PMID:1768720

Castillo, O., & Melin, P. (2002). Hybrid intelligent systems for time series prediction using neural networks, fuzzy logic, and fractal theory. *IEEE Transactions on Neural Networks*, *13*(6), 1395–1408. doi:10.1109/TNN.2002.804316 PMID:18244536

Chen, J. H., & Asch, S. M. (2017). Machine learning and prediction in medicine– beyond the peak of inflated expectations. *The New England Journal of Medicine*, *376*(26), 2507–2509. doi:10.1056/NEJMp1702071 PMID:28657867

Chen, Y., Yang, B., & Dong, J. (2006). Time-series prediction using a local linear wavelet neural network. *Neurocomputing*, *69*(4), 449–465. doi:10.1016/j.neucom.2005.02.006

Choubin, B., Khalighi-Sigaroodi, S., Malekian, A., & Kişi, Ö. (2016). Multiple linear regression, multi-layer perceptron network and adaptive neuro-fuzzy inference system for forecasting precipitation based on large-scale climate signals. *Hydrological Sciences Journal*, *61*(6), 1001–1009. doi:10.1080/02626667.2014.966721

Connor, J. T., Martin, R. D., & Atlas, L. E. (1994). Recurrent neural networks and robust time series prediction. *IEEE Transactions on Neural Networks*, *5*(2), 240–254. doi:10.1109/72.279188 PMID:18267794

Davidor, Y. (1991). *Genetic Algorithms and Robotics: A heuristic strategy for optimization* (Vol. 1). World Scientific. doi:10.1142/1111

Deb, C., Zhang, F., Yang, J., Lee, S. E., & Shah, K. W. (2017). A review on time series forecasting techniques for building energy consumption. *Renewable & Sustainable Energy Reviews*, *74*, 902–924. doi:10.1016/j.rser.2017.02.085

Dounis, A. I., Leftheriotis, G., Stavrinidis, S., & Syrrokostas, G. (2016). Electrochromic device modeling using an adaptive neuro-fuzzy inference system: A model-free approach. *Energy and Building*, *110*, 182–194. doi:10.1016/j.enbuild.2015.10.045

Faruk, D. Ö. (2010). A hybrid neural network and ARIMA model for water quality time series prediction. *Engineering Applications of Artificial Intelligence*, *23*(4), 586–594. doi:10.1016/j.engappai.2009.09.015

Forney, E. M., & Anderson, C. W. (2011). Classification of EEG during imagined mental tasks by forecasting with Elman recurrent neural networks. In *Neural Networks (IJCNN), The 2011 International Joint Conference on* (pp. 2749-2755). IEEE. doi:10.1109/IJCNN.2011.6033579

Francis, H., & Kusiak, A. (2017). Prediction of engine demand with a data-driven approach. *Procedia Computer Science*, *103*, 28–35. doi:10.1016/j.procs.2017.01.005

Froelich, W., & Pedrycz, W. (2017). Fuzzy cognitive maps in the modeling of granular time series. *Knowledge-Based Systems*, *115*, 110–122. doi:10.1016/j.knosys.2016.10.017

Giles, C. L., Lawrence, S., & Tsoi, A. C. (2001). Noisy time series prediction using recurrent neural networks and grammatical inference. *Machine Learning*, *44*(1), 161–183. doi:10.1023/A:1010884214864

Gupta, P., Batra, S. S., & Jayadeva, . (2017). Sparse short-term time series forecasting models via minimum model complexity. *Neurocomputing*, *243*, 1–11. doi:10.1016/j.neucom.2017.02.002

He, X., & Xu, S. (2010). Artificial neural networks. *Process Neural Networks: Theory and Applications*, 20-42.

Ho, S. Y., Shu, L. S., & Chen, J. H. (2004). Intelligent evolutionary algorithms for large parameter optimization problems. *IEEE Transactions on Evolutionary Computation*, *8*(6), 522–541. doi:10.1109/TEVC.2004.835176

Jang, J. S. (1993). ANFIS: Adaptive-network-based fuzzy inference system. *IEEE Transactions on Systems, Man, and Cybernetics*, *23*(3), 665–685. doi:10.1109/21.256541

Jonas, E., Bobra, M. G., Shankar, V., Hoeksema, J. T., & Recht, B. (2017). Flare prediction using photospheric and coronal image data. *arXiv preprint arXiv:1708.01323.*

Kasabov, N. K., & Song, Q. (2002). DENFIS: Dynamic evolving neural-fuzzy inference system and its application for time-series prediction. *IEEE Transactions on Fuzzy Systems*, *10*(2), 144–154. doi:10.1109/91.995117

Kashan, A. H., & Karimi, B. (2010). A new algorithm for constrained optimization inspired by the sport league championships. In *Evolutionary Computation (CEC), 2010 IEEE Congress on* (pp. 1-8). IEEE. doi:10.1109/CEC.2010.5586364

Kashan, A. H., Karimiyan, S., Karimiyan, M., & Kashan, M. H. (2012). A modified League Championship Algorithm for numerical function optimization via artificial modeling of the "between two halves analysis". In *Soft Computing and Intelligent Systems (SCIS) and 13th International Symposium on Advanced Intelligent Systems (ISIS), 2012 Joint 6th International Conference on* (pp. 1944-1949). IEEE.

Khashei, M., & Hajirahimi, Z. (2017). Performance evaluation of series and parallel strategies for financial time series forecasting. *Financial Innovation*, *3*(1), 24. doi:10.1186/s40854-017-0074-9

Kim, S. H., Faloutsos, C., & Yang, H. J. (2013). Coercively adjusted auto regression model for forecasting in epilepsy EEG. *Computational and Mathematical Methods in Medicine*, 2013. PMID:23710252

Latiff, M. S. A., Abdul-Salaam, G., & Madni, S. H. H. (2016). Secure scientific applications scheduling technique for cloud computing environment using global league championship algorithm. *PLoS One*, *11*(7), e0158102. doi:10.1371/journal.pone.0158102 PMID:27384239

Lukasiewycz, M., Glaß, M., Reimann, F., & Teich, J. (2011). Opt4J: a modular framework for metaheuristic optimization. In *Proceedings of the 13th annual conference on Genetic and evolutionary computation* (pp. 1723-1730). ACM. doi:10.1145/2001576.2001808

Mansouri, F., Dunlop, K., Giacobbe, P., Downar, J., & Zariffa, J. (2017). A Fast EEG Forecasting Algorithm for Phase-Locked Transcranial Electrical Stimulation of the Human Brain. *Frontiers in Neuroscience*, *11*, 401. doi:10.3389/fnins.2017.00401 PMID:28775678

Martinetz, T. M., Berkovich, S. G., & Schulten, K. J. (1993). 'Neural-gas' network for vector quantization and its application to time-series prediction. *IEEE Transactions on Neural Networks*, *4*(4), 558–569. doi:10.1109/72.238311 PMID:18267757

Masters, T. (1995). *Neural, novel and hybrid algorithms for time series prediction.* John Wiley & Sons, Inc.

Menezes, J. M. P. Jr, & Barreto, G. A. (2008). Long-term time series prediction with the NARX network: An empirical evaluation. *Neurocomputing*, *71*(16), 3335–3343. doi:10.1016/j.neucom.2008.01.030

Mousavi-Avval, S. H., Rafiee, S., Sharifi, M., Hosseinpour, S., & Shah, A. (2017). Combined application of Life Cycle Assessment and Adaptive Neuro-Fuzzy Inference System for modeling energy and environmental emissions of oilseed production. *Renewable & Sustainable Energy Reviews*, *78*, 807–820. doi:10.1016/j.rser.2017.05.002

Mukherjee, S., Osuna, E., & Girosi, F. (1997). Nonlinear prediction of chaotic time series using support vector machines. In *Neural Networks for Signal Processing [1997] VII. Proceedings of the 1997 IEEE Workshop* (pp. 511-520). IEEE. doi:10.1109/NNSP.1997.622433

Müller, K. R., Smola, A. J., Rätsch, G., Schölkopf, B., Kohlmorgen, J., & Vapnik, V. (1997). Predicting time series with support vector machines. In *International Conference on Artificial Neural Networks* (pp. 999-1004). Springer, Berlin, Heidelberg. doi:10.1007/BFb0020283

Osman, I. H., & Kelly, J. P. (1996). Meta-heuristics: an overview. In Meta-heuristics (pp. 1-21). Springer US. doi:10.1007/978-1-4613-1361-8_1

Pardey, J., Roberts, S., & Tarassenko, L. (1996). A review of parametric modelling techniques for EEG analysis. *Medical Engineering & Physics*, *18*(1), 2–11. doi:10.1016/1350-4533(95)00024-0 PMID:8771033

Pham, D. T., Packianather, M. S., & Afify, A. A. (2007). Artificial neural networks. In Computational Intelligence (pp. 67-92). Springer US. doi:10.1007/0-387-37452-3_3

Rardin, R. L., & Uzsoy, R. (2001). Experimental evaluation of heuristic optimization algorithms: A tutorial. *Journal of Heuristics*, *7*(3), 261–304. doi:10.1023/A:1011319115230

Salles, R., Assis, L., Guedes, G., Bezerra, E., Porto, F., & Ogasawara, E. (2017). A framework for benchmarking machine learning methods using linear models for univariate time series prediction. In *Neural Networks (IJCNN), 2017 International Joint Conference on* (pp. 2338-2345). IEEE. doi:10.1109/IJCNN.2017.7966139

Sancar, N., & Tabrizi, S. S. (2017). Body Mass Index Estimation by Using an Adaptive Neuro Fuzzy Inference System. *Procedia Computer Science*, *108*, 2501–2506. doi:10.1016/j.procs.2017.05.137

Santos, A. R., Cordeiro, G. N., Becchi, G. C., Ayala, H. V., Mariani, V. C., & dos Santos Coelho, L. (2017). Quadcopter Control Based on Electroencephalogram Headset and Hybrid Fuzzy Classifier. In *Brain Function Assessment in Learning: First International Conference, BFAL 2017, Patras, Greece, September 24-25, 2017* [). Springer.]. *Proceedings*, *10512*, 209.

Sapankevych, N. I., & Sankar, R. (2009). Time series prediction using support vector machines: A survey. *IEEE Computational Intelligence Magazine*, *4*(2), 24–38. doi:10.1109/MCI.2009.932254

Stam, C. J. (2005). Nonlinear dynamical analysis of EEG and MEG: Review of an emerging field. *Clinical Neurophysiology*, *116*(10), 2266–2301. doi:10.1016/j.clinph.2005.06.011 PMID:16115797

Su, Z., Wang, J., Lu, H., & Zhao, G. (2014). A new hybrid model optimized by an intelligent optimization algorithm for wind speed forecasting. *Energy Conversion and Management*, *85*, 443–452. doi:10.1016/j.enconman.2014.05.058

Taieb, S. B., Yu, J., Barreto, M. N., & Rajagopal, R. (2017). *Regularization in Hierarchical Time Series Forecasting with Application to Electricity Smart Meter Data* (pp. 4474–4480). AAAI.

Tavakkoli, P., Souran, D. M., Tavakkoli, S., Hatamian, M., Mehrabian, A., & Balas, V. E. (2015). Classification of the liver disorders data using Multi-Layer adaptive Neuro-Fuzzy inference system. In *Computing, Communication and Networking Technologies (ICCCNT), 2015 6th International Conference on* (pp. 1-4). IEEE. doi:10.1109/ICCCNT.2015.7395182

Van Gestel, T., Suykens, J. A., Baestaens, D. E., Lambrechts, A., Lanckriet, G., Vandaele, B., & Vandewalle, J. et al. (2001). Financial time series prediction using least squares support vector machines within the evidence framework. *IEEE Transactions on Neural Networks*, *12*(4), 809–821. doi:10.1109/72.935093 PMID:18249915

Verleysen, M., & François, D. (2005). *The Curse of Dimensionality in Data Mining and Time Series Prediction* (Vol. 5, pp. 758–770). IWANN. doi:10.1007/11494669_93

Voß, S., Martello, S., Osman, I. H., & Roucairol, C. (Eds.). (2012). *Meta-heuristics: Advances and trends in local search paradigms for optimization*. Springer Science & Business Media.

Wang, L. (2001). Intelligent optimization algorithms with applications. Tsinghua University & Springer Press, Beijing.

Wang, Z., & Koprinska, I. (2017). Solar power prediction with data source weighted nearest neighbors. In *Neural Networks (IJCNN), 2017 International Joint Conference on* (pp. 1411-1418). IEEE. doi:10.1109/IJCNN.2017.7966018

Woodward, W. A., Gray, H. L., & Elliott, A. C. (2017). *Applied Time Series Analysis with R*. CRC Press.

Xu, W., Wang, R., & Yang, J. (2015). An improved league championship algorithm with free search and its application on production scheduling. *Journal of Intelligent Manufacturing*, 1–10.

Yaïci, W., & Entchev, E. (2016). Adaptive Neuro-Fuzzy Inference System modelling for performance prediction of solar thermal energy system. *Renewable Energy*, *86*, 302–315. doi:10.1016/j.renene.2015.08.028

Yourui, H. (2008). *Intelligent optimization algorithm and application* (p. 1). Beijing: National Defence Industry Press.

Yu, L., Zhao, Y., & Tang, L. (2017). Ensemble forecasting for complex time series using sparse representation and neural networks. *Journal of Forecasting*, *36*(2), 122–138. doi:10.1002/for.2418

Zhao, L., & Yang, Y. (2009). PSO-based single multiplicative neuron model for time series prediction. *Expert Systems with Applications*, *36*(2), 2805–2812. doi:10.1016/j.eswa.2008.01.061

Zhou, D., Gao, F., Ravey, A., Al-Durra, A., & Simões, M. G. (2017). Online energy management strategy of fuel cell hybrid electric vehicles based on time series prediction. In *Transportation Electrification Conference and Expo (ITEC)*, 2017 *IEEE* (pp. 113-118). IEEE. doi:10.1109/ITEC.2017.7993256

## KEY TERMS AND DEFINITIONS

**Electroencephalogram:** An electrophysiological tracking approach of electrical activity occurred in the brain.

**Intelligent Optimization:** A type of optimization solution approach done thanks to artificial-intelligence-based techniques.

**Neuro-Fuzzy:** A terms used for defining hybrid artificial intelligence systems formed by using both artificial neural networks and fuzzy logic.

**Time Series:** A regular, complex, or chaotic observed series of data considered in a determined/set time period.

**Time Series Prediction/Predicting:** An approach of predicting next flows/data points of an objective time series, thanks to some previously observed features like past data points.

# Chapter 12
# Determining Headache Diseases With Genetic Algorithm

**Gaffari Celik**
*Agri Ibrahim Cecen University, Turkey*

## ABSTRACT

*Currently, medical diagnosis has a strong relation with the artificial-intelligence-oriented approaches. Because it is practical to employ intelligent mechanisms over some input data-expert knowledge and design effective solution ways, even the biomedical engineering field is interested in taking support from artificial intelligence. If applications in this manner are taken into consideration, we can see that medical diagnoses have a big percentage. In the sense of the explanations, the objective of this chapter is to use genetic algorithm (GA) for diagnosing headache diseases. As a popular and essential technique benefiting from evolutionary mechanisms, GA can deal with many different types of real-world problems. So, it has been chosen as the solution way/algorithm over the headache disease detection problem, which shapes the research framework of the study. The chapter content gives information about the performed diagnosis application and the results.*

## INTRODUCTION

Our life is currently run over many technological developments directing the scientific literature in a fast way. This fast way of improvements and changes are of course associated with some past discoveries and inventions, which have important roles on changing our life. Today, individuals can experience very different lives according to the ones experienced by their past ancestors. It is clear that this situation is connected with innovative developments occurred as in fields of especially electrics, electronics and computers. When it is thought about current life standards, it can be easily expressed that the standards are highly directed by computer systems supported with strong environmental technologies. In this sense, especially software and hardware components have remarkable roles to make everything better for individuals while they are interacting with computer systems. At this point, there are currently also some important scientific fields and technologies having great roles on both changing the ways of problem solving within computer systems and introducing new research literatures to the scientific community. Artificial Intelligence is known widely as one of them.

DOI: 10.4018/978-1-5225-4769-3.ch012

Artificial Intelligence has been always a remarkable, strong scientific field to solve different kinds of real-world based problems. In detail, it is possible to indicate that it has reached to a multidisciplinary scope as a result of successful results achieved so far for different problems of different fields (Pannu, 2015). Today, this field is even separated into different research areas – topics such as Machine Learning, Swarm Intelligence and even Cybernetics (Alpaydin, 2014; Jordan & Mitchell, 2015; Kline, 2015; Magin, 2017; Merkle & Middendorf, 2013; Michalski et al., 2013; Sayre, 2014; Waldner, 2013; Yang et al., 2013). At this point, it is also associated with all fields in our life by having stronger connections with also some of them. Medical is one the most remarkable fields in which we can see many problem solution applications of Artificial Intelligence. Although there are different kinds of problem areas of medical as Artificial Intelligence is often employed, diagnosis oriented studies are among the most popular ones nowadays.

Medical diagnosis has always been a remarkable problem solution approach of the field of medical. As a result of changed technologies and problem-solving approaches, even it has improved and evaluated in time. It has been even connected with the latest technological factors appeared in time within the scientific literature. Currently, medical diagnosis has a strong relation with the Artificial Intelligence oriented approaches. Because it is practical to employ intelligent mechanisms over some input data-expert knowledge and design effective solution ways, even Biomedical Engineering field is interested in taking support from Artificial Intelligence. If applications in this manner are taken into consideration, we can see that medical diagnosis have a big percentage.

In the sense of the explanations done so far, objective of this study is to use Genetic Algorithm (GA) for diagnosing headache diseases. As a popular and essential technique benefiting from evolutionary mechanisms, GA can deal with many different types of real world problems. So, it has been chosen as the solution way - algorithm over the headache disease detection problem, which shapes the research framework of the study. Briefly, the formed approach has used a simple function to run from codes of GA particles having weights for different disease symptoms, as similar to an interesting way followed by Tezel and Kose (2011) before. On the other hand, the research done here will be of course a contribution to the associated literature including different examples of medical diagnosis (Some remarkable, recent ones are: Amato et al., 2013; Chattopadhyay et al., 2013; Chikh et al., 2012; Ghumbre et al., 2011; Kharya, 2012; Lavanya & Rani, 2011; Oniśko & Druzdzel, 2013; Prasadl et al., 2011; Qasem & Shamsuddin, 2011; Shankaracharya et al., 2010; Shiraishi et al., 2011; Tripathy et al., 2013; Wall et al., 2012). In detail, the paper content gives information about the performed diagnosis application and the results.

Based on the chapter subject and the research scope, the remaining content is organized as follows: The next section is focused on the Genetic Algorithms (GA) and some information on headache diagnosis. Following that section, the third section is devoted to some brief information about the developed headache diagnosis approach with the support by GA in this study. This section also provides some information about diagnosing performances (classification success) of different GA systems with different specific parameters to enable readers for understanding how the GA based system was good at diagnosing headache disease. Finally, the chapter is ended by focusing on conclusions and some future work ideas under the last section.

## GENETIC ALGORITHMS AND HEADACHE DIAGNOSIS

In the research work – study considered in this chapter, the technique of Genetic Algorithms (GA) was used to deal with a medical diagnosis problem over headache disease. In order to provide more information about working mechanism of GA and the objective headache diseases problem, this section has been devoted to these important two actors of the performed research.

### Genetic Algorithms

Genetic Algorithms (GA) is a popular Artificial Intelligence technique, which is used for many different kinds of real-world based problems. GA is briefly structured as inspiring from the Theory of Evolution (Pham & Karaboga, 2012). From this perspective, it is known as an Evolutionary Computation based technique but also included in the Swarm Intelligence research interest – topic of Artificial Intelligence (Goldberg & Koza, 1994; Yang et al., 2013). In detail, standard GA uses some mechanisms like crossover and mutation in order to improve effectiveness of the particles (individuals) searching for the solution values in the objective solution space (Gen & Cheng, 2000; Man et al., 2012; Sastry et al., 2014). From this perspective, this technique is accepted as an intelligent optimization technique in the related literature.

A typical GA employs the following algorithmic features and mechanisms to use some intelligent problem solution processes (Mitchell, 1998; Sastry et al., 2014):

- GA is initially set with a defined number of particles, which can be called as also individuals.
- Each individual briefly consists of genes forming a whole chromosome. In this sense, an individual is coded with some genes, which can be binary values, hexadecimal values, or even alphanumeric characters or some numbers (Figure 1 – Fisher, 2016).
- An individual code is used within an objective function or a set of functions to calculate some values or decision components associated with the objective problem.
- A typical problem solution flow of a GA is based on an iterative evaluation of each particle and performing crossover and mutation based operators on the successful ones for having newer generations thought to be leading to better solutions in each iteration flow.
- For determining which individuals – particles to be applied crossover, some alternative approaches are used. That's similar for also mutation processes.
- Regarding iterative problem solution flow of GA, Figure represents a simple problem solution steps of the technique (Liao & Sun, 2001).

In the literature, there are many different, improved GA variations (Some recent, remarkable ones: Deng et al., 2015; Hsiao, 2015; Jain & Deb, 2013; Li et al., 2014; Lin et al., 2014; Qu et al., 2013; Shang et al., 2013; Yan et al., 2016; Zuo et al., 2015). But in the research expressed in this chapter, the standard model of GA has been used for diagnosis purposes over headache disease.

*Figure 1. Coding individuals in Genetic Algorithm*
*Fisher, 2016.*

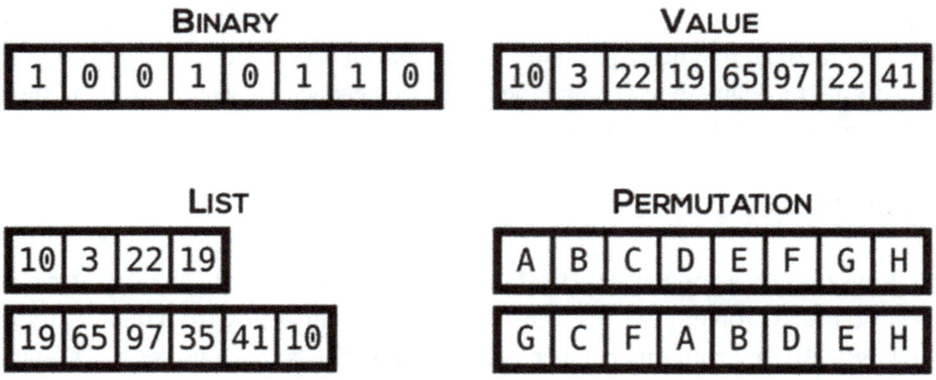


*Figure 2. Problem solution steps of the Genetic Algorithm*
*Liao & Sun, 2001.*

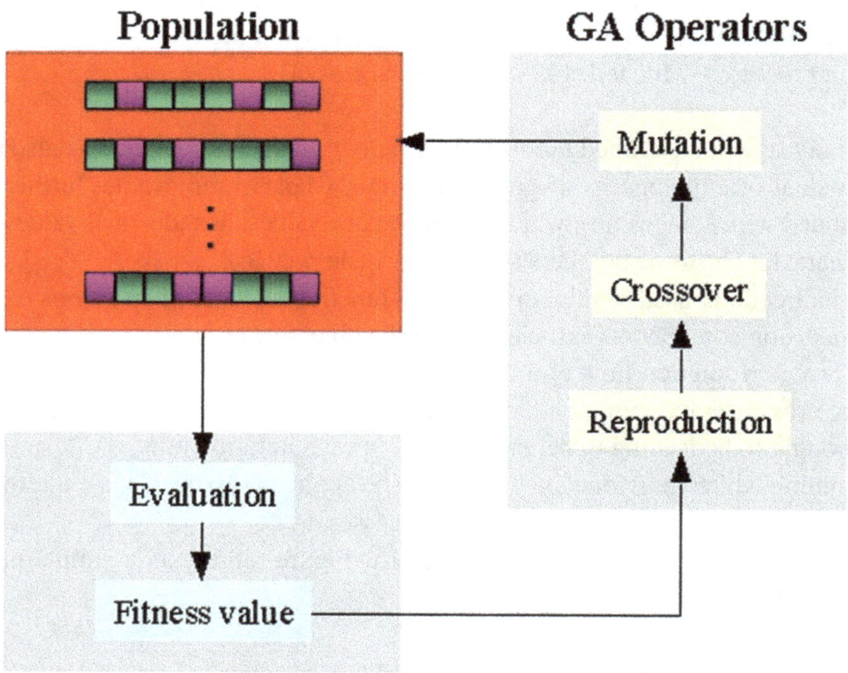


## Headache Disease

Headache diseases is a widely seen disease types as classified by the International Headache Society (HIS) into 13 classes (Olesen, 2000). In this context, it may be seen as a common disease but the literature indicates that 4 of these related classes accepted as primer type headache diseases while 9 ones are within secondary type headache diseases. Although primer ones are common types of headache types, headache diseases in the secondary type are caused by important medical problems such as celebral hemorrhage, neck problems, brain tumors and even eye diseases or some problems on the neck (Tezel & Kose, 2011).

When we focus again generally to the whole headache diseases, it can be seen that the migraine has some certain symptoms associated with this class directly while another type: set headache is an unusual, but a serious headache disease. At this point, because there are different serious headache diseases that should be detected and also majority of headaches are not diseases and just regarding some factors noise, stress, air pollution, it is important to run some diagnosis oriented studies for patients – individuals (Tezel & Kose, 2011).

Considering the whole headache classes, it is possible to derive some symptom types and classify the objective individuals as ill or healthy according to just having information about if these certain symptoms are observed or not. The most essential symptoms in this manner are presented in Table 1 (Tezel & Kose, 2011).

Considering the symptom types and the approach of classifying individuals as ill or healthy according to that, an intelligent diagnosis system over GA has been designed within the scope of the research explained in this chapter. The next section provides some brief information regarding system and some results on samples of symptoms.

## DIAGNOSING APPLICATIONS WITH THE DEVELOPED SYSTEM

The explained GA has been used in the research work – study expressed in this chapter, by using its default structure. In addition, some different parameter values were used at this point, to understand how the formed GA system can be effective with different values of parameters. The next paragraphs provide some information in this manner.

### Some Brief Information About the Diagnosis Problem

The diagnosis system considered here, has been formed over a different use of GA in the context of optimization. This diagnosis approach is similar to the one done by Tezel and Kose (2011) by using the

*Table 1. Symptoms for diagnosing headache disease*

| Symptom No. | Symptom Explanation |
|---|---|
| 1 | *Is there a vomitus?* |
| 2 | *Is there a nausea?* |
| 3 | *Is there a photophobia?* |
| 4 | *Is there an aura symptom?* |
| 5 | *Is the attack period is within 4 to 72 hours?* |
| 6 | *Is the headache most appropriate to be a throb?* |
| 7 | *Is the headache on different points?* |
| 8 | *Does the intensity of the headache improve with a move?* |

Tezel & Kose, 2011.

Clonal Selection Algorithm of Artificial Immunity Systems. Some essential information regarding the formed diagnosing approach can be expressed briefly as follows:

- Individuals – particles in the GA has been coded as representing weight values for the related symptoms.
- By using a sample dataset of headache diseases, appropriate codes of GA particles are determined at the end of a typical training process of GA.
- The most optimum particle has given a code set of weights for each symptom to determine if a newly provided symptom leads to an ill or healthy individual (two-sided classifying) over a function.
- The used function is used a simple summation of the weighted symptoms with also some adjustment parameters and in this context, an average value has been used to determine if the individual is ill or healthy.
- In order to obtain exact diagnosing GA, a sample set of symptoms including a total of 200 symptom value sets has been used. In detail, 150 symptom value sets from the general sample set are for ill individuals while the remaining 50 ones are for healthy individuals.
- After the training phase, the obtained GA structure was then tested over a total of 100 symptom value sets.

## Findings Over Applications

By using the diagnosis set-up indicated under previous paragraphs, some alternative experimental studies have been done to determine which parameter values were effective at performing the diagnosis approach over objective headache sample set. At this point, different crossover rate and mutation rate values for a static population number (which is 100 in this research) were used to see which parameter values are good at classifying the objective test set including 100 symptom value sets. The obtained findings over a total of 10 different applications are reported in Table 2.

*Table 2. Findings over headache diagnosis applications*

| Application No. | Crossover Rate | Mutation Rate | True Classified Symptom Value Sets | False Classified Symptom Value Sets |
|---|---|---|---|---|
| 1 | 0.8 | 0.01 | 49 | 51 |
| 2 | 0.5 | 0.05 | 52 | 48 |
| ***3*** | ***0.6*** | ***0.02*** | ***89*** | ***11*** |
| 4 | 0.7 | 0.08 | 56 | 44 |
| 5 | 0.4 | 0.10 | 75 | 25 |
| 6 | 0.6 | 0.11 | 36 | 64 |
| 7 | 0.8 | 0.05 | 79 | 21 |
| 8 | 0.5 | 0.25 | 74 | 26 |
| 9 | 0.9 | 0.02 | 83 | 17 |
| 10 | 0.8 | 0.03 | 86 | 14 |

The most successful diagnosis performance is in bold and italic.

As it can be seen from the Table 2, the formed GA structure and the followed diagnosing approach is effective enough to detect ill or healthy individuals according to provided symptom situations. On the other hand, it can be understood that different values of critical parameters regarding the technique of GA are effective at directing the diagnosis performance. Here, the third application has provided the most effective performance on the diagnosis approach.

## CONCLUSION

This chapter has provided a research work in which a Genetic Algorithms (GA) approach has been used for diagnosing headache diagnosis. Thanks to flexible, evolutionary based infrastructure of the GA, it has been possible to develop a diagnosis system for detecting if an individual has a serious headache disease or not. In detail, the chosen dataset of headache diseases was used to determine appropriate codes of GA particles and at the end of a typical training process of GA, the most optimum particle has given a code set of weights for each symptom to determine if a newly provided symptom leads to an ill or healthy individual over a function. The objective GA structure has been used for different parameters and the most appropriate parameters for determining the exact headache diagnosis GA system have been reported. The performed diagnosis process introduced here is an alternative work for the related applications on medical diagnosis introduced in the associated literature.

The performed research and the obtained successful results have encouraged the author for thinking about possible future research works on especially medical diagnosis. Some of remarkable future works can be mentioned as follows:

- The used simple function will be improved to include not only two-sided classifying approach but also more advanced classification for all types of headache diagnosis.
- Different types of functions will be modeled and used by the author to see if there is alternative way to improve such interesting way of medical diagnosis.
- In addition to the used GA technique in this research – chapter, there is of course alternative Artificial Intelligence techniques that can deal with classification. So, performances of such different techniques will be evaluated.
- In order to improve medical diagnosis of the GA based approach to include more different types of medical diseases, some experimental studies on forming hybrid systems will be performed in the future.

## REFERENCES

Alpaydin, E. (2014). *Introduction to machine learning*. MIT Press.

Amato, F., López, A., Peña-Méndez, E. M., Vaňhara, P., Hampl, A., & Havel, J. (2013). Artificial neural networks in medical diagnosis. *Journal of Applied Biomedicine*, *11*(2), 47–58. doi:10.2478/v10136-012-0031-x

Chattopadhyay, S., Banerjee, S., Rabhi, F. A., & Acharya, U. R. (2013). A Case-Based Reasoning system for complex medical diagnosis. *Expert Systems: International Journal of Knowledge Engineering and Neural Networks*, *30*(1), 12–20. doi:10.1111/j.1468-0394.2012.00618.x

Chikh, M. A., Saidi, M., & Settouti, N. (2012). Diagnosis of diabetes diseases using an artificial immune recognition system2 (AIRS2) with fuzzy k-nearest neighbor. *Journal of Medical Systems*, *36*(5), 2721–2729. doi:10.1007/s10916-011-9748-4 PMID:21695498

Deng, Y., Liu, Y., & Zhou, D. (2015). An improved genetic algorithm with initial population strategy for symmetric TSP. *Mathematical Problems in Engineering*.

Fisher, J. (2016). *Genetic Algorithm – Programming by the Seat of Your Genes! – SlideShare.Net*. Retrieved from https://www.slideshare.net/JeremyFisher1/genetic-algorithms-programming-by-the-seat-of-your-genes

Gen, M., & Cheng, R. (2000). *Genetic algorithms and engineering optimization* (Vol. 7). John Wiley & Sons.

Ghumbre, S., Patil, C., & Ghatol, A. (2011). Heart disease diagnosis using support vector machine. *International conference on computer science and information technology (ICCSIT') Pattaya.*

Goldberg, D. E., & Koza, J. R. (1994). Genetic Algorithms & Evolutionary Computation. *National Conference on Artificial Intelligence.*

Hsiao, F. H. (2015). Exponential Synchronization of Chaotic Cryptosystems Using an Improved Genetic Algorithm. *The Scientific World Journal*. PMID:26366432

Jain, H., & Deb, K. (2013). An improved adaptive approach for elitist nondominated sorting genetic algorithm for many-objective optimization. In *International Conference on Evolutionary Multi-Criterion Optimization* (pp. 307-321). Springer.

Jordan, M. I., & Mitchell, T. M. (2015). Machine learning: Trends, perspectives, and prospects. *Science*, *349*(6245), 255–260.

Kharya, S. (2012). *Using data mining techniques for diagnosis and prognosis of cancer disease.* arXiv preprint arXiv:1205.1923

Kline, R. R. (2015). *The cybernetics moment: Or why we call our age the information age*. JHU Press.

Lavanya, D., & Rani, K. U. (2011). Performance evaluation of decision tree classifiers on medical datasets. *International Journal of Computers and Applications*, *26*(4).

Li, D., Chen, S., & Huang, H. (2014). Improved genetic algorithm with two-level approximation for truss topology optimization. *Structural and Multidisciplinary Optimization*, *49*(5), 795–814. doi:10.1007/s00158-013-1012-8

Liao, Y. H., & Sun, C. T. (2001). An educational genetic algorithms learning tool. *IEEE Transactions on Education*, *44*(2), 20.

Lin, F., Duan, H., & Qu, X. (2014). PID control strategy for UAV flight control system based on improved genetic algorithm optimization. In *Control and Decision Conference (2014 CCDC), The 26th Chinese* (pp. 92-97). IEEE. doi:10.1109/CCDC.2014.6852124

Magin, R. L. (2017). Bioengineering and Cybernetics: A Modern Caduceus. *IEEE Pulse*, *8*(1), 44–47. doi:10.1109/MPUL.2016.2627461 PMID:28129142

Man, K. F., Tang, K. S., & Kwong, S. (2012). *Genetic algorithms: Concepts and designs*. Springer Science & Business Media.

Merkle, D., & Middendorf, M. (2014). Swarm intelligence. In Search methodologies (pp. 213-242). Springer US.

Michalski, R. S., Carbonell, J. G., & Mitchell, T. M. (Eds.). (2013). *Machine learning: An artificial intelligence approach*. Springer Science & Business Media.

Mitchell, M. (1998). *An introduction to genetic algorithms*. MIT Press.

Olesen, J. (2000). The IHS Members' Handbook. Academic Press.

Oniśko, A., & Druzdzel, M. J. (2013). Impact of precision of Bayesian network parameters on accuracy of medical diagnostic systems. *Artificial Intelligence in Medicine*, *57*(3), 197–206. doi:10.1016/j.artmed.2013.01.004 PMID:23466438

Pannu, A. (2015). Artificial intelligence and its application in different areas. *Artificial Intelligence*, *4*(10).

Pham, D., & Karaboga, D. (2012). *Intelligent optimisation techniques: genetic algorithms, tabu search, simulated annealing and neural networks*. Springer Science & Business Media.

Prasadl, B. D. C. N., Prasad, P. N. K., & Sagar, Y. (2011). An approach to develop expert systems in medical diagnosis using machine learning algorithms (asthma) and a performance study. *International Journal on Soft Computing*, *2*(1), 26–33. doi:10.5121/ijsc.2011.2103

Qasem, S. N., & Shamsuddin, S. M. (2011). Radial basis function network based on time variant multi-objective particle swarm optimization for medical diseases diagnosis. *Applied Soft Computing*, *11*(1), 1427–1438. doi:10.1016/j.asoc.2010.04.014

Qu, H., Xing, K., & Alexander, T. (2013). An improved genetic algorithm with co-evolutionary strategy for global path planning of multiple mobile robots. *Neurocomputing*, *120*, 509–517. doi:10.1016/j.neucom.2013.04.020

Sastry, K., Goldberg, D. E., & Kendall, G. (2014). Genetic algorithms. In Search methodologies (pp. 93-117). Springer US. doi:10.1007/978-1-4614-6940-7_4

Sayre, K. (2014). *Cybernetics and the Philosophy of Mind*. Routledge.

Shang, R., Bai, J., Jiao, L., & Jin, C. (2013). Community detection based on modularity and an improved genetic algorithm. *Physica A*, *392*(5), 1215–1231. doi:10.1016/j.physa.2012.11.003

Shankaracharya, D. O., Samanta, S., & Vidyarthi, A. S. (2010). Computational intelligence in early diabetes diagnosis: A review. *The Review of Diabetic Studies; RDS*, *7*(4), 252–262. doi:10.1900/RDS.2010.7.252 PMID:21713313

Shiraishi, J., Li, Q., Appelbaum, D., & Doi, K. (2011). Computer-aided diagnosis and artificial intelligence in clinical imaging. *Seminars in Nuclear Medicine*, *41*(6), 449–462. doi:10.1053/j.semnuclmed.2011.06.004 PMID:21978447

Tezel, G., & Kose, U. (2011). Headache Disease Diagnosis by Using the Clonal Selection Algorithm. In *6th International Advanced Technologies Symposium (IATS'11)* (pp. 144-148). Academic Press.

Tripathy, B. K., Acharjya, D. P., & Cynthya, V. (2013). *A framework for intelligent medical diagnosis using rough set with formal concept analysis.* arXiv preprint arXiv:1301.6011

Waldner, J. B. (2013). *Nanocomputers and swarm intelligence*. John Wiley & Sons.

Wall, D. P., Dally, R., Luyster, R., Jung, J. Y., & DeLuca, T. F. (2012). Use of artificial intelligence to shorten the behavioral diagnosis of autism. *PLoS One*, *7*(8), e43855. doi:10.1371/journal.pone.0043855 PMID:22952789

Yan, X., Gong, R., & Zhang, Q. (2016). Application of optimization SVM based on improved genetic algorithm in short-term wind speed prediction. *Power Syst. Prot. Control*, *44*(9), 38–42.

Yang, X. S., Cui, Z., Xiao, R., Gandomi, A. H., & Karamanoglu, M. (Eds.). (2013). *Swarm intelligence and bio-inspired computation: theory and applications*. Newnes. doi:10.1016/B978-0-12-405163-8.00001-6

Zuo, X., Chen, C., Tan, W., & Zhou, M. (2015). Vehicle scheduling of an urban bus line via an improved multiobjective genetic algorithm. *IEEE Transactions on Intelligent Transportation Systems*, *16*(2), 1030–1041.

## ADDITIONAL READING

Acuña, G., & Möller, H. (2016). Indirect training of Gray-Box Models using LS-SVM and genetic algorithms. In *Computational Intelligence (LA-CCI), 2016 IEEE Latin American Conference on* (pp. 1-5). IEEE. doi:10.1109/LA-CCI.2016.7885719

Ahmad, R., Akhtar, N., & Choubey, N. S. (2017). Applications of Artificial Bee Colony Algorithms and its Variants in Health Care. *BioChemistry: An Indian Journal, 11*(1).

Al-Shayea, Q. K. (2011). Artificial neural networks in medical diagnosis. *International Journal of Computer Science Issues*, *8*(2), 150–154.

Antonio, L. M., & Coello, C. A. C. (2017). Coevolutionary Multi-objective Evolutionary Algorithms: A Survey of the State-of-the-Art. *IEEE Transactions on Evolutionary Computation*, 1. doi:10.1109/TEVC.2017.2767023

Arsene, O., Dumitrache, I., & Mihu, I. (2015). Expert system for medicine diagnosis using software agents. *Expert Systems with Applications*, *42*(4), 1825–1834. doi:10.1016/j.eswa.2014.10.026

Back, T. (1996). *Evolutionary algorithms in theory and practice: evolution strategies, evolutionary programming, genetic algorithms*. Oxford University Press.

Baxt, W. G. (1995). Application of artificial neural networks to clinical medicine. *Lancet, 346*(8983), 1135–1138. doi:10.1016/S0140-6736(95)91804-3 PMID:7475607

Becker, K., & Gottschlich, J. (2017). AI Programmer: Autonomously Creating Software Programs Using Genetic Algorithms. *arXiv preprint arXiv:1709.05703*.

Bodnar, T., Barclay, V. C., Ram, N., Tucker, C. S., & Salathé, M. (2014). On the ground validation of online diagnosis with Twitter and medical records. In *Proceedings of the 23rd International Conference on World Wide Web* (pp. 651-656). ACM. doi:10.1145/2567948.2579272

Castaneda, C., Nalley, K., Mannion, C., Bhattacharyya, P., Blake, P., Pecora, A., & Suh, K. S. et al. (2015). Clinical decision support systems for improving diagnostic accuracy and achieving precision medicine. *Journal of Clinical Bioinformatics, 5*(1), 4. doi:10.1186/s13336-015-0019-3 PMID:25834725

Charniak, E., Riesbeck, C. K., McDermott, D. V., & Meehan, J. R. (2014). *Artificial intelligence programming*. Psychology Press.

Črepinšek, M., Liu, S. H., & Mernik, M. (2013). Exploration and exploitation in evolutionary algorithms: A survey. [CSUR]. *ACM Computing Surveys, 45*(3), 35. doi:10.1145/2480741.2480752

Dasgupta, D., & Michalewicz, Z. (Eds.). (2013). *Evolutionary algorithms in engineering applications*. Springer Science & Business Media.

Dilsizian, S. E., & Siegel, E. L. (2014). Artificial intelligence in medicine and cardiac imaging: Harnessing big data and advanced computing to provide personalized medical diagnosis and treatment. *Current Cardiology Reports, 16*(1), 441. doi:10.1007/s11886-013-0441-8 PMID:24338557

Dutta, S. (2014). *Knowledge processing and applied artificial intelligence*. Elsevier.

El-Dahshan, E. S. A., Mohsen, H. M., Revett, K., & Salem, A. B. M. (2014). Computer-aided diagnosis of human brain tumor through MRI: A survey and a new algorithm. *Expert Systems with Applications, 41*(11), 5526–5545. doi:10.1016/j.eswa.2014.01.021

Eshelman, L. J., & Schaffer, J. D. (1993). Real-coded genetic algorithms and interval-schemata. In *Foundations of genetic algorithms* (Vol. 2, pp. 187–202). Elsevier.

Fieschi, M. (2013). *Artificial intelligence in medicine: Expert systems*. Springer.

Foster, K. R., Koprowski, R., & Skufca, J. D. (2014). Machine learning, medical diagnosis, and biomedical engineering research-commentary. *Biomedical Engineering Online, 13*(1), 94. doi:10.1186/1475-925X-13-94 PMID:24998888

Furmankiewicz, M., Sołtysik-Piorunkiewicz, A., & Ziuziański, P. (2014). Artificial intelligence systems for knowledge management in e-health: The study of intelligent software agents. *Latest Trends on Systems, 2*, 551–556.

Grefenstette, J. J. (1986). Optimization of control parameters for genetic algorithms. *IEEE Transactions on Systems, Man, and Cybernetics, 16*(1), 122–128. doi:10.1109/TSMC.1986.289288

Hager, G. D., Bryant, R., Horvitz, E., Mataric, M., & Honavar, V. (2017). Advances in Artificial Intelligence Require Progress Across all of Computer Science. *arXiv preprint arXiv:1707.04352*.

Hamet, P., & Tremblay, J. (2017). Artificial intelligence in medicine. *Metabolism: Clinical and Experimental*, *69*, S36–S40. doi:10.1016/j.metabol.2017.01.011 PMID:28126242

Jennings, N. R. (1996). Coordination techniques for distributed artificial intelligence. *Foundations of distributed artificial intelligence*, 187-210.

Khan, I. Y., Zope, P. H., & Suralkar, S. R. (2013). Importance of Artificial Neural Network in Medical Diagnosis disease like acute nephritis disease and heart disease. [IJESIT]. *International Journal of Engineering Science and Innovative Technology*, *2*(2), 210–217.

Kulikowski, C. A. (1980). Artificial intelligence methods and systems for medical consultation. *IEEE Transactions on Pattern Analysis and Machine Intelligence*, *PAMI-2*(5), 464–476. doi:10.1109/TPAMI.1980.6592368

Londhe, V. (2017). Brain MR Image Segmentation for Tumor Detection using Artificial Neural. *Brain*, *6*(1).

Luxton, D. D. (2014). Artificial intelligence in psychological practice: Current and future applications and implications. *Professional Psychology, Research and Practice*, *45*(5), 332–339. doi:10.1037/a0034559

Nahar, J., Imam, T., Tickle, K. S., & Chen, Y. P. P. (2013). Computational intelligence for heart disease diagnosis: A medical knowledge driven approach. *Expert Systems with Applications*, *40*(1), 96–104. doi:10.1016/j.eswa.2012.07.032

Norman, D. (2017). Design, Business Models, and Human-Technology Teamwork: As automation and artificial intelligence technologies develop, we need to think less about human-machine interfaces and more about human-machine teamwork. *Research Technology Management*, *60*(1), 26–30. doi:10.1080/08956308.2017.1255051

Patel, V. L., Shortliffe, E. H., Stefanelli, M., Szolovits, P., Berthold, M. R., Bellazzi, R., & Abu-Hanna, A. (2009). The coming of age of artificial intelligence in medicine. *Artificial Intelligence in Medicine*, *46*(1), 5–17. doi:10.1016/j.artmed.2008.07.017 PMID:18790621

Patil, R. S. (2014). Techniques for diagnostic reasoning in medicine. In *Exploring Artificial Intelligence: Survey Talks from the National Conferences on Artificial Intelligence* (p. 347). Morgan Kaufmann.

Peek, N., Combi, C., Marin, R., & Bellazzi, R. (2015). Thirty years of artificial intelligence in medicine (AIME) conferences: A review of research themes. *Artificial Intelligence in Medicine*, *65*(1), 61–73. doi:10.1016/j.artmed.2015.07.003 PMID:26265491

Poole, D. L., & Mackworth, A. K. (2017). *Artificial intelligence*. Cambridge University Press.

Raedt, L. D., Kersting, K., Natarajan, S., & Poole, D. (2016). Statistical relational artificial intelligence: Logic, probability, and computation. *Synthesis Lectures on Artificial Intelligence and Machine Learning*, *10*(2), 1–189. doi:10.2200/S00692ED1V01Y201601AIM032

Rajan, J. R. (2017). Prognostic system for early diagnosis of pediatric lung disease using artificial intelligence. *Current Pediatric Research, 21*(1).

Reiter, R. (1987). A theory of diagnosis from first principles. *Artificial Intelligence, 32*(1), 57–95. doi:10.1016/0004-3702(87)90062-2

Salman, M., Ahmed, A. W., Khan, O. A., Raza, B., & Latif, K. (2017). Artificial Intelligence in Bio-Medical Domain. *Artificial Intelligence, 8*(8).

Sanz, J. A., Galar, M., Jurio, A., Brugos, A., Pagola, M., & Bustince, H. (2014). Medical diagnosis of cardiovascular diseases using an interval-valued fuzzy rule-based classification system. *Applied Soft Computing, 20*, 103–111. doi:10.1016/j.asoc.2013.11.009

Seera, M., & Lim, C. P. (2014). A hybrid intelligent system for medical data classification. *Expert Systems with Applications, 41*(5), 2239–2249. doi:10.1016/j.eswa.2013.09.022

Sharma, G., & Carter, A. (2017). Artificial Intelligence and the Pathologist: Future Frenemies? *Archives of Pathology & Laboratory Medicine, 141*(5), 622–623. doi:10.5858/arpa.2016-0593-ED PMID:28447905

Shashi, S., & Deep, K. (2016). A Novel Crossover Operator Designed to Exploit Synergies of Two Crossover Operators for Real-Coded Genetic Algorithms.

Sheikhtaheri, A., Sadoughi, F., & Dehaghi, Z. H. (2014). Developing and using expert systems and neural networks in medicine: A review on benefits and challenges. *Journal of Medical Systems, 38*(9), 110. doi:10.1007/s10916-014-0110-5 PMID:25027017

Shrobe, H. E. (Ed.). (2014). *Exploring artificial intelligence: survey talks from the National conferences on artificial intelligence*. Morgan Kaufmann.

Spears, W. M. (2013). *Evolutionary algorithms: The role of mutation and recombination*. Springer Science & Business Media.

Srinivas, M., & Patnaik, L. M. (1994). Adaptive probabilities of crossover and mutation in genetic algorithms. *IEEE Transactions on Systems, Man, and Cybernetics, 24*(4), 656–667. doi:10.1109/21.286385

Szolovits, P. (Ed.). (1982). *Artificial intelligence in medicine* (pp. 1–226). Boulder, CO: Westview Press.

Travé-Massuyès, L. (2014). Bridging control and artificial intelligence theories for diagnosis: A survey. *Engineering Applications of Artificial Intelligence, 27*, 1–16. doi:10.1016/j.engappai.2013.09.018

Wu, T., Ozpineci, B., Chinthavali, M., Wang, Z., Debnath, S., & Campbell, S. (2017, June). Design and optimization of 3D printed air-cooled heat sinks based on genetic algorithms. In *Transportation Electrification Conference and Expo (ITEC)*, 2017 *IEEE* (pp. 650-655). IEEE. doi:10.1109/ITEC.2017.7993346

Ziuziański, P., Furmankiewicz, M., & Sołtysik-Piorunkiewicz, A. (2014). E-health artificial intelligence system implementation: Case study of knowledge management dashboard of epidemiological data in Poland. *International Journal of Biology and Biomedical Engineering, 8*, 164–171.

## KEY TERMS AND DEFINITIONS

**Artificial Intelligence:** A scientific field that is dealing with simulating human and living organism behaviors and actions to develop intelligent machines.

**Genetic Algorithms:** A type of intelligent algorithm inspired by some mechanisms explained under the theory of evolution for solving real-world-based problems over an optimization framework.

**Intelligent Medical Diagnosis:** A type of medical diagnosis done generally thanks to computer systems supported with especially artificial intelligence.

**Medical Diagnosis:** A type of diagnosis used for detecting a disease, medical factor, etc. by using some required input data/information.

**Nature-Inspired Techniques:** A set of techniques that are generally artificial intelligence algorithms inspired by some aspects of the nature in order to deal with real-world-based problems.

# Chapter 13
# Internet of Things and Nature-Inspired Intelligent Techniques for the Future of Biomedical Engineering

**Gur Emre Guraksin**
*Afyon Kocatepe University, Turkey*

## ABSTRACT

*Along with the rise of artificial intelligence (AI), there are many different research fields gaining importance. Because of the growing amount of data and needs for immediate access to information for dealing with the problems, different types of research fields take place within the scientific community. Internet of things (IoT) is one of them, and it enables devices to communicate with each other in order to form a general network of physical, working devices. The objective of this chapter in this manner is to provide a general discussion of using nature-inspired techniques of AI to form the future of biomedical engineering over IoT. Because it is often thought that the medical services of the future will be based on autonomous machines supported with AI and IoT, discussing such a topic by considering biomedical engineering applications will be good for the related literature.*

## INTRODUCTION

Technological developments have always wide-open doors to the dream of the better future. In this context, every technology has a great effort for improving day by day and having a great influence in achieving better living standards for the humankind. In detail, especially computer, electronics and communication oriented technologies had had important roles in meeting with this output at the end. Considering that many different innovative technologies have also taken place in humankind's modern life. Here, more consideration should be given to the Artificial Intelligence (AI) because of its wide popularity and also scope leading to an unstoppable multidisciplinary approach.

DOI: 10.4018/978-1-5225-4769-3.ch013

Since its first introduction to the scientific literature, AI has improved more and more in time and reached to a top place with its detailed sub-research areas that can be applicable to many real world- based problems. In the literature, it is possible to see that AI can be used within almost all fields (Cortés et al., 2000; Nilsson, 2014; Poole & Mackworth, 2010; Remagnino & Foresti, 2005; Russell & Norvig, 2002). Although the AI can be used within almost all fields greatly, some of them are essential for achieving better life for humans and creating a better future also. Here, the field of Biomedical Engineering is one of them as including wide relations with different fields like health, computer engineering, machine engineering, electronics and many more (Bronzino, 1999; Enderle & Bronzino, 2012; Kline, 2012). Thinking about the latest developments in the AI and its early steps, it is possible to see a clear relation among AI based approaches, methods, techniques and also applications that can be examined within Biomedical Engineering.

With the past time over applications of AI, one important thing is that the more improvements in some supportive technologies have enabled AI to rise more in time. Today, AI is more discussed by people because of its more applications within different fields, by providing effective and accurate solutions. When it is thought about supportive technologies, it is possible to focus on i.e. computer or electronics technologies, as important actors in rise of AI. Especially alternative hardware oriented solutions have made it possible to think more about robots or machines, which are not just software systems but also real things that are able to provide actions – feedback according to received triggers from their environment. In detail, there are many different efforts to realize such robots or machines supported with AI (Bakhshipour et al., 2017; Craig, 2005; Geraci, 2012; Liang et al., 2016; Nolfi et al., 2016; Ziemke, 2001). But also, such research efforts have enabled some alternative sub-fields to rise and take active place in the scientific literature. Internet of Things (IoT) is one of these sub-fields and it currently has strong relations with the AI as providing alternative solutions, which are more meaningful when we think about a future with full of autonomous, intelligent systems – machines.

AI is a field with full of detailed approaches, methods, and techniques to provide effective solutions for real world based problems. Considering an intersection of the field of Biomedical Engineering and also IoT, a wide, alternative research interest of future is rapidly designed in the view of scientific community. In addition, it is also possible to think about some specific sides of AI to get an alternative theoretical view over that research interest. Taking a rapid look at to the AI, nature-inspired techniques can be seen as a specific topic to be considered for intelligent, IoT based solutions for Biomedical Engineering and in this way, an alternative discussion for the future can be drawn.

In the context of the explanations above, objective of this chapter is to provide a general discussion over using nature-inspired techniques of AI to form the future of Biomedical Engineering build over IoT. Because it is often thought that the medical services of the future will be based on autonomous machines supported with AI and IoT, discussing about such topic by considering Biomedical Engineering applications will be good for the related literature. In detail, the chapter tries to increase awareness in essential elements of the intersection including AI and IoT and leads the interested readers to have enough idea about how nature-inspired techniques of AI can be employed to achieve IoT oriented, intelligent Biomedical Engineering solutions.

As being associated with the subject of the chapter, the remaining content is organized as follows: The next section is about some brief definitions and explanations regarding main topics under the subject of the chapter. Following to it, the third section discusses some about possible application ways for intelligent IoT based solutions for Biomedical Engineering studies. Next, the fourth section focuses on future state in this manner and finally, the last section ends the chapter with final explanations on the considered subject scope.

## INTERNET OF THINGS NATURE-INSPIRED TECHNIQUES AND THEIR CONNECTIONS WITH BIOMEDICAL ENGINEERING

Research considered in this chapter is in the intersection of IoT and nature-inspired techniques and finds it main meaning in Biomedical Engineering applications. So, before discussing more about applications in this manner, some essential information for them can be given as follows:

### Internet of Things

IoT is today's one of the most remarkable, trendy research interests. As a result of the rapid developments in computer and hardware technologies, IoT has started to rise in the scientific community. It is clear that this rise has been triggered greatly by software oriented innovative technologies including AI effects. IoT is briefly a networked connection of everyday objects and this network is something supported by AI based infrastructure to achieve better form of data – information exchange and appropriate real world based solutions (Gubbi et al., 2013; Kopetz, 2011; Tan & Wang, 2010; Xia et al., 2012; Yang, 2014a). From a general perspective it can be said that IoT based systems are something making people's daily life more practical by ensuring necessary information exchange environment for machines. From this aspect, it can be figured out that such IoT systems require advanced hardware systems including appropriate form of intelligent software solutions for desired communication infrastructure. By considering current state of IoT based systems, we can think about four main elements of IoT as software, hardware, communication, and AI.

Rise of IoT has enabled researchers to think about new concepts like smart objects. Smart objects here are some kind of building elements for a typical IoT system or the whole system of a more general IoT network structure (Atzori et al., 2014; Kortuem et al., 2014). In time the concept of smart objects has become a wider concept affecting people's lives and as a result of this, some concepts like smart homes, smart cities have taken place in the associated literature (Darianian & Michael, 2008; Li et al., 2011; Zanella et al., 2014). As it can be understood, the concepts of smart environments are for resulting effects of IoT in people's living environments. Today IoT is a more comprehensive concept changing the environment of even industries (Da Xu et al., 2014).

### Nature-Inspired Techniques

Inspiration from the nature has resulted in the field of AI to employ many different techniques that can be applied in especially real world based problems. In time, it has been seen that many effective and efficient results can be achieved thanks to employment of nature-inspired techniques of AI. Starting from inspirations from human brain, nature-inspired techniques have included different variations of problem solutions and that state has caused AI to be more multidisciplinary spread within the scientific literature.

The associated literature employs many different types of nature-inspired techniques but only some of them are known popular and widely used in the encountered problems. They can be explained briefly as follows:

*Figure 1. Internet of Things and its main elements*

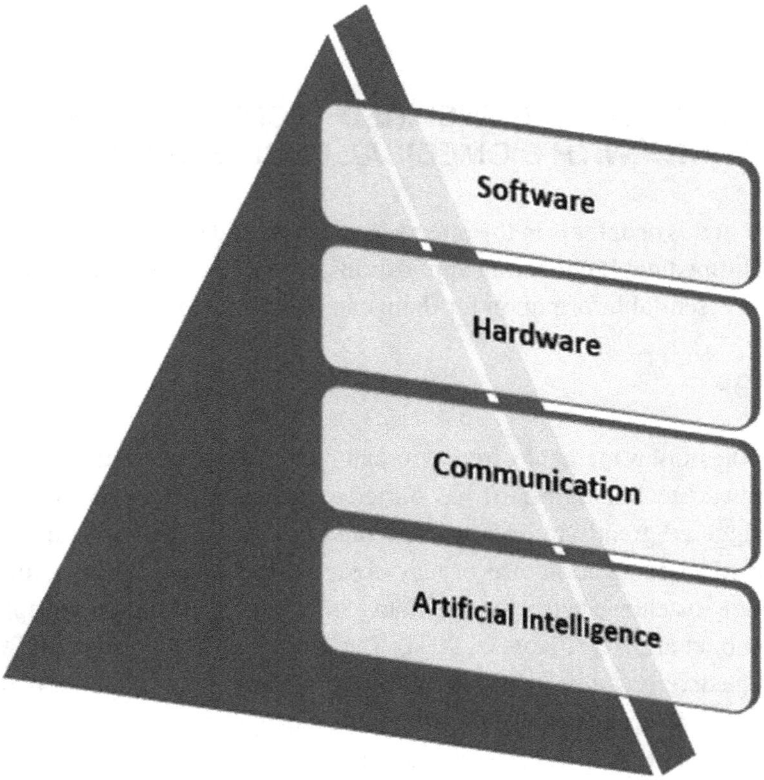


- **Artificial Neural Networks (ANN):** ANN is one of the most important AI techniques as a typical simulation over working mechanism of the human brain. Briefly, a typical ANN is a model of different information processing layers including artificial neurons, which are able to process data by summing their inputs and getting new outputs by feeding some special activation functions with the related inputs (Drew & Monson, 2000; Kubat, 2017; Suzuki, 1997; Zhu, 2017). ANN models can be adjusted according to number of input and output by considering the objective problem. Also, it is important to build optimum number of hidden layers (between input and output layers), number of associated artificial neurons in these hidden layers and also some parameters like learning rate, momentum…etc (Karayiannis & Venetsanopoulos, 2013; Kubat, 2017; Zurada, 1992). As also a Machine Learning technique, ANN models require a training process to be ready for applications. In the literature, there are different types of learning algorithms and even optimization algorithms can be used for this purpose (Jain et al., 1996; Karayiannis & Venetsanopoulos, 2013; White, 2008; Zhu, 2017). Finally, ANN technique has been improved in time and currently there are different types of ANN models introduced to the associated literature (Basheer & Hajmeer, 2000; Krenker et al., 2011; White, 2008).

- **Genetic Algorithms (GA):** GA is another popular nature-inspired technique used for especially problems that can be associated with a fitness function. GA is based on the fundamentals of the Evolution Theory and it employs some problem solution mechanisms like cross-over, mutation, selection to improve solution particles used iteratively along the solution process (Back, 1996; Bakirtzis & Kazarlis, 2016; Gen & Cheng, 2000). Like ANN, GA also has many variations in the associated literature.
- **Fuzzy Logic (FL):** As inspired from human thinking style, FL can be accepted as another nature-inspired technique. FL is briefly based on Zadeh's Fuzzy Set Theory and it is used to eliminate crisp logical values in the Crisp Logic, which is not very well fitting to human thinking process (Kecman, 2001; Ross, 2009; Von Altrock, 1995; Yen & Langari, 1998). In real world cases, FL finds its best application way within intelligent control systems.
- **Swarm Intelligence (SI) Techniques:** As including especially AI based optimization techniques, SI is a remarkable research interest under AI and it employs most accurate examples of nature-inspired techniques for AI, over real-world problems. Briefly, typical SI techniques include problem solution particles, which can act collaboratively to find optimum solutions of objective problems (Kose, 2015; Ma et al., 2017; Vardhini & Sitamahalakshmi, 2016; Yang et al., 2013). With this feature, the nature has always been a main inspiration source for the techniques under SI. At this point, even GA is one of members of SI and especially role or Evolution based mechanisms are very often used solution elements in such techniques of AI. Currently, some other of popular SI techniques are Cuckoo Search, Krill Herd Algorithm, Artificial Bee Colony, Particle Swarm Optimization, Ant Colony Optimization (Brownlee, 2011; Gandomi & Alavi, 2012; Panigrahi et al., 2011; Yang, 2014b; Yang & Deb, 2014).
- **Artificial Immunity:** Inspired from immune systems of living organisms, Artificial Immunity has become another popular research interest under the nature-inspired techniques of AI. As including Evolution and naturel immunity oriented logical – mathematical solution mechanisms, Artificial Immunity employs some widely used nature-inspired techniques like Negative Selection or Clonal Selection algorithms (Brownlee, 2007; Dasgupta et al., 2003; Gao et al., 2008; Jinquan et al., 2009; Tan, 2016; Ulutas & Kulturel-Konak, 2011). Leading the optimization purpose, these techniques are often used for essential problem solution ways within the field of AI.

*Figure 2. A brief view of some widely used nature-inspired techniques of Artificial Intelligence*

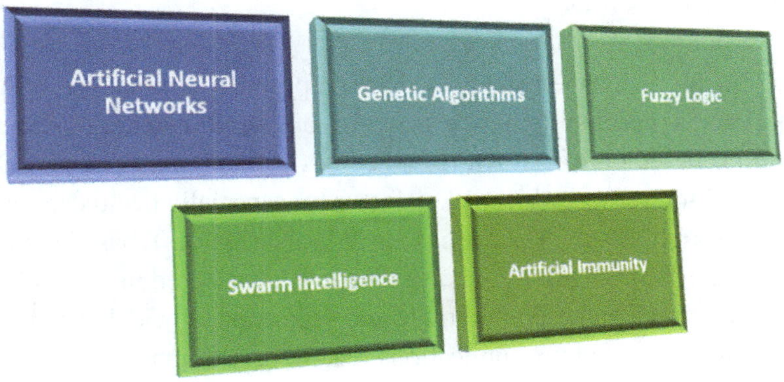

### Internet of Things and Nature-Inspired Techniques for Biomedical Engineering

AI is generally based on the inspirations from nature and because of that, nature-inspired techniques are widely coming in mind when it is considered about developing intelligent IoT systems. Development of intelligent IoT systems can be associated with even specific improvements and more advanced, innovative applications associated with the main IoT system itself. But it is important that combining both IoT and nature-inspired techniques has a great potential in achieving more advanced Biomedical Engineering solutions. Such solutions may include medical diagnosis systems, healthcare systems, specific prediction and / or optimization approaches specific for certain Biomedical Engineering oriented results…etc.

The mentioned use of nature-inspired techniques within IoT for Biomedical Engineering purposes can be realized over the following application approaches:

- Design optimization for IoT based systems,
- Optimization of specific features – functions within IoT based systems,
- Intelligent analysis – evaluation of outputs by IoT based systems,
- More accurate decision making by IoT based systems,
- Intelligent working mechanisms for IoT based systems,
- Innovative, new ideas for Biomedical Engineering applications over IoT,
- More practical and daily application opportunities of intelligent IoT systems for medical – healthcare approaches.

Considering the application approaches expressed above, there are also some current applications of nature-inspired IoT systems for Biomedical Engineering purposes. Some recent ones of them are explained briefly under the next section.

## INTELLIGENT INTERNET OF THINGS BASED APPLICATIONS OF BIOMEDICAL ENGINEERING

If the current literature is examined, it can be seen that there are different research works oriented in the related intersection of intelligent IoT solutions and Biomedical Engineering. At this point, it is possible to inform readers about what have been done and what are possible potentials in this manner, by explaining the performed works briefly:

- In their study, Chiuchisan and Geman (2014) have employed a combination of AI and Data Mining to develop a decision support system for especially patients with neurological disorders. In detail, the whole system is based on IoT structure and comes with not only Decision Support but also an approach of Home Monitoring, which is key factor for especially healthcare purposes.
- Domingo (2012) has provided a review based study in which IoT based applications for people with disabilities are examined briefly. The applications considered in the study include also some intelligent ones, which are generally wearable and supported by widely used AI techniques.
- Qi et al. (2015) have examined the applications regarding daily physical activity, as over IoT systems. In detail, they have provided a survey based research work in which they have also examined the applications with Machine Learning techniques under the support of IoT. As the concept the

main focus of their research is PA (physical activity) monitoring and assessment (PAMA) technologies under the associated literatures of AI and IoT.

- Another example of IoT based monitoring system has been provided by Oniga and Suto (2014). The system introduced here is based on ANN technique and in detail, it is whole an intelligent platform including activity and health monitoring infrastructure, a supportive and telepresence robot over even cloud approaches (Oniga & Suto, 2014).
- The study by Thomas (2013) focuses on smart systems – environments that can be used for at-home care purposes. The discussion provided within this research is focused on especially intelligent use of medical home care systems – machines, which can be examined within monitoring processes indicated under also similar works explained under previous paragraphs.
- Hu et al. (2013) have provided a research in which IoT oriented healthcare and medical approaches were considered. In the research work, they have discussed the applications regarding medication control and medical equipment, management of medical information, mobile medical care, personal health management and many more alternative ones.
- As one of recent works, Dimitrov (2016) has considered two remarkable topics of IoT and Big Data in the context of medical and healthcare applications. The work focuses on wearable and mobile technologies on the background of research and discussion.
- It can be understood from the research works expressed briefly, it is possible to indicate that the related application focus considered in this chapter is one of research interests having great potential for future. So, the associated literature has more applications to be done. On the other hand, when the associated literature is examined, some additional different alternative works in addition to the explained ones above can be suggested for the readers. Some remarkable ones in this manner are: Bui & Zorzi, 2011; Chiuchisan et al., 2014; Islam et al., 2015; Jara et al., 2011; Luo et al., 2009; Perera et al., 2014; Rohokale et al., 2011; Zhao et al., 2011.

## A VIEW ON THE FUTURE OF INTELLIGENT IOT BASED BIOMEDICAL ENGINEERING

By thinking about the relations among essential topics under the subject of this chapter and focusing on current developments in this manner, it is possible to also express some ideas about the future of Biomedical Engineering supported by intelligent IoT in which especially nature-inspired techniques are employed. In this sense, a view on the future can be given under the following sub-titles, briefly:

### More Applications of SI Based IoT

It seems that the more need for developing intelligent materials in the context of intelligent IoT comes with more need for optimization oriented solutions. So, intelligent optimization techniques will probably be always important in the future. As associated with that, the future of Biomedical Engineering with IoT solutions will include more applications of SI. It is clear that the future will bring more alternatives of SI techniques, but it is also clear that they will be always essential elements of successful and effective intelligent IoT solutions within future Biomedical Engineering. Some solution examples in this manner can be listed as follows:

- Better design optimization for the IoT based intelligent healthcare materials, machines of the future.
- Applying optimization for calibrating some functions of the IoT based intelligent healthcare materials, machines of the future.
- Applying optimization for detecting failures or remarkable parameters affecting the performance of the IoT based intelligent healthcare materials, machines of the future and focusing more on possible solution ways in this manner.
- Improving some features or functions of already developed IoT based intelligent healthcare materials, machines taking place in the future.
- More accurate control of the developed IoT based intelligent healthcare materials, machines, thanks to optimization techniques.
- New and innovative IoT based intelligent Biomedical Engineering applications, thanks to applying optimization techniques.
- Possible, separate sub-fields of Biomedical Engineering focused on optimization for better IoT based intelligent solutions.
- More advanced software systems dealing with Biomedical Engineering and intelligent IoT in this manner.

## Alternative ANN Variations for Better IoT Based Biomedical Engineering

As it is already a known fact that there are different variations of ANN, it is possible to imagine that there will be more variations for getting better IoT based Biomedical Engineering in the future. In the literature, there is already a known approach called as Deep Learning (Deng & Yu, 2014; LeCun et al., 2015; Ngiam et al., 2011; Schmidhuber, 2015), which is known with its use over more advanced ANNs (Convotional ANNs) to deal with i.e. Big Data (Chen et al., 2012; Lohr, 2012; Manyika et al., 2011; Marz & Warren, 2015) or more complicated amount of data to solve more advanced problems needing more effective and faster solution processes. It is possible to think that Deep Learning will be a key factor for the intelligent IoT based Biomedical Engineering applications in the future. But it is also possible to indicate that there will be different variations of ANN to deal with more advanced problems including the effects of hardware oriented systems with the features and functions of IoT approach. These ideas can be detailed as follows, by considering essentials of ANNs:

- Future ANN variations may include different models of typical ANNs to deal with IoT oriented system needs having relation with also specific needs of Biomedical Engineering based applications.
- Future ANN variations may use additional parameters as different from typical ANNs to be more effective and faster on working with IoT systems within Biomedical Engineering scope.
- ANN variations in the future may be designed specific for Biomedical Engineering applications including IoT features and functions.
- Future ANN variations may be some type of hybrid ANN models including interaction of big amount of ANNs, in order to deal with IoT based intelligent solutions within healthcare systems.

## New Nature-Inspired Techniques Specific for IoT and Biomedical Engineering

The future of AI and also nature-inspired techniques will be always open to new developments. In this sense, the future of Biomedical Engineering associated with IoT may possibly require design and development of new nature-inspired techniques. That will be because of unstoppable research in roots of the nature and its effects to design new AI based techniques and also possible requirements by IoT and / or Biomedical Engineering to use more specific techniques including more accurate mathematical and logical elements to reach to the desired solution outputs. That view on the future of course can be associated with all aspects of AI techniques in a multidisciplinary manner but here, more consideration has given to the future of intelligent, IoT based Biomedical Engineering applications.

### Full Autonomous Healthcare Systems

With more influence of intelligent IoT in Biomedical Engineering, it is possible to imagine full autonomous healthcare systems needing no doctors, experts or any other staff from a healthcare unit. Current state seems having some early examples of such applications, but the view considered here is about more advanced healthcare systems behaving like they are almost humans and even they are more effective and accurate on providing healthcare services for us. In detail, such systems can be thought as more than a typical software structure by considering the rise robotic applications and also effects of IoT. The idea of such systems is just a triggering factor for building up full autonomous machines – units performing research for improving the literature of Biomedical Engineering. On the other hand, as associated with the topic of nature-inspired techniques, development of such full autonomous systems can be associated with more use of AI based, nature-inspired techniques to form a full intelligent approach for the future.

### Biomedical and Healthcare at Home

The future will probably be associated with more IoT. In this context, also Biomedical Engineering solutions will be available to be used by people at their home, thanks to especially AI and nature-inspired techniques making everything regarding use of such systems easier. In detail, it will be possible for us to receive instant healthcare in case of any injuries or sickness. Additionally, it will be possible to track our health status or prepare something regarding Biomedical solutions thanks to intelligent IoT systems surrounding everywhere at home. So, we can think about smart homes supported by Biomedical Engineering background. Here, we can evaluate the role of AI and nature-inspired techniques as something triggering factors making everything easier and practical to be employed within even daily practical life. Some application examples of the future idea indicated here can be listed briefly as follows:

- Home sets for intelligent IoT supported Biomedical experiments – applications,
- Intelligent healthcare alarm – tracking systems at home,
- Instant intelligent investigation and diagnosis systems easily usable at homes,
- Intelligent medical support software systems at mobile devices and machines at home,
- More advanced traditional machines (i.e. refrigerator, washing machine) that are able to analyze materials in terms of health, to achieve an optimum health level for every individual at home.

### New Software and Hardware Solutions

As a result of continuing rapid developments in intelligent IoT for Biomedical Engineering, the future may have new software and hardware solutions special for designing and developing intelligent IoT systems. Such solutions may be for generally all type of intelligent IoT systems or specific for only Biomedical Engineering applications. But the one key important factor here is to having a potential for more specific software and hardware solutions regarding the future. Furthermore, unstoppable rise of nature-inspired techniques will be more spread in the future by leading more developments of software and hardware solutions to have more effective and advanced systems. If more ideas can be expressed in this manner, new software solutions may include special software components or even special programming languages and hardware solutions may include special IoT materials like special AI CPUs, integrated circuits, management devices…etc.

## CONCLUSION

This chapter has provided a view on applying nature-inspired techniques within Biomedical Engineering supported with IoT. In detail, it is important to indicate that more influence of AI has enabled the scientific community to focus more on hardware oriented intelligent solutions. Eventually, an effective approach called as IoT has attracted interested researchers to employ more effective and efficient communication ways for the machines taking place in our life. Here, the chapter has directed the way of exact subject to a more specific intelligent solution including nature-inspired techniques. From ANNs to SI techniques, it is possible to apply different types of solution approaches to find better IoT application types for Biomedical Engineering research. It is possible to indicate that the future is greatly associated with the AI and some other scientific approaches having strong relations with it. At this point, IoT seems to be having a rapidly increasing popularity when we consider its more intelligent versions in the future. Because healthcare oriented studies are too important in humankind's life cycle, also the field of Biomedical Engineering seems to be having strong interaction and widely open doors to a better future in the context of intelligent solutions for better solutions or solutions for the problems, which have not been solved yet.

## REFERENCES

Atzori, L., Iera, A., & Morabito, G. (2014). From "smart objects" to "social objects": The next evolutionary step of the internet of things. *IEEE Communications Magazine*, *52*(1), 97–105. doi:10.1109/MCOM.2014.6710070

Back, T. (1996). *Evolutionary algorithms in theory and practice: evolution strategies, evolutionary programming, genetic algorithms*. Oxford University Press.

Bakhshipour, M., Ghadi, M. J., & Namdari, F. (2017). Swarm robotics search & rescue: A novel artificial intelligence-inspired optimization approach. *Applied Soft Computing*, *57*, 708–726. doi:10.1016/j.asoc.2017.02.028

Bakirtzis, A., & Kazarlis, S. (2016). Genetic algorithms. *Advanced Solutions in Power Systems: HVDC, FACTS, and Artificial Intelligence: HVDC, FACTS, and Artificial Intelligence*, 845-902.

Basheer, I. A., & Hajmeer, M. (2000). Artificial neural networks: Fundamentals, computing, design, and application. *Journal of Microbiological Methods*, *43*(1), 3–31. doi:10.1016/S0167-7012(00)00201-3 PMID:11084225

Bronzino, J. D. (1999). *Biomedical engineering handbook* (Vol. 2). CRC Press.

Brownlee, J. (2007). *Clonal selection algorithms*. Complex Intelligent Systems Laboratory, Swinburne University of Technology.

Brownlee, J. (2011). *Clever algorithms: nature-inspired programming recipes*. Jason Brownlee.

Bui, N., & Zorzi, M. (2011). Health care applications: a solution based on the internet of things. In *Proceedings of the 4th International Symposium on Applied Sciences in Biomedical and Communication Technologies* (p. 131). ACM. doi:10.1145/2093698.2093829

Chen, H., Chiang, R. H., & Storey, V. C. (2012). Business intelligence and analytics: From big data to big impact. *Management Information Systems Quarterly*, *36*(4).

Chiuchisan, I., Costin, H. N., & Geman, O. (2014). Adopting the internet of things technologies in health care systems. In *Electrical and Power Engineering (EPE), 2014 International Conference and Exposition on* (pp. 532-535). IEEE. doi:10.1109/ICEPE.2014.6969965

Chiuchisan, I., & Geman, O. A. N. A. (2014). An approach of a decision support and home monitoring system for patients with neurological disorders using internet of things concepts. *WSEAS Transations on Systems*, *13*, 460–469.

Cortés, U., Sànchez-Marrè, M., Ceccaroni, L., R-Roda, I., & Poch, M. (2000). Artificial intelligence and environmental decision support systems. *Applied Intelligence*, *13*(1), 77–91. doi:10.1023/A:1008331413864

Craig, J. J. (2005). *Introduction to robotics: mechanics and control* (Vol. 3). Upper Saddle River, NJ: Pearson Prentice Hall.

Da Xu, L., He, W., & Li, S. (2014). Internet of things in industries: A survey. *IEEE Transactions on Industrial Informatics*, *10*(4), 2233–2243. doi:10.1109/TII.2014.2300753

Darianian, M., & Michael, M. P. (2008). Smart home mobile RFID-based Internet-of-Things systems and services. In *Advanced Computer Theory and Engineering, 2008. ICACTE'08. International Conference on* (pp. 116-120). IEEE. doi:10.1109/ICACTE.2008.180

Dasgupta, D., Ji, Z., & Gonzalez, F. (2003). Artificial immune system (AIS) research in the last five years. In Evolutionary Computation, 2003. CEC'03. The 2003 Congress on (Vol. 1, pp. 123-130). IEEE.

Deng, L., & Yu, D. (2014). Deep learning: methods and applications. *Foundations and Trends® in Signal Processing*, *7*(3–4), 197-387.

Dimitrov, D. V. (2016). Medical internet of things and big data in healthcare. *Healthcare Informatics Research*, *22*(3), 156–163. doi:10.4258/hir.2016.22.3.156 PMID:27525156

Domingo, M. C. (2012). An overview of the Internet of Things for people with disabilities. *Journal of Network and Computer Applications*, *35*(2), 584–596. doi:10.1016/j.jnca.2011.10.015

Drew, P. J., & Monson, J. R. (2000). Artificial neural networks. *Surgery*, *127*(1), 3–11. doi:10.1067/msy.2000.102173 PMID:10660751

Enderle, J. D., & Bronzino, J. D. (2012). *Introduction to biomedical engineering*. Academic press.

Gandomi, A. H., & Alavi, A. H. (2012). Krill herd: A new bio-inspired optimization algorithm. *Communications in Nonlinear Science and Numerical Simulation*, *17*(12), 4831–4845. doi:10.1016/j.cnsns.2012.05.010

Gao, X. Z., Ovaska, S. J., & Wang, X. (2008). A GA-based negative selection algorithm. *International Journal of Innovative Computing, Information, & Control*, *4*(4), 971–979.

Gen, M., & Cheng, R. (2000). *Genetic algorithms and engineering optimization* (Vol. 7). John Wiley & Sons.

Geraci, R. M. (2012). *Apocalyptic AI: Visions of heaven in robotics, artificial intelligence, and virtual reality*. Oxford University Press.

Gubbi, J., Buyya, R., Marusic, S., & Palaniswami, M. (2013). Internet of Things (IoT): A vision, architectural elements, and future directions. *Future Generation Computer Systems*, *29*(7), 1645–1660. doi:10.1016/j.future.2013.01.010

Hu, F., Xie, D., & Shen, S. (2013). On the application of the internet of things in the field of medical and health care. In *Green Computing and Communications (GreenCom), 2013 IEEE and Internet of Things (iThings/CPSCom), IEEE International Conference on and IEEE Cyber, Physical and Social Computing* (pp. 2053-2058). IEEE. doi:10.1109/GreenCom-iThings-CPSCom.2013.384

Islam, S. R., Kwak, D., Kabir, M. H., Hossain, M., & Kwak, K. S. (2015). The internet of things for health care: A comprehensive survey. *IEEE Access: Practical Innovations, Open Solutions*, *3*, 678–708. doi:10.1109/ACCESS.2015.2437951

Jain, A. K., Mao, J., & Mohiuddin, K. M. (1996). Artificial neural networks: A tutorial. *Computer*, *29*(3), 31–44. doi:10.1109/2.485891

Jara, A. J., López, P., Fern'ndez, D., Úbeda, B., Zamora, M. A., & Skarmeta, A. F. (2012). Interaction of patients with breathing problems through NFC in Ambient Assisted Living environments. In *Innovative Mobile and Internet Services in Ubiquitous Computing (IMIS), 2012 Sixth International Conference on* (pp. 892-897). IEEE. doi:10.1109/IMIS.2012.150

Jinquan, Z., Xiaojie, L., Tao, L., Caiming, L., Lingxi, P., & Feixian, S. (2009). A self-adaptive negative selection algorithm used for anomaly detection. *Progress in Natural Science*, *19*(2), 261–266. doi:10.1016/j.pnsc.2008.06.008

Karayiannis, N., & Venetsanopoulos, A. N. (2013). *Artificial neural networks: learning algorithms, performance evaluation, and applications* (Vol. 209). Springer Science & Business Media.

Kecman, V. (2001). *Learning and soft computing: support vector machines, neural networks, and fuzzy logic models*. MIT Press.

Kline, J. (Ed.). (2012). *Handbook of biomedical engineering*. Elsevier.

Kopetz, H. (2011). Internet of things. In Real-time systems (pp. 307-323). Springer US. doi:10.1007/978-1-4419-8237-7_13

Kortuem, G., Kawsar, F., Sundramoorthy, V., & Fitton, D. (2010). Smart objects as building blocks for the internet of things. *IEEE Internet Computing*, *14*(1), 44–51. doi:10.1109/MIC.2009.143

Kose, U. (2015). Present state of swarm intelligence and future directions. In *Encyclopedia of Information Science and Technology* (3rd ed.; pp. 239–252). IGI Global. doi:10.4018/978-1-4666-5888-2.ch023

Krenker, A., Bester, J., & Kos, A. (2011). Introduction to the artificial neural networks. In *Artificial neural networks-methodological advances and biomedical applications*. InTech. doi:10.5772/15751

Kubat, M. (2017). Artificial neural networks. In *An Introduction to Machine Learning* (pp. 91–111). Cham: Springer. doi:10.1007/978-3-319-63913-0_5

LeCun, Y., Bengio, Y., & Hinton, G. (2015). Deep learning. *Nature*, *521*(7553), 436–444. doi:10.1038/nature14539 PMID:26017442

Li, X., Lu, R., Liang, X., Shen, X., Chen, J., & Lin, X. (2011). Smart community: An internet of things application. *IEEE Communications Magazine*, *49*(11), 68–75. doi:10.1109/MCOM.2011.6069711

Liang, S. N., Tan, K. O., Clement, T. L., Ng, S. K., Mohammed, A. A., Mailah, M., & Yussof, Z. et al. (2016). Open source hardware and software platform for robotics and artificial intelligence applications. *IOP Conference Series. Materials Science and Engineering*, *114*(1), 012142. doi:10.1088/1757-899X/114/1/012142

Lohr, S. (2012). The age of big data. *New York Times*, *11*(2012).

Luo, J., Chen, Y., Tang, K., & Luo, J. (2009). Remote monitoring information system and its applications based on the Internet of Things. In *BioMedical Information Engineering, 2009. FBIE 2009. International Conference on Future* (pp. 482-485). IEEE.

Ma, H., Ye, S., Simon, D., & Fei, M. (2017). Conceptual and numerical comparisons of swarm intelligence optimization algorithms. *Soft Computing*, *21*(11), 3081–3100. doi:10.1007/s00500-015-1993-x

Manyika, J., Chui, M., Brown, B., Bughin, J., Dobbs, R., Roxburgh, C., & Byers, A. H. (2011). *Big data: The next frontier for innovation, competition, and productivity*. Academic Press.

Marz, N., & Warren, J. (2015). *Big Data: Principles and best practices of scalable realtime data systems*. Manning Publications Co.

Ngiam, J., Khosla, A., Kim, M., Nam, J., Lee, H., & Ng, A. Y. (2011). Multimodal deep learning. In *Proceedings of the 28th international conference on machine learning (ICML-11)* (pp. 689-696). Academic Press.

Nilsson, N. J. (2014). *Principles of artificial intelligence*. Morgan Kaufmann.

Nolfi, S., Bongard, J. C., Husbands, P., & Floreano, D. (2016). *Evolutionary Robotics.* MIT Press / Bradford Books. doi:10.1007/978-3-319-32552-1_76

Oniga, S., & Suto, J. (2014). Human activity recognition using neural networks. In Control Conference (ICCC), 2014 15th International Carpathian (pp. 403-406). IEEE.

Panigrahi, B. K., Shi, Y., & Lim, M. H. (Eds.). (2011). *Handbook of swarm intelligence: concepts, principles and applications* (Vol. 8). Springer Science & Business Media. doi:10.1007/978-3-642-17390-5

Perera, C., Zaslavsky, A., Christen, P., & Georgakopoulos, D. (2014). Context aware computing for the internet of things: A survey. *IEEE Communications Surveys and Tutorials*, *16*(1), 414–454. doi:10.1109/SURV.2013.042313.00197

Poole, D. L., & Mackworth, A. K. (2010). *Artificial Intelligence: Foundations of computational agents.* Cambridge University Press.

Qi, J., Yang, P., Fan, D., & Deng, Z. (2015). A survey of physical activity monitoring and assessment using internet of things technology. In *Computer and Information Technology; Ubiquitous Computing and Communications; Dependable, Autonomic and Secure Computing; Pervasive Intelligence and Computing (CIT/IUCC/DASC/PICOM), 2015 IEEE International Conference on* (pp. 2353-2358). IEEE. doi:10.1109/CIT/IUCC/DASC/PICOM.2015.348

Remagnino, P., & Foresti, G. L. (2005). Ambient intelligence: A new multidisciplinary paradigm. *IEEE Transactions on Systems, Man, and Cybernetics. Part A, Systems and Humans*, *35*(1), 1–6. doi:10.1109/TSMCA.2004.838456

Rohokale, V. M., Prasad, N. R., & Prasad, R. (2011). A cooperative Internet of Things (IoT) for rural healthcare monitoring and control. In *Wireless Communication, Vehicular Technology, Information Theory and Aerospace & Electronic Systems Technology (Wireless VITAE), 2011 2nd International Conference on* (pp. 1-6). IEEE.

Ross, T. J. (2009). *Fuzzy logic with engineering applications.* John Wiley & Sons.

Russell, S. J., & Norvig, P. (2002). *Artificial intelligence: A modern approach* (International Edition). Prentice Hall.

Schmidhuber, J. (2015). Deep learning in neural networks: An overview. *Neural Networks*, *61*, 85–117. doi:10.1016/j.neunet.2014.09.003 PMID:25462637

Suzuki, S. (1997). Artificial neural networks. *Mathematics of Neural Networks: Models. Algorithms and Applications*, *8*, 349.

Tan, L., & Wang, N. (2010). Future internet: The internet of things. In *Advanced Computer Theory and Engineering (ICACTE), 2010 3rd International Conference on* (Vol. 5, pp. V5-376). IEEE.

Tan, Y. (2016). Artificial Immune System. *Artificial Immune System Applications in Computersecurity*, 1-25.

Thomas, A. M., Moore, P., Shah, H., Evans, C., Sharma, M., Xhafa, F., & Chapman, C. et al. (2013). Smart care spaces: Needs for intelligent at-home care. *International Journal of Space-Based and Situated Computing*, *3*(1), 35–44. doi:10.1504/IJSSC.2013.051988

Ulutas, B. H., & Kulturel-Konak, S. (2011). A review of clonal selection algorithm and its applications. *Artificial Intelligence Review*, *36*(2), 117–138. doi:10.1007/s10462-011-9206-1

Vardhini, K. K., & Sitamahalakshmi, T. (2016). A Review on Nature-based Swarm Intelligence Optimization Techniques and its Current Research Directions. *Indian Journal of Science and Technology*, *9*(10).

Von Altrock, C. (1995). *Fuzzy logic and neurofuzzy applications explained.* Prentice-Hall, Inc.

White, H. (2008). Learning in artificial neural networks: A statistical perspective. *Learning, 1*(4).

Xia, F., Yang, L. T., Wang, L., & Vinel, A. (2012). Internet of things. *International Journal of Communication Systems*, *25*(9), 1101–1102. doi:10.1002/dac.2417

Yang, S. H. (2014a). Internet of things. In *Wireless Sensor Networks* (pp. 247–261). Springer London. doi:10.1007/978-1-4471-5505-8_12

Yang, X. S. (2014b). *Nature-inspired optimization algorithms.* Elsevier.

Yang, X. S., Cui, Z., Xiao, R., Gandomi, A. H., & Karamanoglu, M. (Eds.). (2013). *Swarm intelligence and bio-inspired computation: theory and applications.* Newnes. doi:10.1016/B978-0-12-405163-8.00001-6

Yang, X. S., & Deb, S. (2014). Cuckoo search: Recent advances and applications. *Neural Computing & Applications*, *24*(1), 169–174. doi:10.1007/s00521-013-1367-1

Yen, J., & Langari, R. (1998). *Fuzzy logic: intelligence, control, and information.* Prentice-Hall, Inc.

Zanella, A., Bui, N., Castellani, A., Vangelista, L., & Zorzi, M. (2014). Internet of things for smart cities. *IEEE Internet of Things Journal, 1*(1), 22-32.

Zhao, W., Wang, C., & Nakahira, Y. (2011). Medical application on Internet of Things. In *IET Conference Proceedings.* The Institution of Engineering & Technology.

Zhu, A. (2017). *Artificial Neural Networks.* The International Encyclopedia of Geography. doi:10.1002/9781118786352.wbieg0871

Ziemke, T. (2001). The construction of 'reality' in the robot: Constructivist perspectives on situated artificial intelligence and adaptive robotics. *Foundations of Science*, *6*(1), 163–233. doi:10.1023/A:1011394317088

Zurada, J. M. (1992). *Introduction to artificial neural systems* (Vol. 8). St. Paul, MN: West.

## ADDITIONAL READING

Akyildiz, I. F., & Jornet, J. M. (2010). The internet of nano-things. *IEEE Wireless Communications, 17*(6), 58–63. doi:10.1109/MWC.2010.5675779

Akyildiz, I. F., Pierobon, M., Balasubramaniam, S., & Koucheryavy, Y. (2015). The internet of bio-nano things. *IEEE Communications Magazine, 53*(3), 32–40. doi:10.1109/MCOM.2015.7060516

Al-Fuqaha, A., Guizani, M., Mohammadi, M., Aledhari, M., & Ayyash, M. (2015). Internet of things: A survey on enabling technologies, protocols, and applications. *IEEE Communications Surveys and Tutorials, 17*(4), 2347–2376. doi:10.1109/COMST.2015.2444095

Ansari, A. N., Sedky, M., Sharma, N., & Tyagi, A. (2015). An Internet of things approach for motion detection using Raspberry Pi. In *Intelligent Computing and Internet of Things (ICIT), 2014 International Conference on* (pp. 131-134). IEEE. doi:10.1109/ICAIOT.2015.7111554

Antonovici, D. A., Chiuchisan, I., Geman, O., & Tomegea, A. (2014). Acquisition and management of biomedical data using Internet of Things concepts. In *Fundamentals of Electrical Engineering (ISFEE), 2014 International Symposium on* (pp. 1-4). IEEE. doi:10.1109/ISFEE.2014.7050625

Beni, G. (2004). From swarm intelligence to swarm robotics. In *International Workshop on Swarm Robotics* (pp. 1-9). Springer, Berlin, Heidelberg.

Beni, G., & Wang, J. (1993). Swarm intelligence in cellular robotic systems. In *Robots and Biological Systems: Towards a New Bionics?* (pp. 703–712). Berlin, Heidelberg: Springer. doi:10.1007/978-3-642-58069-7_38

Bi, Z., Da Xu, L., & Wang, C. (2014). Internet of things for enterprise systems of modern manufacturing. *IEEE Transactions on Industrial Informatics, 10*(2), 1537–1546. doi:10.1109/TII.2014.2300338

Bonomi, F., Milito, R., Natarajan, P., & Zhu, J. (2014). Fog computing: A platform for internet of things and analytics. In *Big Data and Internet of Things: A Roadmap for Smart Environments* (pp. 169–186). Springer International Publishing. doi:10.1007/978-3-319-05029-4_7

Boric-Lubecke, O., Gao, X., Yavari, E., Baboli, M., Singh, A., & Lubecke, V. M. (2014, June). E-healthcare: Remote monitoring, privacy, and security. In *Microwave Symposium (IMS), 2014 IEEE MTT-S International* (pp. 1-3). IEEE.

Celino, I., & Kotoulas, S. (2013). Smart Cities [Guest editors' introduction]. *IEEE Internet Computing, 17*(6), 8–11. doi:10.1109/MIC.2013.117

Coello, C. C., Dhaenens, C., & Jourdan, L. (Eds.). (2009). *Advances in multi-objective nature inspired computing* (Vol. 272). Springer.

De Guglielmo, D., Anastasi, G., & Seghetti, A. (2014). From IEEE 802.15. 4 to IEEE 802.15. 4e: A step towards the internet of things. In Advances onto the Internet of Things (pp. 135-152). Springer International Publishing.

Di Caro, G., Ducatelle, F., & Gambardella, L. M. (2005). AntHocNet: An adaptive nature-inspired algorithm for routing in mobile ad hoc networks. *Transactions on Emerging Telecommunications Technologies*, *16*(5), 443–455. doi:10.1002/ett.1062

Dlodlo, N., Foko, T. E., Mvelase, P., & Mathaba, S. (2012). *The state of affairs in internet of things research.* Academic Conferences International Ltd.

Dressler, F., & Akan, O. B. (2010). A survey on bio-inspired networking. *Computer Networks*, *54*(6), 881–900. doi:10.1016/j.comnet.2009.10.024

Ducatelle, F., Di Caro, G. A., & Gambardella, L. M. (2010). Principles and applications of swarm intelligence for adaptive routing in telecommunications networks. *Swarm Intelligence*, *4*(3), 173–198. doi:10.1007/s11721-010-0040-x

Fanrong, S., Yuqing, H., & Zhenwen, R. (2013). The multi-sensor Internet of Things wireless monitor system based on ZigBee. *Dianzi Jishu Yingyong*, *3*(13), 96–99.

Fister, I., Jr., Yang, X. S., Fister, I., Brest, J., & Fister, D. (2013). A brief review of nature-inspired algorithms for optimization. *arXiv preprint arXiv:1307.4186.*

Frederix, I. (2009). Internet of things and radio frequency identification in care taking, facts and privacy challenges. In *Wireless Communication, Vehicular Technology, Information Theory and Aerospace & Electronic Systems Technology, 2009. Wireless VITAE 2009. 1st International Conference on* (pp. 319-323). IEEE. doi:10.1109/WIRELESSVITAE.2009.5172467

Garnier, S., Gautrais, J., & Theraulaz, G. (2007). The biological principles of swarm intelligence. *Swarm Intelligence*, *1*(1), 3–31. doi:10.1007/s11721-007-0004-y

Gigli, M., & Koo, S. (2011). Internet of things: Services and applications categorization. *Advances in Internet of Things*, *1*(02), 27–31. doi:10.4236/ait.2011.12004

Holler, J., Tsiatsis, V., Mulligan, C., Avesand, S., Karnouskos, S., & Boyle, D. (2014). *From Machine-to-machine to the Internet of Things: Introduction to a New Age of Intelligence*. Academic Press.

Istepanian, R. S., Hu, S., Philip, N. Y., & Sungoor, A. (2011). The potential of Internet of m-health Things "m-IoT" for non-invasive glucose level sensing. In *Engineering in Medicine and Biology Society, EMBC, 2011 Annual International Conference of the IEEE* (pp. 5264-5266). IEEE. doi:10.1109/IEMBS.2011.6091302

Jara, A. J., Varakliotis, S., Skarmeta, A. F., & Kirstein, P. (2014). Extending the Internet of Things to the Future Internet through IPv6 support. *Mobile Information Systems*, *10*(1), 3–17. doi:10.1155/2014/831974

Jara, A. J., Zamora, M. A., & Skarmeta, A. (2012). Glowbal IP: An adaptive and transparent IPv6 integration in the Internet of Things. *Mobile Information Systems*, *8*(3), 177–197. doi:10.1155/2012/819250

Jeschke, S., Brecher, C., Meisen, T., Özdemir, D., & Eschert, T. (2017). Industrial Internet of Things and Cyber Manufacturing Systems. In *Industrial Internet of Things* (pp. 3–19). Springer International Publishing. doi:10.1007/978-3-319-42559-7_1

Jin, J., Gubbi, J., Marusic, S., & Palaniswami, M. (2014). An information framework for creating a smart city through internet of things. *IEEE Internet of Things Journal*, *1*(2), 112–121. doi:10.1109/JIOT.2013.2296516

Kennedy, J. (2006). Swarm intelligence. In Handbook of nature-inspired and innovative computing (pp. 187-219). Springer US.

Kumar, M., Nallagownden, P., Elamvazuthi, I., Vasant, P., & Rahman, L. H. (2018). Swarm-Intelligence-Based Optimal Placement and Sizing of Distributed Generation in Distribution Network: Swarm-Intelligence-Based Distributed Generation. In S. Khalid (Ed.), *Applied Computational Intelligence and Soft Computing in Engineering* (pp. 29–48). Hershey, PA: IGI Global; doi:10.4018/978-1-5225-3129-6.ch002

Leister, W., & Schulz, T. (2012). Ideas for a Trust Indicator in the Internet of Things. In *The First International Conference on Smart Systems, Devices and Technologies*.

Li, L., Li, S., & Zhao, S. (2014). QoS-aware scheduling of services-oriented internet of things. *IEEE Transactions on Industrial Informatics*, *10*(2), 1497–1505. doi:10.1109/TII.2014.2306782

Liu, C., Zhang, Y., Zeng, J., Peng, L., & Chen, R. (2012). Research on Dynamical Security Risk Assessment for the Internet of Things inspired by immunology. In *Natural Computation (ICNC), 2012 Eighth International Conference on* (pp. 874-878). IEEE. doi:10.1109/ICNC.2012.6234533

Liu, J., & Tsui, K. C. (2006). Toward nature-inspired computing. *Communications of the ACM*, *49*(10), 59–64. doi:10.1145/1164394.1164395

Minoli, D. (2013). *Building the internet of things with IPv6 and MIPv6: The evolving world of M2M communications*. John Wiley & Sons. doi:10.1002/9781118647059

Miranda, J., Mäkitalo, N., Garcia-Alonso, J., Berrocal, J., Mikkonen, T., Canal, C., & Murillo, J. M. (2015). From the Internet of Things to the Internet of People. *IEEE Internet Computing*, *19*(2), 40–47. doi:10.1109/MIC.2015.24

Nebeker, F. (2002). Golden accomplishments in biomedical engineering. *IEEE Engineering in Medicine and Biology Magazine*, *21*(3), 17–47. doi:10.1109/MEMB.2002.1016851 PMID:12119874

Ngai, E. W. T., Moon, K. K., Riggins, F. J., & Candace, Y. Y. (2008). RFID research: An academic literature review (1995–2005) and future research directions. *International Journal of Production Economics*, *112*(2), 510–520. doi:10.1016/j.ijpe.2007.05.004

Ning, H., & Liu, H. (2015). Cyber-physical-social-thinking space based science and technology framework for the Internet of Things. *Science China. Information Sciences*, *58*(3), 1–19. doi:10.1007/s11432-014-5209-2

Paschou, M., Sakkopoulos, E., Sourla, E., & Tsakalidis, A. (2013). Health Internet of Things: Metrics and methods for efficient data transfer. *Simulation Modelling Practice and Theory*, *34*, 186–199. doi:10.1016/j.simpat.2012.08.002

Pencheva, E., & Atanasov, I. (2016). Engineering of web services for internet of things applications. *Information Systems Frontiers*, *18*(2), 277–292. doi:10.1007/s10796-014-9532-3

Perera, C., Liu, C. H., & Jayawardena, S. (2015). The emerging internet of things marketplace from an industrial perspective: A survey. *IEEE Transactions on Emerging Topics in Computing*, *3*(4), 585–598. doi:10.1109/TETC.2015.2390034

Rozenberg, G., Bck, T., & Kok, J. N. (2011). *Handbook of natural computing*. Springer Publishing Company, Incorporated.

Shadbolt, N. (2004). Nature-inspired computing. *IEEE Intelligent Systems*, *19*(1), 2–3. doi:10.1109/MIS.2004.1265875

Sipper, M. (2002). *Machine Nature. The Coming Age of Bio-Inspired Computing* (No. EPFL-BOOK-28714).

Sundmaeker, H., Guillemin, P., Friess, P., & Woelfflé, S. (2010). Vision and challenges for realising the Internet of Things. *Cluster of European Research Projects on the Internet of Things. European Commision*, *3*(3), 34–36.

Tarouco, L. M. R., Bertholdo, L. M., Granville, L. Z., Arbiza, L. M. R., Carbone, F., Marotta, M., & de Santanna, J. J. C. (2012). Internet of Things in healthcare: Interoperatibility and security issues. In *Communications (ICC), 2012 IEEE International Conference on*(pp. 6121-6125). IEEE.

Tsirbas, H., Giokas, K., & Koutsouris, D. (2010). "Internet of Things", an RFID-IPv6 Scenario in a Healthcare Environment. In *XII Mediterranean Conference on Medical and Biological Engineering and Computing 2010* (pp. 808-811). Springer, Berlin, Heidelberg. doi:10.1007/978-3-642-13039-7_204

Vasant, P., Alparslan-Gok, S. Z., & Weber, G. (2018). *Handbook of Research on Emergent Applications of Optimization Algorithms* (Vols. 1–2). pp. 1–812). Hershey, PA: IGI Global; doi:10.4018/978-1-5225-2990-3

Vasant, P., Weber, G., & Dieu, V. N. (2016). *Handbook of Research on Modern Optimization Algorithms and Applications in Engineering and Economics* (pp. 1–960). Hershey, PA: IGI Global; doi:10.4018/978-1-4666-9644-0

Wamba, S. F., & Ngai, E. W. (2013). Internet of Things in healthcare: the case of RFID-enabled asset management. *International Journal of Biomedical Engineering and Technology 45, 11*(3), 318-335.

Wang, W., Li, J., Wang, L., & Zhao, W. (2011). The internet of things for resident health information service platform research.

Wu, J., & Zhao, W. (2013). WInternet: From net of things to internet of things. *Journal of Computer Research and Development*, *50*(6), 1127–1134.

Yun, M., & Yuxin, B. (2010). Research on the architecture and key technology of Internet of Things (IoT) applied on smart grid. In *Advances in Energy Engineering (ICAEE), 2010 International Conference on* (pp. 69-72). IEEE.

Zhang, X. M., & Zhang, N. (2011). An open, secure and flexible platform based on internet of things and cloud computing for ambient aiding living and telemedicine. In *Computer and Management (CAMAN), 2011 International Conference on* (pp. 1-4). IEEE. doi:10.1109/CAMAN.2011.5778905

Zomaya, A. Y. (Ed.). (2006). *Handbook of nature-inspired and innovative computing: integrating classical models with emerging technologies*. Springer Science & Business Media. doi:10.1007/0-387-27705-6

Zorzi, M., Gluhak, A., Lange, S., & Bassi, A. (2010). From today's intranet of things to a future internet of things: A wireless-and mobility-related view. *IEEE Wireless Communications*, *17*(6), 44–51. doi:10.1109/MWC.2010.5675777

## KEY TERMS AND DEFINITIONS

**Artificial Intelligence:** Artificial intelligence is a typical soft computing system that is able to run intelligent operations like reasoning, learning, problem solving, decision making, etc.

**Biomedical Engineering:** Biomedical engineering is an engineering side including the research studies within medicine, biology, and alternative engineering fields to raise healthcare-based solutions/machines/systems.

**Internet of Things:** Internet of things is a type of machine-oriented network in which intelligent machines that are able to communicate with the similar/other ones are employed for the objective problem.

**Nature-Inspired Techniques:** Nature-inspired techniques are artificial-intelligence-based solution techniques as inspired by the natural dynamics in order to get more heuristic and accurate solution ways.

# Chapter 14
# Deep Learning and Biomedical Engineering

**Suraj Sawant**
*College of Engineering Pune, India*

## ABSTRACT

*Deep learning (DL) is a method of machine learning, as running over artificial neural networks, which has a structure above the standards to deal with large amounts of data. That is generally because of the increasing amount of data, input data sizes, and of course, greater complexity of objective real-world problems. Performed research studies in the associated literature show that the DL currently has a good performance among considered problems and it seems to be a strong solution for more advanced problems of the future. In this context, this chapter aims to provide some essential information about DL and its applications within the field of biomedical engineering. The chapter is organized as a reference source for enabling readers to have an idea about the relation between DL and biomedical engineering.*

## INTRODUCTION

It is clear that the World is currently in an age of intelligent innovations. These innovations are all cause of the field of Artificial Intelligence and the rise of this scientific field seems almost infinite. While all improvements – developments by intelligent systems are appeared rapidly, the importance of using the data and processing it becomes more visible. As a result of all technological developments and increasing interaction of the humankind with the data oriented machines – devices, it is now more important to have systems processing data in a fast way and of course accurately. Since its first appearance in the scientific arena, Artificial Intelligence seems the most powerful candidate for this task. Today, we cannot think about any field that is not associated with the approaches, methods, and techniques introduced by Artificial Intelligence. It is a clear situation that we hear news about an innovative intelligent system dealing with a particular real world in each day and it seems that this situation will continue as long as the humankind want to experience the life with a World in which intelligent machines – systems are living. But an important issue regarding to the mentioned reality has become always the amount of data. As the society and technologies rise, the amount of data and complexity of processing them to solve

DOI: 10.4018/978-1-5225-4769-3.ch014

newly encountered real world problems has become a common research problem among scientists. But the most effective answer to that issue has give from the Artificial Intelligence side. 'The science of the future' now has a well-structured, effective approach, which is effective enough to deal with bigger data within complex problems. Called as Deep Learning, this approach of Artificial Intelligence and so Machine Learning has lots of application ways as it is seemed from the reported research results.

The Objective of this chapter is providing some essential information about DL and its applications within the field of Biomedical Engineering. The chapter was organized as a reference source for enabling readers to have idea about relation between DL and Biomedical Engineering. Research works considered – referred here are from the most recent literature to give well-enough information what is actually done in the intersection of Deep Learning and Biomedical Engineering.

Based on the subject of the chapter, the remaining sections are organized as follows: The next section is devoted to some essential information about Deep Learning and its general application types. Following to that, the third section draws the connection between Deep Learning and Biomedical Engineering, and gives information about some remarkable and recent works. Next, the fourth section provides some future ideas by the authors and finally, the chapter ends with the conclusion section.

## DEEP LEARNING

Before discussing about Deep Learning and Biomedical Engineering, it is a good idea to briefly introduce the concept of Deep Learning and take a look at to its application ways.

### The Concept of Deep Learning

In the past few years, Deep Learning (DL) has rapidly evolved into the de-facto approach, showing tremendous improvement in accuracy, robustness, and cross-language generalizability over conventional approaches (Deng & Yu, 2014). DL also termed as Hierarchical Learning or Deep Structured Learning in some of the literature work is a technique of Information Processing. It is a sub-set of machine learning methods. It is the process of training and learning from Artificial Neural Networks (ANNs) containing more than one hidden layer. DL allows computational models that are composed of multiple processing layers to learn representations of data with multiple levels of abstraction. These methods have dramatically improved the state-of-the-art in drug discovery and genomics. DL discovers intricate structure in large data sets by using the backpropagation algorithm to indicate how a machine should change its internal parameters that are used to compute the representation in each layer from the representation in the previous layer. Deep convolutional networks have brought about breakthroughs in processing bio medical images, video, speech and audio, whereas recurrent networks have shone light on sequential data such as text and speech (LeCun et al., 2015). To date, DL has emerged as the leading machine-learning tool in the general imaging and computer vision domains. DL is a growing trend in general data analysis and was termed as one of the 10 breakthrough technologies of 2013 (MIT Tech. Review, 2013). DL has improved day by time till nowadays as an important approach for the future of intelligent systems.

## General Application Types of Deep Learning

There are many different types of application in which DL can enroll easily. Unless it has a big data and an advanced, complex problem structure requiring an effective solution way better than traditional ones, DL can be applied in any real world based problem.

Considering the examinations done over the associated literature, some general application types of the DL can be expressed as follows (Deng, 2014; Deng & Yu, 2014; LeCun et al., 2015):

- Prediction,
- Classification,
- Data (Image, Signal…etc.) Processing,
- Recognition,
- Modeling,
- Diagnosis.

## DEEP LEARNING FOR BIOMEDICAL ENGINEERING PROBLEMS

As it deals with big amount of data, which can also require very detailed examination – evaluation, Biomedical Engineering has become interested widely in Artificial Intelligence and its popular approach of

*Figure 1. General application types of Deep Learning*

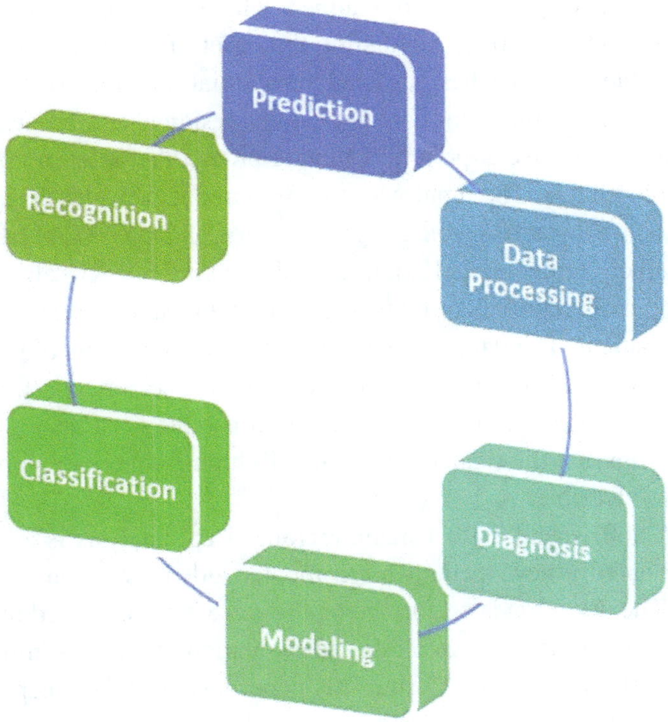

DL. In order to understand more about influence of DL in Biomedical Engineering oriented problems, it is a good way to focus on remarkable and recent works from the related literature.

## Some Recent Works on Deep Learning and Biomedical Engineering

When we examine the associated literature, we can identify that there are many different research works dealing with Biomedical Engineering oriented problems. Research works related to medical – health can be accepted within this scope because of their characteristics in detail. It is possible to indicate that the future of Biomedical Engineering is more and more strong thanks to wide advantages of DL over advanced problems.

Some remarkable research works associated with the applications of DL over Biomedical Engineering problems can be expressed briefly as follows:

Recent results indicate that the generic descriptors extracted from Convolutional Neural Networks (CNNs) are extremely effective in object recognition and localization in natural images. Hence, medical image analysis groups across the world are quickly entering the field and applying CNNs and other DL methodologies to a wide variety of applications (Greenspan et al., 2016). It is possible to say that results are promising.

Accurate diagnosis and/or assessment of a disease in medical imaging, depends on image acquisition and image interpretation. Image acquisition is marked to be improved over past few years, with devices acquiring data with speedy rates and enhanced resolutions. Comparably, image interpretation process, has only recently begun to benefit from advanced computer technology. In general, most of the interpretations of medical images are being performed by physicians which add limits due to the subjectivity, variations in interpreters. In Biomedical engineering, diagnostic tasks require to detect abnormalities, and to quantify measurements and changes over time. Image Analysis and DL if mapped together will improve diagnosis, by identifying the desired treatment which will support the expert's workflow.

According to Petrosian et al., (2000), Predicting the onset of epileptic seizure is an important and difficult biomedical problem, which has attracted substantial attention of the intelligent computing community. They have explored the ability of specifically designed and trained Recurrent Neural Networks (RNN), combined with wavelet preprocessing, to predict the onset of epileptic seizures both on scalp and intracranial recordings. The input was raw EEG and its wavelet-decomposed sub-bands into RNN training/testing, as opposed to specific signal features extracted from EEG. They used network training procedure based on the decoupled extended Kalman xlter (DEKF) algorithm. The data set used consisted of four intracranial seizure recordings of one epileptic patient and six extra-cranial recordings obtained from another patient. The data extracted from selected channels of these multichannel EEGs contained up to 10 min of recordings immediately preceding the seizure onset. Results have shown that the existence of a preictal stage of some minutes duration is quite feasible. As it was demonstrated on intracranial EEG recordings, it is the changes in the signal's high-frequency component ("noisy residue") that makes the prediction feasible.

Boccignone et al. (2000) specifies that mammogram visual analysis is a complex task for expert radiologists. They presented a new method for computer aided detection of micro-calcifications in digital mammograms. The detection is performed on the wavelet transformed image. The calcifications are separated from the background by exploiting the evaluation of Renyi's information at the different decomposition levels of the wavelet transform. This paper has used the Deep Learning for the image analysis and image interpretation part to provide acceptable results.

Shieh et al., (2004) implemented the structure of recurrent network to develop a modified neural network algorithm called a simple recurrent neural network through time (SRNNTT). The proposed recurrent neural network combines Elman architecture of the simple recurrent network structure and backpropagation through time. In order to demonstrate the performance of the proposed model, four kinds of neural-network classifiers have been tested on Mackey–Glass differential-delay equation which is a chaotic time series signal. Finally, they used this SRNNTT model to build the intracranial pressure (ICP) model using data from six head-injured patients. The accuracy of the ICP model was far from ideal, but the methodology used non-invasive vital signs to predict an invasive, dangerous, and expensive signal (i.e., ICP) has achieved this monitoring system more safety and flexibility.

Breast tumor detection in digital mammography is one of the most important methods of breast cancer prevention. Computer-aided diagnosis (CAD) based on extreme learning machine (ELM) has significant meanings for breast tumor detection as it has good generalization abilities and a high learning efficiency, Wang et al. (2014) uses a breast tumor detection algorithm in digital mammography based on ELM. A median filter is used for noise reduction, and contrast enhancement of the digital mammography in data preprocessing. Next, methods of wavelet modulus maxima transform, morphological operation and region growth are used for the breast tumor edge segmentation. Then, five textural features and five morphological features are extracted. Finally, an ELM classifier is used to detect the breast tumor. Comparing breast tumor detection based on Support Vector Machines (SVM), with breast tumor detection based on ELM, it has emerged to have a better classification accuracy than SVM and has a greatly improved training speed. ELM (Huang et al., 2006) is a Single Hidden-Layer Feed-Forward Neural Networks (SLFNs) learning algorithm and can be extended to the generalized SLFNs where the hidden layer need not be neuron like (Huang & Chen, 2007). It first randomly assigns the input weights and hidden layer biases, and then analytically determines the output weights of SLFNs. ELM can achieve better generalization performance than other conventional learning algorithms at an extremely fast learning speed. Besides, ELM is also less sensitive to user-specified parameters, and can be deployed faster and more conveniently.

Characteristic classification of mass plays a role of vital importance in diagnosis of breast cancer. Jiao et al. (2016) designed a deep feature based framework for breast mass classification task. It mainly contains a CNN and a decision mechanism. Combining intensity information and deep features automatically extracted by the trained CNN from the original image, the proposed method could better simulate the diagnostic procedure operated by doctors and achieved state-of-art performance. In this framework, doctors' global and local impressions left by mass images were represented by deep features extracted from two different layers called high-level and middle-level features. The original images were regarded as detailed descriptions of the breast mass. Then, classifiers based on features above were used in combination to predict classes of test images. And outcomes of classifiers based on different features were analyzed jointly to determine the types of test images. With the help of two kinds of feature visualization methods, deep features extracted from different layers illustrate effective in classification performance and diagnosis simulation. The experimental results demonstrated that the CNN model with adjustment for certain data was effective even the scale of specific database was not that large.

CNNs with DL have recently achieved a remarkable success with a superior performance in computer vision applications. Most of CNN-based methods extract image features at the last layer using a single CNN architecture with orderless quantization approaches. Alzu'bi et al. (2017) proposes a new bilinear CNN-based architecture using two parallel CNNs as feature extractors. The activations of convolutional layers are directly used to extract the image features at various image locations and scales. The network architecture is initialized by deep CNNs sufficiently pre-trained on a large generic image dataset then fine-tuned for the Content Based Image Retrieval (CBIR) task. Additionally, an efficient bilinear root pooling is proposed and applied to the low-dimensional pooling layer to reduce the dimension of image features to compact but high discriminative image descriptors. Finally, an end-to-end training with backpropagation is performed to fine-tune the final architecture and to learn its parameters for the image retrieval task. The experimental results achieved on three standard benchmarking image datasets demonstrate the outstanding performance. It achieved a retrieval accuracy of 95.7% (mAP) on Oxford 5K and 88.6% on Oxford 105K dataset.

Regarding to medical image analysis via DL, Shen et al. (2017) provides a remarkable review as a recent source in the associated literature. In detail, the authors express some about applying DL successfully over many image analysis operations like image registration, computer-aided disease diagnosis and prognosis, detection of anatomical and cellular structures, tissue segmentation. As similar, readers can also refer to some alternative works like Zhou et al. (2017), Seebock (2015), and Litjens et al. (2017) in order to understand more about DL and medical image analysis.

A remarkable research work by Kraus et al. (2017) is focused on using DL for automated analysis of high-content microscopy data. In the work, the authors use a deep convolutional neural network (called as DeepLoc) to analyze yeast cell images by obtaining good results regarding to employment of DL for the related data analysis purpose.

Turan et al. (2017) introduces a DL based approach for 6 degree-of-freedom localization of endoscopic capsule robots. By performing real time experiments in a human stomach model, the authors have achieved to predict the pose of the used robot, thanks to a designed 23-layer CNN.

DL has proved its effectiveness many times including innovative research works (Actually, almost all research works done with DL in medical – health and Biomedical Engineering can be accepted as innovative ones). As a remarkable example, DL and 3D-DESI imaging approaches have been used to figure out hidden metabolic heterogeneity of cancer by employing a deep unsupervised neural network based technique through the study done by Inglese et al. (2017).

As an alternative to the traditional methods, Angermueller et al. (2017) employed a DL approach to achieve prediction of single-cell DNA methylation states. In detail, they have developed a DL system called as DeepCpG to obtain a more successful prediction oriented solution way according to previously performed traditional methods.

Heath-care is a remarkable subject that receives support from DL widely. By starting from the idea of using bigger, complex data in order to achieve more health-care for individuals, there has been remarkable research efforts on using DL oriented solutions and evaluating its potentials. As a recent review work in this manner, the paper by Miotto et al. (2017) is a good source that should be examined by the interested readers.

In addition to the mentioned works, the readers are referred to the Additional Reading Section at the end of this chapter and also a few more examples from 2017 as mentioned follows: Chang et al., 2017; Cruz-Roa et al., 2017; Habibi et al., 2017; Heinrich et al., 2017; Lopez et al., 2017; Liu et al., 2017; Peng & Lu, 2017; Qui et al., 2017; Rather et al., 2017; Rawat et al., 2017; Saha et al., 2017; Schirrmeister et al., 2017; Sharma et al., 2017; Trebeschi et al., 2017; Yang et al., 2017; Yi et al., 2017; Zhang et al., 2017.

As it can be seen from the briefly explained research works, there is a great interest and an already formed good literature of DL and its applications in the field of Biomedical Engineering.

## FUTURE RESEARCH DIRECTIONS

Although the ongoing, great research works and innovations appeared often do not leave more things to say, the authors would like to express some ideas regarding to more use of DL (or any new intelligent techniques following the DL) on Biomedical Engineering problems. In this context, some ideas regarding to the future can be listed as below:

- More effective solutions on using more data will clearly cause for rapid development of intelligent, medical robots, which are advanced as like the ones we see in science-fiction movies. In other words, it is not a fiction or dream to think about robot doctors or robot workers dealing with the biomedical side of the medical – health field.
- With the rise of advanced intelligent robots, it will probably be possible for people to receive all medical services without interacting with human doctors.
- Developments and improvements regarding to medical – health, and Biomedical Engineering fields will be done by intelligent systems. At this point, humankind will be responsible for only applying – realizing the investigations or discoveries found and suggested by such systems.
- In the future, there will be more data at the level of we cannot imagine and all of these data will be processed instantly with more advanced machines. There will be a great amount of data flowing instantly through digital systems so there will be many anxieties regarding to privacies and personal life. In this context, a future based on more intelligent systems – machines with unclear issues associated with ethics is waiting for us.
- In the future, there will probably be mobile hospitals, which are reading individuals' health data flow instantly and stepping in if any dangerous cases happen.
- Future of Biomedical Engineering will be based on mostly robotics. But on the background, there will be a great need for Data Processing experts and Artificial Intelligence engineers.
- The future of Biomedical Engineering in connection widely with Artificial Intelligence will possibly require forming new jobs – specialties under the Biomedical Engineering and dealing with only design, development and control of intelligent machines – systems.

## CONCLUSION

This chapter has focused on the subject of Deep Learning (DL) and its applications within especially Biomedical Engineering. Because DL is a strong method of Artificial Intelligence / Machine Learning, it is already favorite among researchers when dealing with advanced problems and also accurate solutions needing more amount of data and complex solution processes. It is clear that the amount of data used for solving real world problems is increasing day by day. So, there is an urgent need for more effective and comprehensive techniques covering problems of nowadays and future as making the field of Artificial Intelligence more robust and wide towards the future. As it was mentioned within the chapter, Data

Processing is another active actor on the background of intelligent systems. As a result of the society busy with using data in even almost every daily life works and great use of more and more data in the scientific fields, Data Processing has become an important factor for the sciences of future and because of that the field of Artificial Intelligence has been improved to deal with bigger amount of data. DL is a wonderful example of that improvement so far. It is still unclear that what the future will bring to us but it will probably be more important to process all types of data for achieving better definable information and solvable advanced problems at the end.

## REFERENCES

Alzu'bi, A., Amira, A., & Ramzan, N. (2017). Content-based image retrieval with compact deep convolutional features. *Neurocomputing*, *249*, 95–105. doi:10.1016/j.neucom.2017.03.072

Angermueller, C., Lee, H. J., Reik, W., & Stegle, O. (2017). DeepCpG: Accurate prediction of single-cell DNA methylation states using deep learning. *Genome Biology*, *18*(1), 67. doi:10.1186/s13059-017-1189-z PMID:28395661

Boccignone, G., Chianese, A., & Picariello, A. (2000). Computer aided detection of microcalcifications in digital mammograms. *Computers in Biology and Medicine*, *30*(5), 267–286. doi:10.1016/S0010-4825(00)00014-7 PMID:10913773

Chang, H., Han, J., Zhong, C., Snijders, A., & Mao, J. H. (2017). Unsupervised Transfer Learning via Multi-Scale Convolutional Sparse Coding for Biomedical Applications. *IEEE Transactions on Pattern Analysis and Machine Intelligence*, 1. doi:10.1109/TPAMI.2017.2656884 PMID:28129148

Cruz-Roa, A., Gilmore, H., Basavanhally, A., Feldman, M., Ganesan, S., Shih, N. N., & Madabhushi, A. et al. (2017). Accurate and reproducible invasive breast cancer detection in whole-slide images: A Deep Learning approach for quantifying tumor extent. *Scientific Reports*, *7*, 46450. doi:10.1038/srep46450 PMID:28418027

Deng, L. (2014). A tutorial survey of architectures, algorithms, and applications for deep learning. *APSIPA Transactions on Signal and Information Processing, 3*.

Deng, L., & Yu, D. (2014). Deep learning: Methods and applications. *Foundations and Trends® in Signal Processing, 7*(3–4), 197-387.

Greenspan, H., van Ginneken, B., & Summers, R. M. (2016). Guest editorial deep learning in medical imaging: Overview and future promise of an exciting new technique. *IEEE Transactions on Medical Imaging*, *35*(5), 1153–1159. doi:10.1109/TMI.2016.2553401

Habibi, M., Weber, L., Neves, M., Wiegandt, D. L., & Leser, U. (2017). Deep learning with word embeddings improves biomedical named entity recognition. *Bioinformatics (Oxford, England)*, *33*(14), i37–i48. doi:10.1093/bioinformatics/btx228 PMID:28881963

Heinrich, L., Bogovic, J. A., & Saalfeld, S. (2017). *Deep Learning for Isotropic Super-Resolution from Non-Isotropic 3D Electron Microscopy*. arXiv preprint arXiv:1706.03142

Huang, G. B., & Chen, L. (2007). Convex incremental extreme learning machine. *Neurocomputing*, *70*(16), 3056–3062. doi:10.1016/j.neucom.2007.02.009

Huang, G. B., Zhu, Q. Y., & Siew, C. K. (2006). Extreme learning machine: Theory and applications. *Neurocomputing*, *70*(1), 489–501. doi:10.1016/j.neucom.2005.12.126

Inglese, P., McKenzie, J. S., Mroz, A., Kinross, J., Veselkov, K., Holmes, E., & Glen, R. C. et al. (2017). Deep learning and 3D-DESI imaging reveal the hidden metabolic heterogeneity of cancer. *Chemical Science (Cambridge)*, *8*(5), 3500–3511. doi:10.1039/C6SC03738K PMID:28507724

Jiao, Z., Gao, X., Wang, Y., & Li, J. (2016). A deep feature based framework for breast masses classification. *Neurocomputing*, *197*, 221–231. doi:10.1016/j.neucom.2016.02.060

Kraus, O. Z., Grys, B. T., Ba, J., Chong, Y., Frey, B. J., Boone, C., & Andrews, B. J. (2017). Automated analysis of high-content microscopy data with deep learning. *Molecular Systems Biology*, *13*(4), 924. doi:10.15252/msb.20177551 PMID:28420678

LeCun, Y., Bengio, Y., & Hinton, G. (2015). Deep learning. *Nature*, *521*(7553), 436–444. doi:10.1038/nature14539 PMID:26017442

Litjens, G., Kooi, T., Bejnordi, B. E., Setio, A. A. A., Ciompi, F., Ghafoorian, M., . . . Sánchez, C. I. (2017). *A survey on deep learning in medical image analysis*. arXiv preprint arXiv:1702.05747

Liu, T., Xie, S., Yu, J., Niu, L., & Sun, W. (2017). Classification of thyroid nodules in ultrasound images using deep model based transfer learning and hybrid features. In *Acoustics, Speech and Signal Processing (ICASSP), 2017 IEEE International Conference on* (pp. 919-923). IEEE. doi:10.1109/ICASSP.2017.7952290

Lopez, A. R., Giro-i-Nieto, X., Burdick, J., & Marques, O. (2017). Skin lesion classification from dermoscopic images using deep learning techniques. In *Biomedical Engineering (BioMed), 2017 13th IASTED International Conference on* (pp. 49-54). IEEE.

Miotto, R., Wang, F., Wang, S., Jiang, X., & Dudley, J. T. (2017). Deep learning for healthcare: Review, opportunities and challenges. *Briefings in Bioinformatics*, bbx044. PMID:28481991

MIT Tech. Rev. (2013). Deep Learning. *MIT Technology Review*. Available at: https://www.technologyreview.com/s/513696/deep-learning

Peng, Y., & Lu, Z. (2017). *Deep learning for extracting protein-protein interactions from biomedical literature*. arXiv preprint arXiv:1706.01556

Petrosian, A., Prokhorov, D., Homan, R., Dasheiff, R., & Wunsch, D. II. (2000). Recurrent neural network based prediction of epileptic seizures in intra-and extracranial EEG. *Neurocomputing*, *30*(1), 201–218. doi:10.1016/S0925-2312(99)00126-5

Qiu, J., Yoon, H. J., Fearn, P. A., & Tourassi, G. D. (2017). Deep Learning for Automated Extraction of Primary Sites from Cancer Pathology Reports. *IEEE Journal of Biomedical and Health Informatics*, 1. doi:10.1109/JBHI.2017.2700722 PMID:28475069

Rather, N. N., Patel, C. O., & Khan, S. A. (2017). Using Deep Learning Towards Biomedical Knowledge Discovery. *International Journal of Mathematical Sciences and Computing*, *3*(2), 1. doi:10.5815/ijmsc.2017.02.01

Rawat, R. R., Ruderman, D., Agus, D. B., & Macklin, P. (2017). *Deep learning to determine breast cancer estrogen receptor status from nuclear morphometric features in H&E images*. Academic Press.

Saha, M., Chakraborty, C., Arun, I., Ahmed, R., & Chatterjee, S. (2017). An Advanced Deep Learning Approach for Ki-67 Stained Hotspot Detection and Proliferation Rate Scoring for Prognostic Evaluation of Breast Cancer. *Scientific Reports*, 7. PMID:28607456

Schirrmeister, R. T., Springenberg, J. T., Fiederer, L. D. J., Glasstetter, M., Eggensperger, K., Tangermann, M., . . . Ball, T. (2017). *Deep learning with convolutional neural networks for brain mapping and decoding of movement-related information from the human EEG*. arXiv preprint arXiv:1703.05051

Seebock, P. (2015). *Deep learning in medical image analysis* (Master's thesis). Vienna University of Technology, Faculty of Informatics.

Sharma, K., Rupprecht, C., Caroli, A., Aparicio, M. C., Remuzzi, A., Baust, M., & Navab, N. (2017). Automatic Segmentation of Kidneys using Deep Learning for Total Kidney Volume Quantification in Autosomal Dominant Polycystic Kidney Disease. *Scientific Reports*, 7. PMID:28515418

Shen, D., Wu, G., & Suk, H. I. (2017). Deep learning in medical image analysis. *Annual Review of Biomedical Engineering*.

Shieh, J. S., Chou, C. F., Huang, S. J., & Kao, M. C. (2004). Intracranial pressure model in intensive care unit using a simple recurrent neural network through time. *Neurocomputing*, *57*, 239–256. doi:10.1016/j.neucom.2003.10.006

Trebeschi, S., van Griethuysen, J. J., Lambregts, D. M., Lahaye, M. J., Parmer, C., Bakers, F. C., & Aerts, H. J. et al. (2017). Deep Learning for Fully-Automated Localization and Segmentation of Rectal Cancer on Multiparametric MR. *Scientific Reports*, 7. PMID:28706185

Turan, M., Almalioglu, Y., Konukoglu, E., & Sitti, M. (2017). *A Deep Learning Based 6 Degree-of-Freedom Localization Method for Endoscopic Capsule Robots*. arXiv preprint arXiv:1705.05435

Wang, Z., Yu, G., Kang, Y., Zhao, Y., & Qu, Q. (2014). Breast tumor detection in digital mammography based on extreme learning machine. *Neurocomputing*, *128*, 175–184. doi:10.1016/j.neucom.2013.05.053

Yang, P., Abo, R., Liu, C., Chen, Z., Wu, H., Cui, J., ... & Chittenden, T. W. (2017). *Novel feature selection strategies for enhanced predictive modeling and deep learning in the biosciences*. Academic Press.

Yi, D., Sawyer, R. L., Cohn, D., III, Dunnmon, J., Lam, C., Xiao, X., & Rubin, D. (2017). *Optimizing and Visualizing Deep Learning for Benign/Malignant Classification in Breast Tumors*. arXiv preprint arXiv:1705.06362

Zhang, J., Xia, Y., Wu, Q., & Xie, Y. (2017). *Classification of Medical Images and Illustrations in the Biomedical Literature Using Synergic Deep Learning*. arXiv preprint arXiv:1706.09092

Zhou, S. K., Greenspan, H., & Shen, D. (Eds.). (2017). *Deep Learning for Medical Image Analysis*. Academic Press.

## ADDITIONAL READING


Amodei, D., Ananthanarayanan, S., Anubhai, R., Bai, J., Battenberg, E., Case, C., & Chen, J. et al. (2016). Deep speech 2: End-to-end speech recognition in english and mandarin. In *International Conference on Machine Learning* (pp. 173-182).

Angermueller, C., Pärnamaa, T., Parts, L., & Stegle, O. (2016). Deep learning for computational biology. *Molecular Systems Biology, 12*(7), 878. doi:10.15252/msb.20156651 PMID:27474269

Arel, I., Rose, D. C., & Karnowski, T. P. (2010). Deep machine learning-a new frontier in artificial intelligence research [research frontier]. *IEEE Computational Intelligence Magazine, 5*(4), 13–18. doi:10.1109/MCI.2010.938364

Baccouche, M., Mamalet, F., Wolf, C., Garcia, C., & Baskurt, A. (2011, November). Sequential deep learning for human action recognition. In *International Workshop on Human Behavior Understanding* (pp. 29-39). Springer, Berlin, Heidelberg. doi:10.1007/978-3-642-25446-8_4

Bengio, Y. (2012). Deep learning of representations for unsupervised and transfer learning. In *Proceedings of ICML Workshop on Unsupervised and Transfer Learning* (pp. 17-36).

Bengio, Y., Goodfellow, I. J., & Courville, A. (2015). Deep learning. *Nature, 521*(7553), 436–444. doi:10.1038/nature14539 PMID:26017442

Chan, T. H., Jia, K., Gao, S., Lu, J., Zeng, Z., & Ma, Y. (2015). PCANet: A simple deep learning baseline for image classification? *IEEE Transactions on Image Processing, 24*(12), 5017–5032. doi:10.1109/TIP.2015.2475625 PMID:26340772

Chen, Y., Lin, Z., Zhao, X., Wang, G., & Gu, Y. (2014). Deep learning-based classification of hyperspectral data. *IEEE Journal of Selected Topics in Applied Earth Observations and Remote Sensing, 7*(6), 2094–2107. doi:10.1109/JSTARS.2014.2329330

Chin, C., & Brown, D. E. (2000). Learning in science: A comparison of deep and surface approaches. *Journal of Research in Science Teaching, 37*(2), 109–138. doi:10.1002/(SICI)1098-2736(200002)37:2<109::AID-TEA3>3.0.CO;2-7

Cho, Y., & Saul, L. K. (2009). Kernel methods for deep learning. In Advances in neural information processing systems (pp. 342-350).

Cohen, N., Sharir, O., & Shashua, A. (2016, June). On the expressive power of deep learning: A tensor analysis. In *Conference on Learning Theory* (pp. 698-728).

Dauphin, G. M. Y., Glorot, X., Rifai, S., Bengio, Y., Goodfellow, I., Lavoie, E., & Courville, A. et al. (2012). Unsupervised and transfer learning challenge: a deep learning approach. In *Proceedings of ICML Workshop on Unsupervised and Transfer Learning* (pp. 97-110).

Deng, L., Li, J., Huang, J. T., Yao, K., Yu, D., Seide, F., . . . Gong, Y. (2013). Recent advances in deep learning for speech research at Microsoft. In *Acoustics, Speech and Signal Processing (ICASSP), 2013 IEEE International Conference on* (pp. 8604-8608). IEEE. doi:10.1109/ICASSP.2013.6639345

Dong, C., Loy, C. C., He, K., & Tang, X. (2014). Learning a deep convolutional network for image super-resolution. In *European Conference on Computer Vision* (pp. 184-199). Springer, Cham. doi:10.1007/978-3-319-10593-2_13

Draelos, T. J., Miner, N. E., Lamb, C. C., Cox, J. A., Vineyard, C. M., Carlson, K. D., . . . Aimone, J. B. (2017). Neurogenesis deep learning: Extending deep networks to accommodate new classes. In *Neural Networks (IJCNN), 2017 International Joint Conference on* (pp. 526-533). IEEE.

Ergün, E., Gürkan, F., Kaplan, O., & Günsel, B. (2017). Video action classification by deep learning. In *Signal Processing and Communications Applications Conference (SIU), 2017 25th* (pp. 1-4). IEEE.

Gal, Y., & Ghahramani, Z. (2016). Dropout as a Bayesian approximation: Representing model uncertainty in deep learning. In international conference on machine learning (pp. 1050-1059).

Gal, Y., Islam, R., & Ghahramani, Z. (2017). Deep Bayesian Active Learning with Image Data. *arXiv preprint arXiv:1703.02910.*

Gan, W. C., & Shu, F. W. (2017). Holography as deep learning. *arXiv preprint arXiv:1705.05750.*

Gebru, T., Krause, J., Wang, Y., Chen, D., Deng, J., Aiden, E. L., & Fei-Fei, L. (2017). Using deep learning and google street view to estimate the demographic makeup of the us. *arXiv preprint arXiv:1702.06683.*

Glorot, X., Bordes, A., & Bengio, Y. (2011). Domain adaptation for large-scale sentiment classification: A deep learning approach. In *Proceedings of the 28th international conference on machine learning (ICML-11)* (pp. 513-520).

Heaton, J. B., Polson, N. G., & Witte, J. H. (2017). Deep learning for finance: Deep portfolios. *Applied Stochastic Models in Business and Industry, 33*(1), 3–12. doi:10.1002/asmb.2209

Jia, S., Lansdall-Welfare, T., & Cristianini, N. (2016). Gender Classification by Deep Learning on Millions of Weakly Labelled Images. In *Data Mining Workshops (ICDMW), 2016 IEEE 16th International Conference on* (pp. 462-467). IEEE. doi:10.1109/ICDMW.2016.0072

Kahou, S. E., Bouthillier, X., Lamblin, P., Gulcehre, C., Michalski, V., Konda, K., & Ferrari, R. C. et al. (2016). Emonets: Multimodal deep learning approaches for emotion recognition in video. *Journal on Multimodal User Interfaces, 10*(2), 99–111. doi:10.1007/s12193-015-0195-2

Kawaguchi, K. (2016). Deep learning without poor local minima. In Advances in Neural Information Processing Systems (pp. 586-594).

Kendall, A., & Cipolla, R. (2017). Geometric loss functions for camera pose regression with deep learning. *arXiv preprint arXiv:1704.00390.*

Kim, Y., Lee, H., & Provost, E. M. (2013, May). Deep learning for robust feature generation in audiovisual emotion recognition. In *Acoustics, Speech and Signal Processing (ICASSP), 2013 IEEE International Conference on* (pp. 3687-3691). IEEE. doi:10.1109/ICASSP.2013.6638346

Le, Q. V. (2013). Building high-level features using large scale unsupervised learning. In *Acoustics, Speech and Signal Processing (ICASSP), 2013 IEEE International Conference on* (pp. 8595-8598). IEEE. doi:10.1109/ICASSP.2013.6639343

Lenz, I., Lee, H., & Saxena, A. (2015). Deep learning for detecting robotic grasps. *The International Journal of Robotics Research, 34*(4-5), 705–724. doi:10.1177/0278364914549607

Levine, S., Pastor, P., Krizhevsky, A., Ibarz, J., & Quillen, D. (2016). Learning hand-eye coordination for robotic grasping with deep learning and large-scale data collection. *The International Journal of Robotics Research.*

Ma, L., Stückler, J., Kerl, C., & Cremers, D. (2017). Multi-view deep learning for consistent semantic mapping with rgb-d cameras. *arXiv preprint arXiv:1703.08866.*

Martens, J. (2010). Deep learning via Hessian-free optimization. In *Proceedings of the 27th International Conference on Machine Learning (ICML-10)* (pp. 735-742).

Mason, E., Yonel, B., & Yazici, B. (2017). Deep learning for SAR image formation. In SPIE Defense+ Security (pp. 1020104-1020104). International Society for Optics and Photonics.

Mnih, V., Kavukcuoglu, K., Silver, D., Graves, A., Antonoglou, I., Wierstra, D., & Riedmiller, M. (2013). Playing atari with deep reinforcement learning. *arXiv preprint arXiv:1312.5602.*

Ngiam, J., Khosla, A., Kim, M., Nam, J., Lee, H., & Ng, A. Y. (2011). Multimodal deep learning. In *Proceedings of the 28th international conference on machine learning (ICML-11)* (pp. 689-696).

Papernot, N., McDaniel, P., Jha, S., Fredrikson, M., Celik, Z. B., & Swami, A. (2016). The limitations of deep learning in adversarial settings. In *Security and Privacy (EuroS&P), 2016 IEEE European Symposium on* (pp. 372-387). IEEE. doi:10.1109/EuroSP.2016.36

Pereira, C., & Yang, X. S. (2016). Learning Parameters in Deep Belief Networks Through Firefly Algorithm. In *Artificial Neural Networks in Pattern Recognition: 7th IAPR TC3 Workshop, ANNPR 2016, Ulm, Germany, September 28–30, 2016* [). Springer.]. *Proceedings, 9896*, 138.

Rushton, A. (2005). Formative assessment: A key to deep learning? *Medical Teacher, 27*(6), 509–513. doi:10.1080/01421590500129159 PMID:16199357

Santos, T., & Kern, R. (2016). A Literature Survey of Early Time Series Classification and Deep Learning. In SAMI@ iKNOW.

Schmidhuber, J. (2015). Deep learning in neural networks: An overview. *Neural Networks, 61*, 85–117. doi:10.1016/j.neunet.2014.09.003 PMID:25462637

Socher, R., Huval, B., Bath, B., Manning, C. D., & Ng, A. Y. (2012). Convolutional-recursive deep learning for 3d object classification. In Advances in Neural Information Processing Systems (pp. 656-664).

Sun, Y., Wang, X., & Tang, X. (2013). Hybrid deep learning for face verification. In *Proceedings of the IEEE International Conference on Computer Vision* (pp. 1489-1496).

Sun, Y., Wang, X., & Tang, X. (2014). Deep learning face representation from predicting 10,000 classes. In *Proceedings of the IEEE Conference on Computer Vision and Pattern Recognition* (pp. 1891-1898). doi:10.1109/CVPR.2014.244

Sutskever, I., Martens, J., Dahl, G., & Hinton, G. (2013). On the importance of initialization and momentum in deep learning. In *International conference on machine learning* (pp. 1139-1147).

Tang, Y. (2013). Deep learning using linear support vector machines. *arXiv preprint arXiv:1306.0239.*

Wan, J., Wang, D., Hoi, S. C. H., Wu, P., Zhu, J., Zhang, Y., & Li, J. (2014, November). Deep learning for content-based image retrieval: A comprehensive study. In *Proceedings of the 22nd ACM international conference on Multimedia*(pp. 157-166). ACM. doi:10.1145/2647868.2654948

Wang, N., & Yeung, D. Y. (2013). Learning a deep compact image representation for visual tracking. In Advances in neural information processing systems (pp. 809-817).

Warburton, K. (2003). Deep learning and education for sustainability. *International Journal of Sustainability in Higher Education*, *4*(1), 44–56. doi:10.1108/14676370310455332

Weigel, V. B. (2002). Deep Learning for a Digital Age: Technology's Untapped Potential To Enrich Higher Education. Jossey-Bass, 989 Market Street, San Francisco, CA 94103-1741.

Weston, J., Ratle, F., Mobahi, H., & Collobert, R. (2012). Deep learning via semi-supervised embedding. In *Neural Networks: Tricks of the Trade* (pp. 639–655). Springer Berlin Heidelberg. doi:10.1007/978-3-642-35289-8_34

## KEY TERMS AND DEFINITIONS

**Artificial Intelligence:** (1) Artificial intelligence is a software-based system solving real-world problems with its functions such as perception, learning, reasoning, thinking, inference, and decision making. (2) Artificial intelligence is a scientific research field in which it is tried to simulate human or living organisms' intelligence or natural dynamics by employing a computer or an autonomous system.

**Convolutional Neural Networks:** Convolutional neural networks is a type of artificial neural networks technique that is specifically designed to include some features/functions like convolution, non-linearity, pooling/sub-sampling, and classification in the context of deep learning.

**Data Processing:** Data processing is operations that are done for manipulating the objective any kind of data (i.e., image, signal, text) to give it different meanings, features, functions, etc.

**Deep Learning:** Deep learning (DL) is a method of machine learning that employs artificial neural networks (ANNs) with the structure above of the standards to deal with a large amount of data.

**Machine Learning:** Machine learning is an artificial intelligence approach that is based on the function of training a technique with some data to make it ready for solving objective problems.

# Chapter 15
# Intelligent Biomedical Engineering Operations by Cloud Computing Technologies

**Hasan Armutlu**
*Usak University, Turkey*

## ABSTRACT

*Cloud computing is an effective way of using hardware- and software-oriented resources at optimum levels. Thanks to this technology, it is possible to share large amounts of resources effectively and accurately among target users. Because it is a rapidly growing technology, one cannot deny that it has remarkable relations with alternative research fields having great potential and application scope. It is clear that artificial intelligence is one of these fields. As associated with both these research fields, the purpose of this chapter is to examine artificial-intelligence-based biomedical engineering works supported/connected with cloud computing. Because it has a vital importance with applications regarding the medical/health problems, biomedical engineering needs support from the most recent technologies and research fields in this manner. So, the chapter provides a view over the intersection of these three research fields as trying to improve awareness among interested readers.*

## INTRODUCTION

Scientific developments have always found their way with combinations of different fields. In even today, we can discuss about innovative outputs of many different fields – technologies and benefit from them along our daily life. As a result of increasing need for more use of information and reaching to it via practical ways, we have surrounded by different tools running over the factor of information – data. Developments occurred in this manner are some typical results of computer and electronics oriented developments and improvements and all these developments have caused the society – community to be passed through different eras – stages (Gurstein, 2000; Jones, 1998). At this point, humankind's improving has enabled to develop newer technologies and in this context, even the field of education has taken an important role to meet with both necessary educational technologies and necessary technological developments

DOI: 10.4018/978-1-5225-4769-3.ch015

synchronically (Gurstein, 2000; Katz and Katz, 2004; Kose et al., 2013). Here, influence of educational resources over the revolutionary communication technology: Internet and its Web platform has made it possible to increase speed of community improvement level and eventually, the humankind has created a more practical but fast-experienced world looking to a more technological future. Nowadays, the look to the future has gained more philosophical trend on achieving an autonomous future and in this context, combinations of different fields and technologies in them are key factors to have desired life standards. While it is also widely thought that the future with full of technology will bring or not disadvantages to us (Abdullah et al., 2017; Brosnan, 2002; Dinello, 2005; Etzioni & Etzioni, 2017; Gelbrich & Sattler, 2014; King, 2017; Kjerulff et al., 1992; Meuter et al., 2003; Oh et al., 2017), some research fields are known as essential components, which are strong enough to direct the humankind's future. The field of Artificial Intelligence is one of them.

Artificial Intelligence is one of the most important scientific fields, which is currently building up the future. Briefly, Artificial Intelligence was born with the idea of simulating human behaviors and thinking styles for solving real world problems at computer systems and in time, its inspiration scope has improved from just human to generally nature (Buchanan, 2005; Kose, 2015a; Nabiyev, 2003; Nilsson, 2014; Padhy, 2005; Russell et al., 1995). After its success real world problems has proved more and more, Artificial Intelligence has become a multidisciplinary approach to design and develop a more accurate and practical world. Today, there are many different types of intelligent problem solution approaches, methods, and techniques used through problems of different fields and in every time it becomes more possible to develop computer systems that are capable of solving even 'impossible' and 'too difficult' problems encountered. This success is because of mathematically and logically well designed Artificial Intelligence techniques, it is always active literature and also its strong relations with different fields. For example, developments in electronics or communication technologies (i.e. Internet, GSM, mobile systems) are important triggering factors for developing more advanced intelligent machines or software environments to deal with many real-world problems. On the other hand, theoretically and philosophically done new research works have been always key components to make infrastructure of the Artificial Intelligence more and more robust. In addition to all these factors, hybrid use of intelligent systems is also another widely followed problem solving approach that we can see often especially nowadays.

When the scientific literature is examined, it is possible to see relations of Artificial Intelligence with different, alternative and innovative technologies having important role on building the future. Briefly it is possible to see applications over Big Data (Chen et al., 2012; Chen et al., 2016; O'Leary, 2013) or Internet of Things (as build already with the idea of 'Artificial Intelligence based things' communicating) – (Bari et al., 2013; Holler et al., 2014; Katasonov et al., 2008; Madakam et al., 2015), which are currently trend research fields – technologies. In addition to them, it has been a remarkable application approach to employ intelligent problem solution ways in communication technologies and the associated software oriented solutions in this manner. At this point, another innovative technology called as Cloud Computing is a widely known essential factor taking many researchers' attention.

Briefly, Cloud Computing is both a field and technology, which has changed the way of using hardware and software by improving also use of Web and data over it. Cloud Computing finds its meaning as an effective way of using hardware and software oriented resources at optimum levels. Thanks to this technology, it is also possible to share big amount of resources effectively and accurately among target users. Because it is a rapidly growing technology, one cannot deny that it has remarkable relations with alternative research fields having great potential and application scope. It is clear that Artificial Intelligence is one of these fields. As associated with both these research fields - technologies, purpose of this chapter is to examine Artificial Intelligence based Biomedical Engineering works supported - connected with Cloud Computing. Because it has a vital importance with applications regarding to the medical - health problems, Biomedical Engineering needs support from the most recent technologies and research fields in this manner. So, the chapter has provided a view over the intersection of these three research fields as trying to improve awareness among interested readers.

Based on the subject of the chapter, remaining sections are organized as follows: The next section provides essential information about what is Cloud Computing and what main components is it build. This section also provides brief explanations about both Biomedical Engineering and Artificial Intelligence based application possibilities of Cloud Computing. After that section, the third section focuses some on the associated literature to discuss about intelligent Biomedical Engineering applications with the support by Cloud Computing. In detail, the section is a typical literature review to enabled readers for having enough ideas about the 'intersection' of the objective fields considered along this chapter. After the third section, the fourth section is devoted to some discussions about conclusions and future work ideas.

## CLOUD COMPUTING

As a result of need for using software and hardware resources, the approach of Cloud Computing has been introduced as a supportive technology regarding especially communication environments. Today, it is a widely used approach to improve efficient use of resources and having opportunities to create – edit – share data over the Web environment. In order to have more idea about how it is related to intelligent – bio-inspired solutions and Biomedical Engineering problems, essential sides of the Cloud Computing should be explained briefly.

### The Concept of Cloud Computing

The concept of Cloud Computing can be defined briefly as an approach of sharing software and hardware oriented resources effectively and efficiently by using the Internet communication infrastructure and even its Web environment (Kose & Armutlu, 2015). Thanks to the Cloud Computing, it is possible to design and develop solutions for sharing hardware resources virtually among users and benefit from also fast, sharable software applications to make everything more practical for users. Virtualization of the resources in order to make sharable them among users has been one key factor in popularity of Cloud Computing and its rise in the scientific community. Today, we can even think about a concept of 'renting computers' virtually by needing only a screen (or screen viewing component), mouse and / or keyboard and running other essential infrastructure over the Web (Armutlu, 2017). Essential advantages of Cloud Computing can be mentioned briefly as follows (Dinh et al., 2013; Hayes, 2008; Kose & Armutlu, 2015; Vouk, 2008):

- Cloud Computing provides efficient share and use of especially hardware resources.
- Cloud Computing allows developing more effective software solutions to create, edit and share data over Web.
- Cloud Computing allows more calculation and work power as a result of combining resources in a practical way.

- Cloud Computing allows optimum share and use of software and hardware resources.
- Cloud Computing allows more collaborative and synchronic working processes for users.
- Thanks to Cloud Computing, users can experience flexible, platform-free, practical and fast working environments.

All the mentioned advantages are provided under some specific application types of Cloud Computing, under also some system formations.

## Application Types of Cloud Computing

As a result of the developments in the field of Cloud Computing, there are many different application types that we can mention. Such application types are generally associated with how a Cloud Computing solution is applied for objective users or objective problem considered to be solved via 'cloud' based processes. In other words, these are some typical 'services' done over Cloud Computing. By eliminating detailed application types introduced in time, we can list five different types of applications – services done with Cloud Computing (Kose & Armutlu, 2015):

- **Software as a Service – SaaS:** This type of application is a common use of Cloud Computing currently. Applications such as Google Drive, DropBox can be given as main examples for SaaS oriented Cloud Computing. Here, users can benefit from software environments for different purposes by not dealing with details on the background as associated with essential features and functions of Cloud Computing.
- **Communication as a Service – CaaS:** CaaS mainly focuses on the mechanism of ensuring communication sessions over Cloud Computing infrastructure. Out of the software environments, this application type deals with how to form a communication structure for building i.e. a Voice IP communication network, teleconferencing solutions.
- **Platform as a Service – PaaS:** This type of application is mostly for multi-purpose works needing design and development of objective solutions, databases to store some data or any other things associated with developing specific software solutions by benefiting from essential advantages of a Cloud Computing infrastructure.

*Figure 1. Advantages of Cloud Computing briefly*

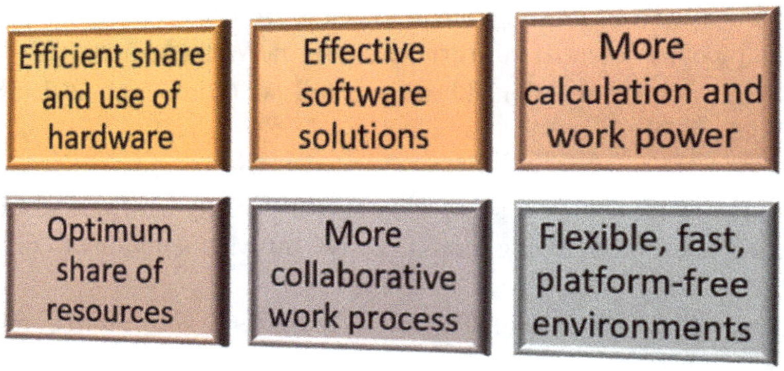

- **Infrastructure as a Service – IaaS:** In this application type, it is possible to benefit from exact physical hardware resources and also some supportive software solutions to build a general Cloud Computing structure according to specific needs. The resources are somehow based on Cloud Computing features – functions, so it is possible to adapt a flexible system running over a 'cloud' solution.
- **Network as a Service – NaaS:** NaaS corresponds to use of Cloud Computing oriented resources as a network management approach. In detail, some formed systems like VPN or dynamic bandwidth management are common examples of this application type.

In addition to the mentioned application types, it is also possible to talk about system formations for Cloud Computing oriented solutions. According to scope – limitations of the formed Cloud Computing system, it is possible to list four different system formations (Kose and Armutlu, 2015):

- **Public Cloud:** It is a system formation in which the related working mechanisms are employed to achieve a cloud approach open to public use.
- **Private Cloud:** It is the cloud system formation, which is private for use within only i.e. a company environment to achieve better security or management processes.
- **Community Cloud:** That is the cloud system formation, which is shared by more than one service receivers. In this way, costs are reduced as well as common works – applications done by different i.e. companies are done in a more practical and manageable way.

*Figure 2. Application types of Cloud Computing*

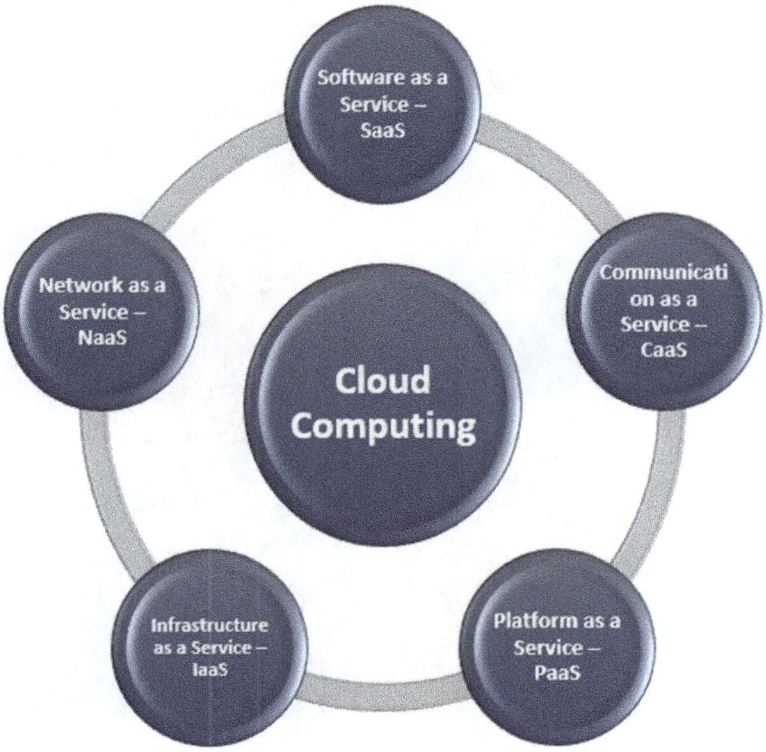

- **Hybrid Cloud:** Hybrid Cloud is another system formation type in which service receivers benefit from both Public Cloud and Private Cloud structures at the same time for different purposes, under a common Cloud Computing system approach.

As the field of Cloud Computing improves in time, many different application orientations, discussions over its different ways, and even issues in management it and also achieving security has been remarkable interests in the associated literature (Rittinghouse and Ransome, 2016; Krutz and Vines, 2010; Zhang et al., 2010). Today, more consideration for Cloud Computing is generally given for especially business oriented works because of its effectiveness on optimum use and share of resources to speed up and reduce costs (Chang et al., 2010; Chang et al., 2013; Marston et al., 2011; Weinhardt et al., 2009). Optimum use to get more speed and reduced costs are important for especially a recent, popular, data oriented research topic called as Big Data (Hashem et al., 2015). On the other hand, its applications within different fields is also another trendy research way followed. In this chapter, the field of Biomedical Engineering has been considered in this manner.

## Biomedical Engineering and Cloud Computing

Biomedical Engineering is an important field with its role on health oriented solutions. Although this field has lots of detailed research interests, some application types have come onto the foreground. At this point, Cloud Computing oriented applications have also been considered in the associated literature. According to some popular research works reported in the literature, it is possible to mention the following solution ways of Cloud Computing for Biomedical Engineering problems (Chen et al., 2013; Doukas et al., 2010; Doukas & Maglogiannis, 2012; Fusaro et al., 2011; Gunarathne et al., 2011; Kuo, 2011; O'Driscoll et al., 2013; Penhaker et al., 2012; Rolim et al., 2010; Rosenthal et al., 2010):

*Figure 3. System formation types of Cloud Computing*

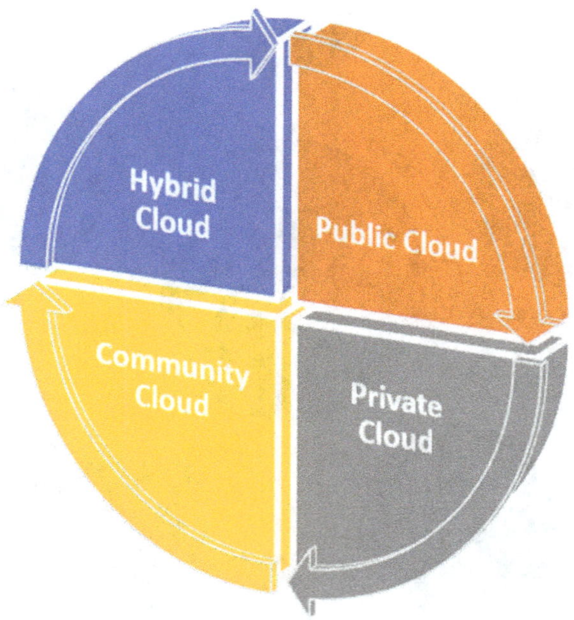

- More effective share of biomedical data – information,
- Practical and effective mobile software solutions for healthcare,
- Better scientific computation solutions for biomedical problems,
- Special healthcare oriented Web platforms – environments,
- Special healthcare oriented Web services,
- Solutions for especially Big Data oriented biomedical applications.

With recent focuses on effective research fields – technologies, Cloud Computing has gone through many changes. In this context, Artificial Intelligence based Cloud Computing applications are remarkable research interests nowadays.

## Intelligent Cloud Computing

While traditional Cloud Computer was in the form of software and hardware based innovative ideas behind, there has been still something more to do for achieving something 'intelligent' in operations – processes done along a typical Cloud Computing system. That's briefly something like supporting innovative technologies with some other innovative technologies to gain more effective technological solutions. This way has enabled researchers to think about possibilities on applying Artificial Intelligence to Cloud Computing.

Relation between Artificial Intelligence and Cloud Computing can be achieved via three different ways: Artificial Intelligence supported by Cloud Computing, Cloud Computing supported by Artificial Intelligence, and finally a mutual support done between both Artificial Intelligence and Cloud Computing. Because Cloud Computing focuses on optimum use of resources and Artificial Intelligence has enough features and functions to achieve such solutions, both these fields – technologies have had common developments for providing intelligent solutions. Currently, one important application type of Artificial Intelligence and Cloud Computing is associated with use of intelligent agents. Because also popular programming approaches like Object Oriented Programming can be widely used to develop intelligent agents (Kose, 2015b), it has been a remarkable research interest in solving real world problems with problem solution elements acting intelligently. At this point, optimizing such agents and running them fast and efficiently over Cloud Computing approaches have been a research interest recently (Li et al., 2011; Sim, 2012; Talia, 2011; Talia, 2012; Zhang and Zhang, 2009). On the other hand, there is a current increasing trend in developing Internet of Thing (IoT) systems supported with Cloud Computing solutions. Because Internet of Things is a typical hardware based intelligent network of machines (Kopetz, 2011; Wortmann & Flüchter, 2015), use of Cloud Computing to achieve efficient and accurate systems is a remarkable interest in Internet of Things research (Botta et al., 2014; Gubbi et al., 2013; Rao et al., 2012; Soliman et al., 2013; Zhou et al., 2013). Some other intelligent Cloud Computing applications are based on also ensuring robots benefiting from advantages of resource share – use supported by cloud solutions and also optimizing some aspects like load balancing, operation schedule…etc. of Cloud Computing systems (Arumugam et al., 2010; Chen et al., 2010; Hu et al., 2012; Lorencik & Sincak, 2013; Mishra & Jaiswal, 2012; Pandey et al., 2010; Ramezani et al., 2014; Xue & Wu, 2012; Zhu et al., 2011).

*Figure 4. Some intelligent Cloud Computing application interests*

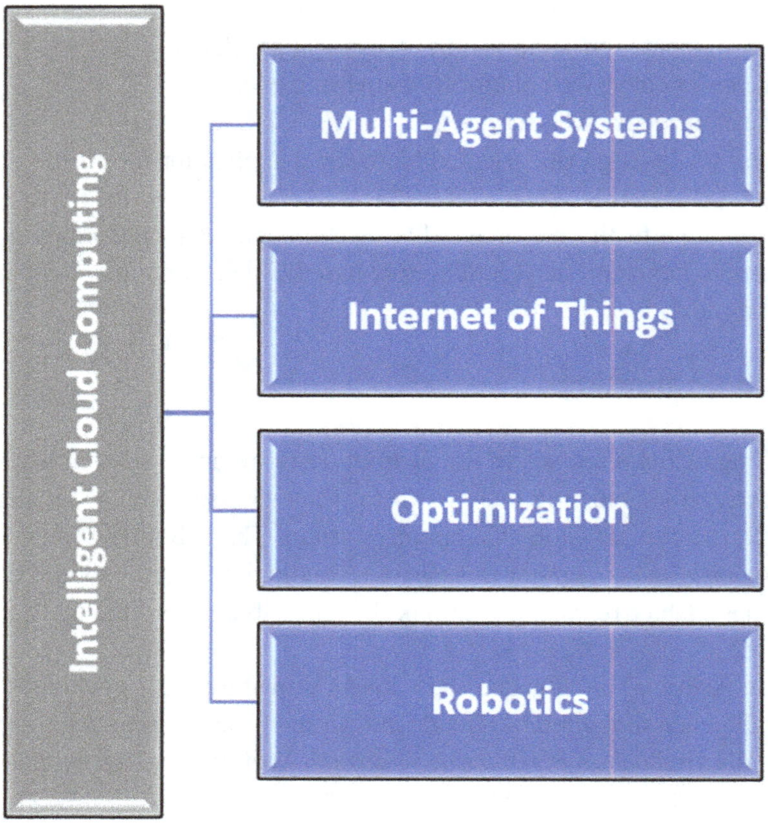


## INTELLIGENT CLOUD COMPUTING APPLICATIONS IN BIOMEDICAL ENGINEERING

The literature associated with intelligent Cloud Computing applications in Biomedical Engineering seems young as research works in the mentioned scope are introduced still to improve it among other literatures. In this context, it is also possible to focus on intelligent use of Cloud Computing to even derive ideas about Biomedical Engineering applications.

When the literature is examined, it is possible to see that there are different types of research works benefiting from especially general purpose and also optimization oriented techniques, which are mostly examined under 'bio-inspired techniques'. Considering their popularities, some remarkable research works can be explained briefly as follows:

- The research work performed by Zhang and Zhang (2011) focused on a development of Internet of Things and Cloud Computing based system to run a health support and telemedicine system for supporting individuals over an open source, flexible platform.
- In a work by Doukas and Maglogiannis (2011), wearable sensors have been employed to get some health data from individuals in order to support them over a cloud system in terms of patient monitoring and emergency response approaches.

- As another work focused on body sensors and monitoring, Fortino et al. (2012) have designed and developed a Cloud Computing system over a network of body sensors. Called as BodyCloud, the system has been able to use data from individuals for monitoring purposes.
- In terms of patient monitoring systems, it is possible to see also examples of Internet of Things oriented systems to run over Cloud Computing for monitoring processes. The research work by Abawajy and Hassan (2017) is a recent example of that development approach.
- Jemal et al. (2015) have performed a research work in which they have designed and developed a mobile Cloud Computing system benefiting from also multi-agent systems to run an intelligent healthcare support approach.
- Sung and Chiang (2012) provided a research work and in that study they have developed a medical care system build over an Internet of Things system structure optimized by also an improved Particle Swarm Optimization (PSO) algorithm as a bio-inspired technique.
- Lee et al. (2009) developed a context-aware system for healthcare purposes as including intelligent techniques like Swarm Intelligence oriented ones and also Bayesian Networks. Data received over the formed system have been gathered thanks to sensor based technologies and also Cloud Computing.
- In the work by Zhang et al. (2014), a data replica placement scheme has been introduced under a healthcare oriented Internet of Things system. In detail, a cloud storage has been run along the study and the bio-inspired techniques of Mosquitoes Oviposition Mating (MOX) and Simulated Annealing (SA) were run within the Artificial Intelligence solution way.
- The work done by Ganapathy (2014) has provided a system of healthcare approach as based on a 'sensor cloud' environment. In detail, the author has employed some techniques like Neuro-Fuzzy Inference System (ANFIS) and Hidden Markov Models to run intelligent solution mechanisms within the introduced approach.
- Thangaraj et al. (2016) provided a research work, which benefits from Internet of Things to build a smart healthcare approach. In detail, a semantic framework has also been used over an agent using infrastructure.
- The research performed by Alamri (2016) run an optimization process done by nature-inspired techniques over a healthcare system. As based on a cloud-media approach, the objective in this research has been to achieve a multimedia service composition in a general manner.
- As another work over healthcare purpose, Hanen et al. (2016) have designed and developed a mobile cloud oriented system. Under essential features – functions of a Web service channel, the system has benefited from also a multi-agent based solution approach.
- In a recent work by Do (2017), application potentials of robotics and Biomedical Engineering problems has been discussed by considering 'predictive models'. As another recent work at this point, by considering also Internet of Things to achieve robots for healthcare, Patel et al. (2017) have provided a remarkable research work.
- In their work, Firorini et al. (2017) provided a personalized intelligent system for chronic disease management. In detail, the system has been built over a robot and cloud oriented problem solution environment.
- In the context of robotics and healthcare purposes, Bachkosky et al. (2017) have focused on a detailed assistive robot design and development process benefiting from Artificial Intelligence and cloud oriented solutions.

## CONCLUSION

This chapter has provided a general view on using Cloud Computing within Artificial Intelligence based Biomedical Engineering applications. In detail, more consideration has been given to especially bio-inspired techniques under the field of Artificial Intelligence. Because it is thought that the main infrastructure of the Artificial Intelligence is associated with mostly bio-inspired elements. In this context, the chapter found its way within the intersection of Cloud Computing and Artificial Intelligence with its bio-inspired techniques. Along the chapter content, essential focus was also regarding Cloud Computing.

It is clear that more influence of the Internet and its 'interface' Web have opened many doors of innovation. As a result, some practical approaches of improved communication have taken active place in our lives. As one of these approaches, Cloud Computing made it possible to run both hardware and software oriented resources in an optimum level and achieve more effectiveness and efficiency in the works of modern life. Among all of these works, health based works have always gained vital importance because of their roles in improving life standards. In time, especially Biomedical Engineering has gained popularity in terms of research works and that increase in popularity has directed the researchers to seek for also alternative application platforms. As a powerful application environment, computer and communication technologies have taken an important place in this scope and finally, one of the most remarkable approaches of today's innovative world, Cloud Computing has been started to be applied in the works of Biomedical Engineering. Following that, need for more 'intelligent' solution ways has increased interest in also using intelligent Cloud Computing applications towards Biomedical Engineering.

Through this chapter, the readers were enabled to increase their awareness towards applications of Cloud Computing in achieving intelligent Biomedical oriented works. In detail, the content has focused some on the associated literature and tried to find answers to the question of 'How can Cloud Computing and Artificial Intelligence be combined for getting better Biomedical Engineering?'. It is important to indicate that the future has many things that were not answered yet and most of answers will be probably given by the innovative technologies like Artificial Intelligence and Cloud Computing. Rise of some solution ways and techniques like Internet of Things Big Data or Deep Learning are some signs of that flow. On the other hand, the approach of Cloud Computing is not the same approach on its first steps to the scientific community. In time, it has divided into many different research interests and we can see the results through the past time in the related literatures.

Research considered here has also enabled the author to think about some future works. The ideas here may be useful also interested readers. It is possible to indicate the thought future work ideas briefly as follows:

- Application potentials and examples expressed here are thought to be realized in the future over appropriate development environments.
- Potentials of Internet of Things supported with Intelligent Cloud Computing for Biomedical Engineering will be studied in the future.
- More focus on developments within the literature, by considering the topic of this chapter, will be given to continue to improve awareness.

## REFERENCES

Abawajy, J. H., & Hassan, M. M. (2017). Federated internet of things and cloud computing pervasive patient health monitoring system. *IEEE Communications Magazine*, *55*(1), 48–53. doi:10.1109/MCOM.2017.1600374CM

Abdullah, L. M., Azuddin, M., Malik, S. A., & Mahmud, M. (2017). Rising Ageing Population: A Preliminary Study of Malaysian Older People Expectations in Information and Communication Technology. In *International Conference of Reliable Information and Communication Technology* (pp. 796-803). Springer.

Alamri, A. (2016). Nature-inspired multimedia service composition in a media cloud-based healthcare environment. *Cluster Computing*, *19*(4), 2251–2260. doi:10.1007/s10586-016-0647-9

Armutlu, H. (2017). A Model for Cloud Computing Based 'Rent A Computer'. *Journal of Multidisciplinary Developments*, *2*(1), 22–24.

Arumugam, R., Enti, V. R., Bingbing, L., Xiaojun, W., Baskaran, K., Kong, F. F., . . . Kit, G. W. (2010). DAvinCi: A cloud computing framework for service robots. In *Robotics and Automation (ICRA), 2010 IEEE International Conference on* (pp. 3084-3089). IEEE. doi:10.1109/ROBOT.2010.5509469

Bachkosky, J. V., Boukhvalova, A., Chou, K., Gunnarsson, W., Ledwell, J., McTaggart, B., . . . Yon, J. (2017). *Sensory Computing and Object Processing Entity: Assistive Robotics for Healthcare* (Doctoral dissertation).

Bari, N., Mani, G., & Berkovich, S. (2013). Internet of things as a methodological concept. In *Computing for Geospatial Research and Application (COM. Geo), 2013 Fourth International Conference on* (pp. 48-55). IEEE. doi:10.1109/COMGEO.2013.8

Botta, A., De Donato, W., Persico, V., & Pescapé, A. (2014). On the integration of cloud computing and internet of things. In *Future Internet of Things and Cloud (FiCloud), 2014 International Conference on* (pp. 23-30). IEEE. doi:10.1109/FiCloud.2014.14

Brosnan, M. J. (2002). *Technophobia: The psychological impact of information technology*. Routledge.

Buchanan, B. G. (2005). A (very) brief history of artificial intelligence. *AI Magazine*, *26*(4), 53.

Chang, V., Bacigalupo, D., Wills, G., & De Roure, D. (2010). A categorisation of cloud computing business models. In *Proceedings of the 2010 10th IEEE/ACM international conference on cluster, cloud and grid computing* (pp. 509-512). IEEE Computer Society. doi:10.1109/CCGRID.2010.132

Chang, V., Walters, R. J., & Wills, G. (2013). The development that leads to the Cloud Computing Business Framework. *International Journal of Information Management*, *33*(3), 524–538. doi:10.1016/j.ijinfomgt.2013.01.005

Chen, J., Qian, F., Yan, W., & Shen, B. (2013). Translational biomedical informatics in the cloud: Present and future. *BioMed Research International*. PMID:23586054

Chen, Y., Du, Z., & García-Acosta, M. (2010). Robot as a service in cloud computing. In *Service Oriented System Engineering (SOSE), 2010 Fifth IEEE International Symposium on* (pp. 151-158). IEEE. doi:10.1109/SOSE.2010.44

Dinello, D. (2005). *Technophobia! Science fiction visions of posthuman technology*. University of Texas Press.

Dinh, H. T., Lee, C., Niyato, D., & Wang, P. (2013). A survey of mobile cloud computing: Architecture, applications, and approaches. *Wireless Communications and Mobile Computing*, *13*(18), 1587–1611. doi:10.1002/wcm.1203

Do, H. N. (2017). *Predictive models for robotics and biomedical applications* (Doctoral dissertation). Michigan State University.

Doukas, C., & Maglogiannis, I. (2011). Managing wearable sensor data through cloud computing. In *Cloud Computing Technology and Science (CloudCom), 2011 IEEE Third International Conference on* (pp. 440-445). IEEE. doi:10.1109/CloudCom.2011.65

Doukas, C., & Maglogiannis, I. (2012). Bringing IoT and cloud computing towards pervasive healthcare. In *Innovative Mobile and Internet Services in Ubiquitous Computing (IMIS), 2012 Sixth International Conference on* (pp. 922-926). IEEE. doi:10.1109/IMIS.2012.26

Doukas, C., Pliakas, T., & Maglogiannis, I. (2010). Mobile healthcare information management utilizing Cloud Computing and Android OS. In *Engineering in Medicine and Biology Society (EMBC), 2010 Annual International Conference of the IEEE* (pp. 1037-1040). IEEE. doi:10.1109/IEMBS.2010.5628061

Etzioni, A., & Etzioni, O. (2017). Incorporating Ethics into Artificial Intelligence. *The Journal of Ethics*, 1–16.

Fiorini, L., Esposito, R., Bonaccorsi, M., Petrazzuolo, C., Saponara, F., Giannantonio, R., & Cavallo, F. et al. (2017). Enabling personalised medical support for chronic disease management through a hybrid robot-cloud approach. *Autonomous Robots*, *41*(5), 1263–1276. doi:10.1007/s10514-016-9586-9

Fortino, G., Pathan, M., & Di Fatta, G. (2012). Bodycloud: Integration of cloud computing and body sensor networks. In *Cloud Computing Technology and Science (CloudCom), 2012 IEEE 4th International Conference on* (pp. 851-856). IEEE.

Fusaro, V. A., Patil, P., Gafni, E., Wall, D. P., & Tonellato, P. J. (2011). Biomedical cloud computing with amazon web services. *PLoS Computational Biology*, *7*(8), e1002147. doi:10.1371/journal.pcbi.1002147 PMID:21901085

Ganapathy, K. (2014). *Soft Computing Approaches for Healthcare in Sensor Cloud Environment* (Thesis). Anna University, Faculty of Information and Communication Engineering.

Gelbrich, K., & Sattler, B. (2014). Anxiety, crowding, and time pressure in public self-service technology acceptance. *Journal of Services Marketing*, *28*(1), 82–94. doi:10.1108/JSM-02-2012-0051

Gubbi, J., Buyya, R., Marusic, S., & Palaniswami, M. (2013). Internet of Things (IoT): A vision, architectural elements, and future directions. *Future Generation Computer Systems*, *29*(7), 1645–1660. doi:10.1016/j.future.2013.01.010

Gunarathne, T., Wu, T. L., Choi, J. Y., Bae, S. H., & Qiu, J. (2011). Cloud computing paradigms for pleasingly parallel biomedical applications. *Concurrency and Computation*, *23*(17), 2338–2354. doi:10.1002/cpe.1780

Gurstein, M. (Ed.). (2000). *Community Informatics: Enabling Communities with Information and Communications Technologies: Enabling Communities with Information and Communications Technologies*. IGI Global. doi:10.4018/978-1-878289-69-8

Hanen, J., Kechaou, Z., & Ayed, M. B. (2016). An enhanced healthcare system in mobile cloud computing environment. *Vietnam Journal of Computer Science*, *3*(4), 267–277. doi:10.1007/s40595-016-0076-y

Hashem, I. A. T., Yaqoob, I., Anuar, N. B., Mokhtar, S., Gani, A., & Khan, S. U. (2015). The rise of "big data" on cloud computing: Review and open research issues. *Information Systems*, *47*, 98–115. doi:10.1016/j.is.2014.07.006

Hayes, B. (2008). Cloud computing. *Communications of the ACM*, *51*(7), 9–11. doi:10.1145/1364782.1364786

Holler, J., Tsiatsis, V., Mulligan, C., Avesand, S., Karnouskos, S., & Boyle, D. (2014). *From Machine-to-machine to the Internet of Things: Introduction to a New Age of Intelligence*. Academic Press.

Hu, G., Tay, W. P., & Wen, Y. (2012). Cloud robotics: Architecture, challenges and applications. *IEEE Network*, *26*(3), 21–28. doi:10.1109/MNET.2012.6201212

Jemal, H., Kechaou, Z., Ayed, M. B., & Alimi, A. M. (2015). Mobile cloud computing in healthcare system. In *Computational Collective Intelligence* (pp. 408–417). Cham: Springer. doi:10.1007/978-3-319-24306-1_40

Jones, S. (Ed.). (1998). *Cybersociety 2.0: Revisiting computer-mediated community and technology* (Vol. 2). Sage Publications.

Katasonov, A., Kaykova, O., Khriyenko, O., Nikitin, S., & Terziyan, V. Y. (2008). Smart Semantic Middleware for the Internet of Things. *ICINCO-ICSO*, *8*, 169–178.

Katz, R. N., & Katz, R. N. (2004). *Information technology leadership in higher education: The condition of the community*. EDUCAUSE Center for Applied Research.

King, B. (2017). *Frankenstein's Legacy: Four Conversations about Artificial Intelligence, Machine Learning, and the Modern World*. Repository.cmu.edu.

Kjerulff, K. H., Pillar, B., Mills, M. E., & Lanigan, J. (1992). Technology anxiety as a potential mediating factor in response to medical technology. *Journal of Medical Systems*, *16*(1), 7–13. doi:10.1007/BF01674093 PMID:1645046

Kopetz, H. (2011). Internet of things. In Real-time systems (pp. 307-323). Springer US. doi:10.1007/978-1-4419-8237-7_13

Kose, U. (2010). *Web 2.0 Technologies in E-learning. In Free and Open Source Software for E-learning: Issues, Successes and Challenges* (pp. 1–23). IGI Global.

Kose, U. (2015a). Present state of swarm intelligence and future directions. In *Encyclopedia of Information Science and Technology* (3rd ed.; pp. 239–252). IGI Global. doi:10.4018/978-1-4666-5888-2.ch023

Kose, U. (2015b). *Algoritma Kavramı ve Bilgisayar Programlama Öğreniminde Pratik Yaklaşımlar [Practical Approaches in Learning Algorithm Concept and Computer Programming]*. Usak: AKY Press. (in Turkish)

Kose, U., & Armutlu, H. (2015). *Bulut Bilişim: Temel Konular ve Amazon Web Services [Cloud Computing: Essential Subjects and Amazon Web Services]*. Ankara: Detay Press. (in Turkish)

Kose, U., Koc, D., & Yucesoy, S. A. (2013). Design and Development of a Sample" Computer Programming" Course Tool via Story-Based E-Learning Approach. *Educational Sciences: Theory and Practice, 13*(2), 1235–1250.

Krutz, R. L., & Vines, R. D. (2010). *Cloud security: A comprehensive guide to secure cloud computing*. Wiley Publishing.

Kuo, A. M. H. (2011). Opportunities and challenges of cloud computing to improve health care services. *Journal of Medical Internet Research, 13*(3), e67. doi:10.2196/jmir.1867 PMID:21937354

Lee, K., Lunney, T., Curran, K., & Santos, J. (2009). A Context-Aware Platform for Computer Assisted Healthcare. In *PGNet 2009, The 10th Annual Conference on the Convergence of Telecommunications, Networking & Broadcasting*. Liverpool John Moores.

Li, Z., Chen, C., & Wang, K. (2011). Cloud computing for agent-based urban transportation systems. *IEEE Intelligent Systems, 26*(1), 73–79. doi:10.1109/MIS.2011.10

Lorencik, D., & Sincak, P. (2013). Cloud Robotics: Current trends and possible use as a service. In *Applied Machine Intelligence and Informatics (SAMI), 2013 IEEE 11th International Symposium on* (pp. 85-88). IEEE.

Madakam, S., Ramaswamy, R., & Tripathi, S. (2015). Internet of Things (IoT): A literature review. *Journal of Computer and Communications, 3*(05), 164–173. doi:10.4236/jcc.2015.35021

Marston, S., Li, Z., Bandyopadhyay, S., Zhang, J., & Ghalsasi, A. (2011). Cloud computing—The business perspective. *Decision Support Systems, 51*(1), 176–189. doi:10.1016/j.dss.2010.12.006

Meuter, M. L., Ostrom, A. L., Bitner, M. J., & Roundtree, R. (2003). The influence of technology anxiety on consumer use and experiences with self-service technologies. *Journal of Business Research, 56*(11), 899–906. doi:10.1016/S0148-2963(01)00276-4

Mishra, R., & Jaiswal, A. (2012). Ant colony optimization: A solution of load balancing in cloud. *International Journal of Web & Semantic Technology, 3*(2), 33–50. doi:10.5121/ijwest.2012.3203

Nabiyev, V. V. (2003). *Yapay Zeka [Artificial Intelligence]*. Ankara: Seçkin Press. (in Turkish)

Nilsson, N. J. (2014). *Principles of artificial intelligence*. Morgan Kaufmann.

O'Driscoll, A., Daugelaite, J., & Sleator, R. D. (2013). 'Big data', Hadoop and cloud computing in genomics. *Journal of Biomedical Informatics, 46*(5), 774–781. doi:10.1016/j.jbi.2013.07.001 PMID:23872175

O'Leary, D. E. (2013). Artificial intelligence and big data. *IEEE Intelligent Systems, 28*(2), 96–99. doi:10.1109/MIS.2013.39 PMID:25505373

Oh, C., Lee, T., Kim, Y., Park, S., & Suh, B. (2017). Us vs. Them: Understanding Artificial Intelligence Technophobia over the Google DeepMind Challenge Match. In *Proceedings of the 2017 CHI Conference on Human Factors in Computing Systems* (pp. 2523-2534). ACM. doi:10.1145/3025453.3025539

Padhy, N. P. (2005). *Artificial intelligence and intelligent systems*. Oxford University Press.

Pandey, S., Wu, L., Guru, S. M., & Buyya, R. (2010). A particle swarm optimization-based heuristic for scheduling workflow applications in cloud computing environments. In *Advanced information networking and applications (AINA), 2010 24th IEEE international conference on* (pp. 400-407). IEEE. doi:10.1109/AINA.2010.31

Patel, A. R., Patel, R. S., Singh, N. M., & Kazi, F. S. (2017). Vitality of Robotics in Healthcare Industry: An Internet of Things (IoT) Perspective. In *Internet of Things and Big Data Technologies for Next Generation Healthcare* (pp. 91–109). Springer International Publishing. doi:10.1007/978-3-319-49736-5_5

Penhaker, M., Krejcar, O., Kasik, V., & Snášel, V. (2012). Cloud computing environments for biomedical data services. *Intelligent Data Engineering and Automated Learning-IDEAL, 2012*, 336–343.

Ramezani, F., Lu, J., & Hussain, F. K. (2014). Task-based system load balancing in cloud computing using particle swarm optimization. *International Journal of Parallel Programming*, *42*(5), 739–754. doi:10.1007/s10766-013-0275-4

Rao, B. P., Saluia, P., Sharma, N., Mittal, A., & Sharma, S. V. (2012). Cloud computing for Internet of Things & sensing based applications. In *Sensing Technology (ICST), 2012 Sixth International Conference on* (pp. 374-380). IEEE. doi:10.1109/ICSensT.2012.6461705

Rittinghouse, J. W., & Ransome, J. F. (2016). *Cloud computing: implementation, management, and security*. CRC Press.

Rolim, C. O., Koch, F. L., Westphall, C. B., Werner, J., Fracalossi, A., & Salvador, G. S. (2010). A cloud computing solution for patient's data collection in health care institutions. In *eHealth, Telemedicine, and Social Medicine, 2010. ETELEMED'10. Second International Conference on* (pp. 95-99). IEEE. doi:10.1109/eTELEMED.2010.19

Rosenthal, A., Mork, P., Li, M. H., Stanford, J., Koester, D., & Reynolds, P. (2010). Cloud computing: A new business paradigm for biomedical information sharing. *Journal of Biomedical Informatics*, *43*(2), 342–353. doi:10.1016/j.jbi.2009.08.014 PMID:19715773

Russell, S., Norvig, P., & Intelligence, A. (1995). A modern approach. *Artificial Intelligence. Prentice-Hall. Egnlewood Cliffs*, *25*, 27.

Sim, K. M. (2012). Agent-based cloud computing. *IEEE Transactions on Services Computing*, *5*(4), 564–577. doi:10.1109/TSC.2011.52

Soliman, M., Abiodun, T., Hamouda, T., Zhou, J., & Lung, C. H. (2013). Smart home: Integrating internet of things with web services and cloud computing. In *Cloud Computing Technology and Science (CloudCom), 2013 IEEE 5th International Conference on* (Vol. 2, pp. 317-320). IEEE.

Sung, W. T., & Chiang, Y. C. (2012). Improved particle swarm optimization algorithm for android medical care IOT using modified parameters. *Journal of Medical Systems*, *36*(6), 3755–3763. doi:10.1007/s10916-012-9848-9 PMID:22492176

Talia, D. (2011). Cloud Computing and Software Agents: Towards Cloud Intelligent Services. In WOA (Vol. 11, pp. 2-6). Academic Press.

Talia, D. (2012). Clouds meet agents: Toward intelligent cloud services. *IEEE Internet Computing*, *16*(2), 78–81. doi:10.1109/MIC.2012.28

Thangaraj, M., Ponmalar, P. P., Sujatha, G., & Anuradha, S. (2016). Agent based Semantic Internet of Things (IoT) in Smart Health care. In *Proceedings of the The 11th International Knowledge Management in Organizations Conference on The changing face of Knowledge Management Impacting Society* (p. 41). ACM. doi:10.1145/2925995.2926023

Vouk, M. A. (2008). Cloud computing–issues, research and implementations. *CIT. Journal of Computing and Information Technology*, *16*(4), 235–246. doi:10.2498/cit.1001391

Weinhardt, C., Anandasivam, A., Blau, B., Borissov, N., Meinl, T., Michalk, W., & Stößer, J. (2009). Cloud computing–a classification, business models, and research directions. *Business & Information Systems Engineering*, *1*(5), 391–399. doi:10.1007/s12599-009-0071-2

Wortmann, F., & Flüchter, K. (2015). Internet of things. *Business & Information Systems Engineering*, *57*(3), 221–224. doi:10.1007/s12599-015-0383-3

Xue, S. J., & Wu, W. (2012). Scheduling workflow in cloud computing based on hybrid particle swarm algorithm. *Indonesian Journal of Electrical Engineering and Computer Science*, *10*(7), 1560–1566.

Zhang, B., Wang, X., & Huang, M. (2014). A data replica placement scheme for cloud storage under healthcare iot environment. In *Fuzzy Systems and Knowledge Discovery (FSKD), 2014 11th International Conference on* (pp. 542-547). IEEE. doi:10.1109/FSKD.2014.6980892

Zhang, Q., Cheng, L., & Boutaba, R. (2010). Cloud computing: State-of-the-art and research challenges. *Journal of Internet Services and Applications*, *1*(1), 7–18. doi:10.1007/s13174-010-0007-6

Zhang, X. M., & Zhang, N. (2011). An open, secure and flexible platform based on internet of things and cloud computing for ambient aiding living and telemedicine. In *Computer and Management (CAMAN), 2011 International Conference on* (pp. 1-4). IEEE. doi:10.1109/CAMAN.2011.5778905

Zhang, Z., & Zhang, X. (2009). Realization of open cloud computing federation based on mobile agent. In Intelligent computing and intelligent systems, 2009. ICIS 2009. IEEE international conference on (Vol. 3, pp. 642-646). IEEE. doi:10.1109/ICICISYS.2009.5358085

Zhou, J., Leppanen, T., Harjula, E., Ylianttila, M., Ojala, T., Yu, C., . . . Yang, L. T. (2013). Cloudthings: A common architecture for integrating the internet of things with cloud computing. In *Computer Supported Cooperative Work in Design (CSCWD), 2013 IEEE 17th International Conference on* (pp. 651-657). IEEE. doi:10.1109/CSCWD.2013.6581037

Zhu, K., Song, H., Liu, L., Gao, J., & Cheng, G. (2011). Hybrid genetic algorithm for cloud computing applications. In *Services Computing Conference (APSCC), 2011 IEEE Asia-Pacific* (pp. 182-187). IEEE. doi:10.1109/APSCC.2011.66

## ADDITIONAL READING

An, B., Lesser, V., Irwin, D., & Zink, M. (2010). Automated negotiation with decommitment for dynamic resource allocation in cloud computing. In *Proceedings of the 9th International Conference on Autonomous Agents and Multiagent Systems: volume 1-Volume 1* (pp. 981-988). International Foundation for Autonomous Agents and Multiagent Systems.

Armbrust, M., Fox, A., Griffith, R., Joseph, A. D., Katz, R. H., Konwinski, A., . . . Zaharia, M. (2009). *Above the clouds: A berkeley view of cloud computing* (Vol. 17). Technical Report UCB/EECS-2009-28, EECS Department, University of California, Berkeley.

Behera, A., Ratha, B. K., & Sethi, S. (2017). Green cloud computing: A survey. *IJSEAT, 4*(12), 763–767.

Buyya, R., Yeo, C. S., & Venugopal, S. (2008). Market-oriented cloud computing: Vision, hype, and reality for delivering it services as computing utilities. In *High Performance Computing and Communications, 2008. HPCC'08. 10th IEEE International Conference on* (pp. 5-13). IEEE.

Calheiros, R. N., Ranjan, R., Beloglazov, A., De Rose, C. A., & Buyya, R. (2011). CloudSim: A toolkit for modeling and simulation of cloud computing environments and evaluation of resource provisioning algorithms. *Software, Practice & Experience, 41*(1), 23–50. doi:10.1002/spe.995

Cao, B. Q., Li, B., & Xia, Q. M. (2009). A service-oriented qos-assured and multi-agent cloud computing architecture. *Cloud computing*, 644-649.

Chen, G., Wang, E., Sun, X., & Lu, Y. (2016). An Intelligent Approval System for City Construction based on Cloud Computing and Big Data. *International Journal of Grid and High Performance Computing, 8*(3), 57–69. doi:10.4018/IJGHPC.2016070104

Chen, H., Chiang, R. H., & Storey, V. C. (2012). Business intelligence and analytics: From big data to big impact. *Management Information Systems Quarterly, 36*(4).

Delic, K. A., & Walker, M. A. (2008). Emergence of the academic computing clouds. *Ubiquity, 2008*(August), 1. doi:10.1145/1459229.1414664

Endo, P. T., de Almeida Palhares, A. V., Pereira, N. N., Goncalves, G. E., Sadok, D., Kelner, J., & Mangs, J. E. et al. (2011). Resource allocation for distributed cloud: Concepts and research challenges. *IEEE Network, 25*(4), 42–46. doi:10.1109/MNET.2011.5958007

Feng, D. G., Zhang, M., Zhang, Y., & Xu, Z. (2011). Study on cloud computing security. *Journal of Software, 22*(1), 71–83. doi:10.3724/SP.J.1001.2011.03958

Fernández, A., del Río, S., López, V., Bawakid, A., del Jesus, M. J., Benítez, J. M., & Herrera, F. (2014). Big Data with Cloud Computing: An insight on the computing environment, MapReduce, and programming frameworks. *Wiley Interdisciplinary Reviews. Data Mining and Knowledge Discovery*, *4*(5), 380–409. doi:10.1002/widm.1134

Fortino, G., Guerrieri, A., Russo, W., & Savaglio, C. (2014). Integration of agent-based and cloud computing for the smart objects-oriented iot. In *Computer Supported Cooperative Work in Design (CSCWD), Proceedings of the 2014 IEEE 18th International Conference on* (pp. 493-498). IEEE. doi:10.1109/CSCWD.2014.6846894

Foster, I., Zhao, Y., Raicu, I., & Lu, S. (2008). Cloud computing and grid computing 360-degree compared. In *Grid Computing Environments Workshop, 2008. GCE'08* (pp. 1-10). IEEE. doi:10.1109/GCE.2008.4738445

Fu, J., Hao, W., Tu, M., Ma, B., Baldwin, J., & Bastani, F. B. (2010, July). Virtual services in cloud computing. In *Services (SERVICES-1), 2010 6th World Congress on* (pp. 467-472). IEEE. doi:10.1109/SERVICES.2010.125

Gutierrez-Garcia, J. O., & Sim, K. M. (2010). Self-organizing agents for service composition in cloud computing. In *Cloud Computing Technology and Science (CloudCom), 2010 IEEE Second International Conference on* (pp. 59-66). IEEE. doi:10.1109/CloudCom.2010.10

Hwang, K., & Li, D. (2010). Trusted cloud computing with secure resources and data coloring. *IEEE Internet Computing*, *14*(5), 14–22. doi:10.1109/MIC.2010.86

Jennings, N. R. (2000). On agent-based software engineering. *Artificial Intelligence*, *117*(2), 277–296. doi:10.1016/S0004-3702(99)00107-1

Ji, C., Li, Y., Qiu, W., Awada, U., & Li, K. (2012). Big data processing in cloud computing environments. In *Pervasive Systems, Algorithms and Networks (ISPAN), 2012 12th International Symposium on* (pp. 17-23). IEEE. doi:10.1109/I-SPAN.2012.9

Jula, A., Sundararajan, E., & Othman, Z. (2014). Cloud computing service composition: A systematic literature review. *Expert Systems with Applications*, *41*(8), 3809–3824. doi:10.1016/j.eswa.2013.12.017

Kaur, V., Tyagi, A., Kritika, M., Kumari, P., & Salvi, S. (2017). Crowd-sourcing based android application for structural health monitoring and data analytics of roads using cloud computing. In *Innovative Mechanisms for Industry Applications (ICIMIA), 2017 International Conference on* (pp. 354-358). IEEE. doi:10.1109/ICIMIA.2017.7975635

Khan, Z., Ludlow, D., McClatchey, R., & Anjum, A. (2012). An architecture for integrated intelligence in urban management using cloud computing. *Journal of Cloud Computing: Advances. Systems and Applications*, *1*(1), 1.

Kovachev, D., & Klamma, R. (2012). Framework for Computation Offloading in Mobile Cloud Computing. *IJIMAI*, *1*(7), 6–15. doi:10.9781/ijimai.2012.171

Leavitt, N. (2009). Is cloud computing really ready for prime time. *Growth*, *27*(5), 15–20.

Li, B. H., Zhang, L., Ren, L., Chai, X. D., Tao, F., Luo, Y. L., & Zhao, X. et al. (2011). Further discussion on cloud manufacturing. *Jisuanji Jicheng Zhizao Xitong*, *17*(3), 449–457.

Li, D., & Du, Y. (2007). *Artificial intelligence with uncertainty*. CRC Press. doi:10.1201/9781584889991

Li, D., Liu, C., & Gan, W. (2009). A new cognitive model: Cloud model. *International Journal of Intelligent Systems*, *24*(3), 357–375. doi:10.1002/int.20340

Li, H., Dai, Y., Tian, L., & Yang, H. (2009). Identity-based authentication for cloud computing. *Cloud computing*, 157-166.

Liu, J., Ma, J., Wu, W., Chen, X., Huang, X., & Xu, L. (2017). Protecting mobile health records in cloud computing: A secure, efficient, and anonymous design. *ACM Transactions on Embedded Computing Systems*, *16*(2), 57. doi:10.1145/2983625

Mesbahi, M., & Rahmani, A. M. (2016). Load balancing in cloud computing: A state of the art survey. *International Journal of Modern Education and Computer Science*, *8*(3), 64–78. doi:10.5815/ijmecs.2016.03.08

Mohit, P., Amin, R., Karati, A., Biswas, G. P., & Khan, M. K. (2017). A standard mutual authentication protocol for cloud computing based health care system. *Journal of Medical Systems*, *41*(4), 50. doi:10.1007/s10916-017-0699-2 PMID:28213882

Mondal, B., Dasgupta, K., & Dutta, P. (2012). Load balancing in cloud computing using stochastic hill climbing-a soft computing approach. *Procedia Technology*, *4*, 783–789. doi:10.1016/j.protcy.2012.05.128

Nakashima, H., Aghajan, H., & Augusto, J. C. (Eds.). (2009). *Handbook of ambient intelligence and smart environments*. Springer Science & Business Media.

Nurmi, D., Wolski, R., Grzegorczyk, C., Obertelli, G., Soman, S., Youseff, L., & Zagorodnov, D. (2009). The eucalyptus open-source cloud-computing system. In *Proceedings of the 2009 9th IEEE/ACM International Symposium on Cluster Computing and the Grid* (pp. 124-131). IEEE Computer Society. doi:10.1109/CCGRID.2009.93

Ostermann, S., Iosup, A., Yigitbasi, N., Prodan, R., Fahringer, T., & Epema, D. (2009). A performance analysis of EC2 cloud computing services for scientific computing. In *International Conference on Cloud Computing* (pp. 115-131). Springer, Berlin, Heidelberg.

Patel, A., Taghavi, M., Bakhtiyari, K., & Celestino Júnior, J. (2013). An intrusion detection and prevention system in cloud computing: A systematic review. *Journal of Network and Computer Applications*, *36*(1), 25–41. doi:10.1016/j.jnca.2012.08.007

Patel, V. L., Shortliffe, E. H., Stefanelli, M., Szolovits, P., Berthold, M. R., Bellazzi, R., & Abu-Hanna, A. (2009). The coming of age of artificial intelligence in medicine. *Artificial Intelligence in Medicine*, *46*(1), 5–17. doi:10.1016/j.artmed.2008.07.017 PMID:18790621

Petre, R. S. (2012). Data mining in cloud computing. *Database Systems Journal*, *3*(3), 67–71.

Qiu, L., Gai, K., & Qiu, M. (2016). Optimal Big Data Sharing Approach for Tele-Health in Cloud Computing. In *Smart Cloud (SmartCloud), IEEE International Conference on* (pp. 184-189). IEEE. doi:10.1109/SmartCloud.2016.21

Ren, F. (2010). From cloud computing to language engineering affective computing and advanced intelligence. *International Journal of Advanced Intelligence*, *2*(1), 1–14.

Rimal, B. P., Choi, E., & Lumb, I. (2009). A Taxonomy and Survey of Cloud Computing Systems. *NCM*, *9*, 44–51.

Santos, N., Gummadi, K. P., & Rodrigues, R. (2009). Towards Trusted Cloud Computing. *HotCloud*, *9*(9), 3.

Scale, M. S. E. (2009). Cloud computing and collaboration. *Library Hi Tech News*, *26*(9), 10–13. doi:10.1108/07419050911010741

Sheu, P. C. Y., Wang, S., Wang, Q., Hao, K., & Paul, R. (2009). Semantic computing, cloud computing, and semantic search engine. In *Semantic Computing, 2009. ICSC'09. IEEE International Conference on* (pp. 654-657). IEEE. doi:10.1109/ICSC.2009.51

Shoham, Y. (1987). *Reasoning about change: time and causation from the standpoint of artificial intelligence*. New Haven, CT, USA: Yale Univ.

Subashini, S., & Kavitha, V. (2011). A survey on security issues in service delivery models of cloud computing. *Journal of Network and Computer Applications*, *34*(1), 1–11. doi:10.1016/j.jnca.2010.07.006

Sucupira, M., Correa, F., Maia, L. C., Cardoso, A. M., & Parreiras, F. S. (2016). Utilization characteristics of software as a service: A study on operators of health plans. In *Proceedings of the XII Brazilian Symposium on Information Systems on Brazilian Symposium on Information Systems: Information Systems in the Cloud Computing Era-Volume 1* (p. 59). Brazilian Computer Society.

Sultan, N. (2010). Cloud computing for education: A new dawn? *International Journal of Information Management*, *30*(2), 109–116. doi:10.1016/j.ijinfomgt.2009.09.004

Voorsluys, W., Broberg, J., & Buyya, R. (2011). Introduction to cloud computing. *Cloud computing: Principles and paradigms*, 1-41.

Wang, L., Tao, J., Kunze, M., Castellanos, A. C., Kramer, D., & Karl, W. (2008, September). Scientific cloud computing: Early definition and experience. In *High Performance Computing and Communications, 2008. HPCC'08. 10th IEEE International Conference on* (pp. 825-830). IEEE.

Wang, L., Von Laszewski, G., Younge, A., He, X., Kunze, M., Tao, J., & Fu, C. (2010). Cloud computing: A perspective study. *New Generation Computing*, *28*(2), 137–146. doi:10.1007/s00354-008-0081-5

Xu, B., Xu, L., Cai, H., Jiang, L., Luo, Y., & Gu, Y. (2017). The design of an m-Health monitoring system based on a cloud computing platform. *Enterprise Information Systems*, *11*(1), 17–36. doi:10.1080/17517575.2015.1053416

Xu, K., Wang, X., Wei, W., Song, H., & Mao, B. (2016). Toward software defined smart home. *IEEE Communications Magazine*, *54*(5), 116–122. doi:10.1109/MCOM.2016.7470945

Xu, X. (2012). From cloud computing to cloud manufacturing. *Robotics and Computer-integrated Manufacturing*, *28*(1), 75–86. doi:10.1016/j.rcim.2011.07.002

Yao, Y. (2016). Emerging Cloud Computing Services: A Brief Opinion Article. *International Journal of Organizational and Collective Intelligence*, *6*(4), 98–102. doi:10.4018/IJOCI.2016100105

Zhang, Y., Qiu, M., Tsai, C. W., Hassan, M. M., & Alamri, A. (2017). Health-CPS: Healthcare cyber-physical system assisted by cloud and big data. *IEEE Systems Journal*, *11*(1), 88–95. doi:10.1109/JSYST.2015.2460747

Zhao, C., Zhang, S., Liu, Q., Xie, J., & Hu, J. (2009). Independent tasks scheduling based on genetic algorithm in cloud computing. In *Wireless Communications, Networking and Mobile Computing, 2009. WiCom'09. 5th International Conference on* (pp. 1-4). IEEE. doi:10.1109/WICOM.2009.5301850

Zou, G., Chen, Y., Yang, Y., Huang, R., & Xu, Y. (2010). AI planning and combinatorial optimization for web service composition in cloud computing. In *Proc international conference on cloud computing and virtualization* (pp. 1-8). doi:10.5176/978-981-08-5837-7_166

## KEY TERMS AND DEFINITIONS

**Artificial Intelligence:** Artificial intelligence is a research field interested in developing intelligent machines and/or software systems that can provide effective problem solutions inspired by human or natural factors.

**Biomedical Engineering:** Biomedical engineering is an engineering field focused on designing and developing biomedical-oriented solutions by combining advantages of different fields like computer, electronics, and even mechanics.

**Cloud Computing:** A technology over combination of hardware and software resources to have more accurate and fast use of such resources to improve communication opportunities for information sharing, creating, and processing.

**Nature-Inspired Techniques:** Nature-inspired techniques are solution algorithms of artificial intelligence inspired by elements of nature.

# Related References


Acevedo, E., Acevedo, A., Felipe, F., & Avilés, P. (2017, June). Artificial intelligence tools for pattern recognition. In *Second International Workshop on Pattern Recognition* (pp. 1044302-1044302). International Society for Optics and Photonics.

Adam, N. R., Wieder, R., & Ghosh, D. (2017). Data science, learning, and applications to biomedical and health sciences. *Annals of the New York Academy of Sciences*, *1387*(1), 5–11. doi:10.1111/nyas.13309 PMID:28122121

Affenzeller, M., Beham, A., Kofler, M., Kronberger, G., Wagner, S. A., & Winkler, S. (2009). Metaheuristic Optimization. In *Hagenberg Research* (pp. 103–155). Springer Berlin Heidelberg.

Aha, D. W., Kibler, D., & Albert, M. K. (1991). Instance-based learning algorithms. *Machine Learning*, *6*(1), 37–66. doi:10.1007/BF00153759

Ahmed, H., & Glasgow, J. (2012). *Swarm intelligence: concepts, models and applications.* School Of Computing, Queens University Technical Report.

Akay, M. (1995). Wavelets in biomedical engineering. *Annals of Biomedical Engineering*, *23*(5), 531–542. doi:10.1007/BF02584453 PMID:7503456

Akay, M. (2000). *Nonlinear Biomedical Signal Processing: Fuzzy logic, neural networks, and new algorithms.* Wiley-IEEE Press. doi:10.1109/9780470545362

AlMuhaideb, S., & Menai, M. E. B. (2014). HColonies: A new hybrid metaheuristic for medical data classification. *Applied Intelligence*, *41*(1), 282–298. doi:10.1007/s10489-014-0519-z

Alpaydın, E. (2011). Machine learning. *Wiley Interdisciplinary Reviews: Computational Statistics*, *3*(3), 195–203. doi:10.1002/wics.166

Alpaydin, E. (2014). *Introduction to machine learning.* MIT Press.

Amato, F., López, A., Peña-Méndez, E. M., Vaňhara, P., Hampl, A., & Havel, J. (2013). Artificial neural networks in medical diagnosis. *Journal of Applied Biomedicine*, *11*(2), 47–58. doi:10.2478/v10136-012-0031-x

Amato, P., Masserini, M., Mauri, G., & Cerofolini, G. (2010, September). Early-Stage Diagnosis of Endogenous Diseases by Swarms of Nanobots: An Applicative Scenario. In *ANTS Conference* (pp. 408-415). Academic Press. doi:10.1007/978-3-642-15461-4_38

Arenas, C., Irigoien, I., Mestres, F., Toma, C., & Cormand, B. (2017). Extreme Observations in Biomedical Data. In *Extended Abstracts Fall 2015* (pp. 3–8). Cham: Birkhäuser. doi:10.1007/978-3-319-55639-0_1

Aribarg, T., Supratid, S., & Lursinsap, C. (2012). Optimizing the modified fuzzy ant-miner for efficient medical diagnosis. *Applied Intelligence*, *37*(3), 357–376. doi:10.1007/s10489-011-0332-x

Azar, A. T., Ali, H. S., Balas, V. E., Olariu, T., & Ciurea, R. (2016). Boosted decision trees for vertebral column disease diagnosis. In *Soft computing applications* (pp. 319–333). Springer International Publishing. doi:10.1007/978-3-319-18296-4_27

Babič, F., Jančuš, A., & Melišová, K. (2016, September). Customized web-based system for elderly people using elements of artificial intelligence. In *Computer Science and Information Systems (FedCSIS), 2016 Federated Conference on* (pp. 277-280). IEEE. doi:10.15439/2016F165

Bahill, T. (1981). *Bioengineering--biomedical, Medical, and Clinical Engineering*. Prentice Hall.

Baillie, J. C. (2016). Artificial Intelligence: The Point of View of Developmental Robotics. In Fundamental Issues of Artificial Intelligence (pp. 413-422). Springer International Publishing.

Beam, A. L., & Kohane, I. S. (2016). Translating artificial intelligence into clinical care. *Journal of the American Medical Association*, *316*(22), 2368–2369. doi:10.1001/jama.2016.17217 PMID:27898974

Beetz, M., Chatila, R., Hertzberg, J., & Pecora, F. (2016). AI Reasoning Methods for Robotics. In *Springer Handbook of Robotics* (pp. 329–356). Springer International Publishing. doi:10.1007/978-3-319-32552-1_14

Begg, R., Lai, D. T., & Palaniswami, M. (2007). *Computational intelligence in biomedical engineering*. CRC Press.

Bench-Capon, T. J., & Dunne, P. E. (2007). Argumentation in artificial intelligence. *Artificial Intelligence*, *171*(10-15), 619–641. doi:10.1016/j.artint.2007.05.001

Bennett, C. C., & Hauser, K. (2013). Artificial intelligence framework for simulating clinical decision-making: A Markov decision process approach. *Artificial Intelligence in Medicine*, 57(1), 9–19. doi:10.1016/j.artmed.2012.12.003 PMID:23287490

Bergholt, M. S., Zheng, W., Lin, K., Ho, K. Y., Teh, M., Yeoh, K. G., & Huang, Z. et al. (2011). In vivo diagnosis of gastric cancer using Raman endoscopy and ant colony optimization techniques. *International Journal of Cancer*, *128*(11), 2673–2680. doi:10.1002/ijc.25618 PMID:20726002

Birattari, M., & Kacprzyk, J. (2009). *Tuning metaheuristics: a machine learning perspective* (Vol. 197). Berlin: Springer. doi:10.1007/978-3-642-00483-4

Boden, M. A. (1998). Creativity and artificial intelligence. *Artificial Intelligence*, *103*(1), 347–356. doi:10.1016/S0004-3702(98)00055-1

Bonato, P. (2003). Wearable sensors/systems and their impact on biomedical engineering. *IEEE Engineering in Medicine and Biology Magazine*, *22*(3), 18–20. doi:10.1109/MEMB.2003.1213622 PMID:12845812

Bostrom, N. (2016). *Fundamental issues of artificial intelligence* (Vol. 376; V. C. Müller, Ed.). Berlin: Springer.

Brier, M. E., & Gaweda, A. E. (2016). Artificial intelligence for optimal anemia management in end-stage renal disease. *Kidney International, 90*(2), 259–261. doi:10.1016/j.kint.2016.05.018 PMID:27418093

Bronzino, J. D. (1999). *Biomedical engineering handbook* (Vol. 2). CRC Press.

Brooks, R. A. (1991). Intelligence without representation. *Artificial Intelligence, 47*(1-3), 139–159. doi:10.1016/0004-3702(91)90053-M

Brundage, M. (2016). Artificial Intelligence and Responsible Innovation. In *Fundamental Issues of Artificial Intelligence* (pp. 541–552). Springer International Publishing.

Budin, G., Gavrilova, M. L., Shell, D. F., Wang, Y., Fiorini, R. A., Widrow, B., & Chan, C. et al. (2016). Cognitive Intelligence: Deep Learning, Thinking, and Reasoning by Brain-Inspired Systems. *International Journal of Cognitive Informatics and Natural Intelligence, 10*(4), 1–20. doi:10.4018/IJCINI.2016100101

Burton, E., Goldsmith, J., Koenig, S., Kuipers, B., Mattei, N., & Walsh, T. (2017). *Ethical considerations in artificial intelligence courses.* arXiv preprint arXiv:1701.07769

Carbonell, J. G., Michalski, R. S., & Mitchell, T. M. (1983). An overview of machine learning. In *Machine learning* (pp. 3–23). Springer Berlin Heidelberg.

Cerofolini, G., Amato, P., Masserini, M., & Mauri, G. (2010). A surveillance system for early-stage diagnosis of endogenous diseases by swarms of nanobots. *Advanced Science Letters, 3*(4), 345–352. doi:10.1166/asl.2010.1138

Chakraborty, A., & Kar, A. K. (2017). Swarm intelligence: A review of algorithms. In *Nature-Inspired Computing and Optimization* (pp. 475–494). Springer International Publishing. doi:10.1007/978-3-319-50920-4_19

Chakraborty, S., Samanta, S., Biswas, D., Dey, N., & Chaudhuri, S. S. (2013). Particle swarm optimization based parameter optimization technique in medical information hiding. In *Computational Intelligence and Computing Research (ICCIC), 2013 IEEE International Conference on* (pp. 1-6). IEEE. doi:10.1109/ICCIC.2013.6724173

Chandiok, A., & Chaturvedi, D. K. (2017). Biologically Inspired Techniques for Cognitive Decision-Making. In *Proceedings of the International Conference on Data Engineering and Communication Technology* (pp. 353-361). Springer Singapore. doi:10.1007/978-981-10-1678-3_34

Chandrasekaran, S., & Hougen, D. F. (2006). Swarm intelligence for cooperation of bio-nano robots using quorum sensing. In *Bio Micro and Nanosystems Conference, 2006. BMN'06* (pp. 104-104). IEEE. doi:10.1109/BMN.2006.330901

Chang, A. C. (2016). Big data in medicine: The upcoming artificial intelligence. *Progress in Pediatric Cardiology, 43*, 91–94. doi:10.1016/j.ppedcard.2016.08.021

Chang, P. C., Lin, J. J., & Liu, C. H. (2012). An attribute weight assignment and particle swarm optimization algorithm for medical database classifications. *Computer Methods and Programs in Biomedicine, 107*(3), 382–392. doi:10.1016/j.cmpb.2010.12.004 PMID:21194784

Charniak, E., Riesbeck, C. K., McDermott, D. V., & Meehan, J. R. (2014). *Artificial intelligence programming*. Psychology Press.

Chaves-González, J. M., Vega-Rodríguez, M. A., & Granado-Criado, J. M. (2013). A multiobjective swarm intelligence approach based on artificial bee colony for reliable DNA sequence design. *Engineering Applications of Artificial Intelligence*, *26*(9), 2045–2057. doi:10.1016/j.engappai.2013.04.011

Chebil, W., Soualmia, L. F., Omri, M. N., & Darmoni, S. J. (2016). Indexing biomedical documents with a possibilistic network. *Journal of the Association for Information Science and Technology*, *67*(4), 928–941. doi:10.1002/asi.23435

Chen, H. L., Huang, C. C., Yu, X. G., Xu, X., Sun, X., Wang, G., & Wang, S. J. (2013). An efficient diagnosis system for detection of Parkinson's disease using fuzzy k-nearest neighbor approach. *Expert Systems with Applications*, *40*(1), 263–271. doi:10.1016/j.eswa.2012.07.014

Chen, H. L., Yang, B., Wang, G., Wang, S. J., Liu, J., & Liu, D. Y. (2012). Support vector machine based diagnostic system for breast cancer using swarm intelligence. *Journal of Medical Systems*, *36*(4), 2505–2519. doi:10.1007/s10916-011-9723-0 PMID:21537848

Choi, S. P. (2016). An Experimental Study on the Relation Extraction from Biomedical Abstracts using Machine Learning. *Journal of the Korean Society for Library and Information Science*, *50*(2), 309–336. doi:10.4275/KSLIS.2016.50.2.309

Christ, M. J., Sivagowri, S., & Babu, P. G. (2014). Segmentation of brain tumors using meta heuristic algorithms. *Open Journal of Communications and Software*, *1*(1), 1–10. doi:10.15764/CS.2014.01001

Cleophas, T. J., & Zwinderman, A. H. (2016). Artificial Intelligence Using Distribution Free Data (90 Patients). In SPSS for Starters and 2nd Levelers (pp. 171-174). Springer International Publishing.

Cohen, P. R., & Feigenbaum, E. A. (Eds.). (2014). *The handbook of artificial intelligence* (Vol. 3). Butterworth-Heinemann.

Conitzer, V., Sinnott-Armstrong, W., Borg, J. S., Deng, Y., & Kramer, M. (2017). *Moral Decision Making Frameworks for Artificial Intelligence* (pp. 4831–4835). AAAI.

Cortes, C., & Vapnik, V. (1995). Support-vector networks. *Machine Learning*, *20*(3), 273–297. doi:10.1007/BF00994018

Cromwell, L., Weibell, F. J., & Pfeiffer, E. A. (1980). *Biomedical instrumentation and measurements*. Prentice Hall.

Darwiche, A. (2001). Recursive conditioning. *Artificial Intelligence*, *126*(1-2), 5–41. doi:10.1016/S0004-3702(00)00069-2

Dawes, T., de Marvao, A., Shi, W., Fletcher, T., Watson, G., Wharton, J., & Cook, S. et al. (2016). Use of artificial intelligence to predict survival in pulmonary hypertension. *Lancet*, *387*, S35. doi:10.1016/S0140-6736(16)00422-0

de Araújo, A. B. (2016). Semantic information and artificial intelligence. In *Fundamental Issues of Artificial Intelligence* (pp. 127–138). Springer International Publishing.

de Moura Oliveira, P. B., & Cunha, J. B. (2016). Blending Artificial Intelligence into PID Controller Design: A Biomedical Engineering Experiment. *IFAC-PapersOnLine*, *49*(6), 366–371. doi:10.1016/j.ifacol.2016.07.205

Deep, G., Kaur, L., & Gupta, S. (2016). Biomedical Image Indexing and Retrieval Descriptors: A Comparative Study. *Procedia Computer Science*, *85*, 954–961. doi:10.1016/j.procs.2016.05.287

Deepa, S. N., & Devi, B. A. (2011). A survey on artificial intelligence approaches for medical image classification. *Indian Journal of Science and Technology*, *4*(11), 1583–1595.

Dembski, W. (2016). *Darwin in mind: 'intelligent design'meets artificial intelligence*. Skeptical Inquirer.

Desjardins-Proulx, P., Poisot, T., & Gravel, D. (2017). Scientific Theories and Artificial Intelligence. *bioRxiv*, 161125.

Dheeba, J., Singh, N. A., & Selvi, S. T. (2014). Computer-aided detection of breast cancer on mammograms: A swarm intelligence optimized wavelet neural network approach. *Journal of Biomedical Informatics*, *49*, 45–52. doi:10.1016/j.jbi.2014.01.010 PMID:24509074

Dilsizian, S. E., & Siegel, E. L. (2014). Artificial intelligence in medicine and cardiac imaging: Harnessing big data and advanced computing to provide personalized medical diagnosis and treatment. *Current Cardiology Reports*, *16*(1), 441. doi:10.1007/s11886-013-0441-8 PMID:24338557

Diprose, W., & Buist, N. (2016). Artificial intelligence in medicine: Humans need not apply? *The New Zealand Medical Journal*, *129*(1434), 73–76. PMID:27349266

Eberhart, R. C. (1998). Overview of computational intelligence [and biomedical engineering applications]. In *Engineering in Medicine and Biology Society, 1998. Proceedings of the 20th Annual International Conference of the IEEE* (*Vol. 3*, pp. 1125-1129). IEEE.

Enderle, J. D., & Bronzino, J. D. (2012). *Introduction to biomedical engineering*. Academic press.

Eremia, M., Liu, C. C., & Edris, A. A. (Eds.). (2016). *Advanced Solutions in Power Systems: HVDC, FACTS, and Artificial Intelligence* (Vol. 52). John Wiley & Sons. doi:10.1002/9781119175391

Esposito, A., Bassis, S., Morabito, F. C., & Pasero, E. (2016). Some Notes on Computational and Theoretical Issues in Artificial Intelligence and Machine Learning. In *Advances in Neural Networks* (pp. 3–12). Springer International Publishing. doi:10.1007/978-3-319-33747-0_1

Ferreira, C. (2006). *Gene expression programming: mathematical modeling by an artificial intelligence* (Vol. 21). Springer.

Fink, M., Montaldo, G., & Tanter, M. (2003). Time-reversal acoustics in biomedical engineering. *Annual Review of Biomedical Engineering*, *5*(1), 465–497. doi:10.1146/annurev.bioeng.5.040202.121630 PMID:14527319

Flasiński, M. (2016). *Introduction to artificial intelligence*. Springer. doi:10.1007/978-3-319-40022-8

Flasiński, M. (2016). History of Artificial Intelligence. In *Introduction to Artificial Intelligence* (pp. 3–13). Springer International Publishing. doi:10.1007/978-3-319-40022-8_1

Fournier, R. L. (2011). *Basic transport phenomena in biomedical engineering*. CRC Press.

Freitag, D. (2000). Machine learning for information extraction in informal domains. *Machine Learning*, *39*(2), 169–202. doi:10.1023/A:1007601113994

Gaggioli, A. (2017). Bringing More Transparency to Artificial Intelligence. *Cyberpsychology, Behavior, and Social Networking*, *20*(1), 68–68. doi:10.1089/cyber.2016.29060.csi PMID:28080147

Galinska, M., & Badura, P. (2016). Swarm intelligence approach to 3D medical image segmentation. In *Information Technologies in Medicine* (pp. 15–24). Cham: Springer. doi:10.1007/978-3-319-39796-2_2

Ganji, M. F., & Abadeh, M. S. (2010). Using fuzzy ant colony optimization for diagnosis of diabetes disease. In *Electrical Engineering (ICEE), 2010 18th Iranian Conference on* (pp. 501-505). IEEE.

Ganji, M. F., Abadeh, M. S., Hedayati, M., & Bakhtiari, N. (2010, November). Fuzzy classifcation of imbalanced data sets for medical diagnosis. In *Biomedical Engineering (ICBME), 2010 17th Iranian Conference of* (pp. 1-5). IEEE.

Garro, B. A., Rodríguez, K., & Vázquez, R. A. (2016). Classification of DNA microarrays using artificial neural networks and ABC algorithm. *Applied Soft Computing*, *38*, 548–560. doi:10.1016/j.asoc.2015.10.002

Ghannad-Rezaie, M., Soltanain-Zadeh, H., Siadat, M. R., & Elisevich, K. V. (2006). Medical data mining using particle swarm optimization for temporal lobe epilepsy. In *Evolutionary Computation, 2006. CEC 2006. IEEE Congress on* (pp. 761-768). IEEE. doi:10.1109/CEC.2006.1688388

Gobinathan, B., Neduncheliyan, S., & Satish, D. (2016). Fuzzy Bee Segmentation-Meta-Heuristic Approach for the Medical Image Segmentation Problem. *Current Signal Transduction Therapy*, *11*(2), 130–136. doi:10.2174/1574362411666160822122831

Göranzon, B., & Josefson, I. (Eds.). (2012). *Knowledge, skill and artificial intelligence*. Springer Science & Business Media.

Guimard, N. K., Gomez, N., & Schmidt, C. E. (2007). Conducting polymers in biomedical engineering. *Progress in Polymer Science*, *32*(8), 876–921. doi:10.1016/j.progpolymsci.2007.05.012

Hart, P., Sacerdoti, E., & Untulis, C. (2017). Research In Progress at the Artificial Intelligence Center, SRI International. *AI Magazine*, *1*(1), 30. doi:10.1609/aimag.v1i1.90

Hassani, K., & Jafarian, K. (2016). An Intelligent Method for Breast Cancer Diagnosis Based on Fuzzy ART and Metaheuristic Optimization. In *XIV Mediterranean Conference on Medical and Biological Engineering and Computing 2016: MEDICON 2016, March 31st-April 2nd 2016, Paphos, Cyprus* (Vol. 57, p. 200). Springer. doi:10.1007/978-3-319-32703-7_41

Hassanien, A. E., Moftah, H. M., Azar, A. T., & Shoman, M. (2014). MRI breast cancer diagnosis hybrid approach using adaptive ant-based segmentation and multilayer perceptron neural networks classifier. *Applied Soft Computing*, *14*, 62–71. doi:10.1016/j.asoc.2013.08.011

Haux, R., Kulikowski, C. A., Bakken, S., de Lusignan, S., Kimura, M., Koch, S., & Moen, A. et al. (2017). Research Strategies for Biomedical and Health Informatics. *Methods of Information in Medicine*, 56. PMID:28119991

Hedeshi, N. G., & Abadeh, M. S. (2011, December). An expert system working upon an ensemble PSO-based approach for diagnosis of coronary artery disease. In *Biomedical Engineering (ICBME), 2011 18th Iranian Conference of* (pp. 249-254). IEEE. doi:10.1109/ICBME.2011.6168566

Helbing, D., Frey, B. S., Gigerenzer, G., Hafen, E., Hagner, M., Hofstetter, Y., & Zwitter, A. et al. (2017). Will Democracy Survive Big Data and Artificial Intelligence. *Scientific American*, (Feb): 25.

Hengstler, M., Enkel, E., & Duelli, S. (2016). Applied artificial intelligence and trust—The case of autonomous vehicles and medical assistance devices. *Technological Forecasting and Social Change*, *105*, 105–120. doi:10.1016/j.techfore.2015.12.014

Hewitt, C. (1991). Open information systems semantics for distributed artificial intelligence. *Artificial Intelligence*, *47*(1-3), 79–106. doi:10.1016/0004-3702(91)90051-K

Holland, J. H. (1992). *Adaptation in natural and artificial systems: An introductory analysis with applications to biology, control, and artificial intelligence*. MIT Press.

Horng, M. H., Lee, Y. X., Lee, M. C., & Liou, R. J. (2012). Firefly meta-heuristic algorithm for training the radial basis function network for data classification and disease diagnosis. In *Theory and new applications of swarm intelligence*. InTech. doi:10.5772/39084

Huang, M. L., Hung, Y. H., & Chen, W. Y. (2010). Neural network classifier with entropy based feature selection on breast cancer diagnosis. *Journal of Medical Systems*, *34*(5), 865–873. doi:10.1007/s10916-009-9301-x PMID:20703622

Hudson, D. L., & Cohen, M. E. (2000). *Neural networks and artificial intelligence for biomedical engineering* (pp. i–xxiii). New York: IEEE.

Iantovics, B. (2006). A novel diagnosis system specialized in difficult medical diagnosis problems solving. *Emergent Properties in Natural and Artificial Dynamical Systems*, 185-195.

Ibrikci, T., Karabulut, E. M., & Uwisengeyimana, J. D. (2016). Meta Learning on Small Biomedical Datasets. In *Information Science and Applications (ICISA) 2016* (pp. 933–939). Singapore: Springer. doi:10.1007/978-981-10-0557-2_89

Iliadis, L., Maglogiannis, I., Papadopoulos, H., Sioutas, S., & Makris, C. (2016). *Artificial Intelligence Applications and Innovations*. Springer International Pu. doi:10.1007/978-3-319-44944-9

Istepanian, R., Laxminarayan, S., & Pattichis, C. S. (2014). *M-health*. John Wiley & Sons.

Jaeger, H. (2016). Artificial intelligence: Deep neural reasoning. *Nature*, *538*(7626), 467–468. doi:10.1038/nature19477 PMID:27732576

Jameel, S. M. (2014). Investigation of Differential Evaluation Optimization Algorithm for Medical Data Classifications. *International Journal of Computer Science and Network Security*, *14*(1), 29.

Jaya, J., & Thanushkodi, K. (2010). Detection of Suspicious Region in Medical Images using Meta-heuristic Algorithms. *Artificial Intelligent Systems and Machine Learning*, *2*(9), 219–222.

Jaya, J., & Thanushkodi, K. (2011). K. certain investigation on mri segmentation for the implementation of cad system. *WSEAS Transactions on Computers*, *10*(6), 189–198.

Jemal, H., Kechaou, Z., & Ayed, M. B. (2014, August). Swarm intelligence and multi agent system in healthcare. In *Soft Computing and Pattern Recognition (SoCPaR), 2014 6th International Conference of* (pp. 423-427). IEEE. doi:10.1109/SOCPAR.2014.7008044

Jha, S., & Topol, E. J. (2016). Adapting to artificial intelligence: Radiologists and pathologists as information specialists. *Journal of the American Medical Association*, *316*(22), 2353–2354. doi:10.1001/jama.2016.17438 PMID:27898975

Jin, L. P., & Dong, J. (2016). Ensemble deep learning for biomedical time series classification. *Computational Intelligence and Neuroscience*. PMID:27725828

Khanmohammadi, S., Adibeig, N., & Shanehbandy, S. (2017). An improved overlapping k-means clustering method for medical applications. *Expert Systems with Applications*, *67*, 12–18. doi:10.1016/j.eswa.2016.09.025

Kharrat, A., Halima, M. B., & Ayed, M. B. (2015, December). MRI brain tumor classification using support vector machines and meta-heuristic method. In *Intelligent Systems Design and Applications (ISDA), 2015 15th International Conference on* (pp. 446-451). IEEE. doi:10.1109/ISDA.2015.7489271

Kitano, H. (2016). Artificial intelligence to win the nobel prize and beyond: Creating the engine for scientific discovery. *AI Magazine*, *37*(1), 39–50. doi:10.1609/aimag.v37i1.2642

Kline, J. (Ed.). (2012). *Handbook of biomedical engineering*. Elsevier.

Kononenko, I. (2001). Machine learning for medical diagnosis: History, state of the art and perspective. *Artificial Intelligence in Medicine*, *23*(1), 89–109. doi:10.1016/S0933-3657(01)00077-X PMID:11470218

Köse, U., & Arslan, A. (2016). Forecasting chaotic time series via anfis supported by vortex optimization algorithm: Applications on electroencephalogram time series. *Arabian Journal for Science and Engineering*, 1–12.

Köse, U., Güraksın, G. E., & Deperlioğlu, Ö. (2015). Diabetes determination via vortex optimization algorithm based support vector machines. In *Medical Technologies National Conference (TIPTEKNO)* (pp. 1-4). IEEE. doi:10.1109/TIPTEKNO.2015.7374614

Köse, U., Güraksın, G. E., & Deperlioğlu, Ö. (2016). Cognitive development optimization algorithm based support vector machines for determining diabetes. *BRAIN. Broad Research in Artificial Intelligence and Neuroscience*, *7*(1), 80–90.

Kruse, R., Borgelt, C., Braune, C., Mostaghim, S., & Steinbrecher, M. (2016). Computational Swarm Intelligence. In *Computational Intelligence* (pp. 299–325). Springer London. doi:10.1007/978-1-4471-7296-3_14

Kruse, R., Borgelt, C., Braune, C., Mostaghim, S., & Steinbrecher, M. (2016). Introduction to Neural Networks. In *Computational Intelligence* (pp. 9–13). Springer London. doi:10.1007/978-1-4471-7296-3_2

Kumar, J. (2016). Applications of artificial intelligence. *International Journal of Research in Engineering and Applied Sciences*, *6*(4), 42–49.

Kutz, M. (2003). *Standard handbook of biomedical engineering and design*. McGraw-Hill.

Kutz, M. (2009). *Biomedical Engineering and Design Handbook, Volume 2: Volume 2: Biomedical Engineering Applications* (Vol. 2). McGraw Hill Professional.

Labovitz, D. L., Shafner, L., Virmani, D., Hanina, A., & Hale, T. (2016). Using Artificial Intelligence to Measure and Optimize Adherence in Patients on Anticoagulation Therapy. *Journal of Medical Internet Research, 18*(12).

Lamy, J. (2016). Ontology-oriented programming for biomedical informatics. *Studies in Health Technology and Informatics, 221*, 64–68. PMID:27071878

Lantada, A. D., & Morgado, P. L. (2012). Rapid prototyping for biomedical engineering: Current capabilities and challenges. *Annual Review of Biomedical Engineering, 14*(1), 73–96. doi:10.1146/annurev-bioeng-071811-150112 PMID:22524389

Lapizco-Encinas, G. C., & Reggia, J. A. (2005, June). Diagnostic problem solving using swarm intelligence. In *Swarm Intelligence Symposium, 2005. SIS 2005. Proceedings 2005 IEEE* (pp. 365-372). IEEE. doi:10.1109/SIS.2005.1501644

Lemaignan, S., Warnier, M., Sisbot, E. A., Clodic, A., & Alami, R. (2017). Artificial cognition for social human–robot interaction: An implementation. *Artificial Intelligence, 247*, 45–69. doi:10.1016/j.artint.2016.07.002

Li, D., & Du, Y. (2007). *Artificial intelligence with uncertainty*. CRC press. doi:10.1201/9781584889991

Li, Q., Chen, H., Huang, H., Zhao, X., Cai, Z., Tong, C., & Tian, X. et al. (2017). An enhanced grey wolf optimization based feature selection wrapped kernel extreme learning machine for medical diagnosis. *Computational and Mathematical Methods in Medicine*. PMID:28246543

Luger, G. F. (2005). *Artificial intelligence: structures and strategies for complex problem solving*. Pearson Education.

Luo, W., Phung, D., Tran, T., Gupta, S., Rana, S., Karmakar, C., & Venkatesh, S. et al. (2016). Guidelines for Developing and Reporting Machine Learning Predictive Models in Biomedical Research: A Multidisciplinary View. *Journal of Medical Internet Research, 18*(12), e323. doi:10.2196/jmir.5870 PMID:27986644

Ma, V. (2016). The Ethics and Implications of Modern Warfare: Robotic Systems and Human Optimization. *Harvard International Review, 37*(4), 43.

Maglogiannis, I., Zafiropoulos, E., & Anagnostopoulos, I. (2009). An intelligent system for automated breast cancer diagnosis and prognosis using SVM based classifiers. *Applied Intelligence, 30*(1), 24–36. doi:10.1007/s10489-007-0073-z

Makas, H., & Yumusak, N. (2013, November). A comprehensive study on thyroid diagnosis by neural networks and swarm intelligence. In *Electronics, Computer and Computation (ICECCO), 2013 International Conference on* (pp. 180-183). IEEE. doi:10.1109/ICECCO.2013.6718258

Mamoun, M., Ibrahim, A. O., & Fatouh, S. E. (2017). *Hybrid Algorithm to Enhance Artificial Neural Network for Parkinson' S Disease Diagnosis: A Framework*. Academic Press.

Marinakis, Y., Dounias, G., & Jantzen, J. (2009). Pap smear diagnosis using a hybrid intelligent scheme focusing on genetic algorithm based feature selection and nearest neighbor classification. *Computers in Biology and Medicine*, *39*(1), 69–78. doi:10.1016/j.compbiomed.2008.11.006 PMID:19147127

McCarthy, T., Rosenblum, L. P., Johnson, B. G., Dittel, J., & Kearns, D. M. (2016). An Artificial Intelligence Tutor: A Supplementary Tool for Teaching and Practicing Braille. *Journal of Visual Impairment & Blindness*, *110*(5), 309–322.

Medrano, I. H., Guijarro, J. T., Belda, C., Ureña, A., Salcedo, I., Espinosa-Anke, L., & Saggion, H. (2017). (in press). Savana: Re-using Electronic Health Records with Artificial Intelligence. *International Journal of Interactive Multimedia and Artificial Intelligence*, *4*.

Michalski, R. S., Carbonell, J. G., & Mitchell, T. M. (Eds.). (2013). *Machine learning: An artificial intelligence approach*. Springer Science & Business Media.

Michie, D., Spiegelhalter, D. J., & Taylor, C. C. (1994). *Machine learning, neural and statistical classification*. Academic Press.

Miller, P. L. (Ed.). (2012). *Selected topics in medical artificial intelligence*. Springer Science & Business Media.

Mittelstadt, B. D., & Floridi, L. (Eds.). (2016). *The ethics of biomedical big data* (Vol. 29). Springer. doi:10.1007/978-3-319-33525-4

Moein, S. (Ed.). (2014). *Medical diagnosis using artificial neural networks*. IGI Global. doi:10.4018/978-1-4666-6146-2

Mohammed, B. H., & Omar, N. (2016). A back propagation neural network for identifying multi-word biomedical named entities. *International Review on Computers and Software*, *11*(8), 682–690.

Montani, S. (2008). Exploring new roles for case-based reasoning in heterogeneous AI systems for medical decision support. *Applied Intelligence*, *28*(3), 275–285. doi:10.1007/s10489-007-0046-2

Moore, J. H., & Holmes, J. H. (2016). The golden era of biomedical informatics has begun. *BioData Mining*, *9*(1), 15. doi:10.1186/s13040-016-0092-6 PMID:27069509

Moravčík, M., Schmid, M., Burch, N., Lisý, V., Morrill, D., Bard, N., & Bowling, M. et al. (2017). DeepStack: Expert-level artificial intelligence in heads-up no-limit poker. *Science*, *356*(6337), 508–513. doi:10.1126/science.aam6960 PMID:28254783

Müller, V. C., & Bostrom, N. (2016). Future progress in artificial intelligence: A survey of expert opinion. In *Fundamental issues of artificial intelligence* (pp. 553–570). Springer International Publishing. doi:10.1007/978-3-319-26485-1_33

Murphy, K. P. (2012). *Machine learning: a probabilistic perspective*. MIT press.

Musib, M., Wang, F., Tarselli, M. A., Yoho, R., Yu, K. H., Andrés, R. M., & Dutton-Regester, K. et al. (2017). Artificial intelligence in research. *Science*, *357*(6346), 28–30. doi:10.1126/science.357.6346.28 PMID:28684488

Nair, S. S. K., & Ganesh, N. (2016, March). An exploratory study on big data processing: A case study from a biomedical informatics. In *Big Data and Smart City (ICBDSC), 2016 3rd MEC International Conference on* (pp. 1-4). IEEE. doi:10.1109/ICBDSC.2016.7460338

Nakib, A., & Talbi, E. G. (Eds.). (2017). *Metaheuristics for Medicine and Biology* (Vol. 704). Springer. doi:10.1007/978-3-662-54428-0

Neshat, M., Sargolzaei, M., Nadjaran Toosi, A., & Masoumi, A. (2012). Hepatitis disease diagnosis using hybrid case based reasoning and particle swarm optimization. *ISRN Artificial Intelligence*.

Ng, E. Y. (2017). A Special Section on Biomedical Imaging in Diagnosis and Treatment–Part 3. *Journal of Medical Imaging and Health Informatics*, *7*(1), 126–128. doi:10.1166/jmihi.2017.1994

Nilsson, N. J. (1991). Logic and artificial intelligence. *Artificial Intelligence*, *47*(1-3), 31–56. doi:10.1016/0004-3702(91)90049-P

Nilsson, N. J. (2014). *Principles of artificial intelligence*. Morgan Kaufmann.

Norris, D. J. (2017). Introduction to Artificial Intelligence. In *Beginning Artificial Intelligence with the Raspberry Pi* (pp. 1–15). Apress. doi:10.1007/978-1-4842-2743-5_1

Ogiela, M. R., & Tadeusiewicz, R. (2003). Artificial intelligence structural imaging techniques in visual pattern analysis and medical data understanding. *Pattern Recognition*, *36*(10), 2441–2452. doi:10.1016/S0031-3203(03)00089-X

Paragios, N., Ferrante, E., Glocker, B., Komodakis, N., Parisot, S., & Zacharaki, E. I. (2016). *(Hyper)-graphical models in biomedical image analysis*. Academic Press.

Parveen, R., Nabi, M., Memon, F. A., Zaman, S., & Ali, M. (2016). A review and survey of artificial neural network in medical science. *J Adv Res Comput Appl*, *3*(1), 8–17.

Patel, V. L., Shortliffe, E. H., Stefanelli, M., Szolovits, P., Berthold, M. R., Bellazzi, R., & Abu-Hanna, A. (2009). The coming of age of artificial intelligence in medicine. *Artificial Intelligence in Medicine*, *46*(1), 5–17. doi:10.1016/j.artmed.2008.07.017 PMID:18790621

Pfister, B. J., Gordon, T., Loverde, J. R., Kochar, A. S., Mackinnon, S. E., & Cullen, D. K. (2011). Biomedical engineering strategies for peripheral nerve repair: surgical applications, state of the art, and future challenges. *Critical Reviews™ in Biomedical Engineering, 39*(2).

Pham, H. N. A., & Triantaphyllou, E. (2009). An application of a new meta-heuristic for optimizing the classification accuracy when analyzing some medical datasets. *Expert Systems with Applications*, *36*(5), 9240–9249. doi:10.1016/j.eswa.2008.12.007

Poole, D. (1997). The independent choice logic for modelling multiple agents under uncertainty. *Artificial Intelligence*, *94*(1), 7–56. doi:10.1016/S0004-3702(97)00027-1

Poole, D. L., & Mackworth, A. K. (2010). *Artificial Intelligence: foundations of computational agents*. Cambridge University Press. doi:10.1017/CBO9780511794797

Punitha, K., & Latha, B. (2016). Validation of Medical Imaging Software Using Metaheuristic Knowledge Discovery. *Journal of Medical Imaging and Health Informatics*, *6*(8), 1966–1971. doi:10.1166/jmihi.2016.1958

Qasem, S. N., & Shamsuddin, S. M. (2011). Radial basis function network based on time variant multi-objective particle swarm optimization for medical diseases diagnosis. *Applied Soft Computing*, *11*(1), 1427–1438. doi:10.1016/j.asoc.2010.04.014

Quinlan, J. R. (1990). Learning logical definitions from relations. *Machine Learning*, *5*(3), 239–266. doi:10.1007/BF00117105

Rahman, M. M., & Bhattacharya, P. (2016, December). Biomedical Image Classification with Multi Response Linear Regression (MLR) as Meta-Learner Combiner and Its Effectiveness on Small to Large Data Sets. In *Computational Science and Computational Intelligence (CSCI), 2016 International Conference on* (pp. 110-115). IEEE.

Rana, S., & Kaur, S. (2016). A Optimized Binary Approach for Heart Disease Decision Model in Biomedical Data Mining. *International Journal of Computers and Applications*, *156*(12).

Reddy, C. V., & Siddaiah, P. (2015). Hybrid LWT-SVD watermarking optimized using metaheuristic algorithms along with encryption for medical image security. *Signal and Image Processing : an International Journal*, *6*(1), 75–95. doi:10.5121/sipij.2015.6106

Rosé, C. P. (2017). Artificial intelligence: A social spin on language analysis. *Nature*, *545*(7653), 166–167. doi:10.1038/545166a PMID:28492252

Russell, S., Norvig, P., & Intelligence, A. (1995). A modern approach. *Artificial Intelligence. Prentice-Hall. Egnlewood Cliffs*, *25*, 27.

Sacchi, L., & Holmes, J. H. (2016). *Progress in biomedical knowledge discovery: a 25-year retrospective*. IMIA Yearbook.

Sahoo, A., & Chandra, S. (2014). Meta-heuristic approaches for active contour model based medical image segmentation. *International Journal of Advances in Soft Computing and Its Applications*, *6*(2).

Sahu, B., & Mishra, D. (2012). A novel feature selection algorithm using particle swarm optimization for cancer microarray data. *Procedia Engineering*, *38*, 27–31. doi:10.1016/j.proeng.2012.06.005

Sandler, R. (Ed.). (2016). *Ethics and emerging technologies*. Springer.

Satishkumar, K., Mahalakshmi, T. S., & Katneni, V. (2016). Computational Intelligence Approach for Prediction of Breast Cancer using Particle Swarm Optimization: A Comparative Study of the Results with Reduced Set of Attributes. In *Computational Intelligence Techniques in Health Care* (pp. 31–44). Springer Singapore. doi:10.1007/978-981-10-0308-0_3

Shachter, R. D., Kanal, L. N., Henrion, M., & Lemmer, J. F. (Eds.). (2017). *Uncertainty in Artificial Intelligence 5* (Vol. 10). Elsevier.

Sharma, A., Virvilis, V., Lekka, T., & Andronis, C. (2016). Binary Relation Extraction from Biomedical Literature using Dependency Trees and SVMs. *bioRxiv*, 082479.

Shukla, S., Lakhmani, A., & Agarwal, A. K. (2016, April). Approaches of artificial intelligence in biomedical image processing: A leading tool between computer vision & biological vision. In *Advances in Computing, Communication, & Automation (ICACCA)(Spring), International Conference on* (pp. 1-6). IEEE. doi:10.1109/ICACCA.2016.7578900

Singh, C., & Singh, J. (2016). Biomedical Signal Processing, Artificial Neural Network: A Review. *Indian Journal of Science and Technology*, *9*(47). doi:10.17485/ijst/2015/v8i1/106889

Slaney, J., & Thiébaux, S. (2001). Blocks world revisited. *Artificial Intelligence*, *125*(1-2), 119–153. doi:10.1016/S0004-3702(00)00079-5

Soliman, M. M., Hassanien, A. E., Ghali, N. I., & Onsi, H. M. (2012). An adaptive watermarking approach for medical imaging using swarm intelligent. *International Journal of Smart Home*, *6*(1), 37–50.

Spitzer, E. (2016). Tacit Representations and Artificial Intelligence: Hidden Lessons from an Embodied Perspective on Cognition. In Fundamental Issues of Artificial Intelligence (pp. 423-439). Springer International Publishing.

Spyropoulos, C. D. (2000). AI planning and scheduling in the medical hospital environment. *Artificial Intelligence in Medicine*, *2*(20), 101–111. doi:10.1016/S0933-3657(00)00059-2 PMID:10936748

Stocker, M. (2016). Decision-making: Be wary of'ethical'artificial intelligence. *Nature*, *540*(7634), 525–525. doi:10.1038/540525b

Stoitsis, J., Valavanis, I., Mougiakakou, S. G., Golemati, S., Nikita, A., & Nikita, K. S. (2006). Computer aided diagnosis based on medical image processing and artificial intelligence methods. *Nuclear Instruments & Methods in Physics Research. Section A, Accelerators, Spectrometers, Detectors and Associated Equipment*, *569*(2), 591–595. doi:10.1016/j.nima.2006.08.134

Streba, C. T., Ionescu, M., Vere, C. C., & Rogoveanu, I. (2017). Artificial Intelligence and Automatic Image Interpretation in Modern Medicine. In *Translational Bioinformatics and Its Application* (pp. 371–407). Springer Netherlands. doi:10.1007/978-94-024-1045-7_16

Subasi, A. (2013). Classification of EMG signals using PSO optimized SVM for diagnosis of neuromuscular disorders. *Computers in Biology and Medicine*, *43*(5), 576–586. doi:10.1016/j.compbiomed.2013.01.020 PMID:23453053

Suganya, P., & Sumathi, C. P. (2015). A novel metaheuristic data mining algorithm for the detection and classification of parkinson disease. *Indian Journal of Science and Technology*, *8*(14). doi:10.17485/ijst/2015/v8i14/72685

Taghavifar, H., & Mardani, A. (2017). Application of Artificial Intelligence on Modeling and Optimization. In *Off-road Vehicle Dynamics* (pp. 133–177). Springer International Publishing. doi:10.1007/978-3-319-42520-7_5

ten Teije, A., Popow, C., Holmes, J. H., & Sacchi, L. (Eds.). (2017). *Artificial Intelligence in Medicine: 16th Conference on Artificial Intelligence in Medicine, AIME 2017, Vienna, Austria,* June 21-24, 2017, *Proceedings* (*Vol. 10259*). Springer.

Thangavel, K., & Karnan, M. (2005). Computer aided diagnosis in digital mammograms: Detection of microcalcifications by meta heuristic algorithms. *GVIP Journal*, 5(7), 41–55.

Thórisson, K. R., Bieger, J., Thorarensen, T., Sigurðardóttir, J. S., & Steunebrink, B. R. (2016, July). Why artificial intelligence needs a task theory. In *International Conference on Artificial General Intelligence* (pp. 118-128). Springer International Publishing. doi:10.1007/978-3-319-41649-6_12

Tien, J. M. (2017). Internet of Things, Real-Time Decision Making, and Artificial Intelligence. *Annals of Data Science*, 1-30.

Tiwana, M. I., Redmond, S. J., & Lovell, N. H. (2012). A review of tactile sensing technologies with applications in biomedical engineering. *Sensors and Actuators. A, Physical*, *179*, 17–31. doi:10.1016/j.sna.2012.02.051

Topalovic, M., Aerts, J. M., Decramer, M., Troosters, T., & Janssens, W. (2017). Artificial Intelligence Detects Lung Diseases Using Pulmonary Function Tests. In C47. COPD: PHYSIOLOGIC ASSESSMENT (pp. A5678-A5678). American Thoracic Society.

Trippi, R. R., & Turban, E. (1992). *Neural networks in finance and investing: Using artificial intelligence to improve real world performance*. McGraw-Hill, Inc.

Tsai, C. W., Chiang, M. C., Ksentini, A., & Chen, M. (2016). Metaheuristic algorithms for healthcare: Open issues and challenges. *Computers & Electrical Engineering*, *53*, 421–434. doi:10.1016/j.compeleceng.2016.03.005

Ulieru, R., & Unland, M. (2006, May). A stigmergic approach to medical diagnosis. In *Proceedings of the 2nd International Workshop on Multi-Agent Systems for Medicine and Computational Biology* (pp. 87-103). Academic Press.

Vassev, E. (2016, October). Safe artificial intelligence and formal methods. In *International Symposium on Leveraging Applications of Formal Methods* (pp. 704-713). Springer International Publishing.

Waal, I. (2017). Skin cancer diagnosed using artificial intelligence on clinical images. *Oral Diseases*. PMID:28326656

Wachowiak, M. P., Smolíková, R., Zheng, Y., Zurada, J. M., & Elmaghraby, A. S. (2004). An approach to multimodal biomedical image registration utilizing particle swarm optimization. *IEEE Transactions on Evolutionary Computation*, *8*(3), 289–301. doi:10.1109/TEVC.2004.826068

Watkins, C. J., & Dayan, P. (1992). Q-learning. *Machine Learning*, *8*(3-4), 279–292. doi:10.1007/BF00992698

Weiss, G. (Ed.). (1999). *Multiagent systems: a modern approach to distributed artificial intelligence*. MIT Press.

Wenger, E. (2014). *Artificial intelligence and tutoring systems: computational and cognitive approaches to the communication of knowledge*. Morgan Kaufmann.

Wiesel-Kapah, I., Kaminka, G. A., Hachmon, G., Agmon, N., & Bachelet, I. (2016, July). *Rule-Based Programming of Molecular Robot Swarms for Biomedical Applications* (pp. 3505–3512). IJCAI.

Witten, I. H., Frank, E., Hall, M. A., & Pal, C. J. (2016). *Data Mining: Practical machine learning tools and techniques*. Morgan Kaufmann.

Yamakawa, H., Osawa, M., & Matsuo, Y. (2016, October). Whole brain architecture approach is a feasible way toward an artificial general intelligence. In *International Conference on Neural Information Processing* (pp. 275-281). Springer International Publishing. doi:10.1007/978-3-319-46687-3_30

Zhang, C. Z., Kareem Abdullah, A., & Abdullabs Abdullah, A. (2016). Electroencephalogram-Artifact Extraction Enhancement Based on Artificial Intelligence Technique. Journal of Biomimetics Biomaterials and Biomedical Engineering, 27, 77-91. doi:10.4028/www.scientific.net/JBBBE.27.77

# Compilation of References

Abawajy, J. H., & Hassan, M. M. (2017). Federated internet of things and cloud computing pervasive patient health monitoring system. *IEEE Communications Magazine*, *55*(1), 48–53. doi:10.1109/MCOM.2017.1600374CM

Abbass, H. A. (2002). An evolutionary artificial neural networks approach for breast cancer diagnosis. *Artificial Intelligence in Medicine*, *25*(3), 265–281. doi:10.1016/S0933-3657(02)00028-3 PMID:12069763

Abdullah, L. M., Azuddin, M., Malik, S. A., & Mahmud, M. (2017). Rising Ageing Population: A Preliminary Study of Malaysian Older People Expectations in Information and Communication Technology. In *International Conference of Reliable Information and Communication Technology* (pp. 796-803). Springer.

Abraham, A., & Pedrycz, W. (2008). *Hybrid Artificial Intelligence Systems. Lecture Notes in Artificial Intelligence, 5271. Retrieved from* http://crina.softcomputing.net/hais2008-preface.pdf

Agrawal, M. R., Kaya, C. Y., & Filar, J. A. (2004). *A Class of Models and Budget Strategies for the Control of Heroin Epidemic in Australia*. Academic Press.

Ahmad, Z., & Saleem, A. (2014). Impact of Governance on Human Development. *Pakistan Journal of Commerce & Social Sciences*, *8*(3).

Ahn, S. H. (2015). *Digital Signal Processing - Convolution.* Retrieved from http://www.songho.ca/dsp/convolution/convolution.html

Akay, M. F. (2009). Support vector machines combined with feature selection for breast cancer diagnosis. *Expert Systems with Applications*, *36*(2), 3240–3247. doi:10.1016/j.eswa.2008.01.009

Akbal, E., & Ergen, B. (2007). A Performance Comparison of User and Access Point based Artificial Immune Systems for intrusion detection on a Wireless Local Area Network. *International Journal of Computer Science and Network Security*, *7*, 119–124.

Akerkar, R. (2014). *Introduction to artificial intelligence*. PHI Learning Pvt. Ltd.

Alamri, A. (2016). Nature-inspired multimedia service composition in a media cloud-based healthcare environment. *Cluster Computing*, *19*(4), 2251–2260. doi:10.1007/s10586-016-0647-9

Aler, R., Vega, A., Galván, I. M., & Nebro, A. J. (2012). Multi-objective metaheuristics for preprocessing EEG data in brain–computer interfaces. *Engineering Optimization*, *44*(3), 373–390. doi:10.1080/0305215X.2011.641542

Algarin, A. D. R. (2011). *Clinical Decision Support Systems in Biomedical Informatics and Their Limitations.* Retrieved November 25, 2016, from http://www.engr.uconn.edu/~steve/Cse5810/delarosa.pdf

Allender, S., Scarborough, P., Peto, V., Rayner, M., Leal, J., Luengo-Fernandez, R., & Gray, A. (2008). *European cardiovascular disease statistics*. Brussels: European Heart Network.

Alm, C. O., Roth, D., & Sproat, R. (2005). Emotions from text: machine learning for text-based emotion prediction. In *Proceedings of the conference on human language technology and empirical methods in natural language processing* (pp. 579-586). Association for Computational Linguistics. doi:10.3115/1220575.1220648

Al-Muhaideb, S., & El Bachir Menai, M. (2013). Hybrid metaheuristics for medical data classification. *Hybrid Metaheuristics*, 187-217.

Alpaydin, E. (2014). *Introduction to machine learning*. MIT Press.

Al-Shayea, Q. K. (2011). Artificial neural networks in medical diagnosis. *International Journal of Computer Science Issues*, *8*(2), 150–154.

Alzu'bi, A., Amira, A., & Ramzan, N. (2017). Content-based image retrieval with compact deep convolutional features. *Neurocomputing*, *249*, 95–105. doi:10.1016/j.neucom.2017.03.072

Amasyalı, S. (2011). *Operasyon İşlemlerinin Hızlandırılmasında Veri Madenciliğinin Kullanılması*. İstanbul: Yıldız Teknik Üniversitesi Fen Bilimleri Enstitüsü.

Amato, F., López, A., Peña-Méndez, E. M., Vaňhara, P., Hampl, A., & Havel, J. (2013). *Artificial neural networks in medical diagnosis*. Academic Press.

Amato, F., López, A., Peña-Méndez, E. M., Vaňhara, P., Hampl, A., & Havel, J. (2013). Artificial neural networks in medical diagnosis. *Journal of Applied Biomedicine*, *11*(2), 47–58. doi:10.2478/v10136-012-0031-x

Anderson, D. M. (2009). *Mosby's Dictionary Of Medicine, Nursing & Health Professions (4th ed.)*. Elsevier.

Anderson, D., & McNeill, G. (1992). Artificial neural networks technology. *Kaman Sciences Corporation*, *258*(6), 1–83.

Andrzejak, R. G., Lehnertz, K., Mormann, F., Rieke, C., David, P., & Elger, C. E. (2001). Indications of nonlinear deterministic and finite-dimensional structures in time series of brain electrical activity: Dependence on recording region and brain state. *Physical Review. E*, *64*(6), 061907. doi:10.1103/PhysRevE.64.061907 PMID:11736210

Angermueller, C., Lee, H. J., Reik, W., & Stegle, O. (2017). DeepCpG: Accurate prediction of single-cell DNA methylation states using deep learning. *Genome Biology*, *18*(1), 67. doi:10.1186/s13059-017-1189-z PMID:28395661

Ari, A. A. A., Yenke, B. O., Labraoui, N., Damakoa, I., & Gueroui, A. (2016). A power efficient cluster-based routing algorithm for wireless sensor networks: Honeybees swarm intelligence based approach. *Journal of Network and Computer Applications*, *69*, 77–97. doi:10.1016/j.jnca.2016.04.020

Ari, S., Hembram, K., & Saha, G. (2010). Detection of cardiac abnormality from PCG signal using LMS based least square SVM classifier. *Expert Systems with Applications*, *37*(12), 8019–8026. doi:10.1016/j.eswa.2010.05.088

Armutlu, H. (2017). A Model for Cloud Computing Based 'Rent A Computer'. *Journal of Multidisciplinary Developments*, *2*(1), 22–24.

Arumugam, R., Enti, V. R., Bingbing, L., Xiaojun, W., Baskaran, K., Kong, F. F., . . . Kit, G. W. (2010). DAvinCi: A cloud computing framework for service robots. In *Robotics and Automation (ICRA), 2010 IEEE International Conference on* (pp. 3084-3089). IEEE. doi:10.1109/ROBOT.2010.5509469

Askarzadeh, A. (2014). Bird mating optimizer: An optimization algorithm inspired by bird mating strategies. *Communications in Nonlinear Science and Numerical Simulation*, *19*(4), 1213–1228. doi:10.1016/j.cnsns.2013.08.027

Asuncion, A., & Newman, D. (2007). *UCI machine learning repository*. Retrieved from http://mlearn.ics.uci.edu/MLRepository.html

Atashpaz-Gargari, E., & Lucas, C. (2007). Imperialist competitive algorithm: an algorithm for optimization inspired by imperialistic competition. In *Evolutionary computation, 2007. CEC 2007. IEEE Congress on* (pp. 4661-4667). IEEE. doi:10.1109/CEC.2007.4425083

Atzori, L., Iera, A., & Morabito, G. (2014). From "smart objects" to "social objects": The next evolutionary step of the internet of things. *IEEE Communications Magazine*, *52*(1), 97–105. doi:10.1109/MCOM.2014.6710070

Auria, L., & Moro, R. A. (2008). *Support vector machines (SVM) as a technique for solvency analysis*. DIW Berlin Discussion Paper No. 811. 10.2139/ssrn.1424949

Austin, W., Lemermeyer, G., Goldberg, L., Bergum, V., & Johnson, M. S. (2005). Moral distress in healthcare practice: The situation of nurses. *HEC Forum*, *17*(1), 33–48. doi:10.1007/s10730-005-4949-1 PMID:15957267

Aybek Yılmaz, A. (2016). *Aile Hekimlerinin Birinci Basamakta Sik Görülen Dermatolojik Lezyonlar Hakkindaki Bilgi Düzeylerinin İncelenmesi*. İzmir: İzmir Katip Çelebi Üniversitesi Atatürk Eğitim Ve Araştirma Hastanesi.

Aydın, İ., Karaköse, M., & Akın, E. (2009). a fault diagnosis work using fuzzy logic and negative selection algorithm. *Journal of the Faculty of Engineering and Architecture of Gazi University*, *24*(4), 745–753.

Aydın, İ., Karaköse, M., & Akın, E. (2009). Artificial Immune System Based Diagnostic Model Using Genetic Algorithm. *Dokuz Eylul Unıversıty Faculty Of Engıneerıng Journal Of Scıence And Engıneerıng*, *11*(31), 58–72.

Azmy, M. M. (2015, December). Classification of normal and abnormal heart sounds using new mother wavelet and support vector machines. In *Electrical Engineering (ICEE), 2015 4th International Conference on* (pp. 1-3). IEEE. doi:10.1109/INTEE.2015.7416684

Bachir, B., Ali, A., & Abdellah, M. (2012). Multiobjective optimization of an operational amplifier by the ant colony optimisation algorithm. *Electrical and Electronic Engineering*, *2*(4), 230–235. doi:10.5923/j.eee.20120204.09

Bachkosky, J. V., Boukhvalova, A., Chou, K., Gunnarsson, W., Ledwell, J., McTaggart, B., . . . Yon, J. (2017). *Sensory Computing and Object Processing Entity: Assistive Robotics for Healthcare* (Doctoral dissertation).

Back, T. (1996). *Evolutionary algorithms in theory and practice: evolution strategies, evolutionary programming, genetic algorithms*. Oxford University Press.

Bahaa, K., Noor, G., & Yousif, Y. (2011). *The Artificial Intelligence Approach for Diagnosis, Treatment and Modelling in Orthodontic*. INTECH Open Access Publisher. doi:10.5772/19597

Ba-Karait, N. O. S., Shamsuddin, S. M., & Sudirman, R. (2009). Swarm Negative Selection Algorithm for Electroencephalogram Signals Classification. *Journal of Computational Science*, *5*(12), 998–1005.

Bakhshipour, M., Ghadi, M. J., & Namdari, F. (2017). Swarm robotics search & rescue: A novel artificial intelligence-inspired optimization approach. *Applied Soft Computing*, *57*, 708–726. doi:10.1016/j.asoc.2017.02.028

Bakirtzis, A., & Kazarlis, S. (2016). Genetic algorithms. *Advanced Solutions in Power Systems: HVDC, FACTS, and Artificial Intelligence: HVDC, FACTS, and Artificial Intelligence*, 845-902.

Balestra, G., Foglietta, C., & Gaetano, L. (2010). A Decision Support System based on Meta-heuristics and MCDA for Healthcare Service and Technology Optimal Location. *Studies in Health Technology and Informatics*, *160*, 1470–1470.

Bari, N., Mani, G., & Berkovich, S. (2013). Internet of things as a methodological concept. In *Computing for Geospatial Research and Application (COM. Geo), 2013 Fourth International Conference on* (pp. 48-55). IEEE. doi:10.1109/COMGEO.2013.8

Baseri, H., Rahmani, B., & Bakhshi-Jooybari, M. (2011). Selection of bending parameters for minimal spring-back using an ANFIS model and simulated annealing algorithm. *Journal of Manufacturing Science and Engineering*, *133*(3), 031010. doi:10.1115/1.4004139

Basheer, I. A., & Hajmeer, M. (2000). Artificial neural networks: Fundamentals, computing, design, and application. *Journal of Microbiological Methods*, *43*(1), 3–31. doi:10.1016/S0167-7012(00)00201-3 PMID:11084225

Başkan, E. B., & Tunalı, Ş. (2004). Sık Rastlanan Pediatrik Deri Hastalıkları - Derleme. *The Journal of Current Pediatrics*, 2-4.

Baum, E. B. (1988). On the capabilities of multilayer perceptrons. *Journal of Complexity*, *4*(3), 193–215. doi:10.1016/0885-064X(88)90020-9

Bayram, B. (2017). *Digital Image Processing*. Retrieved June 16, 2017, from http://www.yildiz.edu.tr/~bayram/sgi/saygi.htm

Bedbiab, E., & Kholladi, M. (2010). The Negative Selection Algortihm A Supervised Learning Approach For Skin Detection And Clasification. *IJCSNS International Journal of Computer Science and Network Security*, *10*(11), 48–54.

Beni, G., & Wang, J. (1993). Swarm intelligence in cellular robotic systems. In *Robots and Biological Systems: Towards a New Bionics?* (pp. 703–712). Berlin: Springer. doi:10.1007/978-3-642-58069-7_38

Bennett, C. C., & Hauser, K. (2013). Artificial intelligence framework for simulating clinical decision-making: A Markov decision process approach. *Artificial Intelligence in Medicine*, *57*(1), 9–19. doi:10.1016/j.artmed.2012.12.003 PMID:23287490

Bentley, P., Nordehn, G., Coimbra, M., & Mannor, S. (2011). *The pascal classifying heart sounds challenge 2011 (chsc2011) results*. Retrieved from http://www. peterjbentley. com/heartchallenge/index. html

Bermejo, P., Ossa, L. D., Gamez, J. A., & Puerta, J. M. (2012). Fast Wrapper Feature Subset Selection In High-Dimensional Datasets By Means Of Filter Re-Ranking. *Knowledge-Based Systems*, *25*(1), 35–44. doi:10.1016/j.knosys.2011.01.015

Bertolazzi, P., Felici, G., Festa, P., & Lancia, G. (2008). Logic classification and feature selection for biomedical data. *Computers & Mathematics with Applications (Oxford, England)*, *55*(5), 889–899. doi:10.1016/j.camwa.2006.12.093

Beucher, S. (1992). The watershed transformation applied to image segmentation. Signal and image processing in microscopy and microanalysis. *Scanning Microscopy Supplement 6*, 299-314.

Beucher, S., & Lantuéjoul, C. (1979). Use of watersheds in contour detection. *International Workshop on Image Processing, Real-Time edge and motion detection/estimation*, 2.1-2.12.

Beucher, S., & Meyer, F. (1993). The morphological approach to segmentation: The watershed transform. In E. R. Dougherty (Ed.), *Mathematical Morphology in Image Processing* (pp. 433–481). Marcel Dekker, Inc.

Bhargava, H. K., Power, D. J., & Sun, D. (2007). Progress in Web-based decision support technologies. *Decision Support Systems*, *43*(4), 1083–1095. doi:10.1016/j.dss.2005.07.002

Bhattacharyya, B., & Raj, S. (2016). Swarm intelligence based algorithms for reactive power planning with Flexible AC transmission system devices. *International Journal of Electrical Power & Energy Systems*, *78*, 158–164. doi:10.1016/j.ijepes.2015.11.086

Biomedical Engineering. (2016). *What is Biomedical Engineering?* Florida Institute of Technology. Retrieved November 20, 2016, from http://coe.fit.edu/biomedical-engineering/

Biswas, A., Dasgupta, S., Das, S., & Abraham, A. (2007). Synergy of PSO and bacterial foraging optimization—a comparative study on numerical benchmarks. In *Innovations in Hybrid Intelligent Systems* (pp. 255–263). Springer Berlin Heidelberg. doi:10.1007/978-3-540-74972-1_34

Blum, C., & Li, X. (2008). Swarm intelligence in optimization. In *Swarm Intelligence* (pp. 43–85). Springer Berlin Heidelberg. doi:10.1007/978-3-540-74089-6_2

Boccignone, G., Chianese, A., & Picariello, A. (2000). Computer aided detection of microcalcifications in digital mammograms. *Computers in Biology and Medicine*, *30*(5), 267–286. doi:10.1016/S0010-4825(00)00014-7 PMID:10913773

Bonato, P. (2003). Wearable sensors/systems and their impact on biomedical engineering. *IEEE Engineering in Medicine and Biology Magazine*, *22*(3), 18–20. doi:10.1109/MEMB.2003.1213622 PMID:12845812

Borovykh, A., Bohte, S., & Oosterlee, C. W. (2017). *Conditional Time Series Forecasting with Convolutional Neural Networks.* arXiv preprint arXiv:1703.04691

Bose, J., Jaya, J., Karnan, M., & Thanushkodi, K. (2005). Metaheuristic algorithms for detection of ovarian cancer. *Proceeding of Image Processing and Pattern Recognition*, 35.

Boskovic, D., & Mehra, R. (2000). Intelligent Adaptive Control of a Tailless Advanced Fighter Aircraft under Wing Damage. *Journal of Guidance Control, and Dynamics*, *23*(5), 876–884. doi:10.2514/2.4617

Bostrom, N. (2003). Ethical issues in advanced artificial intelligence. *Science Fiction and Philosophy: From Time Travel to Superintelligence*, 277-284.

Bostrom, N., & Yudkowsky, E. (2014). The ethics of artificial intelligence. The Cambridge Handbook of Artificial Intelligence, 316-334.

Bostrom, N. (2002). The concept of existential risk. *Journal of Evolution and Technology / WTA*, 9.

Bostrom, N. (2014). *Superintelligence: Paths, Dangers, Strategies.* OUP Oxford.

Botta, A., De Donato, W., Persico, V., & Pescapé, A. (2014). On the integration of cloud computing and internet of things. In *Future Internet of Things and Cloud (FiCloud), 2014 International Conference on* (pp. 23-30). IEEE. doi:10.1109/FiCloud.2014.14

Box, G. E., Jenkins, G. M., Reinsel, G. C., & Ljung, G. M. (2015). *Time series analysis: forecasting and control.* John Wiley & Sons.

Brambilla, M., Ferrante, E., Birattari, M., & Dorigo, M. (2013). Swarm robotics: A review from the swarm engineering perspective. *Swarm Intelligence*, *7*(1), 1–41. doi:10.1007/s11721-012-0075-2

Brockwell, P. J., & Davis, R. A. (2016). *Introduction to Time Series and Forecasting.* Springer. doi:10.1007/978-3-319-29854-2

Broekens, J., Heerink, M., & Rosendal, H. (2009). Assistive social robots in elderly care: A review. *Gerontechnology (Valkenswaard)*, *8*(2), 94–103. doi:10.4017/gt.2009.08.02.002.00

Bronzino, J. D. (1999). *Biomedical engineering handbook* (Vol. 2). CRC Press.

Bronzino, J. D., & Peterson, D. R. (2014). *Biomedical engineering fundamentals.* CRC Press.

Brooks, C. A., Thompson, C., & Teasley, S. D. (2014). Towards A General Method for Building Predictive Models of Learner Success using Educational Time Series Data. In LAK Workshops. Academic Press.

Brosnan, M. J. (2002). *Technophobia: The psychological impact of information technology.* Routledge.

Brownlee, J. (2007). *Clonal selection algorithms*. Complex Intelligent Systems Laboratory, Swinburne University of Technology.

Brownlee, J. (2011). *Clever algorithms: nature-inspired programming recipes*. Jason Brownlee.

Brown, R. G. (2004). *Smoothing, forecasting and prediction of discrete time series*. Courier Corporation.

Buchanan, B. G. (2005). A (very) brief history of artificial intelligence. *AI Magazine*, *26*(4), 53.

Bui, N., & Zorzi, M. (2011). Health care applications: a solution based on the internet of things. In *Proceedings of the 4th International Symposium on Applied Sciences in Biomedical and Communication Technologies* (p. 131). ACM. doi:10.1145/2093698.2093829

Burges, C. J. (1998). A tutorial on support vector machines for pattern recognition. *Data Mining and Knowledge Discovery*, *2*(2), 121–167. doi:10.1023/A:1009715923555

Bursa, M., & Lhotska, L. (2007). Ant Colony Inspired Clustering in Biomedical Data Processing. In *3rd European Symposium on Nature-inspired Smart Information Systems.* Aachen: NiSIS.

Bursa, M., Lhotska, L., Chudacek, V., Spilka, J., Janku, P., & Hruban, L. (2015). Information retrieval from hospital information system: Increasing effectivity using swarm intelligence. *Journal of Applied Logic*, *13*(2), 126–137. doi:10.1016/j.jal.2014.11.006

Cagnoni, S. (2014). Evolutionary image analysis and signal processing. In *Proceedings of the Companion Publication of the 2014 Annual Conference on Genetic and Evolutionary Computation* (pp. 795-818). ACM.

Çalış, K., Gazdağı, O., & Yıldız, O. (2013). Reklam İçerikli Epostaların Metin Madenciliği Yöntemleri Ile Otomatik Tespiti. *Bilişim Teknolojileri Dergisi*, 1-7.

Camarillo, D. B., Krummel, T. M., & Salisbury, J. K. Jr. (2004). Robotic technology in surgery: Past, present, and future. *American Journal of Surgery*, *188*(4), 2–15. doi:10.1016/j.amjsurg.2004.08.025 PMID:15476646

Carpenter, G. A. (1989). Neural network models for pattern recognition and associative memory. *Neural Networks*, *2*(4), 243–257. doi:10.1016/0893-6080(89)90035-X

Caruana, R., & Niculescu-Mizil, A. (2006). An empirical comparison of supervised learning algorithms. In *Proceedings of the 23rd international conference on Machine learning* (pp. 161-168). ACM. doi:10.1145/1143844.1143865

Castro, N., & Timmis, J. (2002). *Artificial Immune Systems: A Novel Paradigm to Pattern Recognition*. University of Paisley.

Cawley, G. C., & Talbot, N. L. (2010). On over-fitting in model selection and subsequent selection bias in performance evaluation. *Journal of Machine Learning Research*, *11*(Jul), 2079–2107.

Çetinel, G., & Kamanlı, A. F. (2016). An automatic embryonic stem cell counting method. *Sakarya University Journal of Science*, *20*(2), 83–97.

Chang, H., Han, J., Zhong, C., Snijders, A., & Mao, J. H. (2017). Unsupervised Transfer Learning via Multi-Scale Convolutional Sparse Coding for Biomedical Applications. *IEEE Transactions on Pattern Analysis and Machine Intelligence*, 1. doi:10.1109/TPAMI.2017.2656884 PMID:28129148

Chang, V., Bacigalupo, D., Wills, G., & De Roure, D. (2010). A categorisation of cloud computing business models. In *Proceedings of the 2010 10th IEEE/ACM international conference on cluster, cloud and grid computing* (pp. 509-512). IEEE Computer Society. doi:10.1109/CCGRID.2010.132

Chang, V., Walters, R. J., & Wills, G. (2013). The development that leads to the Cloud Computing Business Framework. *International Journal of Information Management, 33*(3), 524–538. doi:10.1016/j.ijinfomgt.2013.01.005

Chatfield, C. (2016). *The Analysis of Time Series: An Introduction.* CRC Press.

Chattopadhyay, S., Banerjee, S., Rabhi, F. A., & Acharya, U. R. (2013). A Case-Based Reasoning system for complex medical diagnosis. *Expert Systems: International Journal of Knowledge Engineering and Neural Networks, 30*(1), 12–20. doi:10.1111/j.1468-0394.2012.00618.x

Chaurasia, S. N., & Singh, A. (2015). A hybrid swarm intelligence approach to the registration area planning problem. *Information Sciences, 302*, 50–69. doi:10.1016/j.ins.2015.01.012

Chaves-González, J. M., Pérez-Toledano, M. A., & Navasa, A. (2015). Software requirement optimization using a multiobjective swarm intelligence evolutionary algorithm. *Knowledge-Based Systems, 83*, 105–115. doi:10.1016/j.knosys.2015.03.012

Chawla, M., & Duhan, M. (2014). Applications of recent metaheuristics optimisation algorithms in biomedical engineering: A review. *International Journal of Biomedical Engineering and Technology, 16*(3), 268–278. doi:10.1504/IJBET.2014.065807

Chen, T. C., Tsai, P. W., Chu, S. C., & Pan, J. S. (2007). A novel optimization approach: bacterial-GA foraging. In *Innovative Computing, Information and Control, 2007. ICICIC'07. Second International Conference on* (pp. 391-391). IEEE. doi:10.1109/ICICIC.2007.67

Chen, Y., Du, Z., & García-Acosta, M. (2010). Robot as a service in cloud computing. In *Service Oriented System Engineering (SOSE), 2010 Fifth IEEE International Symposium on* (pp. 151-158). IEEE. doi:10.1109/SOSE.2010.44

Chen, H., Chiang, R. H., & Storey, V. C. (2012). Business intelligence and analytics: From big data to big impact. *Management Information Systems Quarterly, 36*(4).

Chen, J. H., & Asch, S. M. (2017). Machine learning and prediction in medicine– beyond the peak of inflated expectations. *The New England Journal of Medicine, 376*(26), 2507–2509. doi:10.1056/NEJMp1702071 PMID:28657867

Chen, J., Qian, F., Yan, W., & Shen, B. (2013). Translational biomedical informatics in the cloud: Present and future. *BioMed Research International.* PMID:23586054

Cherif, L. H., Mostafi, M., & Debbal, S. M. (2014). Digital filters in Heart Sound Analysis. *International Journal of Clinical Medicine Research, 1*(3), 97–108.

Chikh, M. A., Saidi, M., & Settouti, N. (2012). Diagnosis of diabetes diseases using an artificial immune recognition system2 (AIRS2) with fuzzy k-nearest neighbor. *Journal of Medical Systems, 36*(5), 2721–2729. doi:10.1007/s10916-011-9748-4 PMID:21695498

Chiuchisan, I., Costin, H. N., & Geman, O. (2014). Adopting the internet of things technologies in health care systems. In *Electrical and Power Engineering (EPE), 2014 International Conference and Exposition on* (pp. 532-535). IEEE. doi:10.1109/ICEPE.2014.6969965

Chiuchisan, I., & Geman, O. A. N. A. (2014). An approach of a decision support and home monitoring system for patients with neurological disorders using internet of things concepts. *WSEAS Transations on Systems, 13*, 460–469.

Choi, Y. J., Baek, J. H., Park, H. S., Shim, W. H., Kim, T. Y., Shong, Y. K., & Lee, J. H. (2017). A computer-aided diagnosis system using artificial intelligence for the diagnosis and characterization of thyroid nodules on ultrasound: Initial clinical assessment. *Thyroid, 27*(4), 546–552. doi:10.1089/thy.2016.0372 PMID:28071987

Chong, E. K., & Zak, S. H. (2013). *An Introduction to Optimization* (Vol. 76). John Wiley & Sons.

Chu, S. C., Tsai, P. W., & Pan, J. S. (2006). Cat swarm optimization. In *Pacific Rim International Conference on Artificial Intelligence* (pp. 854-858). Springer Berlin Heidelberg.

Chuang, T. T., & Yada, S. B. (1998). The development of an adaptive decision support system. *Decision Support Systems, 24*(2), 73–87. doi:10.1016/S0167-9236(98)00065-7

Chuh, A. (2003). *Clinical Dermatology: A Clinical Guide To Diagnosis And Therapy* (4th ed.). Journal Of The European Academy Of Dermatology And Venereology.

Clancey, W. J., & Shortliffe, E. H. (1984). *Readings in Medical Artificial Intelligence: The First Decade*. Reading, MA: Addison Wesley.

Coates, A., Ng, A., & Lee, H. (2011, June). An analysis of single-layer networks in unsupervised feature learning. In *Proceedings of the fourteenth international conference on artificial intelligence and statistics* (pp. 215-223). Academic Press.

Cochocki, A., & Unbehauen, R. (1993). *Neural networks for optimization and signal processing*. John Wiley & Sons, Inc.

Coelho, I. M., Coelho, V. N., da Luz, E. J. S., Ochi, L. S., Guimarães, F. G., & Rios, E. (2017). A GPU deep learning metaheuristic based model for time series forecasting. *Applied Energy, 201*, 412–418. doi:10.1016/j.apenergy.2017.01.003

Cohen, J. S., & Erickson, J. M. (2006). Ethical dilemmas and moral distress in oncology nursing practice. *Clinical Journal of Oncology Nursing, 10*(6), 775–780. doi:10.1188/06.CJON.775-780 PMID:17193943

Coiera, E. (2003). Guide to Medical Informatics, the Internet and Telemedicine (2nd ed.). London: Arnold.

Colley, M., Kommoss, F., Bibbo, M., Dytch, H. E., Frnklin, W. A., Holt, J. A., & Wied, G. L. (1989). Assessment of hormone receptors in breast carcinoma by immunocytochemistry and image analysis. *Analytical and Quantitative Cytology and Histology, 11*, 307–314. PMID:2803443

Colorni, A., Dorigo, M., & Maniezzo, V. (1991). *Distributed optimization by ant colonies. In Actes de la première conférence européenne sur la vie artificielle* (pp. 134–142). Paris, France: Elsevier.

Çomak, E., Arslan, A., & Türkoğlu, İ. (2007). A decision support system based on support vector machines for diagnosis of the heart valve diseases. *Computers in Biology and Medicine, 37*(1), 21–27. doi:10.1016/j.compbiomed.2005.11.002 PMID:16426598

Consoli, S., Pérez, J. M., Darby-Dowman, K., & Mladenović, N. (2008). Discrete particle swarm optimization for the minimum labelling Steiner tree problem. In Nature inspired cooperative strategies for optimization (NICSO 2007) (pp. 313-322). Springer Berlin Heidelberg. doi:10.1007/978-3-540-78987-1_28

Copeland, J. (2000). *What is Artificial Intelligence?* Retrieved November 20, 2016, from, http://www.alanturing.net/turing_archive/pages/reference articles/what is ai.html

Corchado, E., Graña, M., & Woniak, M. (2012). Editorial: New trends and applications on hybrid artificial intelligence systems. *Neurocomputing, 75*(1), 61–63. doi:10.1016/j.neucom.2011.06.021

Cortes, C., & Vapnik, V. (1995). Support-vector networks. *Machine Learning, 20*(3), 273–297. doi:10.1007/BF00994018

Cortés, U., Sànchez-Marrè, M., Ceccaroni, L., R-Roda, I., & Poch, M. (2000). Artificial intelligence and environmental decision support systems. *Applied Intelligence, 13*(1), 77–91. doi:10.1023/A:1008331413864

Coskun, H., Yigit, T., & Deperlioglu, O. (2016, October). Effect of filter selection on classification of extrasystole heart sounds via mobile devices. In *Application of Information and Communication Technologies (AICT), 2016 IEEE 10th International Conference on* (pp. 1-5). IEEE. doi:10.1109/ICAICT.2016.7991820

Coşkun, C., & Baykal, A. (2011). Veri Madenciliğinde Sınıflandırma Algoritmalarının Bir Örnek Üzerinde Karşılaştırılması. *Akademik Bilişim, 2011*, 1–8.

Costa, M., Goldberger, A. L., & Peng, C. K. (2002). Multiscale entropy analysis of complex physiologic time series. *Physical Review Letters*, *89*(6), 068102. doi:10.1103/PhysRevLett.89.068102 PMID:12190613

Craig, J. J. (2005). *Introduction to robotics: mechanics and control* (Vol. 3). Upper Saddle River, NJ: Pearson Prentice Hall.

Cristianini, N., & Shawe-Taylor, J. (2000). *An introduction to support vector machines and other kernel-based learning methods*. Cambridge University Press. doi:10.1017/CBO9780511801389

Cruz-Roa, A., Gilmore, H., Basavanhally, A., Feldman, M., Ganesan, S., Shih, N. N., & Madabhushi, A. et al. (2017). Accurate and reproducible invasive breast cancer detection in whole-slide images: A Deep Learning approach for quantifying tumor extent. *Scientific Reports*, *7*, 46450. doi:10.1038/srep46450 PMID:28418027

Da San Martino, G., Cardillo, F. A., & Starita, A. (2006). A new swarm intelligence algorithm for image alignment. In *Advances in Medical, Signal and Information Processing, 2006. MEDSIP 2006. IET 3rd International Conference On* (pp. 1-4). IET. doi:10.1049/cp:20060387

Da Xu, L., He, W., & Li, S. (2014). Internet of things in industries: A survey. *IEEE Transactions on Industrial Informatics*, *10*(4), 2233–2243. doi:10.1109/TII.2014.2300753

Dao, V. L., Nguyen, V. D., Nguyen, H. D., & Hoang, V. P. (2016, December). Hardware Implementation of MFCC Feature Extraction for Speech Recognition on FPGA. In *International Conference on Advances in Information and Communication Technology* (pp. 248-254). Springer International Publishing.

Darianian, M., & Michael, M. P. (2008). Smart home mobile RFID-based Internet-of-Things systems and services. In *Advanced Computer Theory and Engineering, 2008. ICACTE'08. International Conference on* (pp. 116-120). IEEE. doi:10.1109/ICACTE.2008.180

Dasgupta, D., & Attoh-Okine, N. 1997). Immunity-Based Systems: A Survey. *Proceedings of the IEEE International Conference on Systems, Man and Cybernetics*. doi:10.1109/ICSMC.1997.625778

Dasgupta, D., & Forrest, S. (1995). *Tool Breakage Detection in Milling Operations using a Negative-Selection Algorithm*. Technical Report CS95-5. Computer Science, University of New Mexico.

Dasgupta, D., & Forrest, S. (1999). Artificial Immune systems in Industrial Applications. In *Proceedings of the Second International Conference on Intelligent Processing and Manufacturing of materials (IPMM)*. doi:10.1109/IPMM.1999.792486

Dasgupta, D., Ji, Z., & Gonzalez, F. (2003). Artificial immune system (AIS) research in the last five years. In Evolutionary Computation, 2003. CEC'03. The 2003 Congress on (Vol. 1, pp. 123-130). IEEE.

Dasgupta, D. (1996). *Using Immunological Principles in Anomaly Detection. In Proceedings of Artificial Neural Networks in Engineering*. ANNIE.

Dasgupta, D. (1998). *Artificial immune systems and their applications. Springer-Verlag.*

Dasgupta, D., & Forrest, S. (1996). Novelty Detection in Time Series Data Using Ideas from Immunology. *Proceedings of The 5th International Conference on Intelligent Systems.*

Dasgupta, D., & Forrest, S. (1996). Novelty Detection in Time Series Data Using Ideas From Immunology. *Proceedings of the ISCA'96.*

Das, R., Turkoglu, I., & Sengur, A. (2009). Diagnosis of valvular heart disease through neural networks ensembles. *Computer Methods and Programs in Biomedicine, 93*(2), 185–191. doi:10.1016/j.cmpb.2008.09.005 PMID:18951649

Daubechies, I. (1990). The wavelet transform, time-frequency localization and signal analysis. *IEEE Transactions on Information Theory, 36*(5), 961–1005. doi:10.1109/18.57199

de Carvalho Gomes, F., & Maia, J. G. R. (2016). Learning Optimal Decision Lists as a Metaheuristic Search for Diagnosis of Parkinson's Disease. In *International Workshop on Machine Learning, Optimization and Big Data* (pp. 391-401). Springer. doi:10.1007/978-3-319-51469-7_32

De Castro, L. N., & Von Zuben, F. J. (2000). Artificial immune systems: Part I- Basic theory and applications. DCA-RT 02/00, Brasil, 23-28.

de Castro, L. N., & Timmis, J. (2002). *Artificial Immune Systems: A New Computational Intelligence Approach.* London: Springer.

de Castro, L. N., & Von Zuben, F. J. (2002). Learning and optimization using the clonal selection principle. *IEEE Transactions on Evolutionary Computation, 6*(3), 239–251. doi:10.1109/TEVC.2002.1011539

De la Torre, A., Peinado, A. M., Segura, J. C., Perez-Cordoba, J. L., Benitez, M. C., & Rubio, A. J. (2005). Histogram equalization of speech representation for robust speech recognition. *IEEE Transactions on Speech and Audio Processing, 13*(3), 355–366. doi:10.1109/TSA.2005.845805

De Momi, E., & Ferrigno, G. (2010). Robotic and artificial intelligence for keyhole neurosurgery: The ROBOCAST project, a multi-modal autonomous path planner. *Proceedings of the Institution of Mechanical Engineers. Part H, Journal of Engineering in Medicine, 224*(5), 715–727. doi:10.1243/09544119JEIM585 PMID:20718272

Deepa, S. N., & Devi, B. A. (2011). A survey on artificial intelligence approaches for medical image classification. *Indian Journal of Science and Technology, 4*(11), 1583–1595.

Deng, L. (2014). A tutorial survey of architectures, algorithms, and applications for deep learning. *APSIPA Transactions on Signal and Information Processing, 3.*

Deng, L., & Yu, D. (2014). Deep learning: methods and applications. *Foundations and Trends® in Signal Processing, 7*(3–4), 197-387.

Deng, L., & Yu, D. (2014). Deep learning: Methods and applications. *Foundations and Trends® in Signal Processing, 7*(3–4), 197-387.

Deng, N. Y., & Tian, Y. (2004). *A new method of data mining-support vector machine.* Science Publication.

Deng, Y., Liu, Y., & Zhou, D. (2015). An improved genetic algorithm with initial population strategy for symmetric TSP. *Mathematical Problems in Engineering.*

Deperlioglu, O., Guraksin, G. E., & Kose, U. (2016). Web Based Clinical Decision Support Systems: Structures and Features. *Academic Informatics Conferences, AB 2016.*

Deperlioglu, O., Kose, U., & Guraksin, G. E. (2015). *Biomedical Decision Making: Reasoning and Decision Support with Artificial Intelligence.* 3rd International Congress on Natural and Engineering Sciences, ICNES 2015, Sarajevo, Bosnia Herzegovina.

Deperlioglu, O., & Kose, U. (2011). An educational tool for artificial neural networks. *Computers & Electrical Engineering, 37*(3), 392–402. doi:10.1016/j.compeleceng.2011.03.010

Deperlioglu, O., & Polat, K. (2014). Biomedical Informatics Systems and Biostatistics. In M. H. Asyali, S. Kara, & B. Yilmaz (Eds.), *Fundamentals of Biomedical Engineering* (pp. 653–698). Ankara, Turkey: Nobel Academic Publishing.

Dessinioti, C., & Katsambas, A. (2013). Seborrheic Dermatitis: Etiology, Risk Factors, And Treatments: Facts And Controversies. *Clinics in Dermatology, 31*(4), 343–351. doi:10.1016/j.clindermatol.2013.01.001 PMID:23806151

Dimitrov, D. V. (2016). Medical internet of things and big data in healthcare. *Healthcare Informatics Research, 22*(3), 156–163. doi:10.4258/hir.2016.22.3.156 PMID:27525156

Dinello, D. (2005). *Technophobia! Science fiction visions of posthuman technology*. University of Texas Press.

Dinh, H. T., Lee, C., Niyato, D., & Wang, P. (2013). A survey of mobile cloud computing: Architecture, applications, and approaches. *Wireless Communications and Mobile Computing, 13*(18), 1587–1611. doi:10.1002/wcm.1203

Dixon, B. E., Dixon, B. E., Simonaitis, L., Goldberg, H. S., Paterno, M. D., Schaeffer, M., & Middleton, B. et al. (2013). A Pilot Study of Distributed Knowledge Management and Clinical Decision Support in the Cloud. *Artificial Intelligence in Medicine, 59*(1), 45–53. doi:10.1016/j.artmed.2013.03.004 PMID:23545327

Do, H. N. (2017). *Predictive models for robotics and biomedical applications* (Doctoral dissertation). Michigan State University.

Domingo, M. C. (2012). An overview of the Internet of Things for people with disabilities. *Journal of Network and Computer Applications, 35*(2), 584–596. doi:10.1016/j.jnca.2011.10.015

Dorigo, M. (1992). *Optimization, learning and natural algorithms* (PhD Thesis). Politecnico di Milano, Italy.

Doukas, C., & Maglogiannis, I. (2011). Managing wearable sensor data through cloud computing. In *Cloud Computing Technology and Science (CloudCom), 2011 IEEE Third International Conference on* (pp. 440-445). IEEE. doi:10.1109/CloudCom.2011.65

Doukas, C., & Maglogiannis, I. (2012). Bringing IoT and cloud computing towards pervasive healthcare. In *Innovative Mobile and Internet Services in Ubiquitous Computing (IMIS), 2012 Sixth International Conference on* (pp. 922-926). IEEE. doi:10.1109/IMIS.2012.26

Doukas, C., Pliakas, T., & Maglogiannis, I. (2010). Mobile healthcare information management utilizing Cloud Computing and Android OS. In *Engineering in Medicine and Biology Society (EMBC), 2010 Annual International Conference of the IEEE* (pp. 1037-1040). IEEE. doi:10.1109/IEMBS.2010.5628061

Drew, P. J., & Monson, J. R. (2000). Artificial neural networks. *Surgery, 127*(1), 3–11. doi:10.1067/msy.2000.102173 PMID:10660751

Dubuisson, M. P., Jain, A. K., & Jain, M. K. (1994). Segmentation and Classification of Bacterial Culture Images. *Journal of Microbiological Methods, 19*(4), 279–295. doi:10.1016/0167-7012(94)90031-0

Duman, S. (2016). Symbiotic organisms search algorithm for optimal power flow problem based on valve-point effect and prohibited zones. *Neural Computing & Applications*, 1–15.

Eberhart, R., & Kennedy, J. (1995). A new optimizer using particle swarm theory. In *Micro Machine and Human Science, 1995. MHS'95., Proceedings of the Sixth International Symposium on* (pp. 39-43). IEEE. doi:10.1109/MHS.1995.494215

El-Dahshan, E. S. A., Hosny, T., & Salem, A. B. M. (2010). Hybrid intelligent techniques for MRI brain images classification. *Digital Signal Processing, 20*(2), 433–441. doi:10.1016/j.dsp.2009.07.002

Elliot, A., Hsu, C. H., & Slodoba, J. (2017). *Time Series Prediction: Predicting Stock Price.* arXiv preprint arXiv:1710.05751

Elmas, C. (2003). *Artificial neural network – Theory, architecture, training, implementation.* Ankara, Turkey: Seçkin Press. (In Turkish)

Enderle, J. D., & Bronzino, J. D. (2012). *Introduction to biomedical engineering.* Academic press.

Enders, W. (2008). *Applied econometric time series.* John Wiley & Sons.

Engelbrecht, A. P. (2006). *Fundamentals of computational swarm intelligence.* John Wiley & Sons.

Engin, O., & Döyen, A. (2004). Artificial Immune Systems And Applications In Industrial Problems. *Journal of Science of Gazi University, 17*(1), 71–84.

Erdogmuş, P., & Yalçın, E. (2015). The Solutions Of Unconstrained Optimization Benchmark Problems With Particle Swarm Optimization. *Journal of Advanced Technology Sciences*, 14-22.

Erkmen, O., & Bozoğlu, T. F. (2008). *Food Microbiology-1* (1st ed.). Ankara: Gazi University Printing Depermant.

Escalona-Vargas, D. I., Lopez-Arevalo, I., & Gutiérrez, D. (2014). Multicompare tests of the performance of different metaheuristics in EEG dipole source localization. *The Scientific World Journal.* PMID:24757424

Eskidere, Ö. (2012). A Comparison Of Feature Selection Methods For Diagnosis Of Parkinson's Disease From Vocal Measurements. *Journal of Engineering and Natural Sciences*, 402-414.

Etzioni, A., & Etzioni, O. (2017). Incorporating Ethics into Artificial Intelligence. *The Journal of Ethics*, 1–16.

Farkas, I. (2003). Artificial intelligence in agriculture. *Computers and Electronics in Agriculture, 1*(40), 1–3. doi:10.1016/S0168-1699(03)00007-3

Fausett, L. (1994). *Fundamentals of neural networks: architectures, algorithms, and applications (No. 006.3).* Prentice-Hall.

Fidanova, S. S., & Roeva, O. N. (2013). Metaheuristic techniques for optimization of an E. coli cultivation model. *Biotechnology, Biotechnological Equipment, 27*(3), 3870–3876. doi:10.5504/BBEQ.2012.0136

Filter, R. A. (2005). *Dynamic HIV/AIDS parameter estimation with applications* (Doctoral dissertation). University of Pretoria.

Fiorini, L., Esposito, R., Bonaccorsi, M., Petrazzuolo, C., Saponara, F., Giannantonio, R., & Cavallo, F. et al. (2017). Enabling personalised medical support for chronic disease management through a hybrid robot-cloud approach. *Autonomous Robots, 41*(5), 1263–1276. doi:10.1007/s10514-016-9586-9

Fisher, J. (2016). *Genetic Algorithm – Programming by the Seat of Your Genes! – SlideShare.Net.* Retrieved from https://www.slideshare.net/JeremyFisher1/genetic-algorithms-programming-by-the-seat-of-your-genes

Fisher, R. A. (1958). *The genetic theory of natural selection.* Oxford University Press.

Fister, I., Jr., Yang, X. S., Fister, I., Brest, J., & Fister, D. (2013). *A brief review of nature-inspired algorithms for optimization.* arXiv preprint arXiv:1307.4186

Flasiński, M. (2016). *Introduction to Artificial Intelligence.* Springer. doi:10.1007/978-3-319-40022-8

Førde, R., & Aasland, O. G. (2008). Moral distress among Norwegian doctors. *Journal of Medical Ethics, 34*(7), 521–525. doi:10.1136/jme.2007.021246 PMID:18591286

Formisano, E., De Martino, F., & Valente, G. (2008). Multivariate analysis of fMRI time series: Classification and regression of brain responses using machine learning. *Magnetic Resonance Imaging*, *26*(7), 921–934. doi:10.1016/j.mri.2008.01.052 PMID:18508219

Forrest, S. (1994). Self-Non self Discrimination in a Computer. *Proceedings of the IEEE Symposium on Security and Privacy*, 202.

Forrest, S., Perelson, A., & Allen, L. (1994). Self-Nonself Discrimination in a Computer. *Proceedings of the IEEE Symposium on Research in Security and Privacy*, 202-212.

Fortino, G., Pathan, M., & Di Fatta, G. (2012). Bodycloud: Integration of cloud computing and body sensor networks. In *Cloud Computing Technology and Science (CloudCom), 2012 IEEE 4th International Conference on* (pp. 851-856). IEEE.

Fraccaroa, P., O'Sullivana, D., Plastirasa, P., & O'Sullivan, H., Dentonec, Biagioe, A., & Wellera, P. (2015). Behind the screens:Clinical decision support methodologies – A review. *Health Policy and Technology*, *4*, 29–38. doi:10.1016/j.hlpt.2014.10.001

Freitas, A. A., & Kaestner, C. A. (2002). Attribute Selection With A Multiobjective Genetic Algorithm. *16th Brazilian Symposium On Artificial Intelligence* (pp. 280-290). Porto De Galinhas/Recife: Springer-Verlag.

Fusaro, V. A., Patil, P., Gafni, E., Wall, D. P., & Tonellato, P. J. (2011). Biomedical cloud computing with amazon web services. *PLoS Computational Biology*, *7*(8), e1002147. doi:10.1371/journal.pcbi.1002147 PMID:21901085

Gacek, A., & Pedrycz, W. (Eds.). (2011). *ECG signal processing, classification and interpretation: a comprehensive framework of computational intelligence*. Springer Science & Business Media.

Ganapathy, K. (2014). *Soft Computing Approaches for Healthcare in Sensor Cloud Environment* (Thesis). Anna University, Faculty of Information and Communication Engineering.

Gandomi, A. H., & Alavi, A. H. (2012). Krill herd: A new bio-inspired optimization algorithm. *Communications in Nonlinear Science and Numerical Simulation*, *17*(12), 4831–4845. doi:10.1016/j.cnsns.2012.05.010

Gandomi, A. H., Kashani, A. R., Roke, D. A., & Mousavi, M. (2015). Optimization of retaining wall design using recent swarm intelligence techniques. *Engineering Structures*, *103*, 72–84. doi:10.1016/j.engstruct.2015.08.034

Ganesan, P., & Rajini, V. (2014). YIQ color space based satellite image segmentation using modified FCM clustering and histogram equalization. In ICAEE, Velore, India. doi:10.1109/ICAEE.2014.6838440

Gao, X. Z., Ovaska, S. J., & Wang, X. (2008). A GA-based negative selection algorithm. *International Journal of Innovative Computing, Information, & Control*, *4*(4), 971–979.

Garcia, F. J. M., & Perez, J. A. M. (2008). Jumping frogs optimization: a new swarm method for discrete optimization. *Documentos de Trabajo del DEIOC*, (3).

Garlapati, V. K., Vundavilli, P. R., & Banerjee, R. (2010). Evaluation of lipase production by genetic algorithm and particle swarm optimization and their comparative study. *Applied Biochemistry and Biotechnology*, *162*(5), 1350–1361. doi:10.1007/s12010-009-8895-2 PMID:20099046

Garnier, S., Gautrais, J., & Theraulaz, G. (2007). The biological principles of swarm intelligence. *Swarm Intelligence*, *1*(1), 3–31. doi:10.1007/s11721-007-0004-y

Gelbrich, K., & Sattler, B. (2014). Anxiety, crowding, and time pressure in public self-service technology acceptance. *Journal of Services Marketing*, *28*(1), 82–94. doi:10.1108/JSM-02-2012-0051

Gen, M., & Cheng, R. (2000). *Genetic algorithms and engineering optimization* (Vol. 7). John Wiley & Sons.

Geraci, R. M. (2012). *Apocalyptic AI: Visions of heaven in robotics, artificial intelligence, and virtual reality*. Oxford University Press.

Ghumbre, S., Patil, C., & Ghatol, A. (2011). Heart disease diagnosis using support vector machine. *International conference on computer science and information technology (ICCSIT') Pattaya.*

Giagkos, A., & Wilson, M. S. (2014). BeeIP–A Swarm Intelligence based routing for wireless ad hoc networks. *Information Sciences*, *265*, 23–35. doi:10.1016/j.ins.2013.12.038

Giarratano, J. C., & Riley, G. (1998). *Expert systems*. PWS Publishing Co.

Gilchrist, J. E., Campbell, J. E., Donnelly, C. B., Peeler, J. T., & Delaney, J. M. (1973). Spiral Plate Method for Bacterial Determination. *Applied Microbiology*, *25*, 244–252. PMID:4632851

Goldberg, D. E., & Koza, J. R. (1994). Genetic Algorithms & Evolutionary Computation. *National Conference on Artificial Intelligence.*

Gonzalez, F. A., & Dasgupta, D. (2002). Anomaly Detection Using Real-Valued Negative Selection. *Genetic Programming and Evolvable Machines.*

Gonzalez, R. C., & Woods, R. E. (2008). *Digital Image Processing* (3rd ed.). Upper Saddle River, NJ: Prentice Hall.

Greenspan, H., van Ginneken, B., & Summers, R. M. (2016). Guest editorial deep learning in medical imaging: Overview and future promise of an exciting new technique. *IEEE Transactions on Medical Imaging*, *35*(5), 1153–1159. doi:10.1109/TMI.2016.2553401

Gubbi, J., Buyya, R., Marusic, S., & Palaniswami, M. (2013). Internet of Things (IoT): A vision, architectural elements, and future directions. *Future Generation Computer Systems*, *29*(7), 1645–1660. doi:10.1016/j.future.2013.01.010

Gudise, V. G., & Venayagamoorthy, G. K. (2003). Comparison of particle swarm optimization and backpropagation as training algorithms for neural networks. In *Swarm Intelligence Symposium, 2003. SIS'03. Proceedings of the 2003 IEEE* (pp. 110-117). IEEE. doi:10.1109/SIS.2003.1202255

Guermoui, M., Mekhalfi, M. L., & Ferroudji, K. (2013, May). Heart sounds analysis using wavelets responses and support vector machines. In *Systems, Signal Processing and their Applications (WoSSPA), 2013 8th International Workshop on* (pp. 233-238). IEEE. doi:10.1109/WoSSPA.2013.6602368

Guha, D., Roy, P., & Banerjee, S. (2017). Quasi-oppositional symbiotic organism search algorithm applied to load frequency control. *Swarm and Evolutionary Computation*, *33*, 46–67. doi:10.1016/j.swevo.2016.10.001

Guimard, N. K., Gomez, N., & Schmidt, C. E. (2007). Conducting polymers in biomedical engineering. *Progress in Polymer Science*, *32*(8), 876–921. doi:10.1016/j.progpolymsci.2007.05.012

Gümüşçü, A., Aydilek, İ. B., & Taşaltın, R. (2016). Mikro-Dizilim Veri Sınıflandırmasında Öznitelik Seçme Algoritmalarının Karşılaştırılması. *Harran Üniversitesi Mühendislik Dergisi*, *1*(1), 1–7.

Gunarathne, T., Wu, T. L., Choi, J. Y., Bae, S. H., & Qiu, J. (2011). Cloud computing paradigms for pleasingly parallel biomedical applications. *Concurrency and Computation*, *23*(17), 2338–2354. doi:10.1002/cpe.1780

Guner, E. (2003). *Adaptive neuro fuzzy inference system applications in chemical processes* (Doctoral dissertation). METU.

Gupta, C. N., Palaniappan, R., Swaminathan, S., & Krishnan, S. M. (2007). Neural network classification of homomorphic segmented heart sounds. *Applied Soft Computing*, *7*(1), 286–297. doi:10.1016/j.asoc.2005.06.006

Gupta, S., Bhardwaj, S., & Bhatia, P. K. (2011). A Reminiscent Study of Nature Inspired Computation. *International Journal of Advances in Engineering and Technology*, *1*(2), 117–125.

Guraksin, G. E., Ergun, U., & Deperlioglu, O. (2010). Classification of the heart sounds via artificial neural network. *International Journal of Reasoning-based Intelligent Systems*, *2*(3-4), 272–278. doi:10.1504/IJRIS.2010.036874

Gurstein, M. (Ed.). (2000). *Community Informatics: Enabling Communities with Information and Communications Technologies: Enabling Communities with Information and Communications Technologies*. IGI Global. doi:10.4018/978-1-878289-69-8

Güvenir, H. A., Demiröz, G., & İlter, N. (1998, July). Learning Differential Diagnosis Of Erythemato-Squamous Diseases Using Voting Feature Intervals. *Artificial Intelligence in Medicine*, *13*(3), 147–165. doi:10.1016/S0933-3657(98)00028-1 PMID:9698151

Habibi, M., Weber, L., Neves, M., Wiegandt, D. L., & Leser, U. (2017). Deep learning with word embeddings improves biomedical named entity recognition. *Bioinformatics (Oxford, England)*, *33*(14), i37–i48. doi:10.1093/bioinformatics/btx228 PMID:28881963

Hai, N. T., Van Thuyen, N., Mai, T. T., & Van Toi, V. (2015). MFCC-DTW Algorithm for Speech Recognition in an Intelligent Wheelchair. In *5th International Conference on Biomedical Engineering in Vietnam* (pp. 417-421). Springer.

Halkman, A. K., & Ayhan, K. (2017). *Microorganism Counting*. Retrieved June 02, 2017, from http://www.mikrobiyoloji.org/TR/Genel/DosyaGoster.aspx?DIL=1&BELGEANAH=2927&DOSYAISIM=210010601.pdf

Hall, L. O., Bensaid, A. M., Clarke, L. P., Velthuizen, R. P., Silbiger, M. S., & Bezdek, J. C. (1992). A comparison of neural network and fuzzy clustering techniques in segmenting magnetic resonance images of the brain. *IEEE Transactions on Neural Networks*, *3*(5), 672–682. doi:10.1109/72.159057 PMID:18276467

Hall, M., & Holmes, G. (2003). Benchmarking Attribute Selection Techniques For Discrete Class Data Mining. *IEEE Transactions on Knowledge and Data Engineering*, *15*(6), 1437–1447. doi:10.1109/TKDE.2003.1245283

Hamric, A. B., Davis, W. S., & Childress, M. D. (2006). Moral distress in health care professionals. *Pharos*, *69*(1), 16–23. PMID:16544460

Hanen, J., Kechaou, Z., & Ayed, M. B. (2016). An enhanced healthcare system in mobile cloud computing environment. *Vietnam Journal of Computer Science*, *3*(4), 267–277. doi:10.1007/s40595-016-0076-y

Hashem, I. A. T., Yaqoob, I., Anuar, N. B., Mokhtar, S., Gani, A., & Khan, S. U. (2015). The rise of "big data" on cloud computing: Review and open research issues. *Information Systems*, *47*, 98–115. doi:10.1016/j.is.2014.07.006

Hawking, S., Russell, S., Tegmark, M., & Wilczek, F. (2014). Stephen Hawking: Transcendence looks at the implications of artificial intelligence-but are we taking AI seriously enough? *The Independent*, *2014*(05-01), 9313474.

Hayes, B. (2008). Cloud computing. *Communications of the ACM*, *51*(7), 9–11. doi:10.1145/1364782.1364786

Hedayat, A., Davilu, H., Barfrosh, A. A., & Sepanloo, K. (2009). Optimization of the core configuration design using a hybrid artificial intelligence algorithm for research reactors. *Nuclear Engineering and Design*, *239*(12), 2786–2799. doi:10.1016/j.nucengdes.2009.08.027

Hedayatzadeh, R., Salmassi, F. A., Keshtgari, M., Akbari, R., & Ziarati, K. (2010). Termite colony optimization: A novel approach for optimizing continuous problems. In *2010 18th Iranian Conference on Electrical Engineering* (pp. 553-558). IEEE. doi:10.1109/IRANIANCEE.2010.5507009

Heinrich, L., Bogovic, J. A., & Saalfeld, S. (2017). *Deep Learning for Isotropic Super-Resolution from Non-Isotropic 3D Electron Microscopy*. arXiv preprint arXiv:1706.03142

Hetmaniok, E., Słota, D., & Zielonka, A. (2015). Using the swarm intelligence algorithms in solution of the two-dimensional inverse Stefan problem. *Computers & Mathematics with Applications (Oxford, England)*, *69*(4), 347–361. doi:10.1016/j.camwa.2014.12.013

Holland, J. (1995). *Hidden order: How adaptation builds complexity*. Basic Books.

Holland, J. H. (2012). Genetic algorithms. *Scholarpedia*, *7*(12), 1482. doi:10.4249/scholarpedia.1482

Holler, J., Tsiatsis, V., Mulligan, C., Avesand, S., Karnouskos, S., & Boyle, D. (2014). *From Machine-to-machine to the Internet of Things: Introduction to a New Age of Intelligence*. Academic Press.

Hoolimath, P. K. G., Kiran, M., & Reddy, G. R. M. (2012). Optimized Termite: A bio-inspired routing algorithm for MANET's. In *2012 International Conference on Signal Processing and Communications (SPCOM)* (pp. 1-5). IEEE. doi:10.1109/SPCOM.2012.6290230

Hoong, N. K. (1988). Medical Information Science - Framework and Potential. In *International Seminar and Exhibition Computerization for Development-the Research Challenge* (pp. 191 – 198). Universiti Pertanian Malaysia.

Horat, B. (2014). *Genetik Algoritma İle Ankara İlinde Mobese (Kent Güvenlik Yönetim Sistemi) Yerleştirme Optimizasyonu*. Ankara: Gazi Üniversitesi Bilişim Enstitüsü.

Hsiao, F. H. (2015). Exponential Synchronization of Chaotic Cryptosystems Using an Improved Genetic Algorithm. *The Scientific World Journal*. PMID:26366432

Hu, F., Xie, D., & Shen, S. (2013). On the application of the internet of things in the field of medical and health care. In *Green Computing and Communications (GreenCom), 2013 IEEE and Internet of Things (iThings/CPSCom), IEEE International Conference on and IEEE Cyber, Physical and Social Computing* (pp. 2053-2058). IEEE. doi:10.1109/GreenCom-iThings-CPSCom.2013.384

Huang, G. B., & Chen, L. (2007). Convex incremental extreme learning machine. *Neurocomputing*, *70*(16), 3056–3062. doi:10.1016/j.neucom.2007.02.009

Huang, G. B., Zhu, Q. Y., & Siew, C. K. (2006). Extreme learning machine: Theory and applications. *Neurocomputing*, *70*(1), 489–501. doi:10.1016/j.neucom.2005.12.126

Hudson, D. L., & Cohen, M. E. (2000). *Neural networks and artificial intelligence for biomedical engineering*. New York: IEEE.

Hu, G., Tay, W. P., & Wen, Y. (2012). Cloud robotics: Architecture, challenges and applications. *IEEE Network*, *26*(3), 21–28. doi:10.1109/MNET.2012.6201212

IEEE EMB (Engineering in Medicine and Biology). (2015). *Designing a Career in Biomedical Engineering*. IEEE.

Igawa, K., & Ohashi, H. (2009). A negative selection algorithm for classification and reduction of the noise effect. *Applied Soft Computing*, *9*(1), 431–438. doi:10.1016/j.asoc.2008.05.003

Ilonen, J., Kamarainen, J. K., & Lampinen, J. (2003). Differential evolution training algorithm for feed-forward neural networks. *Neural Processing Letters*, *17*(1), 93–105. doi:10.1023/A:1022995128597

Ilter, N., & Guvenir, H. A. (1998). *Dermatology Data Set*. Şubat 1, 2017 Tarihinde The Uci Machine Learning Repository: Https://Archive.Ics.Uci.Edu/Ml/Datasets/Dermatology

Ince, H., & Trafalis, T. B. (2017). A hybrid forecasting model for stock market prediction. *Economic Computation and Economic Cybernetics Studies and Research*, *51*(3).

Inglese, P., McKenzie, J. S., Mroz, A., Kinross, J., Veselkov, K., Holmes, E., & Glen, R. C. et al. (2017). Deep learning and 3D-DESI imaging reveal the hidden metabolic heterogeneity of cancer. *Chemical Science (Cambridge)*, *8*(5), 3500–3511. doi:10.1039/C6SC03738K PMID:28507724

Ishak, W. H., & Siraj, F. (2002). Artificial intelligence in medical application: An exploration. *Health Informatics Europe Journal*, *16*. Retrieved November 21, 2016, from, https://www.researchgate.net/publication/240943548

Islam, S. R., Kwak, D., Kabir, M. H., Hossain, M., & Kwak, K. S. (2015). The internet of things for health care: A comprehensive survey. *IEEE Access: Practical Innovations, Open Solutions*, *3*, 678–708. doi:10.1109/ACCESS.2015.2437951

Israel, C. W. (2014). Mechanisms of sudden cardiac death. *Indian Heart Journal*, *66*, S10–S17.

Iza, S. I., Siti, N. S., Muzaimi, M., & Noor, K. A. K. (2017). Automatic contrast enhancement of brain MR images using Average Intensity Replacement based on Adaptive Histogram Equalization (AIR-AHE). *Biocybernetics and Biomedical Engineering*, *37*, 24–34. doi:10.1016/j.bbe.2016.12.003

Jaemoon, L., Minhyeok, M., Chul, L., & Chang-Su, K. (2017). Contrast enhancement of noisy low-light images based on structure-texture-noise decomposition. *Journal of Visual Communication and Image Representation*, *45*, 107–121. doi:10.1016/j.jvcir.2017.02.016

Jain, A. K., Mao, J., & Mohiuddin, K. M. (1996). Artificial neural networks: A tutorial. *Computer*, *29*(3), 31–44. doi:10.1109/2.485891

Jain, H., & Deb, K. (2013). An improved adaptive approach for elitist nondominated sorting genetic algorithm for many-objective optimization. In *International Conference on Evolutionary Multi-Criterion Optimization* (pp. 307-321). Springer.

Jalali-Heravi, M., & Ebrahimi-Najafabadi, H. (2011). The use of ladder particle swarm optimisation for quantitative structure–activity relationship analysis of human immunodeficiency virus-1 integrase inhibitors. *Molecular Simulation*, *37*(15), 1221–1233. doi:10.1080/08927022.2011.586347

Jang, J. S. R. (2010). *Speech and audio processing toolbox.* Available from the link at the author's homepage at: http:// www. cs. nthu. edu. tw/~ jang

Jang, J. S. (1992). Self-learning fuzzy controllers based on temporal backpropagation. *IEEE Transactions on Neural Networks*, *3*(5), 714–723. doi:10.1109/72.159060 PMID:18276470

Jang, J. S. (1993). ANFIS: Adaptive-network-based fuzzy inference system. *IEEE Transactions on Systems, Man, and Cybernetics*, *23*(3), 665–685. doi:10.1109/21.256541

Jara, A. J., López, P., Fern'ndez, D., Úbeda, B., Zamora, M. A., & Skarmeta, A. F. (2012). Interaction of patients with breathing problems through NFC in Ambient Assisted Living environments. In *Innovative Mobile and Internet Services in Ubiquitous Computing (IMIS), 2012 Sixth International Conference on* (pp. 892-897). IEEE. doi:10.1109/IMIS.2012.150

Jegelevicius, D., & Krisciukaitis, A. (2009). Network Based Clinical Decision Support System. *Proceedings of the 9th International Conference on Information Technology and Applications in Biomedicine.*

Jemal, H., Kechaou, Z., Ayed, M. B., & Alimi, A. M. (2015). Mobile cloud computing in healthcare system. In *Computational Collective Intelligence* (pp. 408–417). Cham: Springer. doi:10.1007/978-3-319-24306-1_40

Ji, Z. (2006). *Negative Selection Algorithms: From The Thymus To V-Detector* (PhD dissertation). The Universty of Memphis.

Jiang, F., Jiang, Y., Zhi, H., Dong, Y., Li, H., Ma, S., ... Wang, Y. (2017). Artificial intelligence in healthcare: past, present and future. *Stroke and Vascular Neurology.*

Jiao, Z., Gao, X., Wang, Y., & Li, J. (2016). A deep feature based framework for breast masses classification. *Neurocomputing, 197*, 221–231. doi:10.1016/j.neucom.2016.02.060

Jinquan, Z., Xiaojie, L., Tao, L., Caiming, L., Lingxi, P., & Feixian, S. (2009). A self-adaptive negative selection algorithm used for anomaly detection. *Progress in Natural Science, 19*(2), 261–266. doi:10.1016/j.pnsc.2008.06.008

Ji, Z., & Dasgupta, D. (2007). Revisiting Negative Selection Algorithms. *Evolutionary Computation, Massachusetts Institute of Technology, 15*(2), 223–251. doi:10.1162/evco.2007.15.2.223 PMID:17535140

Jones, S. (Ed.). (1998). *Cybersociety 2.0: Revisiting computer-mediated community and technology* (Vol. 2). Sage Publications.

Jordan, M. I., & Mitchell, T. M. (2015). Machine learning: Trends, perspectives, and prospects. *Science, 349*(6245), 255–260.

Kaelbling, L. P., Littman, M. L., & Moore, A. W. (1996). Reinforcement learning: A survey. *Journal of Artificial Intelligence Research, 4*, 237–285.

Kaliyeva, S. (2013). *Bilimsel Makalelerin Metin İşleme Yöntemleri İle Siniflandirilmasi.* Ankara, Türkiye: Gazi Üniversitesi Fen Bilimler Enstitüsü.

Kälvemark, S., Höglund, A. T., Hansson, M. G., Westerholm, P., & Arnetz, B. (2004). Living with conflicts-ethical dilemmas and moral distress in the health care system. *Social Science & Medicine, 58*(6), 1075–1084. doi:10.1016/S0277-9536(03)00279-X PMID:14723903

Kamalanand, K., & Jawahar, P. M. (2012). Coupled jumping frogs/particle swarm optimization for estimating the parameters of three dimensional HIV model. *BMC Infectious Diseases, 12*(1), 1. PMID:22214291

Kamalanand, K., & Jawahar, P. M. (2013). Resource limited estimation of HIV-1 viral load from CD4 cell count using coupled bacterial foraging/jumping frogs algorithm. *The World Allergy Organization Journal, 6*(S1), 189. doi:10.1186/1939-4551-6-S1-P189

Kamalanand, K., & Jawahar, P. M. (2014). A graphical user interface for resource limited estimation of HIV-1 viral load using swarm intelligence techniques. *Journal of Bioinformatics and Intelligent Control, 3*(2), 110–116. doi:10.1166/jbic.2014.1073

Kamalanand, K., & Jawahar, P. M. (2015). Comparison of Swarm Intelligence Techniques for Estimation of HIV-1 Viral Load. *IETE Technical Review, 32*(3), 188–195. doi:10.1080/02564602.2014.1000981

Kamalanand, K., & Jawahar, P. M. (2016). Comparison of particle swarm and bacterial foraging optimization algorithms for therapy planning in HIV/AIDS patients. *International Journal of Biomathematics, 9*(02), 1650024. doi:10.1142/S1793524516500248

Kamberaj, H. (2014). Q-Gaussian swarm quantum particle intelligence on predicting global minimum of potential energy function. *Applied Mathematics and Computation, 229*, 94–106. doi:10.1016/j.amc.2013.12.036

Kang, F., Xu, Q., & Li, J. (2016). Slope reliability analysis using surrogate models via new support vector machines with swarm intelligence. *Applied Mathematical Modelling, 40*(11), 6105–6120. doi:10.1016/j.apm.2016.01.050

Kang, Q., Liu, S., Zhou, M., & Li, S. (2016). A weight-incorporated similarity-based clustering ensemble method based on swarm intelligence. *Knowledge-Based Systems, 104*, 156–164. doi:10.1016/j.knosys.2016.04.021

Karaboga, D. (2005). *An idea based on honey bee swarm for numerical optimization* (Vol. 200). Technical report-tr06, Erciyes University, Engineering Faculty, Computer Engineering Department.

Karaboga, D., Akay, B., & Ozturk, C. (2007). Artificial bee colony (ABC) optimization algorithm for training feed-forward neural networks. In *International Conference on Modeling Decisions for Artificial Intelligence* (pp. 318-329). Springer Berlin Heidelberg. doi:10.1007/978-3-540-73729-2_30

Karaboga, D. (2010). Artificial bee colony algorithm. *Scholarpedia, 5*(3), 6915. doi:10.4249/scholarpedia.6915

Karaboga, D., & Akay, B. (2009). A comparative study of artificial bee colony algorithm. *Applied Mathematics and Computation, 214*(1), 108–132. doi:10.1016/j.amc.2009.03.090

Karaduman, A. (2000). Psoriazis. *Sürekli Tıp Eğitimi Dergisi*, 12-19.

Karakoyun, M., & Hacıbeyoğlu, M. (2014). Biyomedikal Veri Kümeleri İle Makine Öğrenmesi Siniflandirma Algoritmalarinin İstatistiksel Olarak Karşilaştirilmasi. *Deü Mühendislik Fakültesi Mühendislik Bilimleri Dergisi*, 30-41.

Karakuş, D. (2006). *Determination of the structural properties of rocks with image analysing methods* (Doctoral Thesis). Dokuz Eylül University, Institute of Science, İzmir, Turkey.

Karayiannis, N., & Venetsanopoulos, A. N. (2013). *Artificial neural networks: learning algorithms, performance evaluation, and applications* (Vol. 209). Springer Science & Business Media.

Kashan, A. H. (2009). League championship algorithm: a new algorithm for numerical function optimization. In *Soft Computing and Pattern Recognition, 2009. SOCPAR'09. International Conference of* (pp. 43-48). IEEE. doi:10.1109/SoCPaR.2009.21

Kashan, A. H. (2014). League Championship Algorithm (LCA): An algorithm for global optimization inspired by sport championships. *Applied Soft Computing, 16*, 171–200. doi:10.1016/j.asoc.2013.12.005

Katasonov, A., Kaykova, O., Khriyenko, O., Nikitin, S., & Terziyan, V. Y. (2008). Smart Semantic Middleware for the Internet of Things. *ICINCO-ICSO, 8*, 169–178.

Katz, R. N., & Katz, R. N. (2004). *Information technology leadership in higher education: The condition of the community*. EDUCAUSE Center for Applied Research.

Kaur, B., & Mann, P. S. (2016). Ant Colony based Metaheuristic Technique for Color Image Segmentation. *International Journal of Computers and Applications, 153*(1).

Kaveh, A., Bakhshpoori, T., & Afshari, E. (2014). An efficient hybrid particle swarm and swallow swarm optimization algorithm. *Computers & Structures, 143*, 40–59. doi:10.1016/j.compstruc.2014.07.012

Kawai, M., Yamaguchin, N., & Nasu, M. (1999). Rapid Enumeration of Physiologically Active Bacteria in purified water used in the Pharmaceutical Manufacturing Process. *Journal of Applied Microbiology, 86*(3), 496–504. doi:10.1046/j.1365-2672.1999.00689.x PMID:10196754

Kawamoto, K., Houlihan, C. A., Balas, E. A., & Lobach, D. F. (2005). Improving clinical practice using clinical decision support systems: A systematic review of trials to identify features critical to success. *British Medical Journal, 330*(7494), 765–772. doi:10.1136/bmj.38398.500764.8F PMID:15767266

Kaya, M., & Özel, S. A. (2014). *Açık Kaynak Kodlu Veri Madenciliği Yazılımlarının Karşılaştırılması*. Mersin: Akademik Bilişim.

Kaymaz, E. D. (2007). *Predicting The Cellular Localization Sites Of Protein Patterns With K-nn Classification Algorithm Based Artificial Immune System* (Master Thesis). Institute of Science of Fırat University, Elazığ.

Kechman, V. (2001). *Learning and soft computing*. Academic Press.

Kecman, V. (2001). *Learning and soft computing: support vector machines, neural networks, and fuzzy logic models*. MIT Press.

Kennedy, J. (1997). The particle swarm: social adaptation of knowledge. In *Evolutionary Computation, 1997., IEEE International Conference on* (pp. 303-308). IEEE. doi:10.1109/ICEC.1997.592326

Kennedy, J. (2006). Swarm intelligence. In Handbook of nature-inspired and innovative computing (pp. 187-219). Springer US.

Kennedy, J., Eberhart, R. C. & Shi, Y. (2001). *Swarm Intelligence*. Morgan Kaufmann.

Kennedy, J., & Eberhart, R. (1995). Particle Swarm Optimization. *Neural Networks, Proceedings. Ieee International Conference On*, 1942-1948.

Kennedy, J., Kennedy, J. F., Eberhart, R. C., & Shi, Y. (2001). *Swarm intelligence*. Morgan Kaufmann.

Khalifa, M. (2014). Clinical Decision Support: Strategies for Success. *Procedia Computer Science*, *37*, 422–427. doi:10.1016/j.procs.2014.08.063

Kharya, S. (2012). *Using data mining techniques for diagnosis and prognosis of cancer disease.* arXiv preprint arXiv:1205.1923

Khubchandani, P., Jha, K., Bijani, R., Lala, S., & Saindane, P. (2017). Medical Prediction using Artificial Intelligence. *International Journal of Engineering Science*, 4898.

King, B. (2017). *Frankenstein's Legacy: Four Conversations about Artificial Intelligence, Machine Learning, and the Modern World.* Repository.cmu.edu.

Kittler, J., & Illingworth, J. (1986). Minimum error thresholding. *Pattern Recognition*, *19*(1), 41–47. doi:10.1016/0031-3203(86)90030-0

Kjerulff, K. H., Pillar, B., Mills, M. E., & Lanigan, J. (1992). Technology anxiety as a potential mediating factor in response to medical technology. *Journal of Medical Systems*, *16*(1), 7–13. doi:10.1007/BF01674093 PMID:1645046

Kline, J. (Ed.). (2012). *Handbook of biomedical engineering*. Elsevier.

Kline, R. R. (2015). *The cybernetics moment: Or why we call our age the information age*. JHU Press.

Klir, G., & Yuan, B. (1995). *Fuzzy sets and fuzzy logic* (Vol. 4). Prentice hall.

Kockanat, S., & Karaboga, N. (2017). Medical Image Denoising Using Metaheuristics. In *Metaheuristics for Medicine and Biology* (pp. 155–169). Springer Berlin Heidelberg. doi:10.1007/978-3-662-54428-0_9

Kodaz, H. (2007). *Information Gain Based Artificial Immunity Recognition System* (PhD thesis). Institute of Science of Selcuk University, Konya.

Kodaz, H., Özşen, S., Arslan, A., & Güneş, S. (2009). Medical application of information gain based artificial immune recognition system (AIRS): Diagnosis of thyroid disease. *Expert Systems with Applications*, *36*(2), 3086–3092. doi:10.1016/j.eswa.2008.01.026

Kononenko, I. (2001). Machine learning for medical diagnosis: History, state of the art and perspective. *Artificial Intelligence in Medicine*, *23*(1), 89–109. doi:10.1016/S0933-3657(01)00077-X PMID:11470218

Kopetz, H. (2011). Internet of things. In Real-time systems (pp. 307-323). Springer US. doi:10.1007/978-1-4419-8237-7_13

Korani, W. M., Dorrah, H. T., & Emara, H. M. (2009). Bacterial foraging oriented by particle swarm optimization strategy for PID tuning. In *Computational Intelligence in Robotics and Automation (CIRA), 2009 IEEE International Symposium on* (pp. 445-450). IEEE. doi:10.1109/CIRA.2009.5423165

Kortuem, G., Kawsar, F., Sundramoorthy, V., & Fitton, D. (2010). Smart objects as building blocks for the internet of things. *IEEE Internet Computing*, *14*(1), 44–51. doi:10.1109/MIC.2009.143

Kose, U. (2017). *Development of Artificial Intelligence based Optimization Algorithms* (PhD. Thesis). Selcuk University, Institute of Natural Sciences, Dept. of Computer Engineering. (in Turkish)

Kose, U., & Pavaloiu, A. (2017). Dealing with Machine Ethics in Daily Life: A View with Examples. *International Virtual Conference on Advanced Scientific Results / Online Scientific Conference 2017 (ScieConf 2017)*, 200-205. Doi:10.18638/scieconf.2017.5.1.454

Kose, U. (2010). *Web 2.0 Technologies in E-learning. In Free and Open Source Software for E-learning: Issues, Successes and Challenges* (pp. 1–23). IGI Global.

Kose, U. (2015). Present state of swarm intelligence and future directions. In *Encyclopedia of Information Science and Technology* (3rd ed.; pp. 239–252). IGI Global. doi:10.4018/978-1-4666-5888-2.ch023

Kose, U. (2015b). *Algoritma Kavramı ve Bilgisayar Programlama Öğreniminde Pratik Yaklaşımlar [Practical Approaches in Learning Algorithm Concept and Computer Programming]*. Usak: AKY Press. (in Turkish)

Kose, U., & Armutlu, H. (2015). *Bulut Bilişim: Temel Konular ve Amazon Web Services [Cloud Computing: Essential Subjects and Amazon Web Services]*. Ankara: Detay Press. (in Turkish)

Kose, U., & Arslan, A. (2015). On the Idea of a New Artificial Intelligence Based Optimization Algorithm Inspired From the Nature of Vortex. *BRAIN. Broad Research in Artificial Intelligence and Neuroscience*, *5*(1-4), 60–66.

Kose, U., & Arslan, A. (2016). Forecasting chaotic time series via anfis supported by vortex optimization algorithm: Applications on electroencephalogram time series. *Arabian Journal for Science and Engineering*, 1–12.

Kose, U., & Arslan, A. (2017). Optimization of self-learning in Computer Engineering courses: An intelligent software system supported by Artificial Neural Network and Vortex Optimization Algorithm. *Computer Applications in Engineering Education*, *25*(1), 142–156. doi:10.1002/cae.21787

Kose, U., Guraksin, G. E., & Deperlioglu, O. (2015). Diabetes determination via vortex optimization algorithm based support vector machines. In *Medical Technologies National Conference (TIPTEKNO)* (pp. 1-4). IEEE. doi:10.1109/TIPTEKNO.2015.7374614

Kose, U., Koc, D., & Yucesoy, S. A. (2013). Design and Development of a Sample" Computer Programming" Course Tool via Story-Based E-Learning Approach. *Educational Sciences: Theory and Practice*, *13*(2), 1235–1250.

Kose, U., & Vasant, P. (2018). A Model of Swarm Intelligence Based Optimization Framework Adjustable According to Problems. In I. Zelinka, P. Vasant, V. Duy, & T. Dao (Eds.), *Innovative Computing, Optimization and Its Applications. Studies in Computational Intelligence* (Vol. 741). Cham: Springer. doi:10.1007/978-3-319-66984-7_2

Kostic, P., & Vasilyevic, Z. (2011). Knowledge Management System for Clinical Decision Support - Application in Cardiology. *19th Telecommunications forum TELFOR*, 1261-1264.

Kramer, J. M., Kendall, M., & Gilbert, R. J. (1978). Evaluation of the Spiral Plate and Laser Colony Counting techniques for the Enumeration of Bacteria in foods. *Journal of Applied Microbiology and Biotechnology*, *6*(3), 289–299. doi:10.1007/BF00508101

Krause, D., Hussein, M. A., & Becker, T. (2015). Online monitoring of bioprocesses via multivariate sensor prediction within swarm intelligence decision making. *Chemometrics and Intelligent Laboratory Systems*, *145*, 48–59. doi:10.1016/j.chemolab.2015.04.012

Kraus, O. Z., Grys, B. T., Ba, J., Chong, Y., Frey, B. J., Boone, C., & Andrews, B. J. (2017). Automated analysis of high-content microscopy data with deep learning. *Molecular Systems Biology*, *13*(4), 924. doi:10.15252/msb.20177551 PMID:28420678

Krenker, A., Bester, J., & Kos, A. (2011). Introduction to the artificial neural networks. In *Artificial neural networks-methodological advances and biomedical applications*. InTech. doi:10.5772/15751

Krutz, R. L., & Vines, R. D. (2010). *Cloud security: A comprehensive guide to secure cloud computing*. Wiley Publishing.

Kubat, M. (2017). Artificial neural networks. In *An Introduction to Machine Learning* (pp. 91–111). Cham: Springer. doi:10.1007/978-3-319-63913-0_5

Kulikowski, C. A. (1984). Artificial Intelligence Methods and Systems for Medical Consultation. In W. J. Clancey & E. H. Shortliffe (Eds.), *Readings in Medical Artificial Intelligence: The First Decade*. Reading, MA: Addison Wesley.

Kumar, Y., Dewal, M. L., & Anand, R. S. (2014). Epileptic seizure detection using DWT based fuzzy approximate entropy and support vector machine. *Neurocomputing*, *133*, 271–279. doi:10.1016/j.neucom.2013.11.009

Kundakcı, N. (2011). Eritemli Skuamlı Hastalıklar. Ankara: Academic Press.

Kuo, A. M. H. (2011). Opportunities and challenges of cloud computing to improve health care services. *Journal of Medical Internet Research*, *13*(3), e67. doi:10.2196/jmir.1867 PMID:21937354

Kurzweil, R. (2005). *The Singularity is Near: When Humans Transcend Biology*. Penguin.

Kutz, M. (2003). *Standard handbook of biomedical engineering and design*. McGraw-Hill.

Lalwani, S., Kumar, R., & Gupta, N. (2016). An efficient two-level swarm intelligence approach for RNA secondary structure prediction with bi-objective minimum free energy scores. *Swarm and Evolutionary Computation*, *27*, 68–79. doi:10.1016/j.swevo.2015.09.008

Längkvist, M., Karlsson, L., & Loutfi, A. (2014). A review of unsupervised feature learning and deep learning for time-series modeling. *Pattern Recognition Letters*, *42*, 11–24. doi:10.1016/j.patrec.2014.01.008

Langtangen, H. P. (2016). Object-oriented programming. In *A Primer on Scientific Programming with Python* (pp. 567–643). Berlin: Springer. doi:10.1007/978-3-662-49887-3_9

Latifoğlu, F., Kodaz, H., Kara, S., & Güneş, S. (2007). Medical application of Artificial Immune Recognition System (AIRS): Diagnosis of atherosclerosis from carotid artery Doppler signals. *Computers in Biology and Medicine*, *37*(8), 1092–1099. doi:10.1016/j.compbiomed.2006.09.009 PMID:17156772

Lavanya, D., & Rani, K. U. (2011). Performance evaluation of decision tree classifiers on medical datasets. *International Journal of Computers and Applications*, *26*(4).

Lazzús, J. A., Rivera, M., & López-Caraballo, C. H. (2016). Parameter estimation of Lorenz chaotic system using a hybrid swarm intelligence algorithm. *Physics Letters. [Part A]*, *380*(11), 1164–1171. doi:10.1016/j.physleta.2016.01.040

LeCun, Y., Bengio, Y., & Hinton, G. (2015). Deep learning. *Nature*, *521*(7553), 436–444. doi:10.1038/nature14539 PMID:26017442

Ledley, R. S., & Lusted, L. B. (1959). Reasoning foundations of medical diagnosis. *Science*, *130*(3366), 9–21. doi:10.1126/science.130.3366.9 PMID:13668531

Lee, K., Lunney, T., Curran, K., & Santos, J. (2009). A Context-Aware Platform for Computer Assisted Healthcare. In *PGNet 2009, The 10th Annual Conference on the Convergence of Telecommunications, Networking & Broadcasting*. Liverpool John Moores.

Lee, Y., & Lee, C. K. (2003). Classification of multiple cancer types by multicategory support vector machines using gene expression data. *Bioinformatics (Oxford, England)*, *19*(9), 1132–1139. doi:10.1093/bioinformatics/btg102 PMID:12801874

Liang, B., Yaali, H., Songyang, L., Jianyun, C., & Lingda, W. (2005, October). Feature analysis and extraction for audio automatic classification. In *Systems, Man and Cybernetics, 2005 IEEE International Conference on* (Vol. 1, pp. 767-772). IEEE. doi:10.1109/ICSMC.2005.1571239

Liang, S. N., Tan, K. O., Clement, T. L., Ng, S. K., Mohammed, A. A., Mailah, M., & Yussof, Z. et al. (2016). Open source hardware and software platform for robotics and artificial intelligence applications. *IOP Conference Series. Materials Science and Engineering*, *114*(1), 012142. doi:10.1088/1757-899X/114/1/012142

Liao, Y. H., & Sun, C. T. (2001). An educational genetic algorithms learning tool. *IEEE Transactions on Education*, *44*(2), 20.

Li, D., Chen, S., & Huang, H. (2014). Improved genetic algorithm with two-level approximation for truss topology optimization. *Structural and Multidisciplinary Optimization*, *49*(5), 795–814. doi:10.1007/s00158-013-1012-8

Li, D., Liu, S., & Zhang, H. (2016). A boundary-fixed negative selection algorithm with online adaptive learning under small samples for anomaly detection. *Engineering Applications of Artificial Intelligence*, *50*, 93–105. doi:10.1016/j.engappai.2015.12.014

Lin, C. T., & Lee, C. G. (1996). *Neural fuzzy systems*. PTR Prentice Hall.

Lin, F., Duan, H., & Qu, X. (2014). PID control strategy for UAV flight control system based on improved genetic algorithm optimization. In *Control and Decision Conference (2014 CCDC), The 26th Chinese* (pp. 92-97). IEEE. doi:10.1109/CCDC.2014.6852124

Lipkin, M., & Hardy, J. D. (1958). Mechanical correlation of data in differential diagnosis of hematologic diseases. *Journal of the American Medical Association*, *166*(2), 113–125. doi:10.1001/jama.1958.02990020001001 PMID:13491316

Li, S. Z. (2000). Content-based audio classification and retrieval using the nearest feature line method. *IEEE Transactions on Speech and Audio Processing*, *8*(5), 619–625. doi:10.1109/89.861383

Litjens, G., Kooi, T., Bejnordi, B. E., Setio, A. A. A., Ciompi, F., Ghafoorian, M., . . . Sánchez, C. I. (2017). *A survey on deep learning in medical image analysis*. arXiv preprint arXiv:1702.05747

Liu, T., Xie, S., Yu, J., Niu, L., & Sun, W. (2017). Classification of thyroid nodules in ultrasound images using deep model based transfer learning and hybrid features. In *Acoustics, Speech and Signal Processing (ICASSP), 2017 IEEE International Conference on* (pp. 919-923). IEEE. doi:10.1109/ICASSP.2017.7952290

Li, X., Lu, R., Liang, X., Shen, X., Chen, J., & Lin, X. (2011). Smart community: An internet of things application. *IEEE Communications Magazine*, *49*(11), 68–75. doi:10.1109/MCOM.2011.6069711

Li, Z., Cao, J., Zhao, X., & Liu, W. (2015). Swarm intelligence for atmospheric compensation in free space optical communication—Modified shuffled frog leaping algorithm. *Optics & Laser Technology*, *66*, 89–97. doi:10.1016/j.optlastec.2014.08.012

Li, Z., Chen, C., & Wang, K. (2011). Cloud computing for agent-based urban transportation systems. *IEEE Intelligent Systems*, *26*(1), 73–79. doi:10.1109/MIS.2011.10

Lohr, S. (2012). The age of big data. *New York Times, 11*(2012).

Londhe, V. (2017). Brain MR Image Segmentation for Tumor Detection using Artificial Neural. *Brain*, *6*(1).

Lopez, A. R., Giro-i-Nieto, X., Burdick, J., & Marques, O. (2017). Skin lesion classification from dermoscopic images using deep learning techniques. In *Biomedical Engineering (BioMed), 2017 13th IASTED International Conference on* (pp. 49-54). IEEE.

Lorencik, D., & Sincak, P. (2013). Cloud Robotics: Current trends and possible use as a service. In *Applied Machine Intelligence and Informatics (SAMI), 2013 IEEE 11th International Symposium on* (pp. 85-88). IEEE.

Lubaib, P., & Muneer, K. A. (2016). The heart defect analysis based on PCG signals using pattern recognition techniques. *Procedia Technology*, *24*, 1024–1031. doi:10.1016/j.protcy.2016.05.225

Luo, J., Chen, Y., Tang, K., & Luo, J. (2009). Remote monitoring information system and its applications based on the Internet of Things. In *BioMedical Information Engineering, 2009. FBIE 2009. International Conference on Future* (pp. 482-485). IEEE.

Machine Learning Group At The University Of Waikato. (2017). *Weka.Attributeselection Class Cfssubseteval*. Retrieved from Http://Weka.Sourceforge.Net/Doc.Stable/: Http://Weka.Sourceforge.Net/Doc.Stable/Weka/Attributeselection/Cfssubseteval.Html

Madakam, S., Ramaswamy, R., & Tripathi, S. (2015). Internet of Things (IoT): A literature review. *Journal of Computer and Communications*, *3*(05), 164–173. doi:10.4236/jcc.2015.35021

Magin, R. L. (2017). Bioengineering and Cybernetics: A Modern Caduceus. *IEEE Pulse*, *8*(1), 44–47. doi:10.1109/MPUL.2016.2627461 PMID:28129142

Maglogiannis, I., Loukis, E., Zafiropoulos, E., & Stasis, A. (2009). Support vectors machine-based identification of heart valve diseases using heart sounds. *Computer Methods and Programs in Biomedicine*, *95*(1), 47–61. doi:10.1016/j.cmpb.2009.01.003 PMID:19269056

Magna, G., Casti, P., Jayaraman, S. V., Salmeri, M., Mencattini, A., Martinelli, E., & Di Natale, C. (2016). Identification of mammography anomalies for breast cancer detection by an ensemble of classification models based on artificial immune system. *Knowledge-Based Systems*, *101*, 60–70. doi:10.1016/j.knosys.2016.02.019

Ma, H., Ye, S., Simon, D., & Fei, M. (2017). Conceptual and numerical comparisons of swarm intelligence optimization algorithms. *Soft Computing*, *21*(11), 3081–3100. doi:10.1007/s00500-015-1993-x

Mahmoodabadi, S. Z., Ahmadian, A., Abolhasani, M. D., Eslami, M., & Bidgoli, J. H. (2006, January). ECG feature extraction based on multiresolution wavelet transform. In *Engineering in Medicine and Biology Society, 2005. IEEE-EMBS 2005. 27th Annual International Conference of the* (pp. 3902-3905). IEEE.

Mair, C., Kadoda, G., Lefley, M., Phalp, K., Schofield, C., Shepperd, M., & Webster, S. (2000). An investigation of machine learning based prediction systems. *Journal of Systems and Software*, *53*(1), 23–29. doi:10.1016/S0164-1212(00)00005-4

Makas, H., & Yumusak, N. (2013). A comprehensive study on thyroid diagnosis by neural networks and swarm intelligence. In *Electronics, Computer and Computation (ICECCO), 2013 International Conference on* (pp. 180-183). IEEE. doi:10.1109/ICECCO.2013.6718258

Mallat, S. (1999). *A wavelet tour of signal processing*. Academic Press.

Man, K. F., Tang, K. S., & Kwong, S. (2012). *Genetic algorithms: Concepts and designs.* Springer Science & Business Media.

Manyika, J., Chui, M., Brown, B., Bughin, J., Dobbs, R., Roxburgh, C., & Byers, A. H. (2011). *Big data: The next frontier for innovation, competition, and productivity.* Academic Press.

Marinakis, Y., Dounias, G., & Jantzen, J. (2009). Pap smear diagnosis using a hybrid intelligent scheme focusing on genetic algorithm based feature selection and nearest neighbor classification. *Computers in Biology and Medicine, 39*(1), 69–78. doi:10.1016/j.compbiomed.2008.11.006 PMID:19147127

Markiewicz, T. (2003). Myelogenous leukemia cell image preprocessing for feature generation. *5th International Workshop on Computational Methods in Electrical Engineering*, 70–73.

Marston, S., Li, Z., Bandyopadhyay, S., Zhang, J., & Ghalsasi, A. (2011). Cloud computing—The business perspective. *Decision Support Systems, 51*(1), 176–189. doi:10.1016/j.dss.2010.12.006

Martinez-Torres, R., & Olmedilla, M. (2016). Identification of innovation solvers in open innovation communities using swarm intelligence. *Technological Forecasting and Social Change, 109*, 15–24. doi:10.1016/j.techfore.2016.05.007

Marz, N., & Warren, J. (2015). *Big Data: Principles and best practices of scalable realtime data systems.* Manning Publications Co.

Matworks. (2016). *Discrete Fourier Transform.* Retrieved from https://www.mathworks.com/help/signal/ug/discrete-fourier-transform.html

Mayer-Schonberger, V., & Cukier, K. (2013). *Big Data: A Revolution That Will Transform How We Live.* Hachette.

Mcaleer, M., Flohr, C., & Irvine, A. (2012). Management Of Difficult And Severe Eczema In Childhood. BMJ (Clinical Research Ed.), 35-41. doi:10.1136/bmj.e4770

McArthur, D., Lewis, M., & Bishary, M. (2005). The Roles of Artificial Intelligence in Education: Current Progress and Future Prospects. *Journal of Educational Technology, 1*(4), 42–80.

Melchels, F. P., Feijen, J., & Grijpma, D. W. (2010). A review on stereolithography and its applications in biomedical engineering. *Biomaterials, 31*(24), 6121–6130. doi:10.1016/j.biomaterials.2010.04.050 PMID:20478613

Melin, P., Castillo, O., & Kacprzyk, J. (Eds.). (2017). *Nature-Inspired Design of Hybrid Intelligent Systems.* Springer. doi:10.1007/978-3-319-47054-2

Merkle, D., & Middendorf, M. (2014). Swarm intelligence. In Search methodologies (pp. 213-242). Springer US.

Mesejo, P., Valsecchi, A., Marrakchi-Kacem, L., Cagnoni, S., & Damas, S. (2015). Biomedical image segmentation using geometric deformable models and metaheuristics. *Computerized Medical Imaging and Graphics, 43*, 167–178. doi:10.1016/j.compmedimag.2013.12.005 PMID:24480648

Meshram, S. P., Thakare, A. D., & Gudadhe, S. (2016). Hybrid Swarm Intelligence Method for Post Clustering Content Based Image Retrieval. *Procedia Computer Science, 79*, 509–515. doi:10.1016/j.procs.2016.03.065

Meuter, M. L., Ostrom, A. L., Bitner, M. J., & Roundtree, R. (2003). The influence of technology anxiety on consumer use and experiences with self-service technologies. *Journal of Business Research, 56*(11), 899–906. doi:10.1016/S0148-2963(01)00276-4

Michalski, R. S., Carbonell, J. G., & Mitchell, T. M. (Eds.). (2013). *Machine learning: An artificial intelligence approach.* Springer Science & Business Media.

Miller, P. L. (Ed.). (2012). *Selected topics in medical artificial intelligence*. Springer Science & Business Media.

Miller, R. A. (1994). Medical Diagnostic Decision Support Systems -Past, Present, and Future: A Threaded Bibliography and Brief Commentary. *Journal of the American Medical Informatics Association*, *1*(1), 8–27. doi:10.1136/jamia.1994.95236141 PMID:7719792

Miller, W. T., Werbos, P. J., & Sutton, R. S. (Eds.). (1995). *Neural networks for control*. MIT press. doi:10.1007/978-3-642-57760-4

Min, X., Liu, L., He, Y., Gong, X., Fong, S., Xu, Q., & Wong, K. K. (2016). Benchmarking Swarm Intelligence Clustering Algorithms with Case study of Medical Data. *Computerized Medical Imaging and Graphics*. doi:10.1016/j.compmedimag.2016.06.005 PMID:27666794

Miotto, R., Wang, F., Wang, S., Jiang, X., & Dudley, J. T. (2017). Deep learning for healthcare: Review, opportunities and challenges. *Briefings in Bioinformatics*, bbx044. PMID:28481991

Mishra, R., & Jaiswal, A. (2012). Ant colony optimization: A solution of load balancing in cloud. *International Journal of Web & Semantic Technology*, *3*(2), 33–50. doi:10.5121/ijwest.2012.3203

MIT Tech. Rev. (2013). Deep Learning. *MIT Technology Review*. Available at: https://www.technologyreview.com/s/513696/deep-learning

MIT Technology Review. (2016). Do we need Asimov's laws? *MIT Technology Review*. Retrieved on July 29, 2017 from https://www.technologyreview.com/s/527336/do-we-need-asimovs-laws/

Mitchell, M. (1998). *An introduction to genetic algorithms*. MIT Press.

Montana, D. J., & Davis, L. (1989). Training Feedforward Neural Networks Using Genetic Algorithms. *IJCAI*.

Moor, J. H. (2006). The nature, importance, and difficulty of machine ethics. *IEEE Intelligent Systems*, *21*(4), 18–21. doi:10.1109/MIS.2006.80

Mori, K., Tsukiyama, M., & Fukuda, T. (1997). Artificial Immunity Based Management System for a Semiconducter Production Line. *IEEE International Conference on Systems, Man, and Cybernetics*, *1*, 852-856.

Mosby. (2013). *Mosby's Medical Dictionary* ($9^{th}$ ed.). St. Louis, MO: Elsevier.

Mousavirad, S. J., & Ebrahimpour-Komleh, H. (2014). Population-Based Feature Selection for Biomedical Data Classification. In *Data Mining and Analysis in the Engineering Field* (pp. 296–326). IGI Global. doi:10.4018/978-1-4666-6086-1.ch016

Mukherjee, D., Pal, A., Sharma, S. E., & Majumder, D. D. (1995). Water Quality Analysis: A Pattern Recognition Approach. *Pattern Recognition*, *28*(2), 269–281. doi:10.1016/0031-3203(94)00089-5

Nabiyev, V. V. (2003). *Yapay Zeka [Artificial Intelligence]*. Ankara: Seçkin Press. (in Turkish)

Nahar, J., Imam, T., Tickle, K. S., Ali, A. S., & Chen, Y. P. P. (2012). Computational intelligence for microarray data and biomedical image analysis for the early diagnosis of breast cancer. *Expert Systems with Applications*, *39*(16), 12371–12377. doi:10.1016/j.eswa.2012.04.045

Naït-Ali, A. (Ed.). (2009). *Advanced biosignal processing*. Springer Science & Business Media. doi:10.1007/978-3-540-89506-0

Nakib, A., Naït-Ali, A., Van Wassenhove, V., & Siarry, P. (2009). Parametric Modeling of Some Biosignals Using Optimization Metaheuristics. In *Advanced Biosignal Processing* (pp. 291–305). Springer Berlin Heidelberg. doi:10.1007/978-3-540-89506-0_14

Naqvi, G. (2012). *A Hybrid Filter-Wrapper Approach For Feature Selection*. Örebro University.

Nedjah, N., & de Macedo Mourelle, L. (2006). *Swarm Intelligent Systems* (Vol. 26). Springer. doi:10.1007/978-3-540-33869-7_6

Nedorost, S. T. (2012). Generalized Dermatitis In Clinical Practice. Springer Science & Business Media.

Neelima, E., & Babu, M. P. (2016). MAGED: Metaheuristic Approach on Gene Expression Data: Predicting the Coronary Artery Disease and the Scope of Unstable Angina and Myocardial Infarction. *Global Journal of Computer Science and Technology, 16*(4).

Negnevitsky, M. (2005). *Artificial intelligence: A guide to intelligent systems*. Pearson Education.

Neshat, M., & Sepidname, G. (2015). A new hybrid optimization method inspired from swarm intelligence: Fuzzy adaptive swallow swarm optimization algorithm (FASSO). *Egyptian Informatics Journal*, *16*(3), 339–350. doi:10.1016/j.eij.2015.07.003

Neshat, M., Sepidnam, G., & Sargolzaei, M. (2013). Swallow swarm optimization algorithm: A new method to optimization. *Neural Computing & Applications*, *23*(2), 429–454. doi:10.1007/s00521-012-0939-9

Ngan, V., & Writer, S. (2003). *Pityriasis Rubra Pilaris*. Dermnet New Zealand.

Ngiam, J., Khosla, A., Kim, M., Nam, J., Lee, H., & Ng, A. Y. (2011). Multimodal deep learning. In *Proceedings of the 28th international conference on machine learning (ICML-11)* (pp. 689-696). Academic Press.

Nguyen, L. T., Wu, P., Chan, W., Peng, W., & Zhang, Y. (2012, August). Predicting collective sentiment dynamics from time-series social media. In *Proceedings of the first international workshop on issues of sentiment discovery and opinion mining* (p. 6). ACM. doi:10.1145/2346676.2346682

Nikolopoulos, C. (1997). *Expert systems: introduction to first and second generation and hybrid knowledge based systems*. Marcel Dekker, Inc.

Nilsson, N. J. (2014). *Principles of artificial intelligence*. Morgan Kaufmann.

Nitze, I., Schulthess, U., & Asche, H. (2012). Comparison of machine learning algorithms random forest, artificial neural network and support vector machine to maximum likelihood for supervised crop type classification. *Proceedings of the 4th GEOBIA*, 7-9.

Niu, B., Xie, T., Tan, L., Bi, Y., & Wang, Z. (2016). Swarm intelligence algorithms for Yard Truck Scheduling and Storage Allocation Problems. *Neurocomputing*, *188*, 284–293. doi:10.1016/j.neucom.2014.12.125

Nolfi, S., Bongard, J. C., Husbands, P., & Floreano, D. (2016). *Evolutionary Robotics*. MIT Press / Bradford Books. doi:10.1007/978-3-319-32552-1_76

O'Driscoll, A., Daugelaite, J., & Sleator, R. D. (2013). 'Big data', Hadoop and cloud computing in genomics. *Journal of Biomedical Informatics*, *46*(5), 774–781. doi:10.1016/j.jbi.2013.07.001 PMID:23872175

O'Leary, D. E. (2013). Artificial intelligence and big data. *IEEE Intelligent Systems*, *28*(2), 96–99. doi:10.1109/MIS.2013.39 PMID:25505373

Oakley, A. (1997). *Seborrhoeic Dermatitis*. Dermnet New Zealand.

Oh, C., Lee, T., Kim, Y., Park, S., & Suh, B. (2017). Us vs. Them: Understanding Artificial Intelligence Technophobia over the Google DeepMind Challenge Match. In *Proceedings of the 2017 CHI Conference on Human Factors in Computing Systems* (pp. 2523-2534). ACM. doi:10.1145/3025453.3025539

Olesen, J. (2000). The IHS Members' Handbook. Academic Press.

Oniga, S., & Suto, J. (2014). Human activity recognition using neural networks. In Control Conference (ICCC), 2014 15th International Carpathian (pp. 403-406). IEEE.

Oniśko, A., & Druzdzel, M. J. (2013). Impact of precision of Bayesian network parameters on accuracy of medical diagnostic systems. *Artificial Intelligence in Medicine*, *57*(3), 197–206. doi:10.1016/j.artmed.2013.01.004 PMID:23466438

Openshaw, J. P., Sun, Z. P., & Mason, J. S. (1993, April). A comparison of composite features under degraded speech in speaker recognition. In *Acoustics, Speech, and Signal Processing, 1993. ICASSP-93., 1993 IEEE International Conference on* (Vol. 2, pp. 371-374). IEEE. doi:10.1109/ICASSP.1993.319316

Otsu, N. (1979). A threshold selection method from gray-level histograms. *IEEE Transactions on Systems, Man, and Cybernetics*, *9*(1), 62–66. doi:10.1109/TSMC.1979.4310076

Ozekes, S. (2003). Veri Madenciliği Modelleri Ve Uygulama Alanlari. *İstanbul Ticaret Üniversitesi Dergisi, 2*(3), 65-82.

Ozkaya, A. U. (2003). *Intelligent Arrhythmia Classification Based on Support Vector Machines* (MSc. Thesis). Istanbul Technical University, Institute of Natural Sciences, Dept. of Computer Engineering. (in Turkish)

Özşen, Ş. (2008). *Development Of A Problem-Based Artificial Immune System For Biomedical Classification Problems And Application On Biomedical Classification Problems* (PhD thesis). Institute of Science of Selcuk University, Konya.

Padhy, N. P. (2005). *Artificial intelligence and intelligent systems.* Oxford University Press.

Pala, T. (2013). *Tibbi Karar Destek Sisteminin Veri Madenciliği Yöntemleriyle Gerçekleştirilmesi.* İstanbul: Marmara Üniversitesi Fen Bilimleri Enstitüsü.

Panda, G., Pradhan, P. M., & Majhi, B. (2011). IIR system identification using cat swarm optimization. *Expert Systems with Applications*, *38*(10), 12671–12683. doi:10.1016/j.eswa.2011.04.054

Pandey, S., Wu, L., Guru, S. M., & Buyya, R. (2010). A particle swarm optimization-based heuristic for scheduling workflow applications in cloud computing environments. In *Advanced information networking and applications (AINA), 2010 24th IEEE international conference on* (pp. 400-407). IEEE. doi:10.1109/AINA.2010.31

Panigrahi, B. K., Shi, Y., & Lim, M. H. (Eds.). (2011). *Handbook of swarm intelligence: concepts, principles and applications* (Vol. 8). Springer Science & Business Media. doi:10.1007/978-3-642-17390-5

Panikhom, S., Sarasiri, N., & Sujitjorn, S. (2010). Hybrid bacterial foraging and tabu search optimization (btso) algorithms for lyapunov's stability analysis of nonlinear systems. *International Journal of Mathematics and Computers in Simulation, 3*(4).

Pannu, A. (2015). Artificial intelligence and its application in different areas. *Artificial Intelligence*, *4*(10).

Papaioannou, V., Economou, G. P. K., & Tsakalidis, A. (2016). A robust smart device app assisting medical diagnosis. *Artificial Intelligence Review*, *5*(2), 82.

Papamichail, K. N., & French, S. (2005). Design and evaluation of an intelligent decision support system for nuclear emergencies. *Decision Support Systems*, *41*(1), 84–111. doi:10.1016/j.dss.2004.04.014

Pardalos, P. M. (2005). Metaheuristics in Optimization. In *Workshop on the European Chapter on Metaheuristics and Large Scale Optimization, Vilnius, Lithuania May* (pp. 19-21). Academic Press.

Parpinelli, R. S., & Lopes, H. S. (2011). New inspirations in swarm intelligence: A survey. *International Journal of Bio-inspired Computation, 3*(1), 1–16. doi:10.1504/IJBIC.2011.038700

Patel, A. R., Patel, R. S., Singh, N. M., & Kazi, F. S. (2017). Vitality of Robotics in Healthcare Industry: An Internet of Things (IoT) Perspective. In *Internet of Things and Big Data Technologies for Next Generation Healthcare* (pp. 91–109). Springer International Publishing. doi:10.1007/978-3-319-49736-5_5

Patel, V. L., Shortliffe, E. H., Stefanelli, M., Szolovits, P., Berthold, M. R., Bellazzi, R., & Abu-Hanna, A. (2009). The coming of age of artificial intelligence in medicine. *Artificial Intelligence in Medicine, 46*(1), 5–17. doi:10.1016/j.artmed.2008.07.017 PMID:18790621

Patidar, S., & Pachori, R. B. (2014). Classification of cardiac sound signals using constrained tunable-Q wavelet transform. *Expert Systems with Applications, 41*(16), 7161–7170. doi:10.1016/j.eswa.2014.05.052

Peek, N. (2011). Guideline-Based Decision Support Systems for Prevention and Management of Chronic Diseases. In C. S. Jao (Ed.), *Efficient Decision Support Systems – Practice and Challenges in Biomedical Related Domain* (pp. 67–86). Rijeka, Croatia: InTech. doi:10.5772/16882

Peng, X., Liu, P., Deng, Y., Feng, B., & Liao, B. (2012). A Novel Feature Selection Method Based On Cfs In Cancer Recognition. *2012 IEEE 6th International Conference on Systems Biology (Isb)* (pp. 226-231). Xi'an, China: IEEE.

Peng, Y., & Lu, Z. (2017). *Deep learning for extracting protein-protein interactions from biomedical literature.* arXiv preprint arXiv:1706.01556

Penhaker, M., Krejcar, O., Kasik, V., & Snášel, V. (2012). Cloud computing environments for biomedical data services. *Intelligent Data Engineering and Automated Learning-IDEAL, 2012*, 336–343.

Perelson, A. S., & Oster, G. F. (1979). Theoretical studies of clonal selection: Minimal antibody repertoire size and reliability of self-non-self discrimination. *Journal of Theoretical Biology, 81*(4), 645–670. doi:10.1016/0022-5193(79)90275-3 PMID:94141

Perera, C., Zaslavsky, A., Christen, P., & Georgakopoulos, D. (2014). Context aware computing for the internet of things: A survey. *IEEE Communications Surveys and Tutorials, 16*(1), 414–454. doi:10.1109/SURV.2013.042313.00197

Pérez-Rendón, A. F., & Robles, R. (2004). The convolution theorem for the continuous wavelet tranform. *Signal Processing, 84*(1), 55–67. doi:10.1016/j.sigpro.2003.07.014

Petrosian, A., Prokhorov, D., Homan, R., Dasheiff, R., & Wunsch, D. II. (2000). Recurrent neural network based prediction of epileptic seizures in intra-and extracranial EEG. *Neurocomputing, 30*(1), 201–218. doi:10.1016/S0925-2312(99)00126-5

Pham, D. T., & Pham, P. T. N. (1999). Artificial intelligence in engineering. *International Journal of Machine Tools & Manufacture, 39*(6), 937–949. doi:10.1016/S0890-6955(98)00076-5

Pham, D., & Karaboga, D. (2012). *Intelligent optimisation techniques: genetic algorithms, tabu search, simulated annealing and neural networks.* Springer Science & Business Media.

Phansalkar, V. V., & Sastry, P. S. (1994). Analysis of the back-propagation algorithm with momentum. *IEEE Transactions on Neural Networks, 5*(3), 505–506. doi:10.1109/72.286925 PMID:18267819

Phua, K., Chen, J., Dat, T. H., & Shue, L. (2008). Heart sound as a biometric. *Pattern Recognition, 41*(3), 906–919. doi:10.1016/j.patcog.2007.07.018

Podgorelec, V., & Kokol, P. (2001). Towards more optimal medical diagnosing with evolutionary algorithms. *Journal of Medical Systems*, *25*(3), 195–219. doi:10.1023/A:1010733016906 PMID:11433548

Polat, K. (2004). *Going To Medical Diagnosis By Using Artificial Immune Recognition Systems With Feature Selection* (Master thesis). Institute of Science of Selcuk University, Konya.

Polat, K., & Güneş, S. (2007). Breast cancer diagnosis using least square support vector machine. *Digital Signal Processing*, *17*(4), 694–701. doi:10.1016/j.dsp.2006.10.008

Polat, K., Şahan, S., & Güneş, S. (2006). A new method to medical diagnosis: Artificial immune recognition system (AIRS) with fuzzy weighted pre-processing and application to ECG arrhythmia. *Expert Systems with Applications*, *31*(2), 264–269. doi:10.1016/j.eswa.2005.09.019

Poole, D. L., & Mackworth, A. K. (2010). *Artificial Intelligence: Foundations of computational agents*. Cambridge University Press.

Poole, D. L., & Mackworth, A. K. (2010). *Artificial Intelligence: foundations of computational agents*. Cambridge University Press. doi:10.1017/CBO9780511794797

Pradhan, P. M., & Panda, G. (2012). Solving multiobjective problems using cat swarm optimization. *Expert Systems with Applications*, *39*(3), 2956–2964. doi:10.1016/j.eswa.2011.08.157

Prasadl, B. D. C. N., Prasad, P. N. K., & Sagar, Y. (2011). An approach to develop expert systems in medical diagnosis using machine learning algorithms (asthma) and a performance study. *International Journal on Soft Computing*, *2*(1), 26–33. doi:10.5121/ijsc.2011.2103

Principe, J. C., & Fancourt, C. L. (2013). Artificial neural networks. Handbook of Global Optimization, 2, 363-386.

Proakis, J. G., & Manolakis, D. G. (2007). *Digital signal processing: principles, algorithms, and applications*. Pearson Prentice Hall.

Prudêncio, R. B., & Ludermir, T. B. (2004). Meta-learning approaches to selecting time series models. *Neurocomputing*, *61*, 121–137. doi:10.1016/j.neucom.2004.03.008

Qasem, S. N., & Shamsuddin, S. M. (2011). Radial basis function network based on time variant multi-objective particle swarm optimization for medical diseases diagnosis. *Applied Soft Computing*, *11*(1), 1427–1438. doi:10.1016/j.asoc.2010.04.014

Qi, J., Yang, P., Fan, D., & Deng, Z. (2015). A survey of physical activity monitoring and assessment using internet of things technology. In *Computer and Information Technology; Ubiquitous Computing and Communications; Dependable, Autonomic and Secure Computing; Pervasive Intelligence and Computing (CIT/IUCC/DASC/PICOM), 2015 IEEE International Conference on* (pp. 2353-2358). IEEE. doi:10.1109/CIT/IUCC/DASC/PICOM.2015.348

Qiu, J., Yoon, H. J., Fearn, P. A., & Tourassi, G. D. (2017). Deep Learning for Automated Extraction of Primary Sites from Cancer Pathology Reports. *IEEE Journal of Biomedical and Health Informatics*, 1. doi:10.1109/JBHI.2017.2700722 PMID:28475069

Qiu, X., Ren, Y., Suganthan, P. N., & Amaratunga, G. A. (2017). Empirical mode decomposition based ensemble deep learning for load demand time series forecasting. *Applied Soft Computing*, *54*, 246–255. doi:10.1016/j.asoc.2017.01.015

Qu, H., Xing, K., & Alexander, T. (2013). An improved genetic algorithm with co-evolutionary strategy for global path planning of multiple mobile robots. *Neurocomputing*, *120*, 509–517. doi:10.1016/j.neucom.2013.04.020

Quinlan, J. R. (1993). C4.5: Programs For Machine Learning. *Machine Learning*, 235–240.

Rajan, J. R. (2017). Prognostic system for early diagnosis of pediatric lung disease using artificial intelligence. *Current Pediatric Research, 21*(1).

Raja, N. S. M., Manic, K. S., & Rajinikanth, V. (2013). Firefly algorithm with various randomization parameters: an analysis. In *International Conference on Swarm, Evolutionary, and Memetic Computing* (pp. 110-121). Springer International Publishing. doi:10.1007/978-3-319-03753-0_11

Rajinikanth, V., & Couceiro, M. S. (2015). RGB histogram based color image segmentation using firefly algorithm. *Procedia Computer Science*, *46*, 1449–1457. doi:10.1016/j.procs.2015.02.064

Rajinikanth, V., & Latha, K. (2011). Bacterial foraging optimization algorithm based PID controller tuning for time delayed unstable system. *The Mediterranean Journal of Measurement and Control*, *7*(1), 197–203.

Rajinikanth, V., & Latha, K. (2012). Controller parameter optimization for nonlinear systems using enhanced bacteria foraging algorithm. *Applied Computational Intelligence and Soft Computing*, *2012*, 22. doi:10.1155/2012/214264

Rajinikanth, V., & Latha, K. (2012). I-PD controller tuning for unstable system using bacterial foraging algorithm: A study based on various error criterion. *Applied Computational Intelligence and Soft Computing*, *2012*, 2. doi:10.1155/2012/329389

Ramesh, A. N., Kambhampati, C., Monson, J. R. T., & Drew, P. J. (2004). Artificial intelligence in medicine. *Annals of the Royal College of Surgeons of England*, *86*(5), 334–338. doi:10.1308/147870804290 PMID:15333167

Ramezani, F., Lu, J., & Hussain, F. K. (2014). Task-based system load balancing in cloud computing using particle swarm optimization. *International Journal of Parallel Programming*, *42*(5), 739–754. doi:10.1007/s10766-013-0275-4

Rao, B. P., Saluia, P., Sharma, N., Mittal, A., & Sharma, S. V. (2012). Cloud computing for Internet of Things & sensing based applications. In *Sensing Technology (ICST), 2012 Sixth International Conference on* (pp. 374-380). IEEE. doi:10.1109/ICSensT.2012.6461705

Rapini, R. P., Bolognia, J. L., & Jorizzo, J. L. (2007). *Dermatology: 2-Volume Set*. St. Louis, MO: Mosby.

Rather, N. N., Patel, C. O., & Khan, S. A. (2017). Using Deep Learning Towards Biomedical Knowledge Discovery. *International Journal of Mathematical Sciences and Computing*, *3*(2), 1. doi:10.5815/ijmsc.2017.02.01

Rawat, R. R., Ruderman, D., Agus, D. B., & Macklin, P. (2017). *Deep learning to determine breast cancer estrogen receptor status from nuclear morphometric features in H&E images*. Academic Press.

Rebollo-Ruiz, I., & Graña, M. (2014). An empirical evaluation of Gravitational Swarm Intelligence for graph coloring algorithm. *Neurocomputing*, *132*, 79–84. doi:10.1016/j.neucom.2013.03.064

Remagnino, P., & Foresti, G. L. (2005). Ambient intelligence: A new multidisciplinary paradigm. *IEEE Transactions on Systems, Man, and Cybernetics. Part A, Systems and Humans*, *35*(1), 1–6. doi:10.1109/TSMCA.2004.838456

Rezaei, H., Bozorg-Haddad, O., & Chu, X. (2018). League Championship Algorithm (LCA). In *Advanced Optimization by Nature-Inspired Algorithms* (pp. 19–30). Singapore: Springer.

Ripley, B. D. (1994). Neural networks and related methods for classification. *Journal of the Royal Statistical Society. Series B. Methodological*, 409–456.

Rittinghouse, J. W., & Ransome, J. F. (2016). *Cloud computing: implementation, management, and security*. CRC Press.

Rogai, F., Manfredi, C., & Bocchi, L. (2016). Metaheuristics for specialization of a segmentation algorithm for ultrasound images. *IEEE Transactions on Evolutionary Computation*, *20*(5), 730–741. doi:10.1109/TEVC.2016.2515660

Rohokale, V. M., Prasad, N. R., & Prasad, R. (2011). A cooperative Internet of Things (IoT) for rural healthcare monitoring and control. In *Wireless Communication, Vehicular Technology, Information Theory and Aerospace & Electronic Systems Technology (Wireless VITAE), 2011 2nd International Conference on* (pp. 1-6). IEEE.

Rolim, C. O., Koch, F. L., Westphall, C. B., Werner, J., Fracalossi, A., & Salvador, G. S. (2010). A cloud computing solution for patient's data collection in health care institutions. In *eHealth, Telemedicine, and Social Medicine, 2010. ETELEMED'10. Second International Conference on* (pp. 95-99). IEEE. doi:10.1109/eTELEMED.2010.19

Rosenthal, A., Mork, P., Li, M. H., Stanford, J., Koester, D., & Reynolds, P. (2010). Cloud computing: A new business paradigm for biomedical information sharing. *Journal of Biomedical Informatics*, *43*(2), 342–353. doi:10.1016/j.jbi.2009.08.014 PMID:19715773

Rossini, P. (2000). Using expert systems and artificial intelligence for real estate forecasting. In *Sixth Annual Pacific-Rim Real Estate Society Conference* (pp. 24-27). Academic Press.

Ross, T. J. (2009). *Fuzzy logic with engineering applications*. John Wiley & Sons.

Ruiz, R., Riquelme, J. C., & Aguilar-Ruiz, J. S. (2005). Incremental Wrapper-Based Gene Selection From Microarray Data For Cancer Classification. *Pattern Recognition*, 2383–2392.

Russell, S. J., & Norvig, P. (2002). *Artificial intelligence: A modern approach* (International Edition). Prentice Hall.

Russell, S., Dewey, D., & Tegmark, M. (2015). Research priorities for robust and beneficial artificial intelligence. *AI Magazine*, *36*(4), 105–114. doi:10.1609/aimag.v36i4.2577

Russell, S., Norvig, P., & Intelligence, A. (1995). A modern approach. *Artificial Intelligence. Prentice-Hall. Egnlewood Cliffs*, *25*, 27.

Sabir, M. M., & Ali, T. (2016). Optimal PID controller design through swarm intelligence algorithms for sun tracking system. *Applied Mathematics and Computation*, *274*, 690–699. doi:10.1016/j.amc.2015.11.036

Saha, M., Chakraborty, C., Arun, I., Ahmed, R., & Chatterjee, S. (2017). An Advanced Deep Learning Approach for Ki-67 Stained Hotspot Detection and Proliferation Rate Scoring for Prognostic Evaluation of Breast Cancer. *Scientific Reports*, 7. PMID:28607456

Salman, S. D., & Bahrani, A. A. (2010). Segmentation of tumor tissue in gray medical images using watershed transformation method. *International Journal of Advancements in Computing Technology*, *2*(4), 123–127. doi:10.4156/ijact.vol2.issue4.13

Sastry, K., Goldberg, D. E., & Kendall, G. (2014). Genetic algorithms. In Search methodologies (pp. 93-117). Springer US. doi:10.1007/978-1-4614-6940-7_4

Sayre, K. (2014). *Cybernetics and the Philosophy of Mind*. Routledge.

Schirrmeister, R. T., Springenberg, J. T., Fiederer, L. D. J., Glasstetter, M., Eggensperger, K., Tangermann, M., . . . Ball, T. (2017). *Deep learning with convolutional neural networks for brain mapping and decoding of movement-related information from the human EEG*. arXiv preprint arXiv:1703.05051

Schmidhuber, J. (2015). Deep learning in neural networks: An overview. *Neural Networks*, *61*, 85–117. doi:10.1016/j.neunet.2014.09.003 PMID:25462637

Scholkopf, B., & Smola, A. J. (2001). *Learning with kernels: support vector machines, regularization, optimization, and beyond*. MIT Press.

Scholkopf, B., Sung, K. K., Burges, C. J., Girosi, F., Niyogi, P., Poggio, T., & Vapnik, V. (1997). Comparing support vector machines with Gaussian kernels to radial basis function classifiers. *IEEE Transactions on Signal Processing, 45*(11), 2758–2765. doi:10.1109/78.650102

ScikitLearn. (2017). *Neural network models (supervised).* Retrieved from http://scikit-learn.org/stable/modules/neural_networks_supervised.html

Seebock, P. (2015). *Deep learning in medical image analysis* (Master's thesis). Vienna University of Technology, Faculty of Informatics.

Sengur, A. (2008). An expert system based on principal component analysis, artificial immune system and fuzzy k-NN for diagnosis of valvular heart diseases. *Computers in Biology and Medicine, 38*(3), 329–338. doi:10.1016/j.compbiomed.2007.11.004 PMID:18177849

Şenol, M. (2014). *Reading Optical Form From Web Camera With C#* (M.Sc. Thesis). Afyon Kocatepe University, Graduate School of Natural and Applied Sciences, Afyonkarahisar, Turkey.

Sertkaya, C. (2009). *Use of Artificial Immune System in Computer Security* (Master thesis). Institute of Science of Sakarya University, Sakarya.

Shah, V. H. (2007, Spring). Machine learning techniques for stock prediction. *Foundations of Machine Learning*, 1-19.

Shah-Hosseini, H. (2009). The intelligent water drops algorithm: A nature-inspired swarm-based optimization algorithm. *International Journal of Bio-inspired Computation, 1*(1-2), 71–79. doi:10.1504/IJBIC.2009.022775

Shang, R., Bai, J., Jiao, L., & Jin, C. (2013). Community detection based on modularity and an improved genetic algorithm. *Physica A, 392*(5), 1215–1231. doi:10.1016/j.physa.2012.11.003

Shankaracharya, D. O., Samanta, S., & Vidyarthi, A. S. (2010). Computational intelligence in early diabetes diagnosis: A review. *The Review of Diabetic Studies; RDS, 7*(4), 252–262. doi:10.1900/RDS.2010.7.252 PMID:21713313

Shanthi, D., Sahoo, G., & Saravanan, N. (2009). Designing an Artificial Neural Network Model for the Prediction of Thrombo-embolic Stroke. *International Journals of Biometric and Bioinformatics, 3*(1), 10–18.

Sharma, K., Rupprecht, C., Caroli, A., Aparicio, M. C., Remuzzi, A., Baust, M., & Navab, N. (2017). Automatic Segmentation of Kidneys using Deep Learning for Total Kidney Volume Quantification in Autosomal Dominant Polycystic Kidney Disease. *Scientific Reports*, 7. PMID:28515418

Shen, D., Wu, G., & Suk, H. I. (2017). Deep learning in medical image analysis. *Annual Review of Biomedical Engineering.*

Sh-Hussain, H., Salleh, S. H., Ariff, A. K., Alhamdani, O., Tian-Swee, T., Noor, A. M., . . . Yusoff, K. (2012, July). Application of multipoint auscultation for heart sound diagnostic system (MAHDS). In *Information Science, Signal Processing and their Applications (ISSPA), 2012 11th International Conference on* (pp. 836-841). IEEE.

Shi, Y., & Eberhart, R. (1998). A modified particle swarm optimizer. In *Evolutionary Computation Proceedings, 1998. IEEE World Congress on Computational Intelligence., The 1998 IEEE International Conference on* (pp. 69-73). IEEE. doi:10.1109/ICEC.1998.699146

Shieh, J. S., Chou, C. F., Huang, S. J., & Kao, M. C. (2004). Intracranial pressure model in intensive care unit using a simple recurrent neural network through time. *Neurocomputing, 57*, 239–256. doi:10.1016/j.neucom.2003.10.006

Shim, J. P., Warkentin, M., Courtney, J. F., Power, D. J., Sharda, R., & Carlsson, C. (2002). Past, present, and future of decision support technology. *Decision Support Systems, 33*(2), 111–126. doi:10.1016/S0167-9236(01)00139-7

Shipp, M. A., Ross, K. N., Tamayo, P., Weng, A. P., Kutok, J. L., Aguiar, R. C., & Ray, T. S. et al. (2002). Diffuse large B-cell lymphoma outcome prediction by gene-expression profiling and supervised machine learning. *Nature Medicine*, *8*(1), 68–74. doi:10.1038/nm0102-68 PMID:11786909

Shiraishi, J., Li, Q., Appelbaum, D., & Doi, K. (2011). Computer-aided diagnosis and artificial intelligence in clinical imaging. *Seminars in Nuclear Medicine*, *41*(6), 449–462. doi:10.1053/j.semnuclmed.2011.06.004 PMID:21978447

Shi, Z., & Han, M. (2007). Support vector echo-state machine for chaotic time-series prediction. *IEEE Transactions on Neural Networks*, *18*(2), 359–372. doi:10.1109/TNN.2006.885113 PMID:17385625

Shortliffe, E. H., Buchanan, B. G., & Feigenbaum, E. A. (1984). Knowledge Engineering for Medical Decision Making: A Review of Computer-Based Decision Aids. In W. J. Clancey & E. H. Shortliffe (Eds.), *Readings in Medical Artificial Intelligence: The First Decade*. Reading, MA: Addison Wesley.

Shyr, W. J. (2010). Parameters Determination for Optimum Design by Evolutionary Algorithm. In *Convergence and Hybrid Information Technologies*. InTech. doi:10.5772/9638

Sikchi, S. S., Sikchi, S., & Ali, M. S. (2012). Artificial intelligence in medical diagnosis. *International Journal of Applied Engineering Research*, *7*(11).

Silahtaroğlu, G. (2016). *Veri Madenciliği Kavram Ve Algoritmaları*. İstanbul: Papatya Bilim.

Sim, K. M. (2012). Agent-based cloud computing. *IEEE Transactions on Services Computing*, *5*(4), 564–577. doi:10.1109/TSC.2011.52

Singh, S. (2002). Anomaly Detection Using Negative Selection Based on the R-Contiguous Matching Rule. In *Proceedings of ICARIS 2002*. University of Kent at Canterbury Printing Unit.

Singh, R., & Balasundaram, S. (2007). Application of extreme learning machine method for time series analysis. *International Journal of Intelligent Technology*, *2*(4), 256–262.

Singularity Summit. (2012). *Ray Kurzweil on the singularity*. Retrieved on July 28, 2017 from https://www.youtube.com/watch?v=EXVrTCjetLg

Smith, B. (2015). Object-oriented programming. In *Advanced ActionScript 3* (pp. 1–23). Apress.

Smith, S. W. (1999). *The scientist and engineer's guide to digital signal processing*. California Technical Pub.

Smola, A. J., & Scholkopf, B. (2004). A tutorial on support vector regression. *Statistics and Computing*, *14*(3), 199–222. doi:10.1023/B:STCO.0000035301.49549.88

Soliman, M., Abiodun, T., Hamouda, T., Zhou, J., & Lung, C. H. (2013). Smart home: Integrating internet of things with web services and cloud computing. In *Cloud Computing Technology and Science (CloudCom), 2013 IEEE 5th International Conference on* (Vol. 2, pp. 317-320). IEEE.

Sondak, V. K., & Sondak, N. E. (1990). New directions for medical artificial intelligent. *Computers & Mathematics with Applications (Oxford, England)*, *20*(4-6), 313–319. doi:10.1016/0898-1221(90)90336-I

Song, Z., Ji, Z., Ma, J. G., Sputh, B., Acharya, U. R., & Faust, O. (2012). A systematic approach to embedded biomedical decision making. *Computer Methods and Programs in Biomedicine*, *108*(2), 656–664. doi:10.1016/j.cmpb.2011.11.002 PMID:22136936

Spelt, P. F., Knee, H. E., & Glover, C. W. (1991). Hybrid artificial intelligence architecture for diagnosis and decision-making in manufacturing. *Journal of Intelligent Manufacturing*, *2*(5), 261–268. doi:10.1007/BF01471174

Stahovich, T. F. (2001). Artificial intelligence for design. *Formal engineering design synthesis*, 228-269.

Stivaros, S. M., Gledson, A., Nenadic, G., Zeng, X.-J., Keane, J., & Jackson, A. (2010). Decision support systems for clinical radiological practice - towards the next generation. *The British Journal of Radiology*, *83*(995), 904–914. doi:10.1259/bjr/33620087 PMID:20965900

Stoitsis, J., Valavanis, I., Mougiakakou, S. G., Golemati, S., Nikita, A., & Nikita, K. S. (2006). Computer aided diagnosis based on medical image processing and artificial intelligence methods. *Nuclear Instruments & Methods in Physics Research. Section A, Accelerators, Spectrometers, Detectors and Associated Equipment*, *569*(2), 591–595. doi:10.1016/j.nima.2006.08.134

Storn, R., & Price, K. (1997). Differential evolution–a simple and efficient heuristic for global optimization over continuous spaces. *Journal of Global Optimization*, *11*(4), 341–359. doi:10.1023/A:1008202821328

Su, M. C., Jean, W. F., & Chang, H. T. (1996, September). A static hand gesture recognition system using a composite neural network. In *Fuzzy Systems, 1996., Proceedings of the Fifth IEEE International Conference on* (*Vol. 2*, pp. 786-792). IEEE.

Suganya, P., & Sumathi, C. P. (2015). A novel metaheuristic data mining algorithm for the detection and classification of parkinson disease. *Indian Journal of Science and Technology*, *8*(14). doi:10.17485/ijst/2015/v8i14/72685

Sundaram, M., Ramar, K., Arumugam, N., & Prabin, G. (2011). Histogram modified local contrast enhancement for mammogram images. *Applied Soft Computing*, *11*(8), 5809–5816. doi:10.1016/j.asoc.2011.05.003

Sung, P. H., Thompson, W. R., Wang, J. N., Wang, J. F., & Jang, L. S. (2015). Computer-assisted auscultation: Patent ductus arteriosus detection based on auditory time–frequency analysis. *Journal of Medical and Biological Engineering*, *35*(1), 76–85. doi:10.1007/s40846-015-0008-9

Sung, W. T., & Chiang, Y. C. (2012). Improved particle swarm optimization algorithm for android medical care IOT using modified parameters. *Journal of Medical Systems*, *36*(6), 3755–3763. doi:10.1007/s10916-012-9848-9 PMID:22492176

Sun, S., Wang, H., Jiang, Z., Fang, Y., & Tao, T. (2014). Segmentation-based heart sound feature extraction combined with classifier models for a VSD diagnosis system. *Expert Systems with Applications*, *41*(4), 1769–1780. doi:10.1016/j.eswa.2013.08.076

Sutton, R. S., & Barto, A. G. (1998). Reinforcement learning: An introduction: Vol. 1. *No. 1*. Cambridge, MA: MIT Press.

Suzuki, S. (1997). Artificial neural networks. *Mathematics of Neural Networks: Models. Algorithms and Applications*, *8*, 349.

Talas, D. (2017). *What is bacteria?* Retrieved June 08, 2017, from http://www.bilgiustam.com/bakteri-nedir/

Talia, D. (2011). Cloud Computing and Software Agents: Towards Cloud Intelligent Services. In WOA (Vol. 11, pp. 2-6). Academic Press.

Talia, D. (2012). Clouds meet agents: Toward intelligent cloud services. *IEEE Internet Computing*, *16*(2), 78–81. doi:10.1109/MIC.2012.28

Tan, L., & Wang, N. (2010). Future internet: The internet of things. In *Advanced Computer Theory and Engineering (ICACTE), 2010 3rd International Conference on* (Vol. 5, pp. V5-376). IEEE.

Tan, Y. (2016). Artificial Immune System. *Artificial Immune System Applications in Computersecurity*, 1-25.

Tang, X., Yang, C., & Zhou, J. (2009). Stock price forecasting by combining news mining and time series analysis. In *Web Intelligence and Intelligent Agent Technologies, 2009. WI-IAT'09. IEEE/WIC/ACM International Joint Conferences on* (Vol. 1, pp. 279-282). IEEE. doi:10.1109/WI-IAT.2009.48

Tang, J., Alelyani, S., & Liu, H. (2015). *Feature Selection For Classification: A Review. In Data Classification Algorithms And Applications* (pp. 37–64). Chapman And Hall/CRC.

Tarique, A., & Gabbar, H. A. (2013). Particle swarm optimization (PSO) based turbine control. *Intelligent Control and Automation*, *4*(02), 126–137. doi:10.4236/ica.2013.42018

Taylor, D., & Corne, D. (2003). An Investigation of the Negative Selection Algorithm for Fault Detection in efrigeration Systems. In *Proceeding of Second International Conference on Artificial Immune Systems (ICARIS*). Napier University.

Taylor, R. H., Menciassi, A., Fichtinger, G., & Dario, P. (2008). Medical robotics and computer-integrated surgery. In *Springer handbook of robotics* (pp. 1199–1222). Springer Berlin Heidelberg. doi:10.1007/978-3-540-30301-5_53

Tengeleng, S., & Armand, N. (2014). Performance of using cascade forward back propagation neural networks for estimating rain parameters with rain drop size distribution. *Atmosphere*, *5*(2), 454–472. doi:10.3390/atmos5020454

Tezel, G., & Kose, U. (2011). Headache diagnosis by using the clonal selection algorithm. In *6th International Advanced Technologies Symposium 2011* (pp. 144-148). Elazig, Turkey: Firat University.

Tezel, G., & Kose, U. (2011). Headache Disease Diagnosis by Using the Clonal Selection Algorithm. In *6th International Advanced Technologies Symposium (IATS'11)* (pp. 144-148). Academic Press.

Thangaraj, M., Ponmalar, P. P., Sujatha, G., & Anuradha, S. (2016). Agent based Semantic Internet of Things (IoT) in Smart Health care. In *Proceedings of the The 11th International Knowledge Management in Organizations Conference on The changing face of Knowledge Management Impacting Society* (p. 41). ACM. doi:10.1145/2925995.2926023

Thangavel, K., & Karnan, M. (2005). Computer aided diagnosis in digital mammograms: Detection of microcalcifications by meta heuristic algorithms. *GVIP Journal*, *5*(7), 41–55.

The Open Clinical Web Site. (n.d.). Retrieved November 28, 2016, from, http://www.openclinical.org/

Thomas, A. M., Moore, P., Shah, H., Evans, C., Sharma, M., Xhafa, F., & Chapman, C. et al. (2013). Smart care spaces: Needs for intelligent at-home care. *International Journal of Space-Based and Situated Computing*, *3*(1), 35–44. doi:10.1504/IJSSC.2013.051988

Timmis, J. (2001). aiVIS - Artificial Immune Network Visualisation. *EuroGraphics UK 2001 Conference Proceedings*.

Timmis, J. (2007). Artificial immune systems-Today and tomorrow. *Natural Computing: An International Journa*l, *6*(1), 1-18.

Toriyama, N., Ono, K., & Orito, Y. (2017). Adaptive GA-based AR-hidden Markov model for time series forecasting. In *Evolutionary Computation (CEC), 2017 IEEE Congress on* (pp. 665-672). IEEE. doi:10.1109/CEC.2017.7969374

Toro, C. H. F., Meire, S. G., Gálvez, J. F., & Fdez-Riverola, F. (2013). A hybrid artificial intelligence model for river flow forecasting. *Applied Soft Computing*, *13*(8), 3449–3458. doi:10.1016/j.asoc.2013.04.014

Trebeschi, S., van Griethuysen, J. J., Lambregts, D. M., Lahaye, M. J., Parmer, C., Bakers, F. C., & Aerts, H. J. et al. (2017). Deep Learning for Fully-Automated Localization and Segmentation of Rectal Cancer on Multiparametric MR. *Scientific Reports*, 7. PMID:28706185

Tripathy, B. K., Acharjya, D. P., & Cynthya, V. (2013). *A framework for intelligent medical diagnosis using rough set with formal concept analysis.* arXiv preprint arXiv:1301.6011

Tsai, P. W., Pan, J. S., Chen, S. M., Liao, B. Y., & Hao, S. P. (2008). Parallel cat swarm optimization. In *2008 International Conference on Machine Learning and Cybernetics* (Vol. 6, pp. 3328-3333). IEEE. doi:10.1109/ICMLC.2008.4620980

Turan, M., Almalioglu, Y., Konukoglu, E., & Sitti, M. (2017). *A Deep Learning Based 6 Degree-of-Freedom Localization Method for Endoscopic Capsule Robots.* arXiv preprint arXiv:1705.05435

Turban, E., Sharda, R. E., & Delen, D. (2011). *Decision Support and Business Intelligence Systems* (9th ed.). Prentice Hall.

Turkoglu, I., Arslan, A., & Ilkay, E. (2003). An intelligent system for diagnosis of the heart valve diseases with wavelet packet neural networks. *Computers in Biology and Medicine*, *33*(4), 319–331. doi:10.1016/S0010-4825(03)00002-7 PMID:12791405

Ubeyli, E. D. (2009). Decision support systems for time-varying biomedical signals: EEG signals classification. *Expert Systems with Applications*, *36*(2), 2275–2284. doi:10.1016/j.eswa.2007.12.025

Ugur, A., & Kinaci, A. C. (2010). A web-based tool for teaching neural network concepts. *Computer Applications in Engineering Education*, *18*(3), 449–457. doi:10.1002/cae.20184

Ulutas, B. H., & Konak, S. (2011). A review of clonal selection algorithm and its applications. *Artificial Intelligence Review*, *36*(2), 117–138. doi:10.1007/s10462-011-9206-1

UNODC. (2013). *United nations office on drugs and crime - world drug report 2013.* United Nations Publication.

Valian, E., Mohanna, S., & Tavakoli, S. (2011). Improved cuckoo search algorithm for feedforward neural network training. *International Journal of Artificial Intelligence & Applications*, *2*(3), 36–43. doi:10.5121/ijaia.2011.2304

Valkenhoef, G. V., Tervonen, T., Zwinkels, T., Brock, B. D., & Hillege, H. (2013). ADDIS: A decision support system for evidence-based medicine. *Decision Support Systems*, *55*(2), 459–475. doi:10.1016/j.dss.2012.10.005

Valsecchi, A., Bermejo, E., Damas, S., & Cordón, O. (2017). Metaheuristics for Medical Image Registration. Handbook of Heuristics, 1-22.

Vapnik, V., Golowich, S. E., & Smola, A. J. (1997). Support vector method for function approximation, regression estimation and signal processing. In Advances in neural information processing systems (pp. 281-287). Academic Press.

Vardhini, K. K., & Sitamahalakshmi, T. (2016). A Review on Nature-based Swarm Intelligence Optimization Techniques and its Current Research Directions. *Indian Journal of Science and Technology*, *9*(10).

Vasant, P. M. (Ed.). (2012). *Meta-heuristics optimization algorithms in engineering, business, economics, and finance.* IGI Global.

Vasant, P., & Kose, U. (2017). Moral Dilemma Scenarios Considering Possible Actions by Humans and Artificial Intelligence Based Systems. *Journal of Multidisciplinary Developments*, *2*(2), 39–49.

Vasant, P., Kose, U., & Watada, J. (2017). Metaheuristic Techniques in Enhancing the Efficiency and Performance of Thermo-Electric Cooling Devices. *Energies*, *10*(11), 1703. doi:10.3390/en10111703

Vassiliadis, V., & Dounias, G. (2009). Nature–Inspired Intelligence: A Review of Selected Methods and Applications. *International Journal of Artificial Intelligence Tools*, *18*(04), 487–516. doi:10.1142/S021821300900024X

Vergin, R., O'Shaughnessy, D., & Gupta, V. (1996, May). Compensated mel frequency cepstrum coefficients. In *Acoustics, Speech, and Signal Processing, 1996. ICASSP-96. Conference Proceedings., 1996 IEEE International Conference on* (Vol. 1, pp. 323-326). IEEE. doi:10.1109/ICASSP.1996.541097

Von Altrock, C. (1995). *Fuzzy logic and neurofuzzy applications explained.* Prentice-Hall, Inc.

Vouk, M. A. (2008). Cloud computing–issues, research and implementations. *CIT. Journal of Computing and Information Technology*, *16*(4), 235–246. doi:10.2498/cit.1001391

Vyborny, C. J., & Giger, M. L. (1994). Computer vision and artificial intelligence in mammography. *AJR. American Journal of Roentgenology*, *162*(3), 699–708. doi:10.2214/ajr.162.3.8109525 PMID:8109525

Waal, I. (2017). Skin cancer diagnosed using artificial intelligence on clinical images. *Oral Diseases*. PMID:28326656

Wada, K., Shibata, T., Musha, T., & Kimura, S. (2008). Robot therapy for elders affected by dementia. *IEEE Engineering in Medicine and Biology Magazine*, *27*(4), 53–60. doi:10.1109/MEMB.2008.919496

Waldner, J. B. (2013). *Nanocomputers and swarm intelligence*. John Wiley & Sons.

Wall, D. P., Dally, R., Luyster, R., Jung, J. Y., & DeLuca, T. F. (2012). Use of artificial intelligence to shorten the behavioral diagnosis of autism. *PLoS One*, *7*(8), e43855. doi:10.1371/journal.pone.0043855 PMID:22952789

Wang, D. W., Wang, J. W., Wang, H., Zhang, R., & Guo, Z. (2007). *Intelligent optimization methods*. Higher Education Press.

Wang, L. (Ed.). (2005). *Support vector machines: theory and applications* (Vol. 177). Springer Science & Business Media. doi:10.1007/b95439

Wang, L., Ni, H., Yang, R., Pappu, V., Fenn, M. B., & Pardalos, P. M. (2014). Feature selection based on meta-heuristics for biomedicine. *Optimization Methods & Software*, *29*(4), 703–719. doi:10.1080/10556788.2013.834900

Wang, X., Yang, J., & Li, X. (2011). Dynamics of a heroin epidemic model with very population. *Applied Mathematics*, *2*(06), 732–738. doi:10.4236/am.2011.26097

Wang, Z., Yu, G., Kang, Y., Zhao, Y., & Qu, Q. (2014). Breast tumor detection in digital mammography based on extreme learning machine. *Neurocomputing*, *128*, 175–184. doi:10.1016/j.neucom.2013.05.053

Weber, R. H., & Weber, R. (2010). *Internet of things* (Vol. 12). New York: Springer. doi:10.1007/978-3-642-11710-7

Weigend, A. S. (1994). *Time series prediction: Forescasting the future and understanding the past* (No. 04; QA280, T5.). Academic Press.

Weinhardt, C., Anandasivam, A., Blau, B., Borissov, N., Meinl, T., Michalk, W., & Stößer, J. (2009). Cloud computing–a classification, business models, and research directions. *Business & Information Systems Engineering*, *1*(5), 391–399. doi:10.1007/s12599-009-0071-2

Weiss, S. M., & Kulikowski, C. A. (1991). *Computer systems that learn: classification and prediction methods from statistics, neural nets, machine learning, and expert systems*. Morgan Kaufmann Publishers Inc.

Welsh, R. C., Jelsone-Swain, L. M., & Foerster, B. R. (2013). The utility of independent component analysis and machine learning in the identification of the amyotrophic lateral sclerosis diseased brain. *Frontiers in Human Neuroscience*, 7. PMID:23772210

Wermter, S. (2000). *Hybrid neural systems (No. 1778)*. Springer Science & Business Media. doi:10.1007/10719871

White, H. (2008). Learning in artificial neural networks: A statistical perspective. *Learning, 1*(4).

White, E., & Comiskey, C. (2007). Heroin epidemics, treatment and ODE modelling. *Mathematical Biosciences*, *208*(1), 312–324. doi:10.1016/j.mbs.2006.10.008 PMID:17174346

WHO. (2004). *Global Burden Disease*. Retrieved from http://www.who.int/entity/healthinfo/global_burden_disease/GBD_report_2004update_full.pdf?ua=1

Wiener, N. (1961). *Cybernetics or Control and Communication in the Animal and the Machine* (Vol. 25). MIT Press. doi:10.1037/13140-000

Williams, N., Zander, S., & Armitage, G. (2006). A Preliminary Performance Comparison Of Five Machine. *ACM Sigcomm Computer Communication Review Learning Algorithms For Practical Ip Traffic Flow Classification*, 7-15.

Witten, I. H., Frank, E., Hall, M. A., & Pal, C. J. (2016). *Data Mining: Practical machine learning tools and techniques*. Morgan Kaufmann.

Wortmann, F., & Flüchter, K. (2015). Internet of things. *Business & Information Systems Engineering*, *57*(3), 221–224. doi:10.1007/s12599-015-0383-3

Wu, Y., Giger, M. L., Doi, K., Vyborny, C. J., Schmidt, R. A., & Metz, C. E. (1993). Artificial neural networks in mammography: Application to decision making in the diagnosis of breast cancer. *Radiology*, *187*(1), 81–87. doi:10.1148/radiology.187.1.8451441 PMID:8451441

Xia, F., Yang, L. T., Wang, L., & Vinel, A. (2012). Internet of things. *International Journal of Communication Systems*, *25*(9), 1101–1102. doi:10.1002/dac.2417

Xia, X. (2003). Estimation of HIV/AIDS parameters. *Automatica*, *39*(11), 1983–1988. doi:10.1016/S0005-1098(03)00220-6

Xing, F. Z., Cambria, E., & Zou, X. (2017). Predicting evolving chaotic time series with fuzzy neural networks. In *International Joint Conference on Neural Networks (IJCNN)*. Academic Press. doi:10.1109/IJCNN.2017.7966252

Xue, S. J., & Wu, W. (2012). Scheduling workflow in cloud computing based on hybrid particle swarm algorithm. *Indonesian Journal of Electrical Engineering and Computer Science*, *10*(7), 1560–1566.

Yang, P., Abo, R., Liu, C., Chen, Z., Wu, H., Cui, J., ... & Chittenden, T. W. (2017). *Novel feature selection strategies for enhanced predictive modeling and deep learning in the biosciences*. Academic Press.

Yang, X. S. (2009, October). Firefly algorithms for multimodal optimization. In *International Symposium on Stochastic Algorithms* (pp. 169-178). Springer Berlin Heidelberg.

Yang, X. S. (2010). A new metaheuristic bat-inspired algorithm. In Nature inspired cooperative strategies for optimization (NICSO 2010) (pp. 65-74). Springer Berlin Heidelberg. doi:10.1007/978-3-642-12538-6_6

Yang, X. S. (2010a). Firefly algorithm, Levy flights and global optimization. *Research and Development in Intelligent Systems, 26*, 209-218.

Yang, X. S. (2011). Metaheuristic optimization: Algorithm analysis and open problems. *Experimental Algorithms*, 21-32.

Yang, X. S., & Deb, S. (2009). Cuckoo search via Lévy flights. In *Nature & Biologically Inspired Computing, 2009. NaBIC 2009. World Congress on* (pp. 210-214). IEEE.

Yang, X. S., & Deb, S. (2009, December). Cuckoo search via Lévy flights. In *Nature & Biologically Inspired Computing, 2009. NaBIC 2009. World Congress on* (pp. 210-214). IEEE.

Yang, S. H. (2014a). Internet of things. In *Wireless Sensor Networks* (pp. 247–261). Springer London. doi:10.1007/978-1-4471-5505-8_12

Yang, X. S. (2010b). *Nature-inspired metaheuristic algorithms*. Luniver Press.

Yang, X. S. (2012). *Flower pollination algorithm for global optimization*. UCNC.

Yang, X. S. (2014b). *Nature-inspired optimization algorithms*. Elsevier.

Yang, X. S., Cui, Z., Xiao, R., Gandomi, A. H., & Karamanoglu, M. (Eds.). (2013). *Swarm intelligence and bio-inspired computation: theory and applications*. Newnes. doi:10.1016/B978-0-12-405163-8.00001-6

Yang, X. S., & Deb, S. (2014). Cuckoo search: Recent advances and applications. *Neural Computing & Applications, 24*(1), 169–174. doi:10.1007/s00521-013-1367-1

Yang, X. S., & Hossein Gandomi, A. (2012). Bat algorithm: A novel approach for global engineering optimization. *Engineering Computations, 29*(5), 464–483. doi:10.1108/02644401211235834

Yan, X., Gong, R., & Zhang, Q. (2016). Application of optimization SVM based on improved genetic algorithm in short-term wind speed prediction. *Power Syst. Prot. Control, 44*(9), 38–42.

Yao, B., Chen, C., Cao, Q., Jin, L., Zhang, M., Zhu, H., & Yu, B. (2017). Short-Term Traffic Speed Prediction for an Urban Corridor. *Computer-Aided Civil and Infrastructure Engineering, 32*(2), 154–169. doi:10.1111/mice.12221

Yao, W., & Kumar, A. (2013). CONFlexFlow: Integrating flexible clinical pathways into clinical decision support systems using context and rules. *Decision Support Systems, 55*(2), 499–515. doi:10.1016/j.dss.2012.10.008

Yegnanarayana, B. (2009). *Artificial neural networks*. PHI Learning Pvt. Ltd.

Yen, J., & Langari, R. (1998). *Fuzzy logic: intelligence, control, and information*. Prentice-Hall, Inc.

Yen, J., & Langari, R. (1999). *Fuzzy logic: Intelligence, Control, and Information* (Vol. 1). Upper Saddle River, NJ: Prentice Hall.

Yi, D., Sawyer, R. L., Cohn, D., III, Dunnmon, J., Lam, C., Xiao, X., & Rubin, D. (2017). *Optimizing and Visualizing Deep Learning for Benign/Malignant Classification in Breast Tumors*. arXiv preprint arXiv:1705.06362

Yourui, H. (2008). *Intelligent optimization algorithm and application*. Beijing: National Defence Industry Press.

Yudkowsky, E. (2008). Artificial intelligence as a positive and negative factor in global risk. *Global Catastrophic Risks, 1*(303), 184.

Yuksel, A. S., Cankaya, I. A., & Yuksel, M. E. (2015). An Analysis of RDF Storage Models and Query Optimization Techniques. *International Journal of Information Engineering and Electronic Business, 7*(2), 20–26. doi:10.5815/ijieeb.2015.02.04

Yuksel, A. S., Cankaya, S. F., & Uncu, I. S. (2017). Design of a Machine Learning Based Predictive Analytics System for Spam Problem. *Acta Physica Polonica A, 132*(3), 500–504. doi:10.12693/APhysPolA.132.500

Zadeh, L. A. (1996). Fuzzy logic= computing with words. *IEEE Transactions on Fuzzy Systems, 4*(2), 103–111. doi:10.1109/91.493904

Zahedi, Z. M., Akbari, R., Shokouhifar, M., Safaei, F., & Jalali, A. (2016). Swarm intelligence based fuzzy routing protocol for clustered wireless sensor networks. *Expert Systems with Applications, 55*, 313–328. doi:10.1016/j.eswa.2016.02.016

Zanella, A., Bui, N., Castellani, A., Vangelista, L., & Zorzi, M. (2014). Internet of things for smart cities. *IEEE Internet of Things Journal, 1*(1), 22-32.

Zeger, S. L., Irizarry, R., & Peng, R. D. (2006). On time series analysis of public health and biomedical data. *Annual Review of Public Health, 27*(1), 57–79. doi:10.1146/annurev.publhealth.26.021304.144517 PMID:16533109

Zhang, B., Wang, X., & Huang, M. (2014). A data replica placement scheme for cloud storage under healthcare iot environment. In *Fuzzy Systems and Knowledge Discovery (FSKD), 2014 11th International Conference on* (pp. 542-547). IEEE. doi:10.1109/FSKD.2014.6980892

Zhang, J., Xia, Y., Wu, Q., & Xie, Y. (2017). *Classification of Medical Images and Illustrations in the Biomedical Literature Using Synergic Deep Learning.* arXiv preprint arXiv:1706.09092

Zhang, X. M., & Zhang, N. (2011). An open, secure and flexible platform based on internet of things and cloud computing for ambient aiding living and telemedicine. In *Computer and Management (CAMAN), 2011 International Conference on* (pp. 1-4). IEEE. doi:10.1109/CAMAN.2011.5778905

Zhang, Z., & Zhang, X. (2009). Realization of open cloud computing federation based on mobile agent. In Intelligent computing and intelligent systems, 2009. ICIS 2009. IEEE international conference on (Vol. 3, pp. 642-646). IEEE. doi:10.1109/ICICISYS.2009.5358085

Zhang, C., & Chen, W. B. (2007). An Effective and Robust Method for Automatic Bacterial Colony Enumeration. *Proceedings of the International Conference on Semantic Computing*, 581-588. doi:10.1109/ICSC.2007.20

Zhang, L., Liu, P., & Gulla, J. A. (2017). A Neural Time Series Forecasting Model for User Interests Prediction On Twitter. In *Proceedings of the 25th Conference on User Modeling, Adaptation and Personalization* (pp. 397-398). ACM. doi:10.1145/3079628.3079648

Zhang, Q., Cheng, L., & Boutaba, R. (2010). Cloud computing: State-of-the-art and research challenges. *Journal of Internet Services and Applications, 1*(1), 7–18. doi:10.1007/s13174-010-0007-6

Zhang, S., Lee, C. K., Chan, H. K., Choy, K. L., & Wu, Z. (2015). Swarm intelligence applied in green logistics: A literature review. *Engineering Applications of Artificial Intelligence, 37*, 154–169. doi:10.1016/j.engappai.2014.09.007

Zhang, Y., Wang, Y., Wang, W., & Liu, B. (2001). Doppler ultrasound signal denoising based on wavelet frames. *IEEE Transactions on Ultrasonics, Ferroelectrics, and Frequency Control, 48*(3), 709–716. doi:10.1109/58.920698 PMID:11381694

Zhao, W., Wang, C., & Nakahira, Y. (2011). Medical application on Internet of Things. In *IET Conference Proceedings.* The Institution of Engineering & Technology.

Zhao, Z., Shen, Q., & Ren, F. (2013). Heart sound biometric system based on marginal spectrum analysis. *Sensors (Basel), 13*(2), 2530–2551. doi:10.3390/s130202530 PMID:23429515

Zheng, W. B., Yin, H. T., Fu, J. F., Fu, P., & Liu, B. (2014). Range expansion of cooperative mobile wireless network based on swarm intelligence optimization. *Measurement, 54*, 83–91. doi:10.1016/j.measurement.2014.04.003

Zheng, Y., Guo, X., Qin, J., & Xiao, S. (2015). Computer-assisted diagnosis for chronic heart failure by the analysis of their cardiac reserve and heart sound characteristics. *Computer Methods and Programs in Biomedicine, 122*(3), 372–383. doi:10.1016/j.cmpb.2015.09.001 PMID:26387633

Zhou, J., Leppanen, T., Harjula, E., Ylianttila, M., Ojala, T., Yu, C., . . . Yang, L. T. (2013). Cloudthings: A common architecture for integrating the internet of things with cloud computing. In *Computer Supported Cooperative Work in Design (CSCWD), 2013 IEEE 17th International Conference on* (pp. 651-657). IEEE. doi:10.1109/CSCWD.2013.6581037

Zhou, S. K., Greenspan, H., & Shen, D. (Eds.). (2017). *Deep Learning for Medical Image Analysis.* Academic Press.

Zhou, Z.-G., Liu, F., Jiao, L.-C., Zhou, Z.-J., Yang, J.-B., Gong, M.-G., & Zhang, X.-P. (2013). A bi-level belief rule based decision support system for diagnosis of lymph node metastasis in gastric cancer. *Knowledge-Based Systems, 54*, 128–136. doi:10.1016/j.knosys.2013.09.001

Zhu, A. (2017). *Artificial Neural Networks.* The International Encyclopedia of Geography. doi:10.1002/9781118786352.wbieg0871

Zhu, J. J., Huang, M., & Lu, Z. R. (2017). Bird mating optimizer for structural damage detection using a hybrid objective function. *Swarm and Evolutionary Computation*, *35*, 41–52. doi:10.1016/j.swevo.2017.02.006

Zhu, K., Song, H., Liu, L., Gao, J., & Cheng, G. (2011). Hybrid genetic algorithm for cloud computing applications. In *Services Computing Conference (APSCC), 2011 IEEE Asia-Pacific* (pp. 182-187). IEEE. doi:10.1109/APSCC.2011.66

Ziemke, T. (2001). The construction of 'reality' in the robot: Constructivist perspectives on situated artificial intelligence and adaptive robotics. *Foundations of Science*, *6*(1), 163–233. doi:10.1023/A:1011394317088

Zikopoulos, P., & Eaton, C. (2011). *Understanding big data: Analytics for enterprise class hadoop and streaming data*. McGraw-Hill Osborne Media.

Zuo, X., Chen, C., Tan, W., & Zhou, M. (2015). Vehicle scheduling of an urban bus line via an improved multiobjective genetic algorithm. *IEEE Transactions on Intelligent Transportation Systems*, *16*(2), 1030–1041.

Zurada, J. M. (1992). *Introduction to artificial neural systems* (Vol. 8). St. Paul, MN: West.

# About the Contributors

**Utku Kose** received the B.S. degree in 2008 from computer education of Gazi University, Turkey as a faculty valedictorian. He received M.S. degree in 2010 from Afyon Kocatepe University, Turkey in the field of computer and D.S. / Ph. D. degree in 2017 from Selcuk University, Turkey in the field of computer engineering. Between 2009 and 2011, he has worked as a Research Assistant in Afyon Kocatepe University. Following, he has also worked as a Lecturer and Vocational School - Vice Director in Afyon Kocatepe University between 2011 and 2012 and as a Lecturer and Research Center Director in Usak University between 2012 and 2017. Currently, he is an Assistant Professor in Suleyman Demirel University, Turkey. His research interest includes artificial intelligence, machine ethics, artificial intelligence safety, optimization, the chaos theory, distance education, e-learning, computer education, and computer science.

**Gur Emre Guraksin** received the B.S. degree in 2005 from computer engineering of Atilim University, Turkey. Following that, he had his M.S. degree in 2009 from Afyon Kocatepe University, Turkey in the field of computer and PhD. degree in 2015 from computer engineering of Selcuk University, Turkey. He has worked as RAN/NMC Engineer at Telsim Telecom, Istanbul and as a Research Assistant at Afyon Kocatepe University, Turkey before. Currently, he is an Assistant Professor at the department of biomedical engineering at the same university. His research interest includes artificial intelligence, biomedical engineering, image processing, data mining, and computer science.

**Omer Deperlioglu** received the B.S. degree in 1988 from the electrical education of Gazi University, in Turkey. He received M.S. degree in 1996 from Afyon Kocatepe University, Turkey and he completed Ph. D. degree in 2001 at Gazi University in field of controlling switch-mode dc-dc converters with neuro-fuzzy system. He is currently an Associate Professor in the Afyon Kocatepe University, in Turkey. His research interest includes computer-based control systems, fuzzy logic control, neuro-fuzzy control, power electronics, distance education and e-learning.

* * *

**Gokhan Akarca** received the B.S. degree in 1995 from food engineering department of Atatürk University, Turkey. He received M.Sc. Degrees in 2005 department of veterinary microbiology and 2010 department of food engineering from Afyon Kocatepe University, Turkey. He Received Ph.D. degree at Afyon Kocatepe University in the field of food engineering. Between 2002 and 2014 he was worked as a lecturer of food engineering department of Afyon Kocatepe University. Currently, He is an Assistant Professor in Afyon Kocatepe University, Food Engineering department, in Turkey. His Research interest includes, food microbiology, food poisoning, natural antimicrobials and dairy microbiology.

**Hasan Armutlu** received the B.S. degree in 2006 from computer and control systems education of Marmara University, Turkey. Following that, he received M.S. degree in 2014 from Kutahya Dumlupinar University, Turkey in the field of electric electronics engineering. Currently, he is a Lecturer in Usak University, Turkey as also the chair of computer technologies at Karahalli Vocational School at the same university. His research interest includes cloud computing, computer software, artificial intelligence, and web – mobile programming.

**Orhan Bölükbaş** has received his BSc degrees in electronic and computer education science program from Gazi University in 1995. He has received his MSc degrees in Niğde University in 2005. He is a Ph.D. student at the Selcuk University in the Department of Computer Engineering. He has been a lecturer in the faculty of Education of Aksaray University since 2007. His research interests focus on image processing and machine learning.

**Ibrahim Arda Cankaya** is a Research Assistant in Department of Computer Engineering at Suleyman Demirel University, Turkey (2012). He has completed B.Sc. at Suleyman Demirel University, Turkey in Department of Computer Engineering. Following that, he has completed M. Sc. again in Department of Computer Engineering at Suleyman Demirel University, Turkey (2015). Currently, he is performing his PhD. at the same department and university. His research interests include indoor navigation, beacon devices and machine learning.

**Sadi Fuat Cankaya** is a Research Assistant in Department of Computer Engineering at Suleyman Demirel University, Turkey (2013). He has completed B.Sc. at Suleyman Demirel University, Turkey in Department of Computer Systems. Following that, he has completed M.Sc. again in Department of Computer Engineering at Suleyman Demirel University, Turkey (2016). Currently, he is performing his PhD. at the same department and university. His research interests include microprocessors, FPGA, image processing, and artificial intelligence.

**Nesibe Cekmez** received the B.S. degree in 2017 from biomedical engineering department of Afyon Kocatepe University, Turkey.

**Gaffari Celik** is a Lecturer in Department of Computer Technologies at Vocational School of Agri Ibrahim Cecen University, Turkey. Celik has performed his MSc. at Ataturk University, Turkey, in the field of Computer Engineering, between 2009 and 2013. His research interests include Data Mining, Machine Learning, Artificial Intelligence, Educational Technologies, Database Management System and Computer Programming.

**Aydin Cetin** has received his MSc. degree from Louisiana State University, US in 1996 from Electrical and Computer engineering and Ph.D. degrees from Gazi University, Turkey in 2001 respectively. He is currently with Computer Engineering Department of Gazi University Faculty of Technology. His current research includes artificial intelligence and machine learning applications, smart systems and technologies.

**Huseyin Coskun** received the B.S. from computer education of Fırat University and Computer Engineering of Suleyman Demirel University, Turkey. He received M.S. degree Suleyman Demirel University, Turkey and now, he continues Ph. D. at Suleyman Demirel University, Turkey in the department of computer engineering. Between 2008 and 2012, he has worked as a computer education teacher at certain technical school in Turkey. Currently, he is a Lecturer in Usak University, Turkey since 2012. His research interest includes artificial intelligence, machine learning, biomedical signal processing, human-computer interaction, distance education, e-learning, computer education, and computer science.

**Tuba Gokhan** received the B.Sc. degree from Eskisehir Osmangazi University and M.Sc. degree from Gazi University in computer engineering in 2012 and 2016 respectively. She has worked as a research assistant in the Department of Computer Engineering, Gazi University, Turkey, from 2013 to 2017.Currently she is with the Newcastle University, Newcastle Upon Tyne for her post graduate studies. Her research interests include Software Engineering, Data Mining, Bioinformatics and Mobile Application Development.

**Mehmet Mehdi Karakoc** has completed his BSc. degree in the Department of Computer Education at Gazi University, Turkey in 2008. After that, he has completed his MSc. degree in the Department of Electronics-Computer Education at Firat University, Turkey in 2013. Now, he performs his PhD. at Firat University, Turkey in the Department of Software Engineering. Karakoc is a Lecturer at Agri Ibrahim Cecen University, Turkey and also Research Center Director at the Computer Sciences Application and Research Center of the same university. His research interests include Digital Forensics, Artificial Intelligence, Expert Systems, Mobile Programming, Computer Programing, and Open Source Operating Systems.

**Arif Koyun** received the B.S. degree in 1988 from Mechanical Engineering of Akdeniz University, Turkey and in 1995 from Computer Engineering, Germany. He received M.S. degree in 1991 from Akdeniz University, Turkey and Ph. D. in 2006 from Suleyman Demirel University, Turkey. Between 2001 and 2007, he has worked as Lecturer in Suleyman Demirel University. Currently, he is an Assist. Prof. Dr. in Suleyman Demirel University, Turkey. His research interest includes artificial intelligence, data mining, database management systems, distance education, e-learning, computer education and computer science.

**Kamalanand Krishnamurthy**, Ph.D., is an Assistant Professor at the Department of Instrumentation Engineering, Madras Institute of Technology Campus, Anna University, Chennai. His research focuses on mathematical modeling and analysis of physical and biological systems. He has authored several papers and book chapters and is a Guest Editor for the European Journal for Biomedical Informatics and Current Bioinformatics.

**Merve Nur Onal** received the B.S. degree in 2017 from biomedical engineering department of Afyon Kocatepe University, Turkey. She begins M.S. in 2017 from Afyon Kocatepe.

**Alice Pavaloiu** received the M.S. degree in 2017 from Paris School of Business in the field of International Management and graduated with the dissertation "The Impact of Artificial Intelligence on Global Trends". Her academic collaboration extended with Dr. Utku Kose upon multiple topics of AI, both sharing the same research interests.

**Mannar Jawahar Ponnuswamy**, Ph.D., is a Senior Professor and Former Vice Chancellor of Anna University, Chennai. He completed his Ph.D. at I.I.T. Delhi and subsequently did his Post Doctoral Research work at Warsaw University of Technology, Poland. He joined Anna University, Chennai, as a Faculty in 1978 and has put in more than three decades of service in the teaching profession. He served as the Vice Chancellor of Anna University from June 2008 to June 2012. His current research focuses on mathematical methods in medicine and biology.

**Suraj Sawant** received his Bachelor's in Engineering degree in 2006 and Master's of Technology degree in 2009. He is pursuing his Ph.D. degree in the Geomatics Engineeing from Indian Institute of Technology, Roorkee since 2015 under Quality Improvement Programme undertaken by All India Council of Technical Education (AICTE). He is working as an Assistant Professor in Computer Engineering and Information Technology Department, College of Engineering, Pune, India from 2010 to till date. His research interest includes Artificial Intelligence, Machine Learning, Geographical Information Systems, Remote Sensing, Geo spatial Technologies and Deep Learning.

**Harun Uğuz** graduated from the Department of Computer Engineering at Selçuk University in 1999. He received his MSc degree from the same department and same university in 2002. His PhD degree was received from the department of Electrical and Electronics Engineering at Selcuk University in 2007. He has been an associate professor in the Departmant of Computer Engineering since 2013. His interest areas include Hidden Markov Models, machine learning and artificial intelligence systems.

**Pandian Vasant** is a senior lecturer at Department of Fundamental and Applied Sciences, Faculty of Science and Information Technology, Universiti Teknologi PETRONAS in Malaysia. He holds PhD (UNEM, Costa Rica) in Computational Intelligence, MSc (UMS, Malaysia, Engineering Mathematics) and BSc (2nd Class Upper-Hons in Mathematics, UM, Malaysia) in Mathematics. His research interests include Soft Computing, Hybrid Optimization, Holistic Optimization, Innovative Computing and Applications. He has co-authored research papers and articles in national journals, international journals, conference proceedings, conference paper presentation, and special issues lead guest editor, lead guest editor for book chapters' project, conference abstracts, edited books, keynote lecture and book chapters (178 publications indexed in SCOPUS). In the year 2009, Dr. Pandian Vasant was awarded top reviewer for the journal Applied Soft Computing (Elsevier), awarded outstanding reviewer in the year 2015 for ASOC (Elsevier) journal and Top reviewer for Sentinels of Science: Computer Science (Oct. 2015 - Sept. 2016). He has 26 years of working experience at the various universities from 1989-2017. Currently he is Editor-in-Chief of IJCO, IJSIEC, IEM, IJEOE and Editor of GJTO. H-Index SCOPUS Citations = 36, H-Index Google Scholar = 26, i-10 index = 80.

**Tuncay Yigit** received the B.S. from Technical Education Faculty of Gazi University, Turkey in 1997. He received M.S. in 2000 and D.S. / Ph.D. in 2005 at Gazi University, Turkey. He has worked as a research assistant and lecturer at Computer Education Department of Gazi University, Turkey. Currently, he is a Professor in Computer Engineering Department of Suleyman Demirel University, Turkey. His research interest includes artificial neural network, genetic algorithm, fuzzy logic, computer based control systems, time scheduling problems, distance education, e-learning, web technologies, computer education, and computer science.

# Index

www.ingramcontent.com/pod-product-compliance
Lightning Source LLC
LaVergne TN
LVHW060301011026
849161LV00011B/2750